The 2nd International Laayoune Forum on Biosaline Agriculture

The 2nd International Laayoune Forum on Biosaline Agriculture

Editors

Abdelaziz Hirich
Redouane Choukr-Allah

MDPI • Basel • Beijing • Wuhan • Barcelona • Belgrade • Manchester • Tokyo • Cluj • Tianjin

Editors
Abdelaziz Hirich
Mohammed VI Polytechnic University (UM6P)
Morocco

Redouane Choukr-Allah
Mohammed VI Polytechnic University (UM6P)
Morocco

Editorial Office
MDPI
St. Alban-Anlage 66
4052 Basel, Switzerland

This is a reprint of articles from the Proceedings published online in the open access journal *Environmental Sciences Proceedings* (ISSN 2673-4931) (available at: https://www.mdpi.com/2673-4931/16/1).

For citation purposes, cite each article independently as indicated on the article page online and as indicated below:

LastName, A.A.; LastName, B.B.; LastName, C.C. Article Title. *Journal Name* **Year**, *Volume Number*, Page Range.

ISBN 978-3-0365-6169-1 (Hbk)
ISBN 978-3-0365-6170-7 (PDF)

Contents

About the Editors

Abdelaziz Hirich

Assistant professor of Biosaline Agriculture at the Mohammed VI Polytechnic University, African Sustainable Agriculture Research Institute in Laayoune, Morocco since January 2020. Prior to his current position, he was a scientist at the International Center for Biosaline Agriculture, Dubai from 2017 to 2020 and a Postdoctoral fellow from 2014 to 2016. He has a PhD in Agricultural Sciences (IAV HASSAN II, Morocco), MSc in Land and Water Resources Management (CIHEAM-IAMB, Italy), and an Engineer diploma in Horticulture (IAV HASSAN II, Morocco). His expertise focuses on biosaline agriculture, salinity management, horticulture, irrigation management, alternative crops, fertigation, etc. Hirich led several projects within the context of biosaline agriculture with a total budget more than USD 11 million implemented in several countries including Morocco, Egypt, UAE, and seven Sub-Saharan countries. He has more than 75 publications in journals, book chapters, and international conferences. Prior to his position with ICBA, Hirich worked as a researcher within GLOBAQUA- and SWUP-MED EU-funded projects and as a horticulture engineer consultant with many private agribusiness companies in Morocco.

Redouane Choukr-Allah

Choukr-Allah is a horticultural, soil, and water environmental expert with more than 35 years of experience in the use of saline water and the use of pretreated sewage in Horticulture. He earned a PhD in environment horticulture at the University of Minnesota, St. Paul, USA. He also served as a technical coordinator of a USD 12 million project, financed by USAID, on water resource sustainability in Morocco. He served as the head of the Horticulture Department from 1983 to 1996 and has served as the head of the salinity and plant nutrition laboratory since 1996. He served at ICBA as a senior fellow scientist in horticulture and as a Section Head of Crop Diversification and Genetics. He has produced numerous publications, including edited books, research reports, articles in peer-reviewed international journals, and books in the field of nonconventional water.

Preface to "The 2nd International Laayoune Forum on Biosaline Agriculture"

The scarcity of water resources and their degradation, aggravated by climate change and socio-economic pressure, will affect food security and thus, there will be a need to use increasing amounts of non-conventional water, soil, and plant resources. This calls for perpetual innovations in techniques and approaches. These innovations require the optimization of efforts and the mobilization of additional resources. Saline water sources have the potential to double the amount of water available for agriculture. Capitalizing on the vast potential of biosaline agriculture necessitates major interdisciplinary and collaborative research efforts to inform effective and supportive policy frameworks and to evaluate the most promising methods for developing biosaline agriculture in different salt-affected regions. Globally, the research illustrates the vast and, so far, underrated potential of growing food on soils generally qualified as saline. This means that salt-affected and other degraded lands should be viewed as a valuable resource rather than a liability. Every type of land available for agricultural production should be used, as the United Nations Food and Agriculture Organization projects need to produce 70% more food by 2050.

This book showcases many examples from highly relevant research that are often also validated at the field level. The next step for training farmers includes demonstrations at several farms to showcase the market opportunities and implement an integrated policy. More and more integrated research is emerging that looks at both crops and soil and water management. Moreover, working more with the private sector will be important to turn this knowledge into products and solutions for farmers. This book also shows that alternative crops offer great potential for contributing to food and nutritional security under marginal environments and, therefore, that the upscaling of their value chains should be prioritized in the environments that need it most.

It is necessary to widen the multidisciplinary scope of the research staff by including ecologists and remote sensing experts (at different levels, from ecosystem to crop monitoring and even phenotyping) together with socio-economists, food specialists, etc. Furthermore, the integration of national, multilateral, and international efforts is urgent to mitigate the negative effects of resource degradation. To this end, the creation of a competence database is a step in this process. Finally, some interesting results were presented from several scientists from various countries; therefore, it is necessary to develop a long-term partnership with institutions working in the biosaline agriculture field.

Abdelaziz Hirich and Redouane Choukr-Allah

Editors

environmental sciences proceedings

Editorial

Statement of Peer Review †

Redouane Choukr-Allah [1] and Abdelaziz Hirich [2,*]

1 African Sustainable Agricultue Research Institute (ASARI), Mohammed VI Polytechnic University, Laayoune 70000, Morocco
2 Agricultural Innovation and Technology Transfer Center (AITTC), Mohammed VI Polytechnic University (UM6P), Ben Guerir 43150, Morocco
* Correspondence: abdelaziz.hirich@um6p.ma; Tel.: +212-666-17-15-46
† Presented at the 2nd International Laayoune Forum on Biosaline Agriculture, 14–16 June 2022; Available online: https://lafoba2.sciforum.net/.

In submitting conference proceedings to *Environment Sciences Proceedings*, the volume editors of the proceedings certify to the publisher that all papers published in this volume have been subjected to peer review administered by the volume editors. Reviews were conducted by expert referees to the professional and scientific standards expected of a proceedings journal.

- Type of peer review: single-blind; double-blind; triple-blind; open; other (please describe): double-blind and triple-blind for some papers
- Conference submission management system: Sciforum
- Number of submissions sent for review: 159
- Number of submissions accepted: 124
- Acceptance rate (number of submissions accepted/number of submissions received): 78%
- Average number of reviews per paper: 2.2
- Total number of reviewers involved: 37
- Any additional information on the review process: N/A

Conflicts of Interest: The authors declare no conflict of interest.

Citation: Choukr-Allah, R.; Hirich, A. Statement of Peer Review. *Environ. Sci. Proc.* **2022**, *16*, 75. https://doi.org/10.3390/environsciproc2022016075

Published: 7 December 2022

environmental sciences proceedings

MDPI

Proceeding Paper

An Emulsion-Based Formulation for Increasing the Resistance of Plants to Salinity Stress: US20160302416A1 Patent Evaluation †

Fatima Zahra Hafiane [1] and Ahmed Fatimi [1,2,*]

1 ERSIC, Polydisciplinary Faculty, Sultan Moulay Slimane University, P.O. Box 592 Mghila, Beni-Mellal 23000, Morocco; f.s.h.hafiane@gmail.com
2 Department of Chemistry, Polydisciplinary Faculty, Sultan Moulay Slimane University, P.O. Box 592 Mghila, Beni-Mellal 23000, Morocco
* Correspondence: a.fatimi@usms.ma

† Presented at the 2nd International Laayoune Forum on Biosaline Agriculture, 14–16 June 2022; Available online: https://lafoba2.sciforum.net/.

Abstract: This study describes the use of an emulsion-based formulation to increase the resistance of plants to salinity and other abiotic stress. Inventors have described and claimed, through the US20160302416A1 patent, an emulsion-based formulation for increasing the resistance of a plant to damage caused by one or more abiotic stresses. The earliest priority date of the US20160302416A1 patent was 6 December 2013, with five patent families and four filled jurisdictions. Based on patent classification, the invention covered by the patent concerns the preservation of plants or parts thereof (e.g., inhibiting evaporation, improving the appearance of leaves, etc.). To prove the concept of this invention, studies were initiated in the greenhouse to look at the impact of different formulations on spring wheat and turf grass grown under salt stress conditions. The inventors confirmed that the proposed formulations could increase the tolerance of spring wheat and turf grass to the presence of excess salt. The method of applying the emulsion-based formulation to the spring wheat or to the turf grass at the onset of the salt stress conditions enhanced the tolerance to the stress conditions.

Keywords: emulsion; formulation; biosaline agriculture; salinity stress; patent; innovation

Citation: Hafiane, F.Z.; Fatimi, A. An Emulsion-Based Formulation for Increasing the Resistance of Plants to Salinity Stress: US20160302416A1 Patent Evaluation. *Environ. Sci. Proc.* **2022**, *16*, 4. https://doi.org/10.3390/environsciproc2022016004

Academic Editor: Abdelaziz Hirich

Published: 16 June 2022

1. Introduction

The protection of plants from environmental stresses is one of the major concerns in the field of agriculture. Growing plants are subjected to a variety of environmental stresses of non-biological origins, referred to as abiotic stresses, including cold, water logging, transplant shock, low light, drought, salinity, etc. [1,2]. In agriculture, conventional plant treatments are generally unable to provide plants with resistance to abiotic stresses and are therefore limited to providing benefits to otherwise healthy or flourishing plants. However, commercial agronomical processes require additional plant treatments to reduce plant stress or enhance the plant's ability to resist common abiotic stresses and/or to recover quickly from such stresses. A typical example of common abiotic stress in marginal lands and sea basins includes continuous exposure to saline water which causes soil salinity. Salinity stress results in poor or no yields mainly due to reducing root growth as well as reducing the nutritional status of plants and/or fruits. More specifically, salinity directly affects the morphology, physiology, and metabolism of plants [3,4].

When a plant is exposed to abiotic stress, it develops certain natural defense mechanisms against such stress, but there is a need to provide plants with enhanced abilities to overcome such stress. Salinity causes two major effects in the plant, i.e., ion toxicity and osmotic stress [5].

This study describes the state of the art by introducing what has been patented in relation to formulations for increasing the resistance of plants to abiotic stresses, among others, salinity stress, and describes the patent US20160302416A1. Regarding preparation methods and applications, this study, in the form of a patent evaluation, describes the use of combinations, including an emulsion-based formulation with a pigment, to increase the resistance of plants to salinity stress. Furthermore, to prove the concept of this invention, studies were initiated in the greenhouse to look at the impact of different formulations on spring wheat and turf grass grown under salt stress conditions.

2. Patent Analysis

The studied patent (US20160302416A1) was invented by Fefer Michael and Liu Jun from Canada, and it was filed by the company Suncor Energy Inc. (Calgary, AB, Canada). The earliest priority date of this patent was 6 December 2013, with five patent families. A patent family is a collection of published patent documents relating to the same invention as the initial document (i.e., a priority document), or to several inventions sharing a common aspect that are published at different times in the same country or published in different countries or regions. That means a number of different patent documents whose technical content is considered identical [6]. The jurisdictions of the related five filled patents correspond to Canada [7,8], the United States [9], Uruguay [10], and the global system for filing patent applications [11], known as the Patent Cooperation Treaty (PCT), and administered by the World Intellectual Property Organization (WIPO) [12].

The International Patent Classification (IPC) is a hierarchical system in the form of codes, which divides all technological areas into a range of sections, classes, subclasses, groups, and subgroups. It is an international classification system that provides standard information to categorize inventions and evaluate their technological uniqueness [13]. Concerning these patents, the IPC codes are presented in Table 1.

Table 1. Classifications of the five patent families of this study [7–11].

IPC Codes	Description
A01N3/00	Preservation of plants or parts thereof (e.g., inhibiting evaporation, improving the appearance of leaves, etc.).
A01N27/00	Biocides, pest repellants or attractants, or plant growth regulators containing hydrocarbons.
A01N43/90	Biocides, pest repellants or attractants, or plant growth regulators containing heterocyclic compounds.
A01N61/02	Biocides, pest repellants or attractants, or plant growth regulators containing substances characterized only by the mode of action, such as mineral oils, and tar oils and tar distillates.

3. Patent Evaluation: Materials, Methods, and Results

Inventors described and claimed through these patents a method and an emulsion-based formulation for increasing the resistance of a plant to damage caused by one or more abiotic stresses, among others, salinity stress, which comprises applying an emulsion amount to the plant. The claimed emulsion-based formulation comprises paraffinic oil, emulsifier, pigment, water, and additives. A typical effective agricultural combination of these oil-in-water emulsions is presented in Table 2.

Two studies concerning salt stress were carried out to prove the concept of this invention, and three oil-in-water emulsion compositions were tested:

- Composition A: Paraffinic oil, having a composition of 98% isoparaffin.
- Composition B: 40% polychlorinated Cu(II) phthalocyanine dispersed in water.
- Composition C: A combination of paraffinic oil and polychlorinated Cu(II) phthalocyanine dispersed in water in a ratio of 30:1, having 90% isoparaffin and 2.4% polychlorinated CU(II) phthalocyanine.

Studies were initiated in the greenhouse to look at the impact of the three above-cited compositions on spring wheat and turf grass grown under salt stress conditions. It should be noted that all results obtained are included in the US20160302416A1 patent [9].

Table 2. Detail of the proposed emulsion-based formulation through the studied patent [9].

Emulsion Components	Description
Paraffinic oil	The paraffinic oil can include a paraffin having an average number of carbon atoms of 23. Paraffinic oil with 98% paraffin content: isoparaffin.
Emulsifier	Emulsifiers include natural or synthetic surfactant or alcohol. Silicone surfactant; Silicone polyether; Polyethylene glycol.
Pigment	The pigment is a water-based pigment dispersion (i.e., hydrophilic) or an oil-based pigment dispersion (i.e., hydrophobic). Copper phthalocyanine: 40% polychlorinated Cu(II) phthalocyanine.
Water	The oil-in-water emulsions include water. Distilled water and/or other waters having a low mineral electrolyte content.
Additives	The formulation can further include a combination of additives. Anti-settling agents; Plant growth regulators; Conventional chemical fungicides.

3.1. *Spring Wheat*

The first salt stress study was conducted on spring wheat in the greenhouse. Wheat seeds were grown in the greenhouse under full light for two to three weeks to reach the three-leaf stage. The wheat plants were treated with the compositions A and B by foliar application with a total spray volume of 935.40 L/ha. Salt stress was introduced 24 h later by drenching of a 150 ppm NaCl solution into each pot (500 mL/pot). A second salt solution was applied one week later. The treatments were performed on different samples under the conditions presented in Table 3. The plant continued to grow for two additional weeks, and the plant height and biomass data (fresh and dry weight) were then measured.

Table 3. Conditions of different plants' treatment under salt stress: the case of spring wheat [9].

Condition [1]	Salt Treatment (ppm) [2]	Composition A (%)	Composition B (%)
1	150	0	0
2	150	5	0.312
3	150	5	0.625
4	150	5	1.25
5	0	0	0

[1] **Condition:** (1) Untreated control under salt stress; (2,3,4) Plants were treated with the compositions A and B by foliar application with a total spray volume of 935.40 L/ha; (5) Untreated control without salt stress. [2] **Salt treatment:** NaCl solution into each pot.

As results for this first salt stress study, each treatment showed no significant difference in plant height. However, the measured biomass (above ground) for each treatment by fresh weight and dry weight showed significant differences. The plants treated with the compositions A and B exhibited higher dry weight and fresh weight compared with the untreated control under salt stress. The treated plant under salt stress also exhibited similar biomass as an untreated control which was not subjected to salt stress.

3.2. *Turf Grasses*

The second salt stress study was conducted on turf grass grown in the greenhouse. Turf grass seed was sown in potting soil in the greenhouse under full light approximately four weeks before the trial start date. Upon trial initiation, pots were watered with different NaCl concentrations at a rate of 100 mL/pot, with additional applications occurring 1–2 times weekly thereafter. Half of the pots from each salt concentration were also treated with either a foliar application of water or the equivalent of 5.2 g/m^2 of the composition C at a rate of 935.40 L/ha, with applications repeated every 14 days. Turf quality ratings (NTEP scale 1–9, where 6 = minimally acceptable turf, and 9 = excellent turf quality) were used to assess the effect of salt stress on overall turf health starting at trial initiation, and then repeated every two weeks thereafter (Table 4).

Table 4. Conditions of different plants' treatment under salt stress: the case of turf grass [9].

Salt Treatment (M) [1]	Turf Quality (NTEP Scale 1–9)			
	35 Days after Initial Treatment		49 Days after Initial Treatment	
	Water [2]	Composition C [3]	Water [2]	Composition C [3]
0 (water)	7.3	8.8	6	9
0.03	6.3	8	5	8.3
0.06	5	7.5	3.5	7.8
0.09	4.8	6.8	3	7.8
0.12	5.8	6.3	5	5.3

[1] **Salt treatment:** NaCl solutions were watered into each pot at a rate of 100 mL/pot, 1–2 times weekly. [2] **Water:** Plants were treated with water at a rate of 935.40 L/ha. [3] **Composition C:** Plants were treated with the equivalent to 5.2 g/m^2 of the composition C at a rate of 935.40 L/ha.

As results for this second salt stress study, turf grass showed no significant sign of stress during the first two applications (0 and 0.03 M) and no significant difference between the treatments. Turf quality ratings shown in Table 4 highlight results assessed 35 days after initial foliar applications (seven days after 3rd application) or 49 days after initial foliar applications (seven days after 4th application), as these were the dates when the greatest differences were observed between the composition C treated and water treated plants.

4. Conclusions

Soil salinity stress causes osmotic stress and affects the agronomical, physiological, and nutritional status of the plants. The studied patent (US20160302416A1) uses an emulsion-based formulation to increase the resistance of plants against salinity stress. Through this patent, the inventors claimed preparation methods/processes and applications of oil-in-water emulsions to treat plants under salt stress. The inventors confirmed that the proposed formulations could increase the tolerance of spring wheat and turf grass against salinity. According to the inventors, the method of applying the emulsion-based formulation to the spring wheat or to the turf grass at the onset of the salt stress conditions enhanced the tolerance to the stress conditions by improving the yield and quality of plants as compared to untreated plants subjected to the same salt stress. In particular, the quality of spring wheat and turf grass was not degraded as quickly or to the same extent as the control. Finally, the inventors proposed that the emulsion-based formulation could be applied prior to or during the crop growing period to improve salinity stress.

Author Contributions: Conceptualization, F.Z.H. and A.F.; methodology, F.Z.H. and A.F.; investigation, F.Z.H. and A.F.; writing—original draft preparation, F.Z.H. and A.F.; writing—review and editing, F.Z.H. and A.F. All authors have read and agreed to the published version of the manuscript.

Funding: This research received no external funding.

Institutional Review Board Statement: Not applicable.

Informed Consent Statement: Not applicable.

Data Availability Statement: The data presented in this study is available within the US20160302416A1 patent.

Conflicts of Interest: The authors declare that this article content has no conflict of interest. The authors have no relevant affiliations or financial involvement with any organization or entity with a financial interest in or financial conflict with the subject matter or materials discussed in this article.

References

1. Raza, A.; Razzaq, A.; Mehmood, S.S.; Zou, X.; Zhang, X.; Lv, Y.; Xu, J. Impact of Climate Change on Crops Adaptation and Strategies to Tackle Its Outcome: A Review. *Plants* **2019**, *8*, 34. [CrossRef] [PubMed]
2. Cramer, G.R.; Urano, K.; Delrot, S.; Pezzotti, M.; Shinozaki, K. Effects of abiotic stress on plants: A systems biology perspective. *BMC Plant Biol.* **2011**, *11*, 163. [CrossRef] [PubMed]
3. Khan, M.S.; Hemalatha, S. Biochemical and molecular changes induced by salinity stress in *Oryza sativa* L. *Acta Physiol. Plant.* **2016**, *38*, 167. [CrossRef]
4. Kumar, S.; Beena, A.S.; Awana, M.; Singh, A. Physiological, Biochemical, Epigenetic and Molecular Analyses of Wheat (*Triticum aestivum*) Genotypes with Contrasting Salt Tolerance. *Front. Plant Sci.* **2017**, *8*, 1151. [CrossRef] [PubMed]
5. Shrivastava, P.; Kumar, R. Soil salinity: A serious environmental issue and plant growth promoting bacteria as one of the tools for its alleviation. *Saudi J. Biol. Sci.* **2015**, *22*, 123–131. [CrossRef] [PubMed]
6. World Intellectual Property Organization. *Handbook on Industrial Property Information and Documentation*; WIPO: Geneva, Switzerland, 2013; p. 43.
7. Fefer, M.; Liu, J. Methods for Increasing Resistance of Plants to Abiotic Stresses. Granted Patent CA2836757C, 10 September 2019.
8. Fefer, M.; Liu, J. Methods for Increasing Resistance of Plants to Abiotic Stresses. Granted Patent CA2836757A1, 6 June 2015.
9. Fefer, M.; Liu, J. Methods for Increasing Resistance of Plants to Abiotic Stresses. Patent Application US20160302416A1, 20 October 2016.
10. Fefer, M.; Liu, J. Métodos Para Aumentar la Resistencia de las Plantas al Estrés Abiótico? Patent Application UY35865A, 31 July 2015.
11. Fefer, M.; Liu, J. Methods for Increasing Resistance of Plants to Abiotic Stresses. Patent Application WO2015081441A1, 11 June 2015.
12. World Intellectual Property Organization. Summary of the Patent Cooperation Treaty (PCT) (1970). Available online: https://www.wipo.int/treaties/en/registration/pct/summary_pct.html (accessed on 2 December 2021).
13. World Intellectual Property Organization. IPC Publication. IPCPUB v8.5. Available online: https://www.wipo.int/classifications/ipc/ipcpub (accessed on 14 January 2022).

environmental sciences proceedings

MDPI

Proceeding Paper

Quinoa Vikinga Response to Salt and Drought Stress under Field Conditions in Italy †

Cataldo Pulvento [1,*], Mohamed Houssemeddine Sellami [2,3], Giuseppe De Mastro [1], Davide Calandrelli [2] and Antonella Lavini [2]

1 Department of Agricultural and Environmental Science (DISAAT), University of Bari, 70121 Bari, Italy; giuseppe.demastro@uniba.it

2 Institute for Mediterranean Agricultural and Forestry Systems, National Research Council of Italy (CNR-ISAFOM), 80055 Portici, Italy; mohamed.sellami@isafom.cnr.it (M.H.S.); davide.calandrelli@cnr.it (D.C.); antonella.lavini@cnr.it (A.L.)

3 Interdepartmental Research Centre on the "Earth Critical Zone", University of Naples Federico II, 80055 Portici, Italy

* Correspondence: cataldo.pulvento@uniba.it

† Presented at the 2nd International Laayoune Forum on Biosaline Agriculture, 14–16 June 2022; Available online: https://lafoba2.sciforum.net/.

Abstract: Agriculture in south Europe is facing the negative effect of abiotic stresses such as salinity that mostly affect the seed production and seed quality of traditional crops. Under these conditions, quinoa represent a good alternative to ensure the production of high-protein-quality seeds thanks to its tolerance to abiotic stresses. In 2015–2017, a sweet variety of quinoa, "Vikinga", was tested in Italy within the PROTEIN2FOOD project (EU Horizon2020) as high-quality-protein crop to enhance food protein production in Europe. A field trial was carried out at the experimental farm of CNR-ISAFOM in South Italy, to evaluate the combined effect of drought and salinity on quinoa Vikinga; both freshwater and saline water were used for irrigation. The plots were arranged in a randomized complete block design. The main yield parameters (seed yield, aboveground dry biomass 1000 seed weight), the protein content and other quality traits were analyzed at harvest, to evaluate the effect of applied treatments. The results showed that, in general, different treatments did not affect the main production and quality traits of quinoa "Vikinga".

Keywords: salinity; seed quality; protein crops; quinoa

check for updates

Citation: Pulvento, C.; Sellami, M.H.; De Mastro, G.; Calandrelli, D.; Lavini, A. *Quinoa Vikinga* Response to Salt and Drought Stress under Field Conditions in Italy. *Environ. Sci. Proc.* **2022**, *16*, 5. https://doi.org/10.3390/environsciproc2022016005

Academic Editor: Abdelaziz Hirich

Published: 16 June 2022

1. Introduction

Salinity is an adverse environmental factor that affects the growth of plants in the Mediterranean region [1,2]. Salinization is rapidly increasing on a global scale, with average yields for most major crop plants declining by more than 50 percent [3]. Salinization may result in the loss of 30% of current agricultural land in the coming 25 years, a figure set to rise by up to 50% by 2050, also due to the rapid rate of population growth. It is estimated that salinization affects around 3.8 million ha in Europe [4]; we can distinguish primary salinization due to natural processes and secondary salinization introduced by human interventions such as irrigation with saline water.

A possible approach is the introduction of species capable of tolerating high-soil salinities and guaranteeing acceptable yields.

Quinoa (*Chenopodium quinoa* Willd.) produce high-quality-protein seeds and is well-known for its tolerance to salinity compared to traditional crops.

A multi-annual field trial was planned to evaluate the effect of saline water irrigation on the protein quantity and quality of these crops, which could represent a source of vegetable proteins in salt-affected environments.

2. Materials and Methods

The field experiments were carried out in Vitulazio (Caserta, Italy) at the experimental research station of CNR-ISAFoM (41°12′ N and 14°20′ E, 23 m above sea level) during three growing seasons: 2015, 2016 and 2017. The soil of the experimental site was characterized by a clay loam texture; the volumetric soil water contents at field capacity was 0.38 m^3m^{-3}, while the permanent wilting point was 0.13 m^3m^{-3}.

The Danish quinoa variety "Vikinga" received from University of Copenhagen was sown. A randomized complete block (RCB) design with two treatments (irrigation regime and water quality) per crops and three replicates per treatment was adopted. Each experimental plot consisted of 10 rows, 4 m length.

Quinoa was grown under two irrigation regimes: (I100, restitution of 100% of the water necessary to replenish to field capacity (F.C.) at 40 cm soil layer and I33, corresponding respectively to restitution of 33% of full irrigation). For each irrigation level, a non-saline treatment irrigated with fresh water (100N and 33N) and a treatment irrigated with saline water was performed at a known salt concentration (100S and 33S).

Irrigation was carried out at fixed weekly intervals using a drip irrigation system. The conductivity achieved in the solution 1/1 (seawater/groundwater) was about 22 dS/m.

At harvest the total yield, the 1000 seed weight and the above-ground biomass were determined on each plot.

The seed samples for each harvest and for each treatment being studied were then chemically analyzed to evaluate the principal qualitative components. The data collected during three years of experimental work were analyzed according to an RCB design.

3. Results

Seasonal precipitation (November–June) was 437, 550 and 220 mm in the first, second and third growing season, compared to the historical average (1976–2017). of 377 mm.

After computing the deciles index (DI) designed by Gibbs and Maher (1967), the first (2015) and second (2016) seasons were classified as normal (DI = 9 and 7, respectively) whereas the third (2017) season was weak dry (DI = 4).

The crop cycle length ranged from 110 to 117 days in the three experimental years. Irrigation was applied 7 times during 2015 and 2016, and 5 times in 2017.

In 2015, 2016 and 2017, 256, 240 and 201 mm of freshwater was applied, respectively, for treatment 100N, and 217, 201 and 192 mm of saline water for treatment 100S. Obviously, the amounts of salt supplied by irrigation water to the soil were higher in the first year than in 2016 and 2017.

The initial ECe (0–0.6 m) value showed an increasing trend over three years for both 100S and 33S treatment, ranging from 1.4 to 2.62 dS m^{-1}; The final ECe value was significantly higher compared to the ECe initial value for each considered treatment; it reached values of 9.3 dS m^{-1} in 2017 for 100S treatment.

From the statistical analysis of the main yield components (yield, dry biomass, harvest index) measured during the three experiments, no significant differences were found for both single effects (Irrigation level and water quality) and interaction irrigation treatment × saline treatment. Vikinga registered low seed yields, ranging from 0.30 g $plant^{-1}$ in 2016 to 0.9 g $plant^{-1}$ in 2015.

No significant differences were recorded for each quality parameter (Table 1). The average protein content of quinoa seeds was about 15% of seed weight.

Table 1. Seed-quality parameters as affected by water quality and water supply on Quinoa.

Source of Variation	Starch (%)	Ashes (%)	Total Protein (%)	Fat (%)
Water quality (WQ)	ns	ns	ns	ns
Fresh water	49.00 ± 2.36	4.37 ± 0.62	14.97 ± 0.92	3.70 ± 0.43
Salinity	49.00 ± 2.21	4.53 ± 0.95	15.30 ± 0.70	3.52 ± 1.40
Water supply (WS)	ns	ns	ns	ns
I100	47.63 ± 1.56	4.47 ± 0.73	15.13 ± 0.50	3.55 ± 0.78
I33	50.37 ± 1.87	4.43 ± 0.87	15.13 ± 1.08	3.66 ± 1.25
WQ × WS	ns	ns	ns	ns

4. Discussion

The results confirm, as reported in the literature, a high tolerance of quinoa to drought and salinity. Titicaca, a quinoa bitter Danish variety was previously shown to have high resistance to drought and saline stresses, in terms of seed yield and qualitative parameters, under the same experimental conditions [5]. Vikinga recorded lower seed yield levels than Titicaca, Regalona [6] and Puno, which are well-adapted to Italian pedo-climatic conditions.

5. Conclusions

The field trials carried out from 2015 to 2017 in Italy confirmed that quinoa is tolerant to high levels of salinity; in fact, seed yield did not significantly vary passing from saline to not-saline treatments for a $p \leq 0.05$.

In addition, total protein content in the seeds did not vary between different treatments; since no differences were found in harvested seed weight and in protein content, was assumed that no changes occurred in the amino acidic composition of quinoa. These results suggest that quinoa has good adaptability and a high degree of flexibility regarding tolerance or resistance to salt stress; this crop represents a valid source of high-quality protein for cultivation in marginal European areas affected by primary and secondary salinization problems.

Author Contributions: Conceptualization, A.L. and C.P.; methodology, A.L. and C.P.; formal analysis, M.H.S.; investigation, D.C., A.L. and C.P.; resources, A.L. and C.P.; data curation, M.H.S., D.C., A.L. and C.P.; writing—original draft preparation, C.P.; writing—review and editing, M.H.S., G.D.M., C.P. and A.L.; visualization, C.P.; supervision, C.P. and A.L. All authors have read and agreed to the published version of the manuscript.

Funding: This research was funded by European Union's Horizon 2020 Research and Innovation Program (Protein2Food project).

Institutional Review Board Statement: Not applicable.

Informed Consent Statement: Not applicable.

Data Availability Statement: Not applicable.

Conflicts of Interest: The authors declare no conflict of interest.

References

1. Gregory, P.J. Food production under poor, adverse climatic conditions. In Proceedings of the IX ESA Congress, Warsaw, Poland, 4–7 September 2006; p. 19.
2. Lin, K.H.; Chao, P.Y.; Yang, C.M.; Cheng, W.C.; Lo, H.F.; Chang, T.R. The effects of flooding and drought stresses on the antioxidant constituents in sweet potato leaves. *Bot. Stud.* **2006**, *47*, 417–426.
3. Bray, E.A.; Bailey-Serres, J.; Weretilnyk, E. Responses to abiotic stresses. In *Biochemistry and Molecular Biology of Plants*; Gruissem, W., Buchnnan, B., Jones, R., Eds.; American Society of Plant Physiologists: Rockville, MD, USA, 2000; pp. 1158–1249.
4. Daliakopoulos, I.N.; Tsanis, I.K.; Koutroulis, A.; Kourgialas, N.N.; Varouchakis, A.E.; Karatzas, G.P.; Ritsema, C.J. The threat of soil salinity: A European scale review. *Sci. Total Environ.* **2016**, *573*, 727–739. [CrossRef] [PubMed]
5. Pulvento, C.; Riccardi, M.; Lavini, A.; Iafelice, G.; Marconi, E.; d'Andria, R. Yield and quality characteristicsof quinoa grown in open field under different saline and non-saline irrigation regimes. *J. Agron. Crop Sci.* **2012**, *198*, 254–263. [CrossRef]
6. Pulvento, C.; Riccardi, M.; Lavini, A.; d'Andria, R.; Iafelice, G.; Marconi, E. Field trial evaluation of two Chenopodium quinoa genotypes grown under rain-fed conditions in a typical Mediterranean environment in South Italy. *J. Agron. Crop Sci.* **2010**, *196*, 407–411. [CrossRef]

environmental sciences proceedings

MDPI

Proceeding Paper

Phenological and Biochemical Characteristics of Almond Cultivars in Arid Climate of Central Tunisia †

Samira Maatallah [1,2,*], Monia Guizani [1,2], Olfa Elloumi [3] and Mohamed Ghrab [3]

1 Non-Conventional Water Valuation Research Laboratory (LR VENC), National Research Institute for Rural Engineering, Water, and Forestry (INRGREF), University of Carthage, Hedi EL Karray Street, El Menzah IV, Tunis 1004, Tunisia; guizanimonia2018@gmail.com

2 Regional Center for Agricultural Research, Institution of Research and Higher Education Agriculture (IRESA), Sidi Bouzid 9100, Tunisia

3 Olive Institute, University of Sfax, LR16IO01, LR16IO02, BP 1087, Sfax 3000, Tunisia; olfa.elloumi@io.usf.tn (O.E.); mghrab@gmail.com (M.G.)

* Correspondence: samiraham2003@yahoo.fr; Tel.: +216-97833374

† Presented at the 2nd International Laayoune Forum on Biosaline Agriculture, 14–16 June 2022; Available online: https://lafoba2.sciforum.net/.

Abstract: Water scarcity is the main limiting factor for fruit trees in arid regions of Tunisia. In this area, almond is widespread, but severe conditions are a key issue for nuts production and kernel quality. In this study, phenological features and kernel quality of local and foreign almond cultivars irrigated with low water quality were investigated. Local cultivars presented an early flowering and seemed to be more appropriate to regional conditions and water salinity. All almond cultivars performed respectable kernel nutritional quality. In conclusion, local cultivars showed better adaptation with early bloom and higher fruit quality under warm conditions.

Keywords: *Prunus dulcis*; cultivar; water scarcity; salinity; bloom; phenol compounds; warm area

Citation: Maatallah, S.; Guizani, M.; Elloumi, O.; Ghrab, M. Phenological and Biochemical Characteristics of Almond Cultivars in Arid Climate of Central Tunisia. *Environ. Sci. Proc.* **2022**, *16*, 7. https://doi.org/10.3390/environsciproc2022016007

Academic Editor: Abdelaziz Hirich

Published: 16 June 2022

1. Introduction

Almond (*Prunus dulcis* Mill) is well cultivated in arid and semi-arid regions of Tunisia. It occupies the second place after the olive tree with a cultivated area around 225,000 ha and providing 2.3% of the world production.

In warm area, almond production is highly linked to climatic conditions as precipitation and winter chill. Recent findings reported frequent warm winters with lack of winter chill [1]. These warm conditions caused yield losses and phenology disorders of nuts trees [2]. Moreover, low yield due to water scarcity has often been achieved.

New almond cultivars and rootstocks were planted [3] using various quality of irrigation water due to less water availability. Almonds are appreciated for their benefit to human health thanks to their antioxidant activity [4]. However, horticultural practices can have different effects on kernel quality. In this context, this work focuses on the phenological behavior and kernel quality of almond in Central Tunisia. Local and foreign cultivars irrigated with low water quality were investigated during a warmer year.

2. Materials and Methods

2.1. Experimental Site and Plant Materials

This study was carried out in an experimental almond orchard located 3 km south-west of Sidi Bouzid (central Tunisia) (35°01′21.9″ N, 9°26′31.3″ E; 160 m above sea level).This region is characterized by a typical Mediterranean climate.

Fourteen cultivars of mature almond trees including 8 local cultivars (Zahaf, Achak, Tlili2, Tlili6, Tlili9, Kf2, Bf1 and Am2) and 6 foreign cultivars (Mazzetto, Lauranne, Constanti, Tarraco, Marinada, Vairo) were selected. Trees were planted in 2012 at 5 m × 6 m spacing and drip-irrigated with low irrigation water quality (salinity: 2.7g L^{-1}; EC: 2.26 mS cm^{-1};

ion content: Na^+ 8.28 meq L^{-1}, K^+ 1.45 meq L^{-1}, Cl^- 15.33 meq L^{-1}) and trained according to standard horticultural management practices typical of the region.

2.2. Phenologivcal Surveys

The phenological stages of local and foreign almond cultivars were monitored based on the phenological growth stages defined by Baggiolini for stone fruits during 2021. The budburst, early and late flowering and the young fruit stages were recorded, and consequently the flowering period was determined.

2.3. Phenolic Compounds Determination

The total phenols and the *O*-diphenols contents were determined colorimetrically [5]. Total flavonoids were measured according to [6]. Flavonols were determined based on [7] using ethanol and HCl. Condensed tannins were evaluated according to [8]. The concentration of condensed tannins was expressed in mg catechin equivalent/100g of DW.

2.4. Statistical Analysis

One-way ANOVA was applied on collected data using SPSS 17.0 for Windows (SPSS, Chicago, IL, USA). Means were compared using the Duncan test ($p < 0.05$).

3. Results

3.1. Phenological Features

Based on phenological observations, local cultivars are the earliest one (Table 1). Their bud-swelling stage occurred during the first two weeks of January against one month later for foreign cultivars (Table 1). The flowering period extended from the last week of January to the second week of February for local cultivars. Achak, Zahaf, Tlili1 and Tlili2 had an extended flowering period ranging between 14 and 20 days.

Table 1. Phenological stages of almond cultivars during 2021: Grey: bud swelling; Blue: flowering period; Green: Young fruit stage.

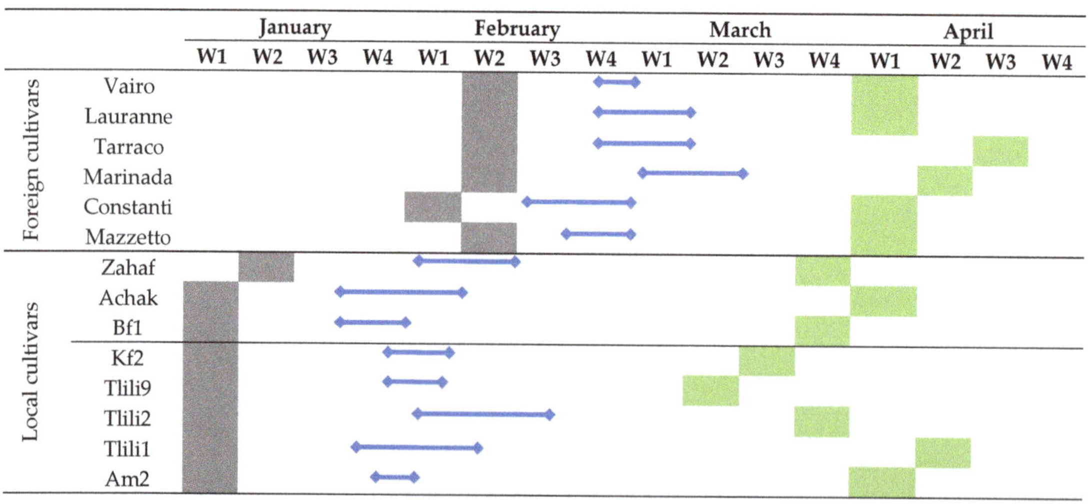

Considering the period in days between buds swelling and young fruit stages, results obtained showed that for local cultivars, this period was in the range of 64–91 days. Tlili9 and Tlili1 had the shortest and largest development cycle, respectively. For foreign cultivars, Terraco and Constanti had the largest development cycle of 70 days, while Vairo and Lauranne seemed to have the shortest one at about 51 days. Local cultivars are the earliest and have the longest development cycle, whereas foreign cultivars presented late flowering and short development period.

3.2. Phenolic Compounds

The results showed that foreign cultivars had the richest kernels in phenols with total content varied from 1937 to 4281 mg/100 g DW for Tarraco and Constanti, respectively. For local cultivars, these contents were ranged between 316 and 4258 mg/100 g DW for Tlili1 and Bf1, respectively (Table 2). Concerning flavonoids, the local cultivars Achak and Zahaf showed the highest contents similarly to both foreign cultivars Mazzetto and Constanti (Table 2). The lowest flavonoids contents were obtained for Tlili2 and Tlili9. Flavonols concentrations varied between 37 and 860 mg/100 g DW in BF1 and Mazzetto, respectively. The highest content of *O*-diphenol was recorded in Mazzetto (1467 mg/100 g DW).The tannin contents varied significantly among cultivar and the highest values were recorded for Mazzetto and BF1cultivars.

Table 2. Variation of the phenolic compounds (mg/100g DW) among almond cultivars.

		Total Phenols	Flavonoids	Flavonols	*O*-Diphenols	Tannins
Foreign cultivars	Vairo	1882.68 ± 60.62 ^{G}	125.28 ± 16.87 ^{FG}	647.16 ± 17.24 ^{C}	335.85 ± 19.61 ^{EF}	140 ± 0.01 ^{F}
	Lauranne	2048.65 ± 51.25 ^{E}	188.88 ± 13.07 ^{E}	652.61 ± 35.34 ^{C}	921.13 ± 10.93	310 ± 0.04 ^{DE}
	Tarraco	1937.39 ± 11.58	190.60 ± 11.52 ^{E}	375.24 ± 20.94 ^{D}	529.59 ± 34.42 ^{D}	290 ± 0.09 ^{DE}
	Marinada	1998.53 ± 72.40 ^{EF}	116.31 ± 26.82 ^{FG}	127.58 ± 40.29 ^{G}	471.94 ± 44.14 ^{D}	357 ± 0.02 ^{D}
	Constanti	4281.29 ± 71.26 ^{A}	473.86 ± 42.09 ^{C}	270.66 ± 7.39 ^{F}	1350.47 ± 39.58 ^{B}	270 ± 0.02 ^{E}
	Mazzetto	3411.45 ± 3.35 ^{B}	864.07 ± 84.29 ^{A}	860.87 ± 31.02 ^{A}	1467.97 ± 64.33 ^{A}	870 ± 0.04 ^{A}
Local cultivars	Zahaf	1700.85 ± 66.59 ^{H}	696.43 ± 7.47 ^{B}	722.55 ± 22.50 ^{B}	391.45 ± 47.31 ^{E}	320 ± 0.08 ^{DE}
	Achak	1992.27 ± 65.28 ^{EF}	517.28 ± 6.53 ^{C}	258.58 ± 39.38 ^{F}	490.92 ± 87.23 ^{D}	170 ± 0.08 ^{F}
	Bf1	4258.15 ± 92.61 ^{A}	205.43 ± 2.20 ^{E}	37.77 ± 1.46 ^{H}	155.07 ± 14.73 ^{G}	660 ± 0.02 ^{B}
	Kf2	2453.28 ± 35.79 ^{D}	292.28 ± 9.63 ^{D}	292.40 ± 0.69 ^{F}	274.22 ± 34.21 ^{F}	20 ± 0.01 ^{G}
	Tlili9	3309.19 ± 40.10 ^{C}	78.08 ± 1.41 ^{GH}	318.05 ± 31.30 ^{DE}	271.24 ± 20.04 ^{F}	470 ± 0.01 ^{C}
	Tlili2	365.51 ± 19.75 ^{IJ}	33.91 ± 1.41 ^{GH}	291.66 ± 15.02 ^{EF}	471.90 ± 21.62 ^{D}	80 ± 0.02 ^{FG}
	Tlili1	316.55 ± 18.34 ^{J}	162.52 ± 25.32 ^{EF}	100.92 ± 25.24 ^{F}	266.69 ± 4.65 ^{F}	110 ± 0.07 ^{F}
	Am2	416.07 ± 68.41 ^{I}	288.60 ± 6.77 ^{D}	335.63 ± 10.20 ^{CD}	133.08 ± 4.6 ^{G}	480 ± 0.02 ^{C}

Values are the means of three different almond samples (n = 3) ± standard deviation. The letters (A, B, C, D, E, F, G, H, I and J) indicate significant differences ($p < 0.05$) between the cultivars.

4. Discussion

Almond flowering and fruit set stages are a key process for yield as the occurrence of hazardous weather conditions during those critical stages has a major impact [9]. Flowering delay and extended duration occurred for both local and foreign cultivars as previously reported [10]. This could be a consequence of the exceptional warm winter with lack of chilling. A negative correlation was reported between flowering initiation and duration as well as winter chills [11]. Early flowering was considered as an interesting trait for almond growing in arid regions, where early blooming could also involve early ripening before the occurrence of extreme summer climatic conditions such as drought and higher temperatures. Local cultivars flowered earlier but had a delayed young fruit stage with the longest fruit development period. Late flowering seemed to be an inappropriate trait in warmer regions. Higher temperatures can affect stigma receptivity and reduce the effective pollination period [12].

Our results revealed rich almond kernels in total phenols content and confirm previous results showing total phenol content varied between 304 and 163715mg/100g [13]. Flavonoids contents recorded in local and foreign cultivars are comparable to previous findings [14]. The flavonoids contents varied with the geographical origin, the harvest period and cultivar [15]. As for results of flavonoids, local cultivars showed the highest values in flavonols with the foreign one Mazzetto, which revealed the highest content of

O-diphenol. These results integrated in part previous report [16]. The tannin contents varied significantly among cultivar and support results obtained by [17].

5. Conclusions

Almond cultivars irrigated with a low quality of water presented different phenological behavior and exhibited various kernel qualities. Local cultivars seemed to be well adapted to warm conditions and presented interesting fruit quality. Foreign cultivars had high kernel quality, while their phenological behavior could be affected by a lack of winter chill.

Author Contributions: S.M. and M.G. (Monia Guizani) wrote the first draft of the manuscript; O.E. and M.G. (Mohamed Ghrab) revised and improved the manuscript. All authors contributed to analyzing and discussing the results. All authors have read and agreed to the published version of the manuscript.

Funding: This research was financially supported by the Tunisian Ministry of Higher Education and Scientific Research (Laboratoire LR VENC).

Institutional Review Board Statement: Not applicable.

Informed Consent Statement: Not applicable.

Data Availability Statement: Not applicable.

Acknowledgments: The authors thank all laboratory members for CRRA-Sidi Bouzid.

Conflicts of Interest: The authors declare that they have no conflict of interest.

References

1. Benmoussa, H.; Luedeling, E.; Ghrab, M.; Ben Mimoun, M. Severe winter chill decline impacts Tunisian fruit and nut or-chards. *Clim. Chang.* **2020**, *162*, 1249–1267. [CrossRef]
2. Elloumi, O.; Ghrab, M.; Kessentini, H.; Ben Mimoun, M. Chilling accumulation effects on performance of pistachio trees cv. Mateur in dry and warm area climate. *Sci. Hortic.* **2013**, *159*, 80–87. [CrossRef]
3. Ghrab, M.; Ben Mimoun, M.; Masmoudi, M.M.; Ben Mechlia, N. Climate change and vulnerability of pistachio and almond crops in Mediterranean arid areas. *Options Medit.* **2016**, *119*, 247–251.
4. Čolić, S.D.; Aksic, M.F.; Lazarević, K.B.; Zec, G.N.; Gašić, U.; Zagorac, D.D.; Natić, M. Fatty acid and phenolic profiles of almond grown in Serbia. *Food Chem.* **2017**, *234*, 455–463. [CrossRef] [PubMed]
5. Montedoro, G.; Servili, M.; Baldioli, M.; Miniati, E. Simple and hydrolyzable phenolic compounds in virgin olive oil. 1. Their extraction, separation, and quantitative and semiquantitative evaluation by HPLC. *J. Agric. Food Chem.* **1992**, *40*, 1571–1576. [CrossRef]
6. Zhishen, J.; Mengcheng, T.; Jianming, W. The determination of flavonoid contents in mulberry and their scavenging effects on superoxide radicals. *Food Chem.* **1999**, *64*, 555–559. [CrossRef]
7. Romani, A.; Mancini, P.; Tatti, S.; Vincieri, F.F. Polyphenols and polysaccharides in Tuscan grapes and wines. *Ital. J. Food Sci.* **1996**, *8*, 13–24.
8. Julkunen-Tiitto, R. Phenolic constituents in the leaves of northern willows: Methods for the analysis of certain phenolics. *J. Agric. Food Chem.* **1985**, *33*, 213–217. [CrossRef]
9. Grab, S.; Craparo, A. Advance of apple and pear tree full bloom dates in response to climate change in the southwestern Cape, South Africa: 1973–2009. *Agric. For. Meteorol.* **2011**, *151*, 406–413. [CrossRef]
10. Pope, K.S.; Dose, V.; Da Silva, D.; Brown, P.; DeJong, T.M. Nut crop yield records show that budbreak-based chilling requirements may not reflect yield decline chill thresholds. *Int. J. Biometeorol.* **2014**, *59*, 707–715. [CrossRef] [PubMed]
11. Benmoussa, H.; Ghrab, M.; Ben Mimoun, M.; Luedeling, E. Chilling and heat requirements for local and foreign almond (*Prunus dulcis* Mill.) cultivars in a warm Mediterranean location based on 30 years of phenology records. *Agric. For. Meteorol.* **2017**, *239*, 34–46. [CrossRef]
12. Segura, J.M.A.; i Company, R.S.; Kodad, O. Late-blooming in almond: A controversial objective. *Sci. Hortic.* **2017**, *224*, 61–67. [CrossRef]
13. Kodad, O. Erratic fruit set in almond under warm climates. *Int. J. Hortic. Sci.* **2014**, *20*, 59–64. [CrossRef]
14. Barreira, J.C.; Ferreira, I.C.; Oliveira, M.B.P.; Pereira, J.A. Antioxidant activity and bioactive compounds of ten Portuguese regional and commercial almond cultivars. *Food Chem. Toxicol.* **2008**, *46*, 2230–2235. [CrossRef] [PubMed]
15. Sfahlan, A.J.; Mahmoodzadeh, A.; Hasanzadeh, A.; Heidari, R.; Jamei, R. Antioxidants and antiradicals in almond hull and shell (*Amygdalus communis* L.) as a function of genotype. *Food Chem.* **2009**, *115*, 529–533. [CrossRef]

16. Prgomet, I.; Gonçalves, B.; Domínguez-Perles, R.; Santos, R.; Saavedra, M.J.; Aires, A.; Pascual-Seva, N.; Barros, A. Irrigation deficit turns almond by-products into a valuable source of antimicrobial (poly)phenols. *Ind. Crops Prod.* **2019**, *132*, 186–196. [CrossRef]
17. Ahrens, S.; Venkatachalam, M.; Mistry, A.; Lapsley, K.; Sathe, S. Almond (*Prunus dulcis* L.) protein quality. *Plant Foods Hum. Nutr.* **2005**, *60*, 123–128. [CrossRef] [PubMed]

environmental sciences proceedings

MDPI

Proceeding Paper

Allometric Model for Predicting Root Biomass of Field Crops in the Salt-Affected Clay Soil: Novel Approach †

Amrakh I. Mamedov [1,2,*] and Emin K. Husiyev [2]

1 Arid Land Research Center, Tottori University, Tottori 680-0001, Japan
2 Institute of Botany, Azerbaijan National Academy of Sciences (ANAS), Baku AZ1043, Azerbaijan; emin9415@gmail.com
* Correspondence: amrakh03@yahoo.com

† Presented at the 2nd International Laayoune Forum on Biosaline Agriculture, 14–16 June 2022; Available online: https://lafoba2.sciforum.net/.

Abstract: Root biomass and phenotyping are vital parameters for studies on crop performance and response to environmental change, as well as abiotic stresses, crop water uptake, nutrient supply, and soil C sequestration and quality. However, root sampling and measurement, including biomass estimation, are laborious and time-consuming tasks. This study developed a novel allometric model to predict the root biomass of annual crop species using root collar diameter, an easy aboveground field measure. The root samples of alfalfa, sorghum and maize were collected (45 from each) at the harvesting stage from the irrigated agricultural field of the semi-arid region (clay soil, salinity: EC = 2–12 dS m^{-1}, 70% of full irrigation). Crops collar diameter (CD) and root biomass (RM) increased in the following order: alfalfa < sorghum < maize. For each crop species, strong power (RM = aCD^b) relations ($R^2 \geq 0.90$) were found between RM and CD (analogous to tree species). The coefficient (a) and exponent (b) of the relations and the soil quality indices (e.g., soil organic carbon, aggregate stability) in the root zone were concomitant with the crop (root) traits. The use of the allometric model was crucial for the fast assessment of the root biomass of the crop species, such as estimating biomass allocation. The approach could be used for evaluation of soil–root–plant interaction under abiotic stresses in the context of the sustainable agriculture (e.g., soil C deposition and respiration, crop transpiration and photosynthesis rate, and selecting the best genotypes-cultivars).

Keywords: root biomass; collar diameter; abiotic stresses; plant allometry; modeling; soil quality; C sequestration

Citation: Mamedov, A.I.; Husiyev, E.K. Allometric Model for Predicting Root Biomass of Field Crops in the Salt-Affected Clay Soil: Novel Approach. *Environ. Sci. Proc.* **2022**, *16*, 11. https://doi.org/10.3390/environsciproc2022016011

Academic Editor: Abdelaziz Hirich

Published: 13 June 2022

1. Introduction

Crop root systems play a crucial role in crop performance and soil C cycle, and affect (i) soil quality parameters, water and nutrient availability; (ii) enzyme activity, microbial biomass and respiration and microbial community structure; and (iii) soil aggregate stability, water retention and hydraulic conductivity and resistance against erosion. Soil organic carbon (SOC) accumulation is related to the interactions of several ecosystem processes (respiration, decomposition, and photosynthesis), yet C sequestration is largely mediated by crops via photosynthesis. The input rate of SOC is governed mostly by the crop root biomass and litter, and indirectly by the transfer of C-enhanced mixtures from roots to soil microbes. The fixation of atmospheric CO_2 into crop biomass occurred by photosynthesis; distribution of photosynthate to roots, stems, and leaves are affected by the environmental factors and ontogenetic drifts. The roots depend on the shoot for carbohydrates, whereas the shoot relies on the root for water and nutrients. In arid and semi-arid regions, environmental abiotic stresses (e.g., drought, salinity, temperature) increase the relative weights of root biomass and its phenotyping. Thus, incorporating the easily assessable parameters of the root system into soil-crop model components will be critical for evaluating the use of the available water and nutrient resources, and climatic change impact on crop productivity and soil quality [1–6].

To quantify crop roots biomass is difficult, since roots are challenging to measure in the soil, and sampling is laborious, especially extraction from clay soil, and is time-consuming [7]. The allometric relations between plant organs (shoot, root or leaf biomass, crop height and diameter, leaf area index, root length) link plant structural development and physiological processes, and are widely used to study the fundamental mechanisms of plant function. In trees, root biomass can be projected indirectly by using allometric relations, such as a root-to-shoot ratio, and by linking the root and shoot biomass with plant height and trynk diameter. It is well established that tree root and shoot biomass can be predicated from the stem or collar diameter [8–10]. Such methods could be useful for agricultural annual crops as well, yet regardless of the approach used, it is necessary to measure the root system directly. The ability of crops to adapt to the environment and abiotic stress conditions affect crop organ mass fraction in both trees and annual crops, and become evident in the parameters of allometric relations [8–14]. Therefore, the objective of the study was to develop an allometric model to predict the root biomass of crop species using a root collar diameter, one of the most easily measured crop parameters.

2. Material and Methods

An experiment was conducted on the semi-arid irrigated and artificially drained field, located in the middle part of the Kur-Aras lowland region of Azerbaijan. Soil was a swelling and crusting clay. Due to the natural condition and farming history, soil salinity distribution was heterogeneous (EC = 2–12 dS m^{-1}) in the upper cultivated layer, and increased considerably with soil depth. Mineralized ground water (4–16 g L^{-1}) was situated 1.2–1.8 m from the soil surface. Three crop species, varying in root system architecture (alfalfa, *Medicago sativa*; sorghum, *Sorghum bicolor*; maize, *Zea mays*), were grown in large plots (40 × 40 m; 50 × 20 spacing) through April–August, over three years. Irrigation rate was traditional (70% of full irrigation frequently applied by farmers) due to the fresh water shortage. Root samples were collected (3 × 45 = 135 samples, 0–40 cm depth) using a monolith excavation method at the harvesting stage [7,14].

The properties of soil samples from root zone were analyzed using standard methods. Root separation was completed carefully by hand and washing (elutriation system) over a sieve (≥0.5 mm); its dry biomass were established by forced-air oven drying at 65 °C. Root collar diameter (CD) was measured by the digital caliper. An ANOVA was conducted to compare the role of crop species on vegetative parameters, and mean comparisons were made by the Tukey HSD ($p < 0.05$). Least squares fitting was used to evaluate the allometric relation (linear, exponential or power) between the RM and the CD [9,10,14].

3. Results and Discussions

The crop species produced varied root biomass in swelling clay soil, associated with salinity and water stresses. Crops RM mostly was clustered within the upper 30 cm of soil layer (>90%). Both the mean of RM (e.g., mean ± se: maize: 12.28 ± 1.36; sorghum: 7.92 ± 0.61, alfalfa 0.92 ± 0.11 g) and CD (13.56 ± 0.65; 11.82 ± 0.38; 3.54 ± 0.25 mm) increased in the following order of crops: control (no crop) < alfalfa < amaranth < maize. There were strong power relations ($RM = aCD^b$; $R^2 \geq 0.90$, $p < 0.05$) between RM and CD for each crop species, as well as power or exponential (not given) relation ($R^2 = 0.93$) for all merged crop species (Figure 1). The differences between the coefficients (a = 0.031, 0.033, 0.105) and exponents (b = 2.22, 2.19, 1.61) of the allometric model was anticipated, since crops with different root types (e.g., tap, fibrous, adventitious; coarse or fine), and widely varying in RM were used. The generalized power or exponential model that merges three crops (Figure 1) may predict the root biomass with reliable accuracy ($R^2 = 0.93$) for comparable soil and environmental condition, particularly when obtaining such data is challenging, and approximate estimation is acceptable for the desired strategy. The study is continued with six annual crop species and a better generalized model is expected.

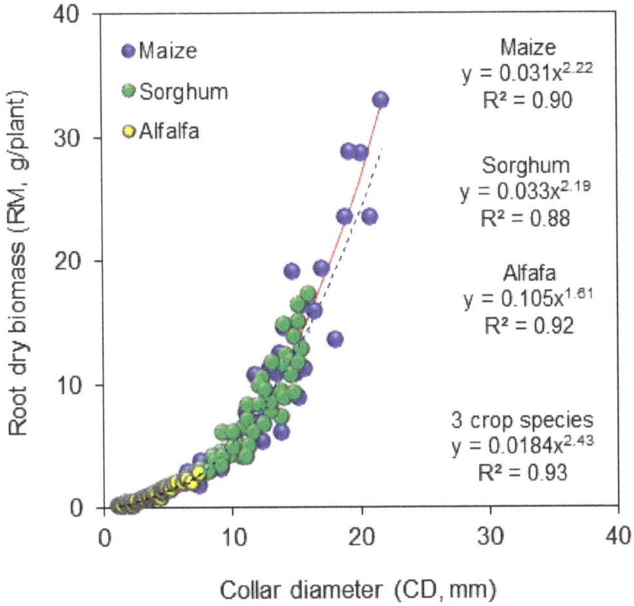

Figure 1. Relation between Root biomass (RM) and collar diameter (CD) of crops.

The largely recognized relations between the height, biomass and CD of the trees could also be used for herbaceous plants with small vertical stems, including used annual field crops [8,10,12–14]. Therefore, if the effect of plant genotype on RM is allocable under various environmental conditions, the allometric models can reflect water and salinity (and nutrient) stress, plant tolerance level, and the input of soil quality [8,10,14]. In the studied region, the root to shoot ratio of annual field crops (0.1–0.3) is mostly related to the abiotic stresses associated with the elevated salinity and scarcity of irrigation water. The order of the RM was somewhat opposite to the root to shoot ratio, showing that sorghum or maize were crops more resistant to the combined abiotic stresses than alfalfa [14].

The cropping improved SOC content and aggregate stability that was significantly higher under crops than in the control (Figure 2). Soil aggregation and pore size distribution in the root zone was frequently modified during the period of active root growth, and by the microbial activity. Roots rhizodeposit (dead fine roots, exudates, mucilages, secretions of insoluble materials, and lysates) differ with crop species, and play a vital role in ecosystem functions; it also form C sources, and enhances SOC and soil structure. However, RM correlated unfairly with SOC and structure stability (Figure 2), which could be explained by the variation in root system attributes, and the rhizodeposition of crop species, and soil texture, since high aggregate stability of clay soil may surpass the role of crops RM [1–6]. Coupled salinity and water stress, and increase in salinity with depth, combined with clay soil texture stress, likely (i) affected the fraction of root system modules, and rooting depth, and water use efficiency triggered by a saline gradient and moisture content; (ii) increased crop stomatal resistance under deficit irrigations, and decreased respiration, transpiration and photosynthetic capacity of crops; and (iii) consequently reduced root and shoot biomass, although roots were less affected by salinity [11,14].

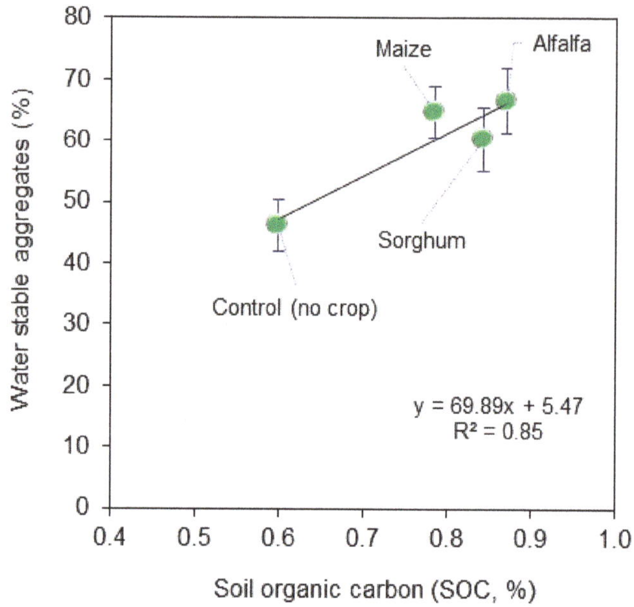

Figure 2. Relation between soil organic carbon (SOC) content and water stable aggregates.

4. Conclusions

Roots play a crucial role in the crop performance and C sequestration, yet selection of the cultivars or crop functional traits are mainly based on shoot-related phenotypes. This study provides a new allometric model for predicting RM from measurements of CD; the relation suggests that various effects of stresses will be laid in the line of the curve. To our knowledge, the study provides the first novel model predictions against direct RM data obtained from field measurements. The RM can be linked to the available data on shoot biomass and crop height (similar to the allometric relations widely used for forest trees) for selecting and breeding of the crop, and functional traits to maximize water and nutrient use efficiency with a consideration of management and climate change scenarios [1,4]. It can be applied to the cropping system for sustainable management, and soil quality improvement associated with C sequestration and enhanced soil processes. Use of this approach in the semi-arid Kur-Araz lowland or other regions with a similar soil type, irrigation requirements, and environmental conditions, seems promising.

Author Contributions: Conceptualization, A.I.M.; methodology, A.I.M. and E.K.H.; software, A.I.M.; validation, A.I.M.; formal analysis, A.I.M. and E.K.H.; investigation, A.I.M. and E.K.H.; resources, A.I.M.; data curation, A.I.M.; writing—original draft preparation, A.I.M.; writing—review and editing, A.I.M. and E.K.H.; visualization, A.I.M.; supervision, A.I.M.; project administration, A.I.M.; funding acquisition, A.I.M. All authors have read and agreed to the published version of the manuscript.

Funding: This research received no external funding.

Acknowledgments: Support by ALRC Tottori University and ANAS is acknowledged.

Conflicts of Interest: The authors declare no conflict of interest.

References

1. Powlson, D.S.; Gregory, P.J.; Whalley, W.R.; Quinton, J.N.; Hopkins, D.W.; Whitmore, A.P.; Hirsch, P.R.; Goulding, K.W.T. Soil management in relation to sustainable agriculture and ecosystem services. *Food Policy* **2011**, *36*, S72–S87. [CrossRef]
2. Daynes, C.N.; Field, D.J.; Saleeba, J.A.; Cole, M.A.; McGee, P.A. Development and stabilisation of soil structure via interactions between organic matter, arbuscular mycorrhizal fungi and plant roots. *Soil Biol. Biochem.* **2013**, *57*, 683–694. [CrossRef]

3. Roumet, C.; Birouste, M.; Picon-Cochard, C.; Ghestem, M.; Osman, N.; Vrignon-Brenas, S.; Cao, K.; Stokes, A. Root structure–function relationships in 74 species: Evidence of a root economics spectrum related to carbon economy. *New Phytol.* **2016**, *210*, 815–826. [CrossRef] [PubMed]
4. Faucon, M.P.; Houben, D.; Lambers, H. Plant functional traits: Soil and ecosystem services. *Trends Plant Sci.* **2017**, *22*, 385–394. [CrossRef] [PubMed]
5. Hirte, J.; Leifeld, J.; Abiven, S.; Oberholzer, H.-R.; Mayer, J. Below ground carbon inputs to soil via root biomass and rhizodeposition of field-grown maize and wheat at harvest are independent of net primary productivity. *Agric. Ecosyst. Environ.* **2018**, *265*, 556–566. [CrossRef]
6. Dijkstra, F.A.; Zhu, B.; Cheng, W. Root effects on soil organic carbon: A double-edged sword. *New Phytol.* **2021**, *230*, 60–65. [CrossRef] [PubMed]
7. Freschet, G.T.; Pagès, L.; Iversen, C.M.; Comas, L.H.; Rewald, B.; Roumet, C.; Klimešová, J.; Zadworny, M.; Poorter, H.; Postma, J.A.; et al. A starting guide to root ecology: Strengthening ecological concepts and standardising root classification, sampling, processing and trait measurements. *New Phytol.* **2021**, *232*, 973–1122. [CrossRef] [PubMed]
8. Poorter, H.; Niklas, K.J.; Reich, P.B.; Oleksyn, J.; Poot, P.; Mommer, L. Biomass allocation to leaves, stems and roots: Meta-analyses of interspecific variation and environmental control. Tansley Review. *New Phytol.* **2012**, *193*, 30–50. [CrossRef] [PubMed]
9. Addo-Danso, S.D.; Prescott, C.E.; Smith, A.R. Methods for estimating root biomass and production in forest and woodland ecosystem carbon studies: A review. *For. Ecol. Manag.* **2015**, *359*, 332–351. [CrossRef]
10. Marden, M.; Lambie, S.; Phillips, C. Biomass and root attributes of eight of New Zealand's most common indigenous evergreen conifer and broadleaved forest species during the first 5 years of establishment. *N. Z. J. For. Sci.* **2018**, *48*, 9. [CrossRef]
11. Fageria, N.K. *The Role of Plant Roots in Crop Production*; CRC Press: Boca Raton, FL, USA, 2012; p. 467. [CrossRef]
12. Reddy, V.R.; Pachepsky, Y.A.; Whisler, F.D. Allometric relationships in field grown soybean. *Ann. Bot.* **1998**, *82*, 125–131. [CrossRef]
13. Song, Y.; Rui, Y.; Birch, C.; Hanan, J. Allometric relationships of maize organ development under different water regimes and plant densities. *Plant Prod. Sci.* **2015**, *18*, 1–10. [CrossRef]
14. Mamedov, A.I.; Gasimova, K.; Husiyev, E.K.; Farzaliyev, V.; Alizade, V.M.; Toderich, K. Root and Shoot Relation of the Quinoa and Forage Plants in Salt-Affected Clay Soil. In *Emerging Research in Alternative Crops. Environment & Policy*; Hirich, A., Choukr-Allah, R., Ragab, R., Eds.; Springer: Cham, Switzerland, 2020; Volume 58, pp. 185–220. [CrossRef]

environmental sciences proceedings

Proceeding Paper

Effect of Seasonal Environmental Changes on Leaf Anatomical Responses of *Limoniastrum guyonianum* in Sabkha Biotope †

Fayçal Boughalleb [1,*], Sameh Maaloul [2] and Raoudha Abdellaoui [1]

1 LR16IRA03 Laboratory of Rangeland Ecosystems and Valorization of Spontaneous Plants, Arid Regions Institute, University of Gabes, Medenine 4119, Tunisia; raoudhamabdellaoui@yahoo.com
2 Faculty of Sciences of Gabes, University of Gabes, Gabes 6072, Tunisia; samah.maaloul@gmail.com
* Correspondence: boughalleb.fay@gmail.com
† Presented at the 2nd International Laayoune Forum on Biosaline Agriculture, 14–16 June 2022; Available online: https://lafoba2.sciforum.net/.

Abstract: Climate change conditions can strongly influence the kinetics of morphogenetic processes. Our study showed that the total thickness of leaf lamina, adaxial palisade parenchyma, abaxial palisade parenchyma and spongy parenchyma increased significantly during the dry period, especially in August (31.4%, 52.1%, 37.6%, 27.69%, respectively). Moreover, the adaxial and abaxial epidermis becomes thicker during the most dry months (July and August). Likewise, the adaxial cuticle thickness increased during the dry period. The stomata density in the adaxial and abaxial leaf sides is 1.36- and 1.4-fold higher than those recorded during the wet periods. However, the salt glands' density showed a much greater increase in the abaxial face (+2.4-fold). The bundle sheath size was unchanged under the seasonal environmental fluctuation in the sabkha. The xylem vessels diameter showed a maximum reduction in August (–63.8%). Likewise, the xylem vessels density increased significantly during the dry period. The closer relationship between the anatomical proprieties with soil salinity allows us to conclude that salt stress is one of the most limiting factors for *Limoniastrum guyonianum* in its natural biotope.

Keywords: anatomical proprieties; *Limoniastrum guyonianum*; seasonal environmental changes; saline biotope

Citation: Boughalleb, F.; Maaloul, S.; Abdellaoui, R. Effect of Seasonal Environmental Changes on Leaf Anatomical Responses of *Limoniastrum guyonianum* in Sabkha Biotope. *Environ. Sci. Proc.* **2022**, *16*, 12. https://doi.org/10.3390/environsciproc2022016012

Academic Editor: Abdelaziz Hirich

Published: 16 June 2022

1. Introduction

Limoniastrum guyonianum Boiss is a wild herb (*Plumbaginaceae*) growing in the deserts of North Africa, especially in Northern Sahara (Algeria, Tunisia) in the salty soils of the great chotts [1]. This halophyte is characterized by the presence of salt glands that contribute to salt excess excretion [2]. Commonly used for dune stabilization and landscaping [3] also played an important role in folk medicine as an anti-dysenteric, antibacterial and antidiabitic [4]. Photochemical studies have shown that this species presents anti-inflammatory and antitumor proprieties [5]. However, no study has been carried out on the impacts of environmental condition changes on *L. guyonianum* grown in their natural habitats yet. Saline habitats could be subjected, in addition to salinity, to high temperature; drought; flooding; active deflation depending on site and seasons [6]. Consequently, this study aimed to evaluate the monthly anatomical behavior of *L. guyonianum* in relation to the environmental conditions of the Sabkha of Aïn Maïder.

2. Materials and Methods

The plant was harvested from the shott of the Sebkha of Aïn Maider–Boughrara, a south-eastern coastal area, located at 33°27′52″ N, 10°43′31″ E, which is 35 km from the city of Medenine, Tunisia. Meteorological data were obtained monthly from the nearest synoptic weather station to the study site. The soil electrical conductivity (EC) was measured using a conductivity meter. Sodium content was determined using an atomic absorption spectrophotometer. The Anatomical observations were performed under a light microscope

3. Results and Discussion

The soil EC showed significant monthly variation, and the highest value was recorded in August which was about 4.68-fold greater than those measured during the wet period (1.27 in January) (Figure 1).The Na^+ value significantly increased during the dry period with a maximum of 769.1 mmol·g^{-1} soil in August, reflecting an increase of about 4.68-fold as compared to those obtained during the rainy period.

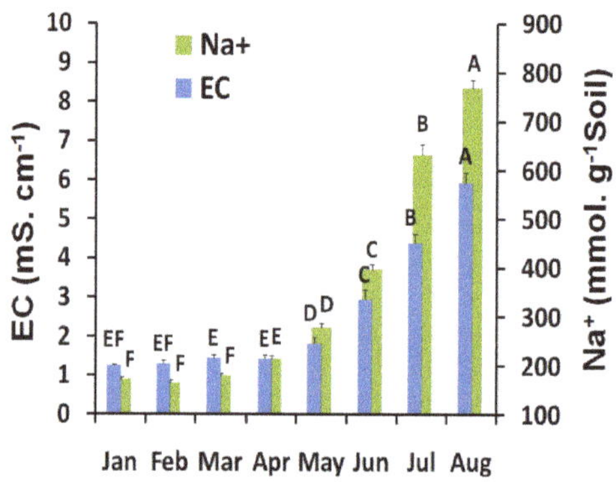

Figure 1. Monthly variations of the soil physicochemical parameters.

Leaf anatomical modifications under limited moisture availability can play an important role under salt stress, and they are an indicator of the degree of tolerance.The increased salinity in Ain Maider's Sabkha significantly increased the leaf lamina thickness, palisade and spongy parenchyma (Table 1 and Figure 2). The necessity to conserve water renders the leaves succulent, thus increasing leaf thickness. These anatomical features may help in storing ions inside the plant body due to increased vacuolar volume [7], thus permitting the plant to cope with higher salt amounts.

Table 1. Anatomical parameters changes in the aerial part of *L. guyonianum* under the variation of climatic conditions in the Sabkha of Ain Maider.

Characters	Treatments							
	January	February	March	April	May	June	July	August
Leaf thickness (μm)	485.4 ± 3.98 F	484.1 ± 7.66 F	491.7 ± 6.98 F	518.8 ± 7.18 E	554.6 ± 9.66 D	581.7 ± 6.45 C	608.2 ± 3.29 B	637.9 ± 6.98 A
Adaxial epidermis (μm)	24.1 ± 0.35 B	23.9 ± 0.40 B	24.0 ± 0.32 B	24.2 ± 0.38 B	24.1 ± 0.42 B	24.3 ± 0.30 B	25.1 ± 0.71 B	26.1 ± 0.26 A
Adaxial stomatal density (nb·mm^{-2})	62.2 ± 1.01 C	60.3 ± 1.52 C	63.1 ± 1.05 C	63.0 ± 2.64 C	63.3 ± 1.55 C	71.3 ± 4.02 B	82.0 ± 1.18 A	84.3 ± 0.69 A
Adaxial salt glandsdensity (nb·mm^{-2})	8.23 ± 0.25 CD	7.73 ± 0.61 D	8.16 ± 0.15 CD	8.32 ± 0.20 CD	8.90 ± 0.36 C	10.56 ± 0.40 B	11.66 ± 0.76 A	11.85 ± 0.33 A
Abaxial epidermis (μm)	23.9 ± 0.15 BC	23.2 ± 0.41 BC	23.1 ± 0.30 C	23.7 ± 0.35 BC	23.9 ± 0.65 BC	23.7 ± 0.40 BC	24.3 ± 0.20 B	25.1 ± 0.61 A
Abaxial stomatal density (nb·mm^{-2})	58.7 ± 1.12 D	57.9 ± 2.83 D	58.8 ± 2.02 D	60.4 ± 2.15 D	58.5 ± 2.50 D	66.7 ± 1.30 C	75.6 ± 1.35 B	82.1 ± 4.37 A
Abaxial salt glandsdensity (nb·mm^{-2})	10.4 ± 0.64 D	10.5 ± 0.71 D	11.7 ± 0.27 D	12.3 ± 0.36 D	13.0 ± 0.30 D	15.7 ± 0.37 C	23.8 ± 2.41 B	25.7 ± 2.83 A
Adaxial palisadeparenchyma (μm)	95.3 ± 3.51 F	92.6 ± 2.08 F	96.3 ± 1.52 F	102.7 ± 2.10 E	113.4 ± 4.49 D	127.1 ± 2.68 C	138.0 ± 3.10 B	144.9 ± 2.57 A
Abaxial palisadeparenchyma (μm)	62.0 ± 1.01 CD	60.3 ± 1.52 D	63.7 ± 2.05 CD	64.6 ± 1.50 CD	68.0 ± 2.04 C	75.3 ± 4.85 B	82.8 ± 1.25 A	85.3 ± 3.51 A

Table 1. *Cont.*

Characters	Treatments							
	January	February	March	April	May	June	July	August
Spongyparenchyma (µm)	269.3 ± 6.94 D	273.0 ± 9.58 D	273.7 ± 8.61 D	292.8 ± 6.25 C	313.8 ± 6.27 B	319.4 ± 3.87 B	326.5 ± 4.11 B	344.6 ± 3.81 A
Xylem vessel diameter (µm)	14.7 ± 0.21 A	14.9 ± 0.15 A	14.1 ± 0.10 B	12.0 ± 0.10 C	11.4 ± 0.25 D	11.0 ± 0.05 D	9.4 ± 0.10 F	9.4 ± 0.06 F
Xylem density (nb·mm^{-2})	1934.3 ± 15.1 E	1909.0 ± 10.6 E	1930.6 ± 24.1 E	2149.5 ± 54.3 D	2662.1 ± 53.8 C	2852.0 ± 50.1 B	3246.2 ± 77.4 A	3317.0 ± 43.8 A

The different letters (A–F) in the same row indicate significantly different values at $p < 0.05$ as described by Duncan's test.

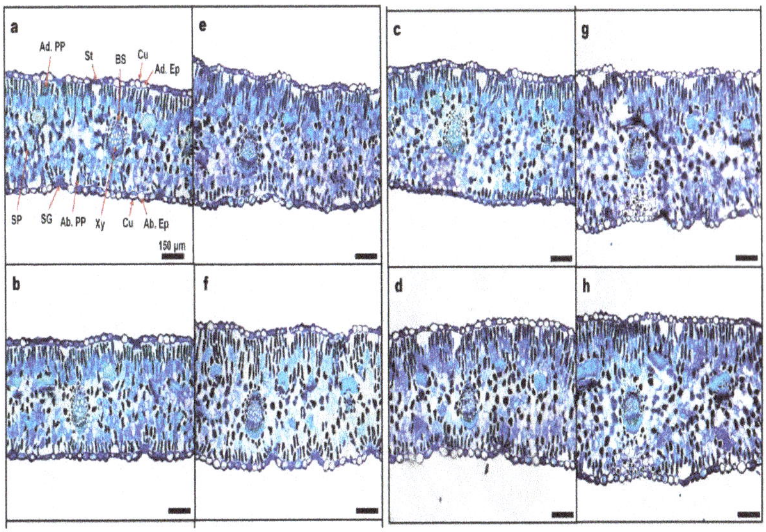

Figure 2. Cross-section showing anatomical changes in the aerial part of *L. guyonianum* subjected to monthly variations of climatic conditions in Ain Maider Sabkha ((**a**) = January, (**b**) = February, (**c**) = March, (**d**) = April, (**e**) = May, (**f**) = June, (**g**) = July and (**h**) = August). Bars = 1500 µm. Ab. Ep, Abaxial epidermis; Ab. PP, abaxial palisade parenchyma; Ad. Ep, Adaxial epidermis; Ad. PP, adaxial palisadeparenchyma; BS, bundle sheath; Cu, cuticle; SP, spongy parenchyma; SG, salt gland; St, stomata; Xy, xylem.

Concerning epidermis size, it is common to observe a thickening of the skin under salt stress, which can be related to salt tolerance.In our results, epidermis thickness was significantly thicker only during the highest dry month (August). A positive correlation between the thickening of epidermal cells and salt tolerance has been demonstrated [8]. This characteristic is critical in conditions where the availability of water is limited; in fact, a thick epidermis can better control the loss of water [9].

Regarding the vascular system, our results exhibited that high salinity (dry period) reduced the xylem vessel diameter, while the xylem density increased (Table 1). Stiller et al. [10] reported that cavitation occurs when the flow of water in the xylem vessels cannot keep the sweat rate. Thus, selection for narrow vessels in response to improved water use efficiency would reduce the risk of xylem embolisms in saline habitats.

This study also revealed that stomata were evenly distributed on both leaf surfaces of *L. guyonianum*. Numerous works linked drought and or salt stress plant adaptation to the increase in stomatal density [11,12]. In the present study, leaf stomatal density increased during the dry season (June–August).

Salt glands play an important role in the regulation of ionic balance, contributing to salt tolerance [13]. In our study, higher salt gland densities were observed in *L. guyoni-*

anum plants subjected to summer climatic conditions as these epidermal structures are important for the plant's survival when grown under stressful conditions. In *Limoniastrum monopetalum* the salt gland is organized as an embedded cup of multiple cells [14].

4. Conclusions

Our study indicated that *Limoniastrum guyonianum* used many anatomical mechanisms to tolerate salinity, especially a decrease of the mesophyll cells area, the promotion of xylem production and the accumulation and excretion of the excess of salt by salt glands.

Author Contributions: F.B. and R.A. conceived and designed the experiments. S.M. and F.B. conducted experiments. F.B. and R.A. co-wrote all drafts of the paper and also approved the final draft for submission. All authors have read and agreed to the published version of the manuscript.

Funding: Tunisian Ministry of Higher Education and Scientific Research (LR16IRA03).

Institutional Review Board Statement: Not applicable.

Informed Consent Statement: Not applicable.

Data Availability Statement: Data is contained within the article.

Conflicts of Interest: The authors declare no conflict of interest.

References

1. Quezel, P.; Santa, S. *Nouvelle Flore de l'Algérie et des Régions Désertiques Méridionales;* CNRS: Paris, France, 1963.
2. Ding, F.; Chen, M.; Sui, N.; Wang, B.S. Ca^{2+} significantly enhanced development and salt-secretion rate of salt glands of Limonium bicolor under NaCl treatment. *S. Afr. J. Bot.* **2010**, *76*, 95–101. [CrossRef]
3. Zouhaier, B.; Najla, T.; Abdallah, A.; Wahbi, D.; Wided, C.; Chedly, A.; Abderrazak, S. Salt stress response in the halophyte Limoniastrum guyonianum Boiss. *Fun. Ecol. Plants* **2015**, *217*, 1–9. [CrossRef]
4. Telli, A.; Esnault, M.A.; Khelil, A.O.E.H. An ethnopharmacological survey of plants used in traditional diabetes treatment in south-eastern Algeria (Ouargla province). *J. Arid Environ.* **2016**, *127*, 82–92. [CrossRef]
5. Krifa, M.; Mustapha, N.; Ghedira, Z.; Ghedira, K.; Pizzi, A.; Chekir-Ghedira, L. Limoniastrum guyonianum methanol extract induces immunomodulatory and anti-inflammatory effects by activating cellular anti-oxidant activity. *Drug Chem. Toxicol.* **2015**, *38*, 84–91. [CrossRef] [PubMed]
6. Al Hassan, M.; Chaura, J.; Donat-Torres, M.P.; Boscaiu, M.; Vicente, O. Antioxidant responses under salinity and drought in three closely related wild monocots with different ecological optima. *AoB Plants* **2017**, *9*, plx009. [CrossRef] [PubMed]
7. Hameed, M.; Ashraf, M.; Naz, N. Anatomical adaptations to salinity in cogon grass (*Imperatacylindrica* (L.) Raeuschel) from the Salt Range, Pakistan. *Plant Soil* **2009**, *322*, 229–238. [CrossRef]
8. Wankhade, S.D.; Comejo, M.J.; Mateu-Andre's, I.; Sanz, A. Morphophysiological variations in response to NaCl stress during vegetative and reproductive development of rice. *Acta Physiol. Plant* **2013**, *35*, 323–333. [CrossRef]
9. Nawazish, S.; Hameed, M.; Naurin, S. Leaf anatomical adaptations of *Oenchrusciliaris* L. From the salt range, pakistan against drought stress. *Pak. J. Bot.* **2006**, *38*, 1723–1730.
10. Stiller, V.; Lafitte, H.R.; Sperry, J.S. Hydraulic properties of rice and the response of gas exchange to water stress. *Plant Physiol.* **2003**, *132*, 1698–1706. [CrossRef] [PubMed]
11. Xu, Z.; Zhou, G. Responses of leaf stomatal density to water status and its relationship with photosynthesis in a grass. *J. Exp. Bot.* **2008**, *59*, 3317–3325. [CrossRef] [PubMed]
12. Laajimi, N.O.; Boussadia, O.; Skhiri, F.H. Teixeira da Silva, J.A., Rezgui, S., Hellali, R. Anatomical Adaptations in Vegetative Structures of Apricot Tree (*Prunus armeniaca* L.) cv. 'Amor El Euch' Grown under Water Stress. *FVCSB* **2011**, *5*, 46–51.
13. Barhoumi, Z.; Djebali, W.; Abdelly, C.; Chaïbi, W.; Smaoui, A. Ultrastructure of Aeluropus littoralis leaf salt glands under NaCl stress. *Protoplasma* **2008**, *233*, 195–202. [PubMed]
14. Ioannidou-Akoumianaki, A.; Spentza, R.P.; Fasseas, C. *Limoniastrum monopetalum* (L.) Boiss, a candidate plant for use in urban and suburban areas with adverse conditions. An anatomical and histochemical study. *Bull. Univ. Agric. Sci. Vet. Med. Cluj-Napoca Hortic.* **2015**, *72*, 438–440.

environmental sciences proceedings

MDPI

Proceeding Paper

Impact of Water Stress on the Productivity and Quality Parameters of Wheat and Their Potential Use in the Breeding Program †

Mohamed Amine Abdellaoui [1,2,*], Mouna Taghouti [1], Ghizlane Diria [1], Mackaye Moussa Hassane [1,3], Houda EL Yacoubi [2] and Sahar Bennani [1]

1 Plant Breeding Department, National Institute of Agricultural Research (INRA), Rabat 10101, Morocco; mouna.taghouti@inra.ma (M.T.); ghizlane.diria@inra.ma (G.D.); mackayemoussa@gmail.com (M.M.H.); sahar.bennani@inra.ma (S.B.)

2 Laboratory of Natural Resources and Sustainable Development, Faculty of Sciences, University Ibn Tofail, Kenitra 14000, Morocco; elyacoubihouda@gmail.com

3 Botany Laboratory Valuation and Conservation of Plant and Fungal Resources, Faculty of Science, Mohamed V University, Rabat 10100, Morocco

* Correspondence: mohamedamine.abdellaoui@uit.ac.ma

† Presented at the 2nd International Laayoune Forum on Biosaline Agriculture, 14–16 June 2022; Available online: https://lafoba2.sciforum.net/.

Abstract: Wheat occupies a prominent place at the national and international levels. This food crop is considered a main source of human and animal nutrition. Water stress is one of the main causes of the decline in bread wheat production in Morocco. Therefore, breeding for resilient varieties is of great importance to overcome this challenge. In this context, the main objective of this study is to assess the impact of drought on the productivity of bread wheat and to determine the key selection criteria for drought tolerance in a Mediterranean environment. To do this, 31 varieties and lines of bread wheat were evaluated at two contrasted INRA experimental stations during the 2020–2021 cropping season. Five productivity and quality parameters were assessed: grain yield, biomass, number of fertile spikes, thousand grain weight, and protein content. The results demonstrate the negative impact of water stress on the various parameters studied, except for the protein content which was higher in arid conditions compared to the favourable ones. Based on correlation analysis, the number of fertile spikes and the biomass could constitute the potential selection criteria for drought resistance. These trials will be repeated for several years under the effect of other intensities of stress and agro-ecosystems and by incorporating other parameters to confirm the results obtained. The line "BT19I5" showed great yield potential under both conditions and can be of great interest for farmers and for potential release at the national program.

Keywords: bread wheat; yield; quality; drought; selection criteria

Citation: Abdellaoui, M.A.; Taghouti, M.; Diria, G.; Hassane, M.M.; EL Yacoubi, H.; Bennani, S. Impact of Water Stress on the Productivity and Quality Parameters of Wheat and Their Potential Use in the Breeding Program. *Environ. Sci. Proc.* **2022**, *16*, 13. https://doi.org/10.3390/environsciproc2022016013

Academic Editor: Abdelaziz Hirich

Published: 16 June 2022

1. Introduction

The cereal sector occupies a decisive place in the Moroccan agriculture. Cereal crops are considered a main source of human and animal nutrition. In Morocco, bread wheat (*Triticum aestivum* L.) is the most consumed cereal, estimated at 258 kg/year/person. About 3 million hectares of wheat are grown annually, with a global production of 48.2 million quintals in 2021.

This cereal crop has been given special attention, particularly in breeding programs in Mediterranean countries, because of its adaptation to semi-arid environments and its unique technological quality, compared to other cereals [1]. Bread wheat production is affected by several abiotic stresses, mainly drought and heat and biotic stresses via attacks by parasites and pests [2], which hinders the achievement of good economic yields of good quality.

The sensitivity of bread wheat to water stress is one of the main causes of the decline in national production over years and even decades. Thus, the orientation of research in the identification of new agricultural techniques and the dissemination of resilient varieties, are part of the adaptation strategies for reducing the effects of climate change and increasing cereal yield in the long term. Grain yield has low heritability (<20%) over variable stress intensities and unpredictable environmental conditions [2]. Thus, the identification of reliable secondary indirect selection criteria can improve the efficiency of selection for drought tolerance [3].

In this perspective, the main objective of this study is to (1) evaluate the impact of drought on the productivity of bread wheat; (2) deduce the key selection criteria for drought tolerance in the Moroccan environment; and (3) determine the best promising lines that are resilient to stress and of interest to farmers in the Mediterranean and arid regions in general.

2. Materials and Methods

The genetic material consists of 31 varieties and lines of bread wheat from the national breeding program at the National Institute for Agronomic Research (INRA). The trials were carried out in two contrasting cereal regions in terms of rainfall at two INRA experimental sites. The first station is located in the region of Rabat "Marchouch" known by humid and relatively favorable climatic conditions (>500 mm). The second station "Jemaat Shaim" is part of the arid agroecosystem in Safi region with an average annual rainfall of less than 300 mm.

The experimental design is a complete randomized block design with three repetitions. Each elementary plot consists of 6 lines of 5 m in length and a spacing of 0.20 m. Five parameters of productivity and quality were analyzed: grain yield, biomass (BM), number of fertile spikes (NFS), thousand grain weight (TGW) and protein content. Statistical analyzes were performed using Genstat 18 software to perform analysis of variance for each site and between sites in addition to Pearson correlation between the various parameters studied.

3. Results and Discussion

The cropping season 2021–2022 has been generally favorable. The national cumulative rainfall amounted to 271.9 mm with a good temporal and spatial distribution against 286.9 mm for the average of the last 30 years, with a slight decrease of 5%. The favorable station of Marchouch (MCH) recorded 367 mm, while the arid station of Jemmat Shaim (JS) accumulated 201 mm.

The average yield at Marchouch was 57.3 qx/ha and 10.7 qx/ha at Jemaat Shaim. The genotypes gave better yields in the Marchouch station (MCH) than in Jemaat Shaim (JS). Indeed, the MCH station represents the favorable environment for the production of cereals where rainfall generally exceeds 400 mm annually. However, the JS station is located towards the south in arid zones where annual rainfall does not exceed 200 mm/year. The genotypes evaluated showed different performances from one station to another (Table 1). The lines "BT19I5", "BT19A21", "CCBT175" and "CCBG" genotypes presented the best results in terms of yield in the Jemaat Shaim station with an average yield of 14.2 qx/ha; while "BT19I5", "BT20A217", in addition to the varieties Resulton and Radia, are the most productive in Marchouch with an average yield of 57.1 qx/ha. The line "BT19I5" remains the most promising as it shows the best performance in both water conditions.

The results demonstrate the negative impact of water stress on the various parameters studied (Table 2), as demonstrated by other previous studies [2,4]. The most influenced traits by water stress were biomass (−61%) and number of fertile spikes (−42%); while the protein content was higher in arid conditions compared to the favourable ones. The analysis of variance showed significant differences for all the parameters studied between the two contrasting stations in terms of rainfall ($p < 0.001$).

Table 1. Mean grain yield of the different genotypes at the two stations JS: Jemaat Shaim, MCH: Marchouch.

Genotype/Line	Yield at JS (qx/ha)	Yield at MCH (qx/ha)	Genotype/Line	Yield at JS (qx/ha)	Yield at MCH (qx/ha)
Achtar	8.56	50.14	T3BT	8.89	50.85
T1BT	8.34	47.08	T4BT	12.89	59.44
T2BT	5.78	49.98	T5BT	11.56	45.03
Arrehane	12.89	62.11	CCBT144	10.22	54.53
Bandera	9.78	60.64	T6BT	12.00	49.00
BT19A04	13.00	59.89	T7BT	10.34	49.25
BT19A21	13.67	62.92	T8BT	9.89	60.11
BT19I5	15.56	70.25	T9BT	10.56	59.00
BT20A217	11.11	64.67	PVT3	10.34	57.61
CCBG	11.78	62.07	Radia	11.00	65.31
CCBT108	10.44	62.53	INRA-10	7.22	50.36
CCBT155	9.67	50.36	PVT4	9.00	53.83
CCBT175	14.56	55.72	Resulton	7.56	65.43
CCBT65	11.56	60.61	BT15-42	14.22	57.5
PVT1	12.44	62.50	INRA-11	7.11	54.22
PVT2	9.33	64.22			

Table 2. Descriptive statistics of the different parameters studied in the two experimental stations.

Criteria	Favourable Station (Marchouch)	Arid Station (Jemaat Shaim)
Yield (qx/ha)	57.33	10.68
BM (g)	275.5	106.2
NEF	77.48	44.60
TGW (mg)	36.13	28.01
Proteins (%)	13.65	17.09

BM: Biomass; NEF: Number of fertile spikes; TGW: Thousand grain weight.

Correlation analysis (Tables 3 and 4) showed a strong association between yield, biomass and the number of fertile spikes in the arid conditions. Biomass is an important site of photosynthesis for the plant, allowing it to generate better productivity [5]. At the same time, the fertility of the spikes has a significant impact on the final yield, making it possible to generate a large number of grains. Therefore, preventing floret mortality at pre-flowering stage can hinder significant reductions in yield [6,7]. These two parameters could constitute potential selection criteria for drought tolerance in Mediterranean conditions. On the other hand, a negative association linked the biomass to the protein rate. This could be linked to the absence of rain during the March–April grain-filling period, which promotes competition between the different organs. At Marchouch, the only positive and significant correlation is that between the biomass and the number of fertile spikes under relatively favourable conditions.

Table 3. Pearson correlation between the characters studied in the arid site of Jemaat Shaim.

Traits	BM	NEF	TGW	Proteins	Yield
BM	1				
NEF	0.24	-			
TGW	0.01	−0.24	-		
Proteins	−0.43 *	0.0007	−0.30	-	
Yield	0.48 **	0.35 *	0.01	−0.15	-

BM: Biomass; NEF: Number of fertile spikes; TGW: Thousand grain weight. *, ** significant correlation at 0.5, 0.01 probabilities respectively.

Table 4. Pearson correlation between the traits studied in the semi-favourable site of Marchouch.

Traits	BM	NEF	TGW	Proteins	Yield
BM	-				
NEF	0.52 **	-			
TGW	−0.07	−0.002	-		
Proteins	0.28	0.23	−0.19	-	
Yield	0.17	−0.14	−0.04	−0.03	-

BM: Biomass; NEF: Number of fertile spikes; TGW: Thousand grain weight. ** significant correlation at 0.01 probabilities respectively.

4. Conclusions

The results obtained from this study demonstrate the significant impact of water stress on wheat productivity in Mediterranean and Moroccan conditions in particular. The number of fertile spikes and the biomass could constitute the potential selection criteria for drought resistance. These trials will be repeated for several years under the effect of other intensities of stress and agro-ecosystems and by incorporating other parameters to confirm the results obtained. The line "BT19I5" showed great yield potential under both conditions, can be of great interest for farmers, and therefore could be presented to the national catalogue for release after further evaluations.

Author Contributions: Conceptualization, S.B. and M.T.; methodology, S.B., G.D. and M.T.; validation, M.A.A., S.B., G.D. and M.T.; formal analysis, M.A.A. and S.B.; investigation, M.A.A. and M.M.H.; data curation, M.A.A.; writing—original draft preparation, M.A.A.; writing—review and editing, S.B. and M.A.A.; supervision, H.E.Y., S.B., G.D. and M.T. All authors have read and agreed to the published version of the manuscript.

Funding: This research received no external funding.

Acknowledgments: The authors gratefully acknowledge valuable contribution and support of the technical staff of the two experimental stations and the wheat breeding laboratory.

Conflicts of Interest: The authors declare no conflict of interest.

References

1. Nazco, R.; Villegas, D.; Ammar, K.; Peña, R.J.; Moragues, M.; Royo, C. Can Mediterranean durum wheat landraces contribute to improved grain quality attributes in modern cultivars? *Euphytica* **2012**, *185*, 1–17. [CrossRef]
2. Bennani, S.; Nsarellah, N.; Jlibene, M.; Tadesse, W.; Birouk, A.; Ouabbou, H. Efficiency of drought tolerance indices under different stress severities for bread wheat selection. *Austr. Agric. Crop Sci.* **2017**, *4*, 395–405. [CrossRef]
3. Monneveux, P.; Jing, R.; Misra, S., II. 1.2 Phenotyping wheat for adaptation to drought using physiological traits. *Front. Physiol.* **2012**, *3*, 429. [CrossRef] [PubMed]
4. Farooq, M.; Wahid, A.; Kobayashi, N.; Fujita, D.; Basra, S.M.A. Plant drought stress: Effects, mechanisms and management. *Agron. Sustain.* **2009**, *29*, 185–212. [CrossRef]
5. Richards, R.A. Selectable traits to increase crop photosynthesis and yield of grain crops. *J. Exp. Bot.* **2000**, *51*, 447–458. [CrossRef] [PubMed]

6. Garcia del Moral, L.F.; Rharrabti, Y.; Villegas, D.; Royo, C. Evaluation of Grain Yield and Its Components in Durum Wheat under Mediterranean Conditions: An Ontogenic Approach. *Agronomy* **2003**, *95*, 266–274. [CrossRef]
7. Slafer, G.A.; Elia, M.; Savin, R.; García, G.A.; Terrile, I.; Ferrante, A.; Miralles, D.J.; González, F.G. Fruiting efficiency, an alternative trait to further rise wheat yield. *Food Energy Secur.* **2015**, *4*, 92–109. [CrossRef]

environmental sciences proceedings

MDPI

Proceeding Paper

Assessment of Morphological and Physiological Traits of Moroccan Barley (*Hordeum vulgare* L.) Varieties Submitted to Severe Salt Stress †

Said Bouhraoua [1,*], Nassira Srhiouar [1], Mohamed Ferioun [1,2], Khalil Hammani [1] and Said Louahlia [1]

1 Natural Resources and Environment Laboratory, Multidisciplinary Faculty of Taza, Sidi Mohammed Ben Abdellah University, Taza 35000, Morocco; nassira.oligo@gmail.com (N.S.); Mohamed.ferioun@usmba.ac.ma (M.F.); khalil.hammani@usmba.ac.ma (K.H.); said.louahlia@usmba.ac.ma (S.L.)

2 Microbial Biotechnology and Bioactive Molecules Laboratory, Faculty of Sciences and Technologies, Fez, Sidi Mohammed Ben Abdellah University, Fez 30000, Morocco

* Correspondence: said.bouhraoua88@gmail.com; Tel.: +212-67-525-0324

† Presented at the 2nd International Laayoune Forum on Biosaline Agriculture, 14–16 June 2022; Available online: https://lafoba2.sciforum.net/.

Abstract: The maintenance of Moroccan barley (*Hordeum vulgare* L.) cultivation under rising saline conditions requires investigations to be performed. In the present work, we aimed to test the effect of salt stress on four Moroccan barley varieties (ADRAR, AMIRA, LAANACEUR, and MASSINE). Salt stress was applied by increasing NaCl concentration gradually in nutrient solution to 300 mM (Severe stress). Our results showed that salt stress induces significant decreases in RDW, SDW, and Chl content. In addition, significant increases of EL% and proline content were recorded. Analysis of variance showed a significant intraspecific variability between varieties and a significant effect of treatment and combination of the varieties factor and treatment factor. Principal component analysis (PCA) showed that under NaCl treatment, MASSINE is the genotype that kept significantly high values of SDW and RDW.

Keywords: *Hordeum vulgare*; salinity; Moroccan barley

Citation: Bouhraoua, S.; Srhiouar, N.; Ferioun, M.; Hammani, K.; Louahlia, S. Assessment of Morphological and Physiological Traits of Moroccan Barley (*Hordeum vulgare* L.) Varieties Submitted to Severe Salt Stress. *Environ. Sci. Proc.* **2022**, *16*, 21. https://doi.org/10.3390/environsciproc2022016021

Academic Editor: Abdelaziz Hirich

Published: 16 June 2022

1. Introduction

In the Mediterranean area, salinity is a serious problem reducing plants growth and crop productivity [1] markedly affecting agricultural land. For instance, more than 500,000 ha of the saline area are considered as damaged soils [2]. Salt stress affects morphological, physiological, and biochemical parameters as well as crop production in plants including cereals [3]. In Morocco, barley (*Hordeum vulgare* L.) is the second cereal used by Moroccan population in their alimentation after wheat [4]. The degree of salt tolerance is not the same at all barley varieties [5], which makes the selection of Moroccan tolerant varieties very important to ensure nutrition security for the next decades. Here, we aimed to screen the effect of severe salt stress on some morphological and physiological traits at four barley Moroccan varieties.

2. Material and Methods

Experiments were conducted in a greenhouse at the experimental station. Six days after sowing, seedlings of four barley varieties were subsequently transferred to a hydroponic culture system (10 L per pot) containing a complete and continuously aerated Hoagland nutrient solution. Two treatments were applied, control without NaCl and stressed with 300 mM of NaCl in the nutrient solution. Plants were grown at 21 °C under natural light supplemented with artificial light with PAR of 300 μmol photons.m^{-2}.s^{-1}. Fresh leaves and root samples used for analysis were harvested 10 days after the application of the treatment.

For morphological parameters, the dry weights of shoots and roots were determined after being dried at 70 °C for 72 h. The electrolyte leakage (EL) was assayed by the technique of [6]. Total chlorophyll (Chl) was determined according to the method of Burnison, B. K. (1980) [7]. The proline accumulation was assessed as described in Bates et al. (1973) [8]. For statistical analyses, we submitted data to a variance analysis (ANOVA) to test the effects of salt stress, genotype, and their interactions on the morphological, physiological and biochemical traits of the Barley. Mean comparisons were made using test (Fisher's comparison) at $p \leq 0.05$. Different statistical approaches were performed using XLSTAT software (XLSTAT Version 2016.02.28451).

3. Results and Discussion

3.1. Effect of Salt Stress on Barley Genotypes

Combined analysis of variance (Table 1), was used for biochemical and physiological traits of the four barley varieties measured after two-week under severe salt stress conditions. ANOVA also showed that salt stress significantly affected the biochemical traits as well as physiological traits. Moreover, highly significant differences were noticed among the genotypes for all the parameters. The ANOVA also disclosed a significant interaction between varieties and treatment for the biochemical and physiological traits. The comparisons of means (Figure 1) showed a significant reduction in root and shoot dry weight as well as Chl content in response to salt stress, AMIRA was found to have the most important reduction of RDW and SDW (79% and 82.2%, respectively) and total chlorophyll content (70%) when submitted to salinity. MASSINE displayed the lowest reduction percentage of RDW and SDW (59% and 71%, respectively), this variety showed also the highest score of proline accumulation (96%). The electrolyte leakage and proline content were significantly higher under salt stress compared to the control. The highest increase of the electrolyte leakage percentage was noted at AMIRA (55%).

Table 1. Mean squares of parameters studied in leaves and roots samples of four barley Moroccan varieties.

Source	Df	RDW	SDW	EL	Chl T	Proline
Variety	3	0.248947 ***	0.3165 ***	45.72 ***	0.2398 ***	22.872 ***
Treatment	2	0.7168 ***	12.2108 ***	6734.57 ***	11.7309 ***	386.290 ***
replicate	2	0.000165	0.0205	3.03	0.0082	0
Variety*trt	6	0.157225 ***	0.2287 ***	15.38 ***	1.2233 ***	17.974 ***
Residual	22	0.000272	0.0136	1.27	0.0039	16
Total	35					

Df: Degree of freedom. ***: significant at 0.001.

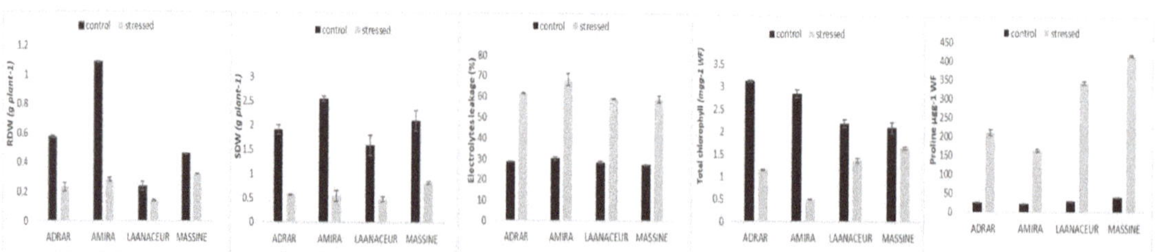

(a) (b) (c) (d) (e)

Figure 1. Effect of 300 mM NaCl treatment on RDW (**a**) SDW (**b**) EL% (**c**) chlorophyll content (**d**) and proline content (**e**) in four barley Moroccan varieties. Values indicate the means of three replicates. Treatment was applied during ten days.

It is widely spread in literature that salt stress induces a great reduction of plant growth and crop yield [5,9,10]. The decrease of RDW and SDW under osmotic stresses indicates that plants maintain resources by reducing their vegetative growth [11], which is well documented in many papers [12,13]. AMIRA in our results was the variety that showed the important decrease of SDW and RDW under salt stress, which indicates that this variety seems to be sensitive against this stress.

Photosynthetic pigments content is considered as good criterion for evaluating the tolerance of plants against abiotic stresses [14]. The reduction of total Chl content is considered as an indicator of abiotic stresses sensitivity [3], that was also recorded in many other studies [13], and could be due to occurrence of increase Chl degradation and pigment photo-oxidation [15]. The most important decrease of Chl content in our case was observed in AMIRA, which makes this variety, based on this parameter changes, more sensitive against salt stress. The increase of EL% under stress is always linked with sensitivity of plant into abiotic stress [16]. As in our case, many studies describe the increase of EL% under salt stress, which indicates cell membrane deterioration.

Under stressful conditions, as in our case, many other studies show a significant increase in leaves proline content [5]. This amino acid is considered as an osmoregulator for membrane stability, also it is known by its ability of buffering cellular redox potential, and for scavenging free radicals [17]. In our results, MASSINE and LAANACEUR are the varieties that accumulated more proline under salt stress, which make them more tolerant based on this criteria.

3.2. Correlations Matrix among Parameters Studied

Table 2 shows the correlations matrix among parameters studied, positive and negative correlations were recorded in our results. In control, a significant positive correlation of RDW with SDW, EL, and Chl content, and a significant negative correlation with proline content. SDW shows a significant positive correlation with EL. For EL, significant positive and negative correlations were recorded with Chl and proline contents, respectively. Furthermore, Chl content shows a significant negative correlation with proline. Under salt stress, RDW was in significant positive correlation with SDW. EL% correlates significantly and negatively with Chl and proline contents. A significant positive correlation between Chl and proline contents was recorded.

Table 2. Correlation matrix (Pearson) Somme parameters studied under control conditions and under salt stress.

Variables	RDW	SDW	EL	Chl T	Proline
Control					
RDW		**0.881** ***	**0.779** ***	**0.588** *	**−0.578** *
SDW			**0.591** *	0.356	−0.248
EL				**0.615** *	**−0.821** ***
Chl					**−0.764** ***
Salt Stress					
RDW		**0.703** *	0.298	−0.091	0.013
SDW			−0.366	0.498	0540
EL				**−0.916** ***	**−0.799** ***
Chl					**0.915** ***

*: Significant at 0.05 probability level; ***: significant at 0.001 probability level.

3.3. Principal Component Analysis (PCA)

Principal component analysis (PCA) was used as a preferment multivariate statistical tool to discriminate between treatments and varieties based on two top first factors (F1 and F2). In control, F1 and F2 accounted, respectively, 72.59% and 20.37%, our biplot (Figure 2a) discriminates AMIRA in the positive side of F1 with high values of SDW, RDW, and EL%. The negative side of F1 shows MASSINE with high value of proline content. On the other hand, F1 and F2 accounted, respectively, 63.38% and 34.23% under salt stress, the biplot created in this case (Figure 2b) discriminates MASSINE with high levels of proline and Chl T contents, and SDW, and RDW in the positive side of F1 and F2. The negative side of F1 shows AMIRA with high score of EL%. Under salt stress, PCA shows MASSINE variety with high values of tolerance indexes (SDW, RDW, and Chl and Proline contents). However, AMIRA is linked with sensitivity marker (EL%).

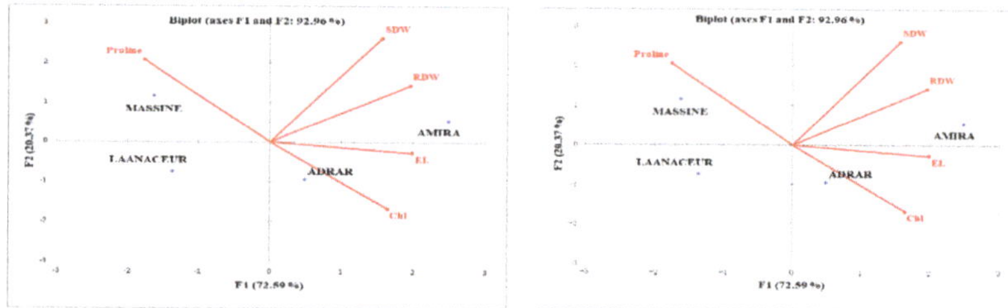

(**a**) (**b**)

Figure 2. The loading biplot of principal component analysis for physiological responses in four barley Moroccan varieties under control conditions (**a**) and under severe salt stress (**b**).

Author Contributions: Conceptualization S.L. and S.B.; methodology, S.L., S.B. and M.F.; analysis, S.B., N.S., M.F. and K.H.; resources, N.S. and K.H.; data elaboration, S.L., S.B. and K.H.; writing—original draft preparation, S.B. and S.L.; writing—review and editing, S.B., N.S., M.F., K.H. and S.L.; supervision, S.L. and K.H. All authors have read and agreed to the published version of the manuscript.

Funding: This research received no external funding.

Institutional Review Board Statement: Not applicable.

Informed Consent Statement: Not applicable.

Data Availability Statement: Rough data can be requested from the corresponding author.

Conflicts of Interest: The authors declare no conflict of interest.

References

1. Sayar, R.; Bchini, H.; Mosbahi, M.; Khemira, H. Response of durum wheat (*Triticum durum Desf.*) growth to salt and drought stresses. *Czech. J. Genet. Plant Breed* **2010**, *46*, 54–63. [CrossRef]
2. Khosravinejad, F.; Heydari, R.; Farboodnia, T. Effects of salinity on photosynthetic pigments, respiration, and water content in two barley varieties. *Pak. J. Biol. Sci. PJBS* **2008**, *11*, 2438–2442. [CrossRef] [PubMed]
3. Sallam, A.; Alqudah, A.M.; Dawood, M.F.; Baenziger, P.S.; Börner, A. Drought stress tolerance in wheat and barley: Advances in physiology, breeding and genetics research. *Int. J. Mol. Sci.* **2019**, *20*, 3137. [CrossRef] [PubMed]
4. El Goumi, Y.; Fakiri, M.; Lamsaouri, O.; Benchekroun, M. Salt stress effect on seed germination and some physiological traits in three Moroccan barley (*Hordeum vulgare* L.) cultivars. *J. Mater. Environ. Sci.* **2014**, *5*, 625–632.
5. Khattabi, D.; Sakar, E.H.; Louahlia, S. Flag Leaf Tolerance Study in Moroccan Barley (*Hordeum vulgare* L.) Varieties Submitted to a Severe Salt Stress. *Biointerface Res. Appl. Chem.* **2021**, *12*, 2787–2799.
6. Szalai, G.; Janda, T.; Páldi, E.; Szigeti, Z. Role of light in the development of post-chilling symptoms in maize. *J. Plant Physiol.* **1996**, *148*, 378–383. [CrossRef]
7. Burnison, B.K. Modified dimethyl sulfoxide (DMSO) extraction for chlorophyll analysis of phytoplankton. *Can. J. Fish. Aquat. Sci.* **1980**, *37*, 729–733. [CrossRef]
8. Bates, L.S.; Waldren, R.P.; Teare, I.D. Rapid determination of free proline for water-stress studies. *Plant Soil* **1973**, *39*, 205–207. [CrossRef]
9. Adjel, F.; Bouzerzour, H.; Benmahammed, A. Salt stress effects on seed germination and seedling growth of barley (*Hordeum Vulgare* L.) genotypes. *J. Sustain. Agric.* **2013**, *3*, 223–237.
10. Wu, D.; Cai, S.; Chen, M.; Ye, L.; Chen, Z.; Zhang, H.; Dai, F.; Wu, F.; Zhang, G. Tissue metabolic responses to salt stress in wild and cultivated barley. *PLoS ONE* **2013**, *8*, e55431. [CrossRef] [PubMed]

11. Nguyen, K.H.; Van Ha, C.; Nishiyama, R.; Watanabe, Y.; Leyva-González, M.A.; Fujita, Y.; Tran, U.T.; Li, W.; Tanaka, M.; Seki, M. Arabidopsis type B cytokinin response regulators ARR1, ARR10, and ARR12 negatively regulate plant responses to drought. *Proc. Natl. Acad. Sci. USA* **2016**, *113*, 3090–3095. [CrossRef] [PubMed]
12. Wang, N.; Cao, F.; Richmond, M.E.A.; Qiu, C.; Wu, F. Foliar application of betaine improves water-deficit stress tolerance in barley (*Hordeum vulgare* L.). *Plant Growth Regul.* **2019**, *89*, 109–118. [CrossRef]
13. Pour-Aboughadareh, A.; Etminan, A.; Abdelrahman, M.; Siddique, K.H.; Tran, L.S.P. Assessment of biochemical and physiological parameters of durum wheat genotypes at the seedling stage during polyethylene glycol-induced water stress. *Plant Growth Regul.* **2020**, *92*, 81–93. [CrossRef]
14. Hussain, H.A.; Men, S.; Hussain, S.; Chen, Y.; Ali, S.; Zhang, S.; Zhang, K.; Li, Y.; Xu, Q.; Liao, C.; et al. Interactive effects of drought and heat stresses on morpho-physiological attributes, yield, nutrient uptake and oxidative status in maize hybrids. *Sci. Rep.* **2019**, *9*, 3890. [CrossRef] [PubMed]
15. Gururani, M.A.; Venkatesh, J.; Tran, L.S.P. Regulation of photosynthesis during abiotic stress-induced photoinhibition. *Mol. Plant* **2015**, *8*, 1304–1320. [CrossRef] [PubMed]
16. Khan, M.N.; Zhang, J.; Luo, T.; Liu, J.; Ni, F.; Rizwan, M.; Fahad, S.; Hu, L. Morpho-physiological and biochemical responses of tolerant and sensitive rapeseed cultivars to drought stress during early seedling growth stage. *Acta Physiol. Plant.* **2019**, *41*, 25. [CrossRef]
17. Mann, A.; Kumar, A.; Saha, M.; Lata, C.; Kumar, A. Stress induced changes in osmoprotectants, ionic relations, antioxidants activities and protein profiling characterize Spo-robolus marginatus Hochst. ex A. Rich. salt tolerance mechanism. *Indian J. Exp. Biol.* **2019**, *57*, 672–679.

environmental sciences proceedings

MDPI

Proceeding Paper

Drought Stress Responses of Four Contrasting Provenances of *Argania spinosa* †

Mohamed Mouafik [1,*], Abdelghani Chakhchar [2], Mohamed Ouajdi [3], Salwa El Antry [3], Ismail Ettaleb [1], Jalila Aoujdad [3] and Ahmed El Aboudi [1]

1 Botany, Mycology and Environment Laboratory, Department of Biology, Faculty of Sciences, Mohammed V University, Rabat 10050, Morocco; ettalib1ismail@gmail.com (I.E.); elaboudi@gmail.com (A.E.A.)

2 École Normale Supérieure Marrakech, Cadi Ayyad University, Marrakech 40000, Morocco; chakhchar.ckr@gmail.com

3 Forest Research Center, Silviculture and Forest Health Service, Rabat 10050, Morocco; ouajdim@gmail.com (M.O.); elantry.salwa@gmail.com (S.E.A.); jaoujdad@gmail.com (J.A.)

* Correspondence: mohamed960@gmail.com; Tel.: +212-650586856

† Presented at the 2nd International Laayoune Forum on Biosaline Agriculture, 14–16 June 2022; Available online: https://lafoba2.sciforum.net/.

Abstract: The Argane tree (*Argania spinosa* (L.) Skeels) is an endemic species of Morocco, widely adapted to the arid and semi-arid climate of the southwest. The Argane may serve as an oilseed crop in desert areas threatened by drought and salinity. Its domestication will open up important economic opportunities for Morocco in the face of the growing demand for Argane oil and will help reduce the pressure on the wild Argane forest. In this research, we aimed to study and characterize some physiological and biochemical traits of Argane tolerance to drought stress. We measured physiological parameters related to the water content (foliar water potential and relative water content of the leaves), biochemical parameters involved in osmoregulation (proline and total sugars), and photosynthesis (chlorophylls) in Argane seedlings from four contrasting provenances (Bouizakarne, Agadir, Essaouira and Berkane) cultivated under drought stress induced by cessation of irrigation. The results showed that the basic and minimal foliar water potential, relative water content as well as chlorophyll content significantly decreased in seedlings under severe drought stress compared to control ones, whereas a significant accumulation of proline and total soluble sugars was noted in stressed seedlings. Nonetheless, inter-provenance differences were recorded for some parameters studied. Provenance effect was determinant for variation in drought stress responses of *A. spinosa*. The study of drought stress-adaptive traits in Argane tree can help to understand the tolerance mechanisms and discriminate between the most drought tolerant provenances in order to rehabilitate degraded Argane forests. It is also relevant for domestication and conservation programs in others abiotic stress conditions resulting from climate change.

Keywords: *Argania spinosa* (L.) Skeels; drought stress; provenance; tolerance; domestication

Citation: Mouafik, M.; Chakhchar, A.; Ouajdi, M.; El Antry, S.; Ettaleb, I.; Aoujdad, J.; El Aboudi, A. Drought Stress Responses of Four Contrasting Provenances of *Argania spinosa*. *Environ. Sci. Proc.* **2022**, *16*, 25. https://doi.org/10.3390/environsciproc2022016025

Academic Editor: Abdelaziz Hirich

Published: 16 June 2022

1. Introduction

The Argane tree [*Argania spinosa* (L.) Skeels] is a horticultural forestry species endemic to Morocco that has multiple uses. It is the only tree representing the tropical family of Sapotaceae listed in the flora of Morocco [1]. Argane tree plays an important role, mainly for the local population, in terms of its botanical, ecological, and economic interest as well as its social value [2]. It contributes to the preservation of the ecosystem and provides an environment conducive to maintaining floristic and faunistic biodiversity [2]. However, the sustainability of this ecosystem is threatened by the effect of global climatic changes such as the successive chronic drought which considerably reduces its natural spread. Tree species cannot escape drought like annuals but they have developed several mechanisms of tolerance against this abiotic constraint reflecting different types of adaptations [3]. Biogeographical distribution of Argane tree in a restricted area in southwestern Morocco is characterized by low water availability and high evapotranspiration demand [2]. Nonetheless, this species can withstand drought by developing adaptive mechanisms through various biochemical and physiological processes [4–7]. In this perspective, we studied some physiological and biochemical mechanisms of drought tolerance in four contrasting provenances of the Argane tree in order to discriminate the most tolerant provenances with a view to rehabilitate the degraded Argane forests.

2. Materials and Methods

Argane seedlings (6-month-old) from the germination of seeds of four contrasting provenances (Essaouira (Ess), Agadir (Agd), Bouizakarne (Bzk) and Berkane (Brk)), were subjected to severe water stress by cessation of irrigation for 25 days under greenhouse conditions. Control was irrigated every two days during the planned experimental period. Physiological and biochemical measurements were collected every five days interval from 3 replications per treatment/provenance. For physiological parameters, we determined basic (Ψ_b) and minimum (Ψ_m) leaf water potential and the relative leaf water content (RWC). For the biochemical parameters, we determined proline, total sugars and total proteins as well as chlorophyll a and b contents.

All the results obtained were statistically analyzed using analysis of variance (ANOVA). The Tukey post-hoc test was used to compare the means at the 5% significance level. A canonical discriminant analysis (CDA) was achieved on the four contrasting *A. spinosa* provenances considering all traits studied. All these statistical analyzes were carried out using SPSS software, version 25.

3. Results and Discussion

The results of the physiological and biochemical parameters studied in this work showed significant differences across observation time and provenance (Tables 1 and 2). At the end of the experimental period, stopping irrigation significantly reduced leaf RWC by 13.6%, 11.1%, 9.2% and 14.6% in Ess, Agd, Bzk and Brk, respectively. Similarly, stress decreased Ψ_b about 2 to 4.6-fold and Ψ_m about 1.5 to 4.6-fold in Argane seedlings compared to the control ones. Significant decrease in RWC and leaf water potential in Argane tree under severe water stress (25 days unirrigated) has been reported in different provenances [6,8]. This reduction in leaf water status has been considered a clear response to drought stress to prevent water loss in *A. spinosa* [6].

Table 1. Analysis of variance (ANOVA) showing the effect of provenance, observation time, and drought stress and their interaction on the physiological parameters studied in the four contrasting provenances of the *A. spinosa* (Berkane, Essaouira, Agadir, and Bouizakarne).

	Ψ_b	Ψ_m	RWC
Berkane (Brk)	−0.50 b	−1.41 b	62.82 a
Essaouira (Ess)	−0.41 a	−1.19 a	56.54 b
Agadir (Agd)	−0.38 a	−1.42 b	65.26 a
Bouizakarne (Bzk)	−0.47 b	−1.11 a	50.70 c
ANOVA ($p \leq 0.05$)			
Provenance (P)	≤0.05	≤0.05	≤0.05
Time (T)	≤0.05	≤0.05	0.15
Stress (S)	≤0.05	≤0.05	≤0.05
P × S	≤0.05	≤0.05	0.18
P × T	≤0.05	≤0.05	≤0.05
T × S	≤0.05	≤0.05	0.78
P × S × T	≤0.05	≤0.05	≤0.05

(Means within a column flanked by the same letter (a–c) are not significantly different at 5% (Tukey test)).

Table 2. Analysis of variance (ANOVA) showing the effect of provenance, observation time, and drought stress and their interaction (ANOVA) on the biochemical parameters studied in the four contrasting provenances of the *A. spinosa* (Berkane, Essaouira, Agadir, and Bouizakarne).

	Proline Content	Sugars Content	Proteins Content	Chl a	Chl b	Chl ab
Berkane (Brk)	32.41 b	103.29 a	214.37 b	11.19 a	4.49 a	15.68 a
Essaouira (Ess)	28.76 c	97.89 ab	266.61 a	9.07 c	3.29 c	12.36 a
Agadir (Agd)	42.42 a	96.43 c	229.89 b	9.30 bc	3.40 bc	14.70 a
Bouizakarne (Bzk)	34.05 b	101.44 ab	273.25 a	10.36 ab	4.04 ab	14.40 a
ANOVA ($p \leq 0.05$)						
Provenance (P)	≤0.05	≤0.05	≤0.05	≤0.05	≤0.05	0.17
Observation time (T)	≤0.05	≤0.05	≤0.05	≤0.05	≤0.05	≤0.05
Stress (S)	≤0.05	≤0.05	≤0.05	≤0.05	≤0.05	≤0.05
P × S	≤0.05	≤0.05	0.74	≤0.05	≤0.05	≤0.05
P × T	≤0.05	≤0.05	≤0.05	≤0.05	≤0.05	≤0.05
T × S	≤0.05	≤0.05	≤0.05	≤0.05	≤0.05	≤0.05
P × S × T	≤0.05	≤0.05	≤0.05	≤0.05	≤0.05	≤0.05

(Means within a column flanked by the same letter (a–c) are not significantly different at 5% (Tukey test)).

At the end of the experimental period, the cessation of irrigation also induced a significant decrease of chlorophylls a (29 to 78%) and b (11 to 86%) contents in all *A. spinosa* provenances. This decrease in chlorophylls under drought stress is mainly attributed to the damage of chloroplasts caused by active oxygen species [5]. However, stressed Argane seedlings showed a significant increase in free proline (46 to 167%), total sugars (60 to 166%) and total protein (17 to 67%) contents compared to control ones. The marked accumulation of osmolytes (proline and sugars) in the leaves of the Argane tree under drought stress conditions has also been reported by other authors [6,7]. These osmolytes are considered as one of the adaptive strategies triggered by the plant in response to environmental constraints. Agd and Brk showed the high accumulation of proline (42.42 mg L^{-1}) and total sugars (103.29 mg L^{-1}), respectively, compared to other provenances (Table 2). Furthermore, protein accumulation in stressed Argane seedlings may be due to the synthesis of new proteins, including antioxidant enzymes, to ensure growth and/or survival under drought stress conditions [6]. The highest protein accumulation was recorded in Bzk (273.25 mg L^{-1}) and Ess (266.61 mg L^{-1}).

Provenance × water (stress) × observation time interaction was significant for all physiological and biochemical parameters studied (Tables 1 and 2). According to CDA, Agd and Brk provenances were mainly separated from Ess and Bzk by a good conservation of the water (high values of RWC and high negative values of Ψ_b). The physiological and biochemical differences noted in our study between the four provenances indicated adaptive genetic diversity of Argane tree for the conservation of its forests.

4. Conclusions

The present study highlighted the inter-provenance variation in the adaptation mechanism of *A. spinosa* to drought stress. The four contrasting provenances of Argane tree showed, under drought conditions, a significant decrease in water potential, as a strategy of water conservation, as well as a significant accumulation of certain osmolytes related to osmoregulation and homeostasis maintenance. These physiological and biochemical responses varied significantly according the provenance of the Argane.

Author Contributions: Conceptualization, A.E.A. and J.A.; methodology, M.M. and I.E.; validation, M.O. and S.E.A.; formal analysis, A.C.; writing—original draft preparation, M.M.; writing—review and editing, A.C.; supervision, A.E.A. All authors have read and agreed to the published version of the manuscript.

Funding: This research received no external funding.

Institutional Review Board Statement: Not applicable.

Informed Consent Statement: Not applicable.

Data Availability Statement: The data presented in this study are available on request from the corresponding author. The data are not publicly available due to privacy restrictions.

Acknowledgments: The authors would like to thank all members and staff of the Forest Research Center in Rabat for technical support and guidance.

Conflicts of Interest: The authors declare no conflict of interest.

References

1. Mhirit, O.; Benzyane, M.; Benchekroun, F.; El Yousfi, S.M.; Bendaanoun, M. *L'arganier, une espèce Fruitière Forestière à Usages Multiples*; Mardaga: Sprimont, Belgium, 1998.
2. Chakhchar, A.; Ben Salah, I.; El Kharrassi, Y.; Filali-Maltouf, A.; El Modafar, C.; Lamaoui, M. Agro-fruit-forest systems based on argan tree in Morocco: A review of recent results. *Front. Plant Sci.* **2022**, *12*, 783615. [CrossRef] [PubMed]
3. Pita, P.; Cañas, I.; Soria, F.; Ruiz, F.; Toval, G. Use of physiological traits in tree breeding for improved yield in drought-prone environments. The case of Eucalyptus globulus. *Investig. Agrar. Sist. Recur.* **2005**, *14*, 383–393. [CrossRef]
4. Chakhchar, A.; Haworth, M.; El Modafar, C.; Lauteri, M.; Mattioni, C.; Wahbi, S.; Centritto, M. An assessment of genetic diversity and drought tolerance in argan tree (*Argania spinosa*) populations: Potential for the development of improved drought tolerance. *Front. Plant Sci.* **2017**, *8*, 276. [CrossRef] [PubMed]
5. Chakhchar, A.; Lamaoui, M.; Aissam, S.; Ferradous, A.; Wahbi, S.; El Mousadik, A.; Ibnsouda, K.S.; Filali, M.A.; El Modafar, C. Using chlorophyll fluorescence, photosynthetic enzymes and pigment composition to discriminate drought tolerant ecotypes of *Argania spinosa. Plant Biosyst.* **2018**, *152*, 356–367. [CrossRef]
6. Chakhchar, A.; Lamaoui, M.; Aissam, S.; Ferradous, A.; Wahbi, S.; El Mousadik, A.; Ibnsouda, K.S.; Filali, M.A.; El Modafar, C. Physiological and biochemical mechanisms of drought stress tolerance in the argan tree. In *Plant Metabolites and Regulation under Environmental Stress*; Ahmad, P., Ahanger, M.A., Singh, V.P., Tripathi, D.K., Alam, P., Eds.; Elsevier: Amsterdam, The Netherlands, 2018; p. 446.
7. Chakhchar, A.; Lamaoui, M.; Aissam, S.; Ferradous, A.; Wahbi, S.; El Mousadik, A.; Ibnsouda, K.S.; Filali, M.A.; El Modafar, C. Physiological and carbohydrate metabolism traits for discrimination of drought-tolerant elite ecotypes of *Argania spinosa. Plant Physiol. Rep.* **2019**, *24*, 388–398. [CrossRef]
8. El Aboudi, A.; Carlier, G.; Peltier, J.P. Régime hydrique de l'Arganier (*Argania spinosa*) dans le Souss (Maroc). In *Physiologie des arbres et arbustes des zones arides et semi-arides*; GEA: Paris, France, 1991.

environmental sciences proceedings

MDPI

Proceeding Paper

Phenotypic Diversity of Agronomical Traits and Nut Phenolic Compounds among Pistachio (*Pistacia vera* L.) Cultivars †

Walid Abidi * and **Rawaa Akrimi**

Regional Center of Agricultural Research of Sidi Bouzid (CRRA) PB 357, Sidi Bouzid 9100, Tunisia; akrimirawaa2017@gmail.com
* Correspondence: walid.abidi@iresa.agrinet.tn
† Presented at the 2nd International Laayoune Forum on Biosaline Agriculture, 14–16 June 2022; Available online: https://lafoba2.sciforum.net/.

Abstract: The phenotypic diversity of agronomical traits and nut phenolic compound content of four pistachio cultivars (Mateur, Elguetar, Kerman, and Ohadi), conducted under regulated deficit irrigation (RDI) during two growing seasons (2017/2018), were evaluated. The experimental orchard was located in the Regional Center of Agricultural Research of Sidi Bouzid (CRRA), Tunisia. Three water treatments were applied from March to September; control (T0; 100% ETc during all the developmental stages), RDI treatment (T1; 50% ETc during stages I and II of fruit development followed by full irrigation 100% ETc during stage III), and stressed treatment (T2; 50% ETc during all the growing season). The results showed that treatments T0 and T1 no presented statistically significant difference in yield and nut biochemical traits, and 50% less irrigation water was used during stages I and II of nut development. The trees under treatment T2 were affected by water stress showing low values of yield and nut phenolic compounds.

Keywords: pistachio; drought stress; phenolic compounds

Citation: Abidi, W.; Akrimi, R. Phenotypic Diversity of Agronomical Traits and Nut Phenolic Compounds among Pistachio (*Pistacia vera* L.) Cultivars. *Environ. Sci. Proc.* **2022**, *16*, 26. https://doi.org/10.3390/environsciproc2022016026

Academic Editor: Abdelaziz Hirich

Published: 16 June 2022

1. Introduction

In semiarid conditions, water availability for crop irrigation has declined as a consequence of global climate change. Pistachio is a nut tree that has the reputation of being a drought tolerant and saline resistant species, cultivated under rainfed conditions in its region of origin [1]. It is well known that pistachio irrigation increases the yield, improves the nut quality, and dampens the normal alternate bearing pattern [2]. Regulated deficit irrigation (RDI) is a system of managing water supply by imposing some water deficits in specific phenological stages, which are found to be less sensitive, with no (or low) reduction in economic benefits [3]. Goldhamer and Beede [4] showed that the same reduction in the irrigation water during stages I (nut growth) and II (shell hardening) did not reduce the total amount of fruits and increased the percentage of shell splitting, although it also increased early splitting.

Pistachio nut has peculiar organoleptic characteristics including a high nutritional value and high unsaturated fatty acids content; it is a rich source of fat (about 50–70%) and a good source of proteins and minerals and bioactive components [5].

Our experiment was conducted to study the effectiveness of RDI irrigation on yield and nut phenolic compounds content in four pistachio cultivars conducted in a high-density orchard under semiarid conditions.

2. Materials and Methods

2.1. Plant Material

The trial was carried out in the Regional Center of Agriculture Research (CRRA, Sidi Bouzid) in west central Tunisia (9°43′ E, 35°01′ N; altitude 354 m). Fifteen-year-old pistachio trees grafted onto *P. atlantica* rootstock were studied.

2.2. *Treatments for Regulated Deficit Irrigation (RDI)*

The phenological stages taken into account in the RDI treatments were those suggested by Goldhamer and Beede (2004): Stage I (from sprouting until the end of rapid nut growth), stage II (from maximum nut size until the beginning of kernel growth), and stage III (from the beginning of kernel growth until harvest). Three irrigations were applied during the two years of the study:

- Treatment (T0): trees received water to cover estimated evapotranspiration (ETc) losses.
- Treatment (T1): trees received 50% of the water received by the control trees during stages I and II and the same amount of water as the control trees during stage III.
- Treatment (T2): trees received water to replace 50% ETc calculated for the control treatment during all the steps.

Drip irrigation was applied 3 days per week and was controlled and adjusted weekly according to soil potential (measured by tensiometers located 25 cm from the drip head at depths of 30 and 60 cm). A drip line was utilized in each tree row, with four self-compensating drippers (4 L h^{-1}) per tree, 0.5 m apart. The amount of provided water was calculated on the basis of the crop evapotranspiration (ETc) and the crop coefficient (Kc) according to the FAO method [6]: ETc = ETo Kc.

2.3. *Agronomical Traits*

The trunk cross-sectional area (TCSA) was measured during the dormant season at 30 cm above the graft union. Yield (kg/tree) was determined per tree, and yield efficiency (YE) of each scion–stock was computed from the harvest data. At harvest, a representative nut sample (100 nuts) was taken for pomological evaluations.

2.4. *Nut Biochemical Quality*

Samples were grounded to powder by a mortar and pestle, separately. Twenty grams of each kernel were extracted by 200 mL of 95% methanol for 48 h at room temperature. Then, the extracts were filtered and evaporated at low pressure; samples were stored at −20 °C until analysis. The total anthocyanin content, flavonoid, total phenolics and antioxidant capacity were determined as described in [7].

2.5. *Statistical Analysis*

Mean values and mean standard error (SE), were calculated for each studied trait. Analysis of variance (ANOVA) was performed using SPSS 20.0 (SPSS Inc., Chicago, IL, USA). Means were compared using the Scheffe test at the 5% significance level.

3. Results and Discussion

3.1. *Agronomical Traits*

The results showed significant differences ($p < 0.05$) among cultivars in yield, TCSA, YE, tree height, fruit weight, and production of blanks (Table 1). The irrigation regimes affected the yield and its components, and significant differences ($p < 0.05$) were observed between the three treatments. The control and treatment T1 presented similar yield values, while treatment T2 showed a significantly ($p < 0.05$) lower yield. These results are similar to those reported by [8], studying the impact of regulated deficit irrigation on pistachio trees, showing that the mean yield for three years of study was not reduced in T1 compared to the control treatment. The cv. Mateur presented the highest yield (7.1 Kg/tree) under the control treatment, whereas the lowest yield was observed in the cv. Elguetar under the treatment T2 (4.0 Kg/tree). The treatment T2 showed low values of split nuts and blanks as compared to the control and RDI treatment.

Table 1. Agronomical and pomological traits of pistachio cultivars. Values are means ($n = 3$) ± SE. Different letters a, b, c, indicate differences ($p < 0.05$) among three irrigation treatments (T0: control; T1: regulated deficit irrigation; and T2: stressed) in the same cultivar. a, b, c indicate differences ($p < 0.05$) among cultivars.

	Cultivar	Yield	TCSA	YE	Height	FW	DW	L	Split	Blanks
T0	Mateur	7.1 ± 1^{a}	35.5 ± 5^{a}	0.20^{a}	2.5 ± 0.1^{a}	0.62 ± 0.1^{b}	0.50 ± 0.1^{b}	17.8 ± 2^{a}	76 ± 5^{a}	7 ± 2^{b}
	Elguetar	5.1 ± 1^{b}	31.8 ± 4^{b}	0.16^{b}	2.3 ± 0.1^{a}	0.44 ± 0.1^{c}	0.50 ± 0.1^{b}	15.8 ± 2^{b}	71 ± 5^{b}	9 ± 3^{b}
	Kerman	5.0 ± 1^{b}	32.5 ± 5^{b}	0.15^{b}	2.2 ± 0.1^{a}	0.58 ± 0.1^{b}	0.46 ± 0.1^{c}	17.2 ± 2^{a}	70 ± 2^{b}	10 ± 3^{b}
	Ohadi	3.1 ± 1^{c}	29.8 ± 2^{c}	0.10^{c}	2.0 ± 0.1^{a}	0.77 ± 0.1^{a}	0.64 ± 0.1^{a}	18.8 ± 2^{a}	55 ± 3^{c}	25 ± 5^{a}
T1	Mateur	6.4 ± 2^{a}	36.5 ± 4^{a}	0.17^{a}	2.3 ± 0.1^{b}	0.60 ± 0.5^{b}	0.51 ± 0.5^{b}	17.4 ± 1^{a}	70 ± 3^{b}	10 ± 2^{b}
	Elguetar	4.7 ± 3^{b}	33.2 ± 3^{b}	0.14^{b}	1.9 ± 0.1^{b}	0.46 ± 0.6^{c}	0.4 ± 0.7^{c}	15.4 ± 1^{b}	75 ± 2^{a}	13 ± 3^{b}
	Kerman	4.1 ± 1^{b}	28.2 ± 5^{b}	0.14^{b}	2.3 ± 0.1^{a}	0.52 ± 0.1^{b}	0.42 ± 0.1^{c}	17.0 ± 2^{a}	65 ± 2^{b}	12 ± 3^{b}
	Ohadi	2.1 ± 1^{c}	22.0 ± 4^{c}	0.10^{c}	1.8 ± 0.1^{b}	0.72 ± 0.1^{a}	0.62 ± 0.1^{a}	18.1 ± 2^{a}	48 ± 2^{c}	29 ± 2^{a}
T2	Mateur	5.1 ± 1^{a}	32.5 ± 3^{b}	0.15^{a}	2.2 ± 0.1^{b}	0.51 ± 0.5^{b}	0.48 ± 0.4^{b}	17.1 ± 1^{a}	68 ± 4^{a}	19 ± 1^{b}
	Elguetar	4.0 ± 2^{b}	38.6 ± 4^{a}	0.10^{b}	1.7 ± 0.1^{b}	0.42 ± 0.1^{c}	0.3 ± 0.1^{c}	15.1 ± 1^{b}	70 ± 3^{a}	16 ± 2^{b}
	Kerman	4.0 ± 1^{b}	30.0 ± 5^{b}	0.13^{b}	2.2 ± 0.1^{a}	0.50 ± 0.1^{b}	0.40 ± 0.1^{b}	17.0 ± 2^{a}	60 ± 3^{b}	14 ± 2^{b}
	Ohadi	2.0 ± 1^{c}	20.5 ± 2^{c}	0.10^{c}	1.7 ± 0.1^{b}	0.70 ± 0.1^{a}	0.60 ± 0.1^{a}	17.8 ± 2^{a}	41 ± 2^{c}	35 ± 2^{a}

Units and Abbreviations: Yield (kg/tree); TCSA = trunk cross-sectional area in cm^2; FW = Fresh weight (g); DW = dry weight (g); YE = Yield Efficiency (kg/cm^2); L = Length (mm).

3.2. Nut Phenolic Compounds Content

The anthocyanins, flavonoids, total phenolics, and relative antioxidant capacity of pistachio nuts under the three water regimes are shown in Figure 1.

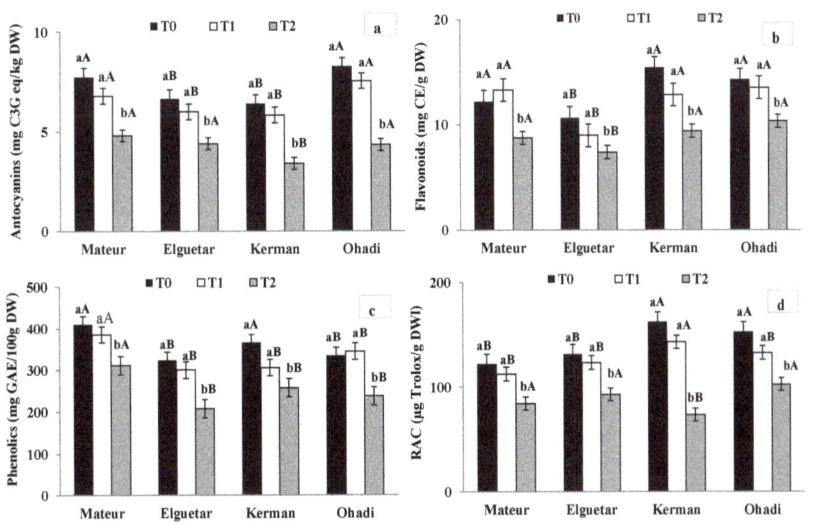

Figure 1. Phenolic compounds in pistachio nuts under RDI. The anthocyanin content are shown in (**a**), flavonoids are shown in (**b**), total phenolic are shown in (**c**), relative antioxidant capacity are showed in (**d**). Values are means ($n = 6$) ± SE. Letters (a, b, c), indicate difference among treatments in the same cultivars. Letters (A, B, C), indicate differences ($p < 0.05$) among cultivars. Abbreviations: C3GE = Cyanidin-3-glucoside equivalents; CE = Catechin equivalents; GAE = Gallic acid equivalents; RAC = Relative Antioxidant Capacity; DW = dry weight.

The irrigation treatments applied significantly ($p < 0.05$) affected the anthocyanins, flavonoids, total phenolics, and RAC values with the T2 treatment presenting lower values. The two treatments T0 and T1 presented similar behavior and showed statistically significant differences ($p < 0.05$) with treatment T2. Our results are in accordance with those presented by [9] showing an anthocyanins content between 1.4 and 8.9 mg/Kg in pistachio whole nut. The flavonoid values obtained in our study are in accordance with the range (3.5 to 7.2 mg CE g^{-1} DW) of a previous study conducted by [10]. The antioxidant capacity values showed great variations among cultivars, but the values are in accordance with the range (45.7 to 122.6 µg TE g^{-1} DW) reported by [10]. The irrigation treatments applied significantly affected the RAC values with the T2 treatment presenting lower values.

4. Conclusions

Our results showed that the studied pistachio cultivars presented differential responses to RDI irrigation. However, treatment T1 led to a similar yield, nut weight, and phenolic compounds as the control treatment. The reduction in irrigation volumes by 50% during stage I and II of nut development over the control (100% ETc) could be a suitable irrigation strategy implemented in high density pistachio orchards in semiarid conditions.

Author Contributions: W.A. performed the experiments, analyzed data and wrote the manuscript. R.A. performed the biochemical analyses. W.A. and R.A. revised the manuscript. All authors have read and agreed to the published version of the manuscript.

Funding: This research received no external funding.

Institutional Review Board Statement: Not applicable.

Informed Consent Statement: Not applicable.

Data Availability Statement: The data that support the findings of this study are available from the corresponding author, upon reasonable request.

Acknowledgments: The authors thank Donya Ammari, Boutheina Fridhi, Chaima Maamri, and Boutheina Hajlaoui for technical assistance and support.

Conflicts of Interest: The authors have no conflict of interest to declare that are relevant to the content of this article.

References

1. Rieger, M. Offsetting effects of reduced root hydraulic conductivity and osmotic adjustment following drought. *Tree Physiol.* **1993**, *15*, 379–385. [CrossRef] [PubMed]
2. Kanber, R.; Yazar, A.; Onder, S.; Koksal, H. Irrigation response of pistachio (*Pistacia vera* L.). *Irrig. Sci.* **1993**, *14*, 7–14. [CrossRef]
3. Carbonell-Barrachina, Á.A.; Memmi, H.; Noguera-Artiaga, L.; Gijón-López, M.C.; Ciapaa, R.; Pérez-López, D. Quality attributes of pistachio nuts as affected by rootstock and deficit irrigation. *J. Sci. Food Agric.* **2015**, *95*, 2866–2873. [CrossRef] [PubMed]
4. Goldhamer, D.A.; Beede, B.H. Regulated deficit irrigation effects on yield, nut quality and water-use efficiency of mature pistachio trees. *J. Hortic. Sci. Biotechnol.* **2004**, *79*, 538–545. [CrossRef]
5. Satil, F.; Azcan, N.; Baser, K.H.C. Fatty acid composition of Pistachio nuts in Turkey. *Chem. Nat. Compd.* **2003**, *39*, 322–324. [CrossRef]
6. Allen, R.G.; Pereira, L.S.; Raes, D.; Smith, M. Crop Evapotranspiration-Guidelines for Computing Crop Water Requirements. In *Irrigation and Drain*; Paper No. 56; FAO: Rome, Italy, 1998; p. 300.
7. Taghizadeh, S.F.; Rezaee, R.; Davarynejad, G.; Karimi, G.; Nemati, S.H.; Asili, J. Phenolic profile and antioxidant activity of Pistaciavera var. Sarakhs hull and kernel extracts: The influence of different solvents. *J. Food Meas. Charact.* **2018**, *12*, 2138–2144. [CrossRef]
8. Memmi, H.; Gijón, M.C.; Couceiro, J.F.; Pérez-López, D. Water stress thresholds for regulated deficit irrigation in pistachio trees: Rootstock influence and effects on yield quality. *Agric. Water Manag.* **2015**, *164*, 58–72. [CrossRef]
9. Ojeda-Amador, R.M.; Salvador Moya, M.D.; Fregapane, G.; Gómez Alonso, S. Comprehensive study of phenolic compounds profile and antioxidant activity of eight pistachio cultivars, their residual cakes and virgin oils. *J. Agric. Food Chem.* **2019**, *67*, 3583–3594. [CrossRef] [PubMed]
10. Tsantili, E.; Konstantinidis, K.; Christopoulos, M.V.; Roussos, P.A. Total phenolics and flavonoids and total antioxidant capacity in pistachio (*Pistacia vera* L.) nuts in relation to cultivars and storage conditions. *Sci. Hortic.* **2011**, *129*, 694–701. [CrossRef]

environmental sciences proceedings

MDPI

Proceeding Paper

Impact of Severe Salt Stress on Morphological, Physiological, and Biochemical Parameters in Alfalfa (*Medicago sativa* L.) †

Nassira Srhiouar [1,*], Mohamed Ferioun [1,2], Said Bouhraoua [1], Khalil Hammani [1] and Said Louahlia [1]

[1] Natural Resources and Environment Laboratory, Multidisciplinary Faculty of Taza, Sidi Mohammed Ben Abdellah University, Fez 30000, Morocco; mohamed.ferioun@usmba.ac.ma (M.F.); said.bouhraoua88@gmail.om (S.B.); khalil.hammani@usmba.ac.ma (K.H.); said.louahlia@usmba.ac.ma (S.L.)

[2] Microbial Biotechnology and Bioactive Molecules Laboratory, Faculty of Sciences and Technologies, Sidi Mohammed Ben Abdellah University, Fez 30000, Morocco

* Correspondence: nassira.oligo@gmail.com; Tel.: +212-61-891-1794

† Presented at the 2nd International Laayoune Forum on Biosaline Agriculture, 14–16 June 2022; Available online: https://lafoba2.sciforum.net/.

Abstract: In the Mediterranean area, climate change induces an important increase in soil salinity, which threatens crop production. Here, we aimed to test the impact of severe salt stress on the Alfalfa (*Medicago sativa* L.) used by Moroccan breeders. Stress was applied by increasing salinity gradually by 8 dS/m. Two weeks later, the root dry weight (RDW) and shoot dry weight (SDW) were measured as morphological parameters. Soluble proteins, total soluble sugars (TSS), and proline contents were quantified in both of leaves and roots parts. Furthermore, Chlorophyll a (Chl a), b (Chl b), and total Chlorophyll (Chl T) contents were quantified in leaves to estimate the salt stress impact on the photosynthetic apparatus. Our results showed a significant decrease in morphological parameters under stress. Additionally, a significant reduction in photosynthetic pigment was recorded. Compared to leaves, important accumulations of proline and soluble protein contents in roots were observed. These results indicate that under salt treatment, alfalfa plants react to salinity by storing many molecules in the roots for planned mobilization after recovery.

Keywords: alfalfa; *Medicago sativa*; salt stress; morphology; biochemical parameters

check for updates

Citation: Srhiouar, N.; Ferioun, M.; Bouhraoua, S.; Hammani, K.; Louahlia, S. Impact of Severe Salt Stress on Morphological, Physiological, and Biochemical Parameters in Alfalfa (*Medicago sativa* L.). *Environ. Sci. Proc.* **2022**, *16*, 27. https://doi.org/10.3390/environsciproc2022016027

Academic Editor: Abdelaziz Hirich

Published: 13 June 2022

1. Introduction

Saline soils cover more than 7% of the world's total land and 70% of all agricultural soils in the world. They have devastating global consequences and are projected to cause 30% land loss within the next 25 years and up to 50% by 2050 [1]. Unfortunately, in Morocco, there is a significant area of land affected by salinity, and about 500,000 hectares are reported to be affected [2]. Salinity is becoming one of the major environmental factors reducing crop yields and threatening the world food balance [3]. To overcome this environmental constraint, different strategies can be adopted, but these methods are very expensive. [4] Therefore, the introduction of resistant varieties to abiotic stress with a high socioeconomic value constitutes one of the approaches to rehabilitating saline soils. Selecting ideal plants for these conditions is the first step to solving the problem of salinity. [5] The objective of this study was to screen the effects of salt stress (induced by NaCl) on physiological and biochemical parameters in alfalfa (*Medicago sativa* L.).

2. Material and Methods

Homogenous seeds of alfalfa were disinfected and sown in pots containing approximately 1 kg of substrate (soil/peat/sand, (1:1:1)). Twenty-one days after sowing (DAS), two homogenous plants were maintained per pot, and salt stress was applied by irrigation with NaCl solution to increase soil conductivity to 8 dS/m. All pots were arranged randomly in greenhouse at 21 °C under natural light supplemented with artificial light with PAR of

300 μmol photons·m^{-2}·s^{-1}. At 35 DAS, root and shoot parts were harvested, dried at 70 °C for three days, and weighted to estimate root (RDW) and shoot (SDW) dry weight. For physiological parameters, the Chlorophyll (Chl a), b (Chl b), and total Chlorophyll (Chl T) contents were determined using Burnison's method (1980) [6]; total soluble sugars (TSSs) content was determined according to Erice et al.'s (2007) method [7]; proline was determined using Bates et al.'s (1973) method [8]; and soluble proteins content was quantified following Lowry's method (1951) [9].

3. Results and Discussion

Our results show that salinity markedly affects all parameters studied. Thus, for morphological parameters, a significant ($p < 0.05$) reduction in RDW and SDW was recorded under salt stress compared to control plants (Figure 1). The reduction was 77% and 42% in roots and shoots, respectively. These results confirm what was previously reported in the literature [10,11]. This biomass reduction indicates that plants are trying to manage their resources by downregulation of their vegetative growth [12]. The measure of photosynthetic pigments content in leaves showed that salt stress significantly reduces Chl a, Chl b, and total Chlorophyll contents (Figure 1). The decrease in photosynthetic pigment under stressful conditions is widely spread in the literature [13], and it is reported to be linked to other metabolism disturbances, such as electrolytes leakage from thylakoids, the dehydration of protoplasm, lowered photo-assimilation levels, and Chlorophyll photo-oxidation [14]. Moreover, the decrease in photosynthetic pigment under stress is considered a sign of sensitivity [15]. An analysis of variance (Table 1) showed that both parts of plant and treatment factors and their interaction significantly influenced ($p < 0.05$) the parameters studied. The treatment factor is the most important source of variability for proline, TSS, and soluble protein contents, where this factor explained 75%, 64%, and 60% of the total variability, respectively. Moreover, part of plant factors explained 37%, 29%, and 12% of the total variability of soluble proteins, TSS, and proline contents, respectively, while the effect of the interaction between the last factors was less than 10% for all parameters studied. The results in Figure 2 showed a significant increase in biochemical parameters studied resulting from NaCl treatment in both leaves and root parts, which is in agreement with other studies [16,17]. A significant increase in soluble protein content was shown in salt treatment compared to the control in both leaves and root parts. Soluble proteins contribute to the osmoregulation process in indoor plant mediums [15]. Furthermore, significant increases in proline content in both root and shoot parts were recorded. The increase in proline in stressful conditions is well noted in various studies [11,16]. This amino acid is involved in many physiological functions, such as osmotic adjustment, free radical scavenging, cellular redox potential buffering, and detoxification pathway activation [18]. In alfalfa plants, the accumulation of proline salt stress is linked to salt tolerance [11].

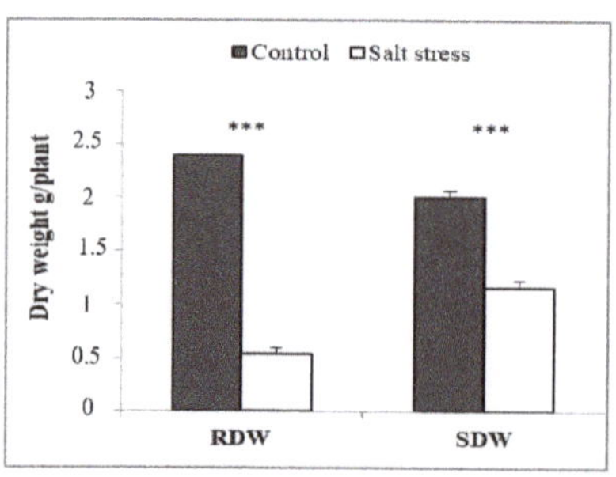

(a)

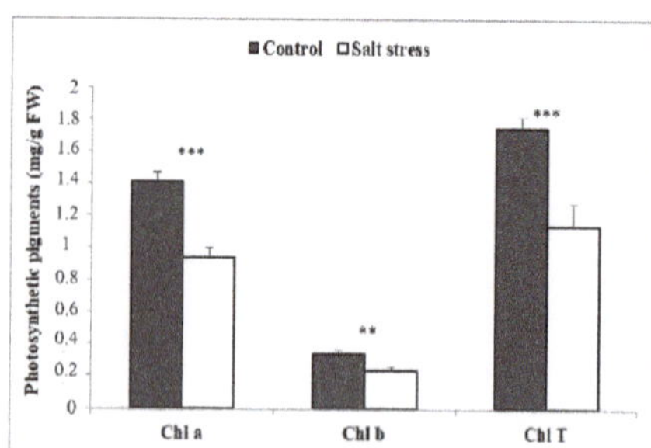

(b)

Figure 1. Effect of salt stress on roots (RDW) and shoot (SDW) dry weight (**a**) and Chlorophyll contents in leaves (**b**) of *Medicago sativa* L. (**: significant at 0.01. ***: significant at 0.005).

Table 1. Analysis of variance for proline and soluble sugar and protein contents in root plant part and area plant part (Df: degree of freedom. *: significant at 0.05, **: significant at 0.01. ***: significant at 0.005.).

Source	Df	Proline	TSS	Proteins
Part of plant (pp)	1	0.89142 *	29.8207 ***	246.385 ***
Treatment (trt)	1	5.38135 ***	64.6927 ***	399.661 ***
Replicates	2	0.05616	0.2490	1.208
pp*trt	1	0.74541 *	6.4557 **	20.663 *
Error	6	0.08834	0.4190	1.759
Total	11			

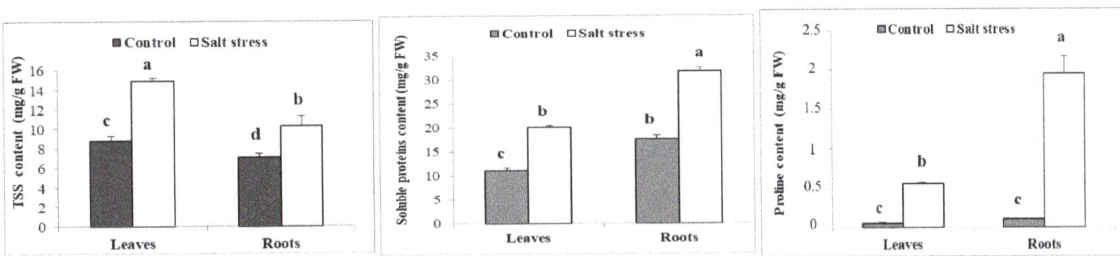

(a) (b) (c)

Figure 2. Effect of salt stress on Total Soluble Sugars (**a**), soluble proteins (**b**), and proline (**c**) contents in roots of *Medicago sativa* subjected to salt stress. Values indicate means of triplicates (scores with the same letter are not significantly different at $p = 0.05$).

The accumulation of soluble proteins and proline under salt stress appears to be more important in roots, which is also noted in many other papers [19]. It is widely known that alfalfa roots accumulate many molecules under stressful conditions to ensure a stock of molecules for mobilization after recovery [15]. Total soluble sugar contents also increased in stressed plants by 41% and 31% in leaves and roots, respectively. It is well documented that salinity increases the production of many sugars in alfalfa leaves, such as sucrose and pinitol [20]. These carbohydrates are considered osmoregulators that decrease the osmotic potential in order to maintain osmotic homeostasis inside of the cell [21]. The lower accumulation level of these molecules in roots compared to leaves might be explained by the storage of these molecules in the starch form in this part of the alfalfa plant [20].

Author Contributions: N.S.: conceptualization, methodology, formal analysis, writing original draft, formal analysis. M.F.: methodology, formal analysis, writing original draft, formal analysis. S.B. and K.H.: methodology. S.L.: conceptualization, supervision, review and editing, validation, resources, and funding acquisition. All authors have read and agreed to the published version of the manuscript.

Funding: This research received no external funding.

Institutional Review Board Statement: Not Applicable.

Informed Consent Statement: Informed consent was obtained from all subjects involved in the study.

Data Availability Statement: The study did not report any data.

Conflicts of Interest: The authors declare no conflict of interest.

References

1. Rengasamy, P. World salinization with emphasis on Australia. *J. Exp. Bot.* **2006**, *57*, 1017–1023. [CrossRef] [PubMed]
2. El Goumi, Y.; Fakiri, M.; Lamsaouri, O.; Benchekroun, M. Salt stress effect on seed germination and some physiological traits in three Moroccan barley (*Hordeum vulgare* L). *cultivars. J. Mat. Env. Sci.* **2014**, *5*, 625–632.
3. Walia, H.; Wilson, C.; Wahid, A.; Condamine, P.; Cui, X.; Close, T.J. Expression analysis of barley (*Hordeum vulgare* L.) during salinity stress. *Funct. Integr. Genomics* **2006**, *6*, 143–156. [CrossRef] [PubMed]
4. Rhodes, J.; Laveday, J. *Salinity in Irrigated Agriculture Riverside*; FAO: Rome, Italy, 1990; pp. 1089–1141.
5. Karoune, S.; Kechebar, M.S.A.; Halis, Y.; Djellouli, A.; Rahmoune, C. Effet du stress salin sur la morphologie, la physiologie et la biochimie de l'Acacia albida. *J. Alg. Rég. Ar.* **2017**, *14*, 60–73.
6. Burnison, B.K. Modified dimethyl sulfoxide (DMSO) extraction for chlorophyll analysis of phytoplankton. *Can. J. Fish. Aquat. Sci.* **1980**, *37*, 729–732. [CrossRef]
7. Erice, G.; Irigoyen, J.J.; Sánchez-Díaz, M.; Avice, J.C.; Ourry, A. Effect of drought, elevated CO2 and temperature on accumulation of N and vegetative storage proteins (VSP) in taproot of nodulated alfalfa before and after cutting. *Pla. sci.* **2007**, *172*, 903–912. [CrossRef]
8. Bates, L.S.; Waldern, S.P.; Teare, I.D. Rapid determination of free proline for water-stress studies. *Plant Soil* **1973**, *39*, 205–207. [CrossRef]
9. Lowry, O.H.; Rosebrough, N.J.; Farr, A.L.; Rose, J.; Randall, A.J. Protein measurement with the folin phenol reagent. *J. Bio. Chem.* **1951**, *193*, 265–275. [CrossRef]
10. El-Sharkawy, M.S.; El-Beshsbeshy, T.R.; Mahmoud, E.K.; Abdelkader, N.I.; Al-Shal, R.M.; Missaoui, A.M. Response of alfalfa under salt stress to the application of potassium sulfate nanoparticles. *AJPS* **2017**, *8*, 1751–1773. [CrossRef]
11. Wang, X.S.; Han, J.G. Changes of proline content, activity, and active isoforms of antioxidative enzymes in two alfalfa cultivars under salt stress. *JAS China* **2009**, *8*, 431–440. [CrossRef]
12. Nguyen, K.H.; Ha, C.V.; Nishiyama, R.; Watanabe, Y.; Leyva-González, M.A.; Fujita, Y.; Tran, U.T.; Li, W.; Tanaka, M.; Seki, M.; et al. Arabidopsis type B cytokinin response regulators ARR1, ARR10, and ARR12 negatively regulate plant responses to drought. *Proc. Natl. Acad. Sci. USA.* **2016**, *113*, 3090–3095. [CrossRef] [PubMed]
13. Ashrafi, E.; Razmjoo, J.; Zahedi, M.; Pessarakli, M. Selecting alfalfa cultivars for salt tolerance based on some physiochemical traits. *Agro. J.* **2014**, *106*, 1758–1764. [CrossRef]
14. Gomes, M.A.D.C.; Suzuki, M.S.; Cunha, M.D.; Tullii, C.F. Effect of salt stress on nutrient concentration, photosynthetic pigments, proline and foliar morphology of Salvinia auriculata Aubl. *Acta Limno Brasiliensia.* **2011**, *23*, 164–176. [CrossRef]
15. Bhattarai, S.; Biswas, D.; Fu, Y.B.; Biligetu, B. Morphological, physiological, and genetic responses to salt stress in alfalfa. *Agronomy* **2020**, *10*, 577. [CrossRef]
16. Fougere, F.; Le Rudulier, D.; Streeter, J.G. Effects of salt stress on amino acid, organic acid, and carbohydrate composition of roots, bacteroids, and cytosol of alfalfa (*Medicago sativa* L.). *Plant physiology.* **1991**, *96*, 1228–1236. [CrossRef] [PubMed]
17. Torabi, M.; Halim, M.R.A. Variation of root and shoot growth and free proline accumulation in Iranian alfalfa ecotypes under salt stress. *J. Food Agric. Environ.* **2010**, *8*, 323–327.
18. Balnška, F.; Ovecka, M.; Hirt, H. Salt stress induces changes in amounts and localization of the mitogen-activated protein kinase SIMK in alfalfa roots. *Protoplasma* **2000**, *212*, 262–267. [CrossRef]
19. Li, J.; Guo, X.; Zhang, M.; Wang, X.; Zhao, Y.; Yin, Z.; Zhang, Z.; Wanga, Y.; Xionga, H.; Zhanga, H.; et al. OsERF71 confers drought tolerance via modulating ABA signaling and proline biosynthesis. *Plant Sci.* **2018**, *270*, 131–139. [CrossRef] [PubMed]
20. Xiong, J.; Sun, Y.; Yang, Q.; Tian, H.; Zhang, H.; Liu, Y.; Chen, M. Proteomic analysis of early salt stress responsive proteins in alfalfa roots and shoots. *Proteome Sci.* **2017**, *15*, 19. [CrossRef] [PubMed]
21. Bertrand, A.; Dhont, C.; Bipfubusa, M.; Chalifour, F.P.; Drouin, P.; Beauchamp, C.J. Improving salt stress responses of the symbiosis in alfalfa using salt-tolerant cultivar and rhizobial strain. *Appl. Soil Ecol.* **2015**, *87*, 108–117. [CrossRef]

environmental sciences proceedings

MDPI

Proceeding Paper

Effect of Seed Priming with ZnO Nanoparticles and Saline Irrigation Water in Yield and Nutrients Uptake by Wheat Plants †

Yasser Al-Salama [1,2]

[1] The Arab Center for the Studies of Arid Zones and Dry Lands—ACSAD, Damscus 963 11, Syria; yassersa@hotmail.com; Tel.: +963-944393708

[2] Faculty of Agriculture, Al-Furat University, Deir Ezzor 963 51, Syria

† Presented at the 2nd International Laayoune Forum on Biosaline Agriculture, 14–16 June 2022; Available online: https://lafoba2.sciforum.net/.

Abstract: Salinity is one of the major abiotic stresses that affect crop production in arid and semiarid areas. The nano-priming can be applied to improve the seeds germination and seedling establishment and enhance plant growth in addition to improving resistance to abiotic stresses. A pot experiment was conducted to study the influence of seed priming with (50 mg L^{-1}) ZnO nanoparticles (particle size < 50 nm) and irrigation with saline water on the yield and uptake of some nutrients (N, P, K, and Zn) by durum wheat variety (*Triticum Durum* L., cv. Acsad1105). Wheat variety was grown in calcareous soil under four levels of saline irrigation water (0.52, 4.42, 6.84 and 9.3 dS m^{-1}). Increasing water salinity caused a gradual and remarkable decrease in the grain yield, where the reduction was the highest (42.8%) at the higher salinity level (9.13 dS m^{-1}). Seed priming with ZnO NPs increased the grain yield by 36.2, 24.1, 13.2 and 5.6% for the investigated salinity levels, respectively, compared with non-primed seeds. The uptake of macronutrients (N, P and K) by both the straw and grains was significantly increased by increasing the salinity level up to 6.84 dS m^{-1}, particularly in the primed seeds. On the other hand, Zn uptake was significantly decreased in the two treatments with an increasing salinity level of irrigation water. However, field studies may further enhance the mechanistic understanding of the pertinence of NPs in seed priming and salinity-tolerant plants.

Keywords: durum wheat; salinity stress; seed priming; ZnO NPs

Citation: Al-Salama, Y. Effect of Seed Priming with ZnO Nanoparticles and Saline Irrigation Water in Yield and Nutrients Uptake by Wheat Plants. *Environ. Sci. Proc.* **2022**, *16*, 37. https://doi.org/10.3390/environsciproc2022016037

Academic Editor: Abdelaziz Hirich

Published: 16 June 2022

1. Introduction

Wheat is a moderately tolerant plant to salinity stress. It is considered, drought, as the major abiotic stress that affects the production capacity in arid and semiarid areas. Seed germination and seedling establishment are the most sensitive stages to salinity. Salt stress can adversely affect seed germination and the initial seedling growth via both the osmotic effect and ionic toxicity, which will induce oxidative stress [1]. It is a well-known fact that world resources of freshwater are getting exhausted through the increasing demand to meet the ever-increasing requirements of the world population. Therefore, the use of low water quality, e.g., groundwater, drainage water and treated seawater, should be considered as complementary sources for agricultural development. Nano-priming can be applied in seeds to provide protection during storage and improve germination and germination synchronization.

The use of nanotechnology for seed priming is a new area of research, although studies have already shown promising results [2,3]. Seed priming with nanomaterials can promote seed germination and enhance initial vegetative growth and biochemical characteristics in various plant species [4]. The treatment is simple and cost-effective. Primed seeds can rapidly absorb and renovate the seed metabolism, resulting in a higher germination rate and decreasing the intrinsic physiological non-uniformity in germination [5,6]. Seed priming can also improve plant growth, development, and the production capacity of crops [7,8]. Different methods are used to apply the zinc oxide (ZnO) nanoparticles on wheat, like foliar spray, soil mixing and seed priming methods [9]. ZnO nanoparticles at low concentrations promote the production of antioxidant enzymic and non-enzymic molecules, phytohormones and the overexpression of new water channels, thereby improving the absorption of water and nutrients with ZnO nanoparticles seed priming which can help the initial growth of seedlings [10]. New studies showed that priming of wheat seeds with zinc oxide nanoparticles (50 mg L^{-1}) and after 20 days of recovery of the plants being exposed to salt stress (200 mM NaCl) for 10 days reduced the concentration of Na^+ and increased water uptake [11]. However, when plants were exposed to salt stress, the nano-ZnO primed plants had significantly higher photosynthetic, stomatal conductance and transpiration rates compared with the non-primed plants [12]. The current work had the objective to study the influence of seed priming with ZnO nanoparticles and saline water irrigation (drainage water) on yield and nutrient uptake (N, P, K, and Zn) of durum wheat variety (Acsad1105) grown in calcareous soil.

2. Materials and Methods

Two varieties of wheat seeds (*Triticum Durum* L., cv. Acsad1105) were used in this experiment, the first one was left without any treatment (non-primed seed) and the second group was primed with 50 ppm ZnO nanoparticles for 4 h before sowing. The ZnO nanoparticles (particle size < 50 nm) were obtained from Begbroke Directorate, University of Oxford Science Park, Oxford OX5 1PF, UK, and both groups were used in a pot experiment under four levels of saline water. The pots were filled with 10 kg calcareous soil (Sandy loam, 24.5% $CaCO_3$ and EC: 2.27 dS m^{-1}) collected from Al-Tabny village in the west of Deir Ezzor Governorate, Syria. The primed and non-primed wheat seeds were sown separately in each pot. The seedlings were irrigated with tap water (W1: 0.52 dS m^{-1}) as a control and three levels of saline drainage water (W2: 4.42, W3: 6.82 and W4: 9.13 dS m^{-1}). After 160 days from planting, plants were harvested and the grain yield per pot, straw per pot and the 1000-kernel weight were recorded. Representative plant samples were digested for N, P, K and Zn determination. The IBM SPSS software (Version 25.0) was used to statistically analyze data at a 5.0% probability ($p \leq 0.05$).

3. Results and Discussion

The results revealed that increasing the salinity of irrigation water in the primed and non-primed seeds significantly ($p < 0.05$) decreased all the studied traits, but the reduction was significantly lower ($p < 0.05$) in ZnO NPs primed seeds (Figure 1a). However, increasing water salinity up to 4.42, 6.84 and 9.13 dS m^{-1} caused a remarkable decline in the grain yield of the non-primed wheat seeds by 12.9, 20.4 and 42.8%, respectively, meanwhile the priming of seeds with 50 ppm ZnO NPs increased grain yield by 36.2, 24.1, 13.2 and 5.6% under saline irrigation water of 0.52, 4.42, 6.84 and 9.3 dS m^{-1}, respectively, compared with the non-primed seeds (Figure 1b).

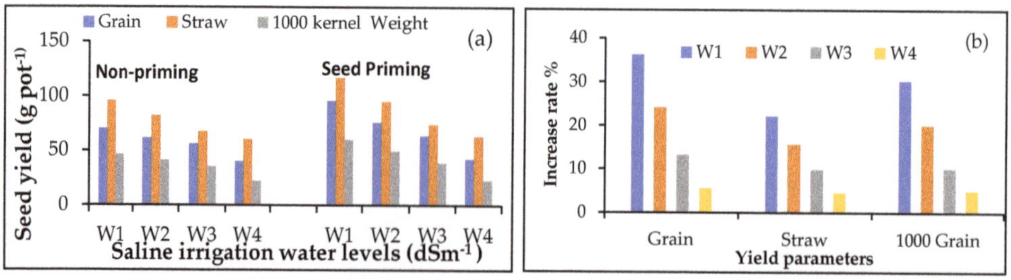

Figure 1. Effect of saline irrigation water and seed priming on seed yield (g pot^{-1}) of durum wheat plants (**a**), and increase rate in yield parameters due to seed priming(**b**).

Concerning the straw yield and 1000-kernel weight, the obtained data showed a similar trend as previously shown for grain yield. Both treatments had significant differences ($p < 0.05$) between primed seeds over non-primed ones (Figure 1b). Macronutrients, N, P and K uptake by grains and straw of wheat plants were significantly affected ($p < 0.05$) by salinity in both the treatments (primed and non-primed seeds) (Figure 2a,b). The maximum uptake values by grains were found to be under the W3 level. However, it has been noticed that the seed priming significantly enhanced ($p < 0.05$) the uptake of these nutrients by grain in non-primed seeds (1455, 412 and 583 mg pot^{-1}) and 1674, 467 and 647 mg pot^{-1} for N, P and K in primed seeds. A similar trend was recorded for the nutrients uptake by the straw; at the same time, the nutrient uptake was the least at the highest salinity level for both treatments. This could be attributed to the effect of higher soluble salts on decreasing the water potential gradient between the soil solution and the root cells, thereby decreasing the water flow and absorption which will adversely affect the nutrient uptake. In addition to the antagonistic effects of the higher concentration of Na^+ and Cl^- on the uptake of other beneficial cations and anions.

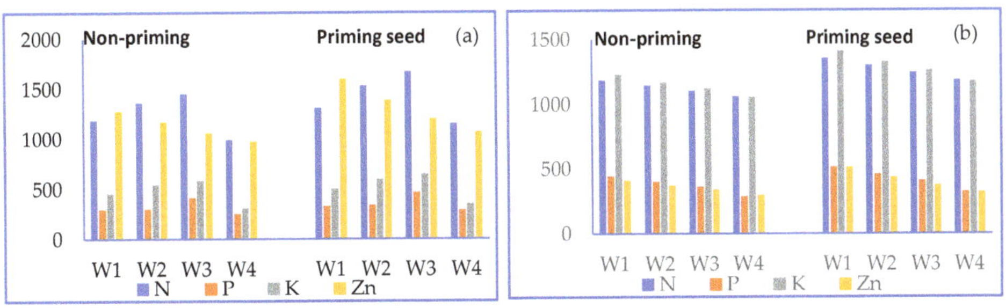

Figure 2. Effect of irrigation water salinity and seed priming on nutrients uptake (N, P, K in mg pot^{-1} and Zn in μg pot^{-1}) by grain (**a**) and straw (**b**).

On the other hand, Zn uptake (μg pot^{-1}) significantly decreased ($p < 0.05$) in both treatments with an increasing salinity level of irrigation water due to the reduction of the dry weight of wheat plants for both straw and grains. The decline was highest in the non-primed seeds compared with primed ones, thus the seed priming with ZnO NPs could improve the Zn uptake from 8 to 25% depending on the level of salinity imposed. These results agree with those obtained by Gaafar et al. (2020) [13], they found that Zn^{+2} content in soybeans was significantly decreased under salinity stress by 76% compared to control, and presoaking of soybean seeds with ZnO NPs (50 mg L^{-1}) increased Zn^{+2} content even under salinity stress 2.7-fold compared to salt treatment alone.

Excessive salinity has a detrimental effect on the growth and wheat yield by restricting nutrients uptake to extent that a deficiency takes place. This may be due to the possibility that plants grown under saline conditions utilize energy for the osmotic adjustment process at the expense of growth and development, in addition to limiting the water flow from soil to plant under saline conditions as a consequence of the decreasing water potential gradient, which will cause a remarkable decline in turgor potential and stomatal conductance (gs). However, seed priming with ZnO NPs increases the root extension and the number of effective root hairs during the early stages of wheat plant growth [14] and improved plant height, chlorophyll content, photosynthetic efficiency, tiller number and, finally, biological and grain yields. Furthermore, ZnO seed priming increases the Zn content in both leaves and seeds [15], and positively affects the growth traits in salt-stressed plants, whereas ZnO NPs stimulated natural auxin (Indole Acetic Acid) and thus activating cell division and enlargement as well as augmentation of photosynthetic pigments, organic solutes, total phenols, ascorbic acid and Zn in stressed plants [16].

4. Conclusions

Salinity stress adversely effects wheat plant growth and grain yield and yield components. However, priming seeds with ZnO nanoparticles was an effective technique to improve the performance of wheat plants, straw and seed yield under salinity stress conditions.

Funding: This research received no external funding.

Institutional Review Board Statement: Not applicable.

Informed Consent Statement: Not applicable.

Acknowledgments: The author would like to thank Ian P. Thompson and Rodrigo Molina from Begbroke Directorate, University of Oxford, UK, for their support in provided and characterized the ZnO nanoparticles.

Conflicts of Interest: The authors declare no conflict of interest.

References

1. Ibrahim, E.A. Seed priming to alleviate salinity stress in germinating seeds. *J. Plant Physiol.* **2016**, *192*, 38–46. [CrossRef] [PubMed]
2. Hussain, A.; Rizwan, M.; Ali, Q.; Ali, S. Seed priming with silicon nanoparticles improved the biomass and yield while reduced the oxidative stress and cadmium concentration in wheat grains. *Environ. Sci. Pollut. Res.* **2019**, *26*, 7579–7588. [CrossRef] [PubMed]
3. Ye, Y.; Cota-Ruiz, K.; Hernández-Viezcas, J.A.; Valdés, C.; Medina-Velo, I.A.; Turley, R.S.; Peralta-Videa, J.R.; Gardea-Torresdey, J.L. Manganese Nanoparticles Control Salinity-Modulated Molecular Responses in *Capsicum annuum* L. through Priming: A Sustainable Approach for Agriculture. ACS Sustain. *Chem. Eng.* **2020**, *8*, 1427–1436. [CrossRef]
4. Waqas, M.; Korres, N.E.; Khan, M.D.; Nizami, A.S.; Deeba, F.; Ali, I.; Hussain, H. Advances in the concept and methods of seed priming. In *Priming and Pretreatment of Seeds and Seedlings*; Hasanuzzaman, M., Fotopoulos, V., Eds.; Springer: Singapore, 2019; pp. 11–41. [CrossRef]
5. Dimkpa, C.O.; McLean, J.E.; Britt, D.W.; Anderson, A. Bioactivity and Biomodification of Ag, ZnO, and CuO Nanoparticles with Relevance to Plant Performance in Agriculture. *Ind. Biotechnol.* **2012**, *8*, 344. [CrossRef]
6. Rowse, H.R. Drum priming: A non-osmotic method of priming seeds. *Seed Sci. Technol.* **1995**, *24*, 281–294.
7. Rico, C.M.; Lee, S.C.; Rubenecia, R.; Mukherjee, A.; Hong, J.; Peralta-Videa, J.; Gardea-Torresdey, J.L. Cerium Oxide Nanoparticles Impact Yield and Modify Nutritional Parameters in Wheat (*Triticum aestivum* L.). *J. Agric. Food Chem.* **2014**, *62*, 9669–9675. [CrossRef] [PubMed]
8. Ali, Q.; Daud, M.K.; Haider, M.Z.; Ali, S.; Rizwan, M.; Aslam, N.; Noman, A.; Iqbal, N.; Shahzad, F.; Deeba, F.; et al. Seed priming by sodium nitroprusside improves salt tolerance in wheat (*Triticum aestivum* L.) by enhancing physiological and biochemical parameters. *Plant Physio. Biochem.* **2017**, *119*, 50–58. [CrossRef] [PubMed]
9. Rizwan, M.; Ali, S.; Qayyum, M.F.; Ok, Y.S.; Adrees, M.; Ibrahim, M.; Rehman, M.Z.; Farid, M.; Abbas, F. Effect of metal and metal oxide nanoparticles on growth and physiology of globally important food crops. *J. Hazard. Mater.* **2017**, *322*, 2–16. [CrossRef] [PubMed]
10. Santo Pereira, A.D.E.; Oliveira, H.C.; Fraceto, L.F.; Santaella, C. Nanotechnology Potential in Seed Priming for Sustainable Agriculture. *Nanomaterials* **2021**, *11*, 267. [CrossRef] [PubMed]
11. Wang, Z.; Li, H.; Li, X.; Xin, C.; Si, J.; Li, S.; Li, Y.; Zheng, X.; Li, H.; Wei, X.; et al. Nano-ZnO priming induces salt tolerance by promoting photosynthetic carbon assimilation in wheat. *Arch. Agron. Soil Sci.* **2020**, *66*, 1259–1273. [CrossRef]
12. Pooja, R.M.; Bhaumik, J.; Kaur, R. Role of Zinc Oxide Nanoparticles in Mitigation of Drought and Salinity-A Review. *Int. J. Curr. Microbiol. App. Sci.* **2020**, *9*, 467–481. [CrossRef]
13. Gaafar, R.M.; Diab, R.H.; Halawa, M.L.; El-Shanshory, A.R.; El-Shaer, A.; Hamouda, M.M. Role of Zinc Oxide Nanoparticles in Ameliorating Salt Tolerance in Soybean. *Egypt. J. Bot.* **2020**, *60*, 733–747. [CrossRef]
14. Larue, C.; Veronesi, G.; Flank, A.M.; Surble, S.; Herlin-Boime, N.; Carrière, M. Comparative Uptake and Impact of TiO_2 Nanoparticles in Wheat and Rapeseed. *J. Toxicol. Environ. Health Part A* **2012**, *75*, 722–734. [CrossRef]
15. Raddy, R.; Pavithra, G.J.; Rajashekar Reddy, B.H.; Mahesh, S.; Geetha, K.N.; Shankar, A.G. Zinc oxide nano particles increases Zn uptake, translocation in rice with positive effect on growth, yield and moisture stress tolerance. *Indian J. Plant Physiol.* **2017**, *22*, 287–294. [CrossRef]
16. Latef, A.A.H.A.; Alhmad, M.F.A.; Fattah, K.E.A. The possible roles of priming with ZnO nanoparticles in mitigation of salinity stress in Lupine (*Lupinus termis*) plants. *J. Plant Growth Regul.* **2017**, *36*, 60–70. [CrossRef]

environmental sciences proceedings

MDPI

Proceeding Paper

Effects of Water Stress on Leaf Photosynthesis and Yield of Melon and Tomato Crops Grown under Mediterranean Conditions of the Northeast of Morocco †

Sara Uald Lamkaddam [1], Salah Ed-Dine Samri [1], Mohamed Hassani Zerrouk [2] and Kamal Aberkani [1,*]

1 Polydisciplinary Faculty of Nador, University Mohammed First, P.O. Box 300, Selouane 62700, Morocco; sara.lamk1007@gmail.com (S.U.L.); salahsamri@gmail.com (S.E.-D.S.)
2 Faculty of Science and Techniques Al Hoceima, University Abdelmalek Essaadi, P.O. Box 34, Ajdir 32003, Morocco; m.hassani@uae.ac.ma
* Correspondence: kamalaberkani@gmail.com; Tel.: +212-613-591-745

† Presented at the 2nd International Laayoune Forum on Biosaline Agriculture, 14–16 June 2022; Available online: https://lafoba2.sciforum.net/.

Abstract: Drought is the most important limiting factor affecting plant yield and photosynthesis and has impacts on reducing yields. The objective of this work is to study the impacts of drought regimes of irrigation on Chlorophyll a fluorescence, stress status and yield of melon and tomato crops. Experiments were conducted under the Mediterranean climate conditions. Two cultivars of each crop and three irrigation strategies (100%, 75% and 50%) were used. Many physiological stress indices based on chlorophyll a fluorescence parameters were measured. Results showed a change in crop light phase of photosynthesis and a decrease in yields according to treatments and cultivars.

Keywords: chlorophyll a fluorescence; drought; irrigation

Citation: Uald Lamkaddam, S.; Samri, S.E.-D.; Zerrouk, M.H.; Aberkani, K. Effects of Water Stress on Leaf Photosynthesis and Yield of Melon and Tomato Crops Grown under Mediterranean Conditions of the Northeast of Morocco. *Environ. Sci. Proc.* **2022**, *16*, 42. https://doi.org/10.3390/environsciproc2022016042

Academic Editor: Abdelaziz Hirich

Published: 16 June 2022

1. Introduction

The United Nations Climate Change Conference (COP26) held recently in Glasgow, Scotland, adopted an agreement to accelerate the process in order to face global warming [1]. Morocco is among the African countries affected by the irregularity and the decrease in rainfall during the last decade [2]. The focus of research on water management is an important step in order to decrease the impact of drought on field production.

Many researchers studied the impact of water scarcity on physiological aspects such as plant growth, photosynthesis, metabolite translocation, plant yield and quality, etc. Yildirim et al. (2009) [3] reported that plants should be irrigated during the whole period of the production. Fabeiro et al. (2002) [4] found that water stress applied during the flowering period negatively influenced the fruit yield and quality of muskmelon. Moreover, low stress levels were found to result in maximum yields and high stress resulted in the lowest yields [5]. Birhanu and Zeleke (2010) [6] also conducted a study to determine the effect of 0%, 25%, 50% and 75% crop evapotranspiration deficits on the fruit yield and quality on drip-irrigated tomatoes. They showed that growth, yield and quality were directly related to water shortage levels.

Moreover, it has been shown that under high temperature and radiation levels, drought stress enhances the inhibition of electron transport [7]. Chlorophyll a fluorescence is a non-destructive measurement for studying the equilibrium between metabolic and energy evolving processes that may be affected by temperature and drought stresses [8]. Chlorophyll a fluorescence indicates the transfer of electrons during the light phase of photosynthesis from the excitation of chlorophyll by light energy to the transfer of electrons for the dark phase [9]. Drought stress reduces variable fluorescence (F_v), initiative fluorescence (F_0) and quantum yield (F_v, F_m) [10]. Chlorophyll a fluorescence was considered as a useful tool for screening and breeding under dry conditions and for high temperature resistance. The ratio F_v/F_m varies between 0.75 and 0.85 in non-stressed plants [11], and it is a good indicator for stress level status. Other parameters of the Chl a fluorescence (Vi, Vj, N, S, ABS, TR, ETR, DI, PI, etc.) also present good indicators for plant stress status.

The objective of this research work is to study, under Mediterranean climate conditions and at a commercial production level, the impacts of different drought strategies of irrigation on chlorophyll a fluorescence, stress status and yield of melon and tomato, which represent two of the main crop productions in the country.

2. Material and Methods

This study was conducted in two sites spaced out at 70 km. The first site is located in the Driouch area, Morocco (34°59′30.3″ N 3°22′30.6″ W), and was intended for melon experiments. However, the second site was located in the ARID area, Nador, Morocco (35°08′52.6″ N 2°57′40.3″ W), and was intended for tomato experiments. Both sites were realized at the commercial farm production scale. For the melon trial, two commercial cultivars were used: "cv. Missoura" and "cv. Wifaq". An area of 0.5 ha was used for each cultivar, at a plant density of 8333 plants per ha. Rows were spaced by 2 m, and the distance between two plants of the same row was 60 cm. Date of plantation was 23 May 2021. The trial ended on 10 September 2021. For tomato trials, two commercial cultivars were used: "cv. Karima" and "cv. Jad". An area of 1200 m was used for each cultivar, at plant density of 14285 plants per ha. Each row was spaced at 1.20 m and the distance between two plants of the same row was 70 cm. Date of plantation was 28 August 2021, and the trial was finished on 25 January 2022. Both crops were managed according to commercial production practices.

Three irrigation treatments were applied: 100%, 75% and 50% water. An irrigation tape (200 micron with dripper of 1.5 L/h, Rolland tape, Roll-Drip, France) was used. The distances of 20 cm, 15 cm and 10 cm between two drippers were used for 100%, 75% and 50% water, respectively. In the case of melon experiments, the length of each irrigation lateral was 100 m and split on 2 semi-lines of 50 m. However, for tomato experiments, the length of each irrigation lateral was 50 m long.

For the melon trial, six repetitions of each treatment were used. Each treatment replicate contains 6 blocks of 6 plants. For tomato experiments, 8 replicates of each treatment were used containing 6 blocks of 3 plants.

The chlorophyll a fluorescence measurements were recorded on the fifth young expanded leaf, during a hot and sunny day. The HPEA device (Handey PEA, Hansatech, King's Lynn, UK) was used. Both cultivars of melon and tomato leaves were adapted to the dark for 30 min using a clip. Then, a light flash of 3000 µmol/m/s (650 nm) was applied for 1 s (gain = ×1) on the leaf adapted to darkness for 30 min. The measurements were taken on 6 plants for each treatment, from 12:00 to 14:00. F_v/F_m and DI0/CS0 were measured (F_v = variable fluorescence; F_m = maximum fluorescence; DI/CS_0 = Dissipated heat per cross section).

The fruit harvests were carried out on 18 July and 3 August 2021 for "cv. Missoura" and for "cv. Wifaq", respectively. However, the last harvests were carried out on 19 September and 27 August 2021, respectively. For tomato trials, the harvest starts on 3 November and ends on 23 January 2022. Total yield was counted for both experiments.

3. Results and Discussion

During the hot period of the day and clear sky conditions (19 July 2021; 12:00–14:00 with Tmax = 35 °C; Tmin = 20 °C; Relative humidity = 58%; and wind speed = 11 km/h), the F_v/F_m ratio of the cv. Missoura melon was lowered and the DI/CS0 ratio was increased by the water shortages (75% and 50%). However, for cv. Wifaq, both ratios were not affected by the irrigation treatment (Table 1). This means that the response of the cultivar was not the same for irrigation treatments and cv. Missoira was more affected by water stress compared to cv. Wifaq. It was shown by Baker and Horton (1987) [11] that the F_v/F_m

ratio varies between 0.75 and 0.85 for non-stressed plants. The DI/CS_0 ratio showed high values, which indicates that plants were stressed and would have dissipated heat energy. A similar occurrence was observed for cv. Missoura, which was more stressed under the 50% and 75% treatments compared to the 100% treatment. The DI/CS_0 values were 86, 114 and 120, for the 50%, 75% and 100% irrigation treatments, respectively.

Table 1. Fluorescence parameters (F_v/F_m and DI/CS_0) of melon and tomato plants grown under three strategies of irrigation (100%, 75% and 50% water).

			100% Water	75% Water	50% Water
Melon 19/7/2021	cv. **Missoura** (12:00–12:45)	F_v/F_m	0.81 ± 0.02	0.77 ± 0.03	0.77 ± 0.03
		DI/CS_0	86 ± 13	114 ± 17	120 ± 36
	cv. Wifaq (13:15–14:00)	F_v/F_m	0.80 ± 0.03	0.81 ± 0.02	0.79 ± 0.01
		DI/CS_0	77 ± 6	70 ± 9	78 ± 10
Tomato 29/9/2021	**cv. Karima** (11:40–12:10)	F_v/F_m	0.77 ± 0.04	0.73 ± 0.05	0.72 ± 0.04
		DI/CS_0	123 ± 26	185 ± 64	192 ± 44
	cv. Jad (12:20–12:50)	F_v/F_m	0.73 ± 0.08	0.73 ± 0.06	0.78 ± 0.03
		DI/CS_0	197 ± 77	181 ± 76	122 ± 37

F_v = variable fluorescence; F_m = maximum fluorescence; DI/CS_0 = Dissipated heat per cross section. Data are mean of 6 repetitions ± standard deviation.

For tomato, during a day of clear sky (29 September 2021; 11:40–12:50 with T_{max} = 29 °C; T_{min} = 22 °C; Relative humidity = 62%; and wind speed = 16 km/h), the same tendencies were observed for cv. Karima than cv. Missoura. For cv. Jad, no differences were recorded for the F_v/F_m and for DI/CS_0 ratios between treatments. This could mean that cv. Karima did not have the same response to water shortage as cv. Jad. This was shown by Salinas-Vargas et al. (2021) [12], which indicates that tomato cultivars could have different responses (tolerance or resistance) for water stress.

The yield of melon was lowered by shortage in irrigation amount (Table 2), for both cultivars (cv. Missoura and cv. Wifaq). This indicates that 100% treatment (adapted by the farmer) was the appropriate irrigation strategy. This is consistent with the findings of Fabeiro et al. (2002) [4], who found that water stress negatively influenced the fruit yield of melon. For tomato, the treatments did not affect the yield of both cultivars. This could be also be explained by the fact that tomato trials were carried out in fall, where the evapotranspiration demand is lower than conditions of the melon experiments in summer. As reported by Agele et al. (2011) [11], the high evapotranspiration rate has a significant effect on decreasing the plant yield for tomato crops.

Table 2. Total yield (kg/plant) of melon and tomato crops grown under three strategies of irrigation (100%, 75% and 50% water).

		100% Water	75% Water	50% Water
Melon (kg/plant)	cv. Missoura	6.0 ± 1.5	4.9 ± 1.1	4.4 ± 1.0
	cv. Wifaq	7.3 ± 1.0	6.9 ± 1.2	5.8 ± 1.1
Tomato (kg/plant)	cv. Karima	2.9 ± 0.7	2.3 ± 0.7	2.2 ± 0.7
	cv. Jad	2.4 ± 0.9	2.2 ± 1.1	2.1 ± 0.7

Data are the mean of 36 repetitions (melon); 48 repetitions (tomato) ± standard deviation.

4. Conclusions

This study showed that reducing the amount of irrigation water decreased the yield of both cultivars of melon (cv. Missoura and cv. Wifaq). This reduction was higher for the 50% irrigation compared to 75% and 100% irrigation. However, this shortage in water did not affect yields of both tomato cultivars (cv. Karima and cv. Jad). For chlorophyll a fluorescence, the results showed that the F_V/F_m ratio was decreased and the DI/CS0 ratio was increased for cv. Missoura (melon) and cv. Karima (tomato) with the 75% and 50% irrigation treatments compared to the control treatments, and this means that plants were more stressed. Economic studies on the amount of water saving and yield decrease will be also interesting in order to determine the best strategy of water management that could be adapted by the grower in such conditions.

Author Contributions: Methodology, S.U.L., K.A. and S.E.-D.S.; software, M.H.Z.; validation, S.U.L. and K.A.; investigation, S.U.L. and K.A.; data curation, S.U.L.; writing—original draft preparation, S.U.L. and K.A.; writing–review and editing, K.A. and S.E.-D.S.; supervision, K.A., S.E.-D.S. and M.H.Z.; project administration, K.A. and S.E.-D.S. All authors have read and agreed to the published version of the manuscript.

Funding: This research was funded by the PRIMA program, which is supported by Horizon 2020, the European Union's Framework Program for Research and Innovation (https://prima-med.org, accessed on 1 March 2021). The project is intitule DATI (Digital Agriculture for Technologies Efficiency).

Institutional Review Board Statement: Not applicable.

Informed Consent Statement: Not applicable.

Data Availability Statement: Not applicable.

Acknowledgments: The authors would like to thank PRIMA for funding the DATI project (Digital Agriculture Technologies for Irrigation), wherein this work is part of the proposal. In addition, they would like to thank the students and staff for their technical help and the growers for giving their field for trials.

Conflicts of Interest: This research work has no conflict of interest.

References

1. Glasgow Climate Change Conference. October-November 2021. Available online: https://unfccc.int/conference/glasgow-climate-change-conference-october-november-2021 (accessed on 14 March 2022).
2. Salhi, A.; Martin-Vide, J.; Benhamrouche, A. Rainfall distribution and trends of the daily precipitation concentrastion index in northern Morocco: A need for an adaptive environmental policy. *SN. Appl. Sci.* **2019**, *1*, 277. [CrossRef]
3. Yildirim, M.; Akçal, A.; Kaynaş, L. The response of *Cyclamen hederifolium* to water stressinduced by different irrigation levels. *Afr. J. Biotechnol.* **2009**, *8*, 1069–1073.
4. Fabeiro, C.; Olalla, F.; Juan, J.A. Production of muskmelon (*Cucumis melo* L.) under controlled deficit irrigation in a semi-arid climate. *Agri. Water Manag.* **2002**, *54*, 93–105. [CrossRef]
5. May, D.M. Moisture stress to maximize processing tomato yield and quality. *Acta Hort.* **1993**, *335*, 547–552. [CrossRef]
6. Birhanu, K.; Zeleke, K. Fruit yield and quality of drip-irrigated tomato under deficit irrigation. *Afr. J. Food Agric. Nutr. Dev.* **2010**, *10*, 2139–2151. [CrossRef]
7. Lu, Q.; Lu, C.; Zhang, G.; Kuang, T. Photosynthesis and chlorophyll a fluorescence during flag leaf senescence of field-grown wheat plants. *J. Plant Physiol.* **2002**, *159*, 1173–1178. [CrossRef]
8. Araus, J.L.; Hogan, K.P. Comparative leaf structure and patterns of photo inhibition of the geotropically palms *Scheelea zonensis* and *Socratea durissima* growing in clearings and forest understory during the dry season in Panama. *Am. J. Bot.* **1994**, *81*, 726–738. [CrossRef]
9. Strasser, R.; Tsimilli-Michael, M.; Srivastava, A. Analysis of the Chlorophyll a Fluorescence Transient. In *Chlorophyll a Fluorescence: A Signature of Photosynthesis*; Papageorgiou, G.C., Govindjee, Eds.; Springer: New York, NY, USA, 2004; pp. 321–362. [CrossRef]
10. Vazan, S. Effects of Chlorophyll Parameters and Photosynthesis Efficiency in Difference Beet. Ph.D. Thesis, Islamic Azad University Science and Research Tehran-Branch, Tehran, Iran, 2002; p. 285.
11. Baker, N.R.; Horton, P. Chlorophyll Fluorescence Quench-ing During Photoinhibition. In *Photoinhibition*; Kyle, D.L., Osmond, C.B., Arntzen, C.J., Eds.; Elsevier Science Publishers: Amsterdam, The Netherlands, 1987; pp. 145–168.
12. Salinas-Vargas, D.; Maldonado Peralta, M.; Rojas García, A.; León-Villanueva, A.; Álvarez-Vázquez, P.; Maldonado-Peralta, R. Response of tomato varieties (*Solanum lycopersicum* L.) to water stress. *Agro Product.* **2021**, *13*, 69–74. [CrossRef]

environmental sciences proceedings

MDPI

Proceeding Paper

Saline Water Irrigation Effect on Oil Yield and Quality of Argan Trees Domesticated in Laâyoune, Morocco †

Chaima Afi [1,2,*], Jamal Hallam [2], Abdelaziz Mimouni [2], Fouad Msanda [1] and Naima Ait Aabd [2]

1 Laboratory of Biotechnology and Valorization of Natural Resources, Faculty of Sciences, Ibn Zohr University, P.O. Box 8106 Agadir, Morocco; f.msanda@uiz.ac.ma
2 National Institute of Agricultural Research of Morocco, P.O. Box 124 Inezgane, Morocco; jamal.hallam@inra.ma (J.H.); abdelaziz.mimouni@inra.ma (A.M.); naima.aitaabd@inra.ma (N.A.A.)
* Correspondence: chaima.afi@edu.uiz.ac.ma
† Presented at the 2nd International Laayoune Forum on Biosaline Agriculture, 14–16 June 2022; Available online: https://lafoba2.sciforum.net/.

Abstract: Salinity is one of the major severe constraints that limit crop productivity in 40% of the land surface, notably in the Mediterranean region. In this study, we worked in an argan orchard located in Laâyoune, Morocco. The orchard is characterized by a saline soil and trees that are irrigated with highly saline water. The study aimed to evaluate the effect of irrigation with saline water on oil yield, total phenolic content, flavonoid content, antioxidant activity, and fatty acid composition. The results show that saline water irrigation has a no significant effect on oil yield and most oil quality parameters, yet has a highly significant effect on total polyphenols, flavonoid content as well as two saturated acids (C16:0 and C20:0).

Keywords: salinity; argan; oil yield; total polyphenol; flavonoid content; antioxidant activity; fatty acids

check for updates

Citation: Afi, C.; Hallam, J.; Mimouni, A.; Msanda, F.; Ait Aabd, N. Saline Water Irrigation Effect on Oil Yield and Quality of Argan Trees Domesticated in Laâyoune, Morocco. *Environ. Sci. Proc.* **2022**, *16*, 45. https://doi.org/10.3390/environsciproc2022016045

Academic Editor: Abdelaziz Hirich

Published: 16 June 2022

1. Introduction

Around the world, it is estimated that approximately 20% of irrigated cultivated lands since 1990 are affected by salinization [1]. The growing demand for water and the climate change shift seem to increasingly limit crop production, making the issues even worse [2]. Agriculture will then increasingly be forced to use low-quality waters such as brackish water or reclaimed effluent which in turn increases the risks of soil salinization and yield reduction [3]. The introduction of plant species that are tolerant to salt stress of great socio-economic values is one of the possible approaches to promote salt-affected soils and tackle this major abiotic factor that reduces agricultural yield [4]. The approach would improve the vegetation cover and solve the regeneration of particular forest species, such as the argan tree (Argania spinosa) that occurs in a restricted arid areas and which certainly represents a threatened ecological wealth in Southwestern Morocco [3].

Adult argan trees are adapted to aridity and are able to survive long drought periods. They can produce leaves, branches and fruits under as little as 100 mm rainfall [5]. It was shown that the argan tree could tolerate water salinity at germination stage and at the first phases of its in vitro growth [4]. Yet, the majority of studies consider the vegetative parameters, germination or oil quality of argan trees. Despite this knowledge, little is known about the effect of salinity on argan oil quality, and such understanding could help mitigate and buffer extreme events caused by climate change. Therefore, we conducted a comprehensive study on the quality of argan oil obtained from trees domesticated under saline constraints in the Saharan climate in Laâyoune.

2. Materials and Methods

2.1. Site and Plant Materials

The study was carried out in a 700 m^2 orchard in Laâyoune, Morocco. Twelve argan trees were investigated, eight of which were at production stage during the sampling period (Figure 1). The trees have been planted since 1996 at a density of 5 × 5 m and have been irrigated twice a month with 40 L of saline well water per tree. The soil is a sandy-loamy-clayey texture on a limestone slab, with a pH of 7.95, an electrical resistivity of 7.64 dS m^{-1} and a 108 ppm Na+. As for the irrigation water, the pH, the electrical conductivity and the sodium content were 7.27, 16.45 dS m^{-1} and 3083 ppm, respectively (experimental INRA-field, Laâyoune data).

Figure 1. The argan orchard design with the twelve domesticated trees at experimental INRA-field, Laâyoune.

The kernels were recovered from the ripped fruits and grinded. The powder was fried down for 24 h at 40 °C to eliminate all traces of humidity

2.2. Extraction

Once oven dried, a 10 g mass was mixed with 80 mL hexane then poured in a 100 mL flask. The mixture was then exposed to an ultrasonic bath set at 50 ± 1 C for 45 min at a frequency of 28 kHz.

2.3. Physico-Chemcial and Biochemical Study

The extraction of phenolic compounds was achieved following the method of polarity [6]. The determination of the total polyphenol and flavonoid content was carried out by colorimetry using the Folin–Ciocalteu reagent [7], and the aluminum trichloride (AlCl3) and sodium nitrite (NaNO2) [8] methods, respectively. To study the anti-radical activity of the different extracts, the method based on diphenyl picryl-hydrayl (DPPH) as a relatively stable free radical was used according to the protocol described by [9]. The International Organization for Standardization (ISO 5509:2009) [10] method was used for fatty acid composition. Data were analyzed using analysis of variance (ANOVA). Statistica software version 10 was used to assess the statistical significance of mean variations in terms of the effect on the quality of oil. The threshold for significance was set at <0.05 *p* values.

3. Results

Table 1 shows that the mean oil yields ranged from 37.04% to 47.60%. The total polyphenols and flavonoids ranged from 0.075 to 0.094 and 1.749 to 2.494, respectively. The antioxidant activity varied between 22.80% and 46.17%. As for the fatty acid profiles, the results show variations between 12.82 and 15.1 in C16:0, 0.1 and 0.14 for C16:1, 5.36 and 7.9 for C18:0, 44.33 and 53.66 for C18:1, 25.27 and 34.78 for C18:2, 0.06 and 0.1 for C20:0 and 0.02 and 0.04 for C20:1. From the table, we observe that the results obtained show an interesting aspect in terms of the several analyses carried out.

Table 1. Analysis of variance data of oil yield and quality under highly saline water irrigation.

Tree Code	Oil Yield (%)	TPC (mg EAG/g oil)	FC (mg EQ/g Oil)	AOA (%)	C16:0	C16:1	C18:0	C18:1	C18:2	C20:0	C20:1
A1	43.02	0.094	2.194	31.95	14.39	0.11	6.44	53.66	25.27	0.08	0.04
A2	37.58	0.075	2.466	22.80	15.10	0.14	6.02	51.67	26.97	0.08	0.02
A3	46.12	0.077	2.449	41.33	13.74	0.1	6.93	44.33	34.78	0.08	0.03
B1	44.73	0.080	2.255	46.12	12.82	0.11	7.90	49.69	29.38	0.06	0.03
B2	43.02	0.077	2.416	40.68	12.94	0.14	5.36	46.65	34.80	0.08	0.04
C1	47.60	0.084	2.494	45.85	14.22	0.12	6.27	50.31	28.96	0.08	0.04
C2	41.80	0.093	1.933	39.08	13.16	0.11	7.30	48.02	31.27	0.10	0.04
C4	37.04	0.081	1.749	46.17	13.92	0.13	5.79	48.5	31.53	0.09	0.04
Mean	42.61	0.083	2.245	39.25	13.79	0.12	6.50	49.10	30.37	0.08	0.04
Marge	37.04 −47.60	0.075– 0.094	1.749 −2.494	22.80 −46.17	12.82 −15.10	0.10 −0.14	5.36 −7.90	44.33 −53.66	25.27 −34.80	0.06 −0.10	0.02 −0.04
F value	6.162	9.130	650.025	0.355	21.792	0.000	2.141	3.265	0.451	1017.679	3.764
LSD (0.05)	0.005 ns	0.004 *	0.000 ***	0.554 ns	0.000 ***	1.000 ns	0.150 ns	0.077 ns	0.505 ns	0.000 ***	0.059 ns

TPC: Total polyphenol content; EAG: Equivalent gallic acid; FC: Flavonoid content; EQ: Equivalent quercetin; AOA: Antioxidant activity; LSD: Least significant differences; * significant at $p < 0.05$; *** significant at $p < 0.001$; ns:no significant.

Table 1 also shows that irrigation with highly saline water has no significant effect on oil yield, a highly significant effect on the content of both total polyphenols and flavonoids, a non-significant effect on the antioxidant activity relative to the control treatments. For the fatty acid, the ANOVA showed a highly significant effect of saline irrigation water on two saturated acids which are palmitic C 16:0 and arachidic acid C20:0, for the rest of the fatty acids, the results shows that there is no significant effect.

4. Discussion

Soil and water of the orchards were classified as moderately and highly saline, respectively [1,11]. We observed that the mean of the oil yield of trees irrigated with saline water is lower than that in the previous findings of Charrouf, Kouidri and Mechqoq [12–14], who reported an oil yield with solvent ranging between 50–55%, a yield of 55.94% and a yield of 57.12%, respectively. This could be explained by the fact that the yields obtained by those authors were extracted using the Soxhlet method, which generally gives better yield than UAE. However, the total polyphenol content for all irrigated trees with highly saline water was higher than the values obtained by Sour and Demnati [15,16], 0.083 mg EGA/g oil vs. 0.0563 and 0.0724 mg EGA/g oil, respectively. The high stress phase of argan trees under high saline conditions in our orchards would stimulate the production of the secondary metabolites and therefore explain their high values. Similar higher values were obtained for flavonoids, relative to the control. The antioxidant activity values were slightly lower than those of the control yet both are lower than the value obtained by [17], who reported 86.45% antioxidant activity for non-irrigated, old trees. Perhaps the antioxidant activity difference is mostly due to the climate or tree age. Chromatographic profiles showed that the results are overly within the ranges given by the Moroccan standers [18] except for some slight variation, for example in A2 for C16:0, B1 and C2 for C18:0, A1, A2, B1 and C1 for C18:1 and C18:2, all trees for C20:0 and C20:1 (except A2 for the last acid).

5. Conclusions

Our results suggest that except the secondary metabolites, all parameters are either genetically controlled by the tree treats or are affected biotic/abiotic factors such as soil type, climate, age, extraction methods, etc. For a better understanding of those interactions, a profound study on different concentrations, different soils and climates should be conducted.

Author Contributions: Conceptualization: C.A. and N.A.A.; Methodology: C.A. and N.A.A.; Investigation and writing—original draft preparation: C.A.; Writing—review and editing: C.A., N.A.A., J.H. and F.M.; Formal analysis: C.A., N.A.A., J.H. and F.M.; Funding acquisition: N.A.A. and A.M. All authors have read and agreed to the published version of the manuscript.

Funding: This work was supported by National Institute of Agronomic Research, Regional Center of Agronomic Research of Agadir, Morocco (Grant id: PRMT Arganier 2021–2024).

Institutional Review Board Statement: Not applicable.

Informed Consent Statement: Not applicable.

Data Availability Statement: Data available by contacting the corresponding author.

Conflicts of Interest: The authors declare no conflict of interest.

References

1. Shahid, S.A.; Heng, L. *Guideline for Salinity Assessment, Mitigation and Adaptation Using Nuclear and Related Techniques*; Springer Nature: Cham, Switzerland, 2018.
2. Stefanoudaki, E.; Williams, M.; Chartzoulakis, K.; Harwood, J. Olive oil qualitative parameters after orchard irrigation with saline water. *J. Agric. Food Chem.* **2009**, *57*, 1421–1425. [CrossRef] [PubMed]
3. Chakhchar, A.; Haworth, M.; El Modafar, C.; Lauteri, M.; Mattioni, C.; Wahbi, S.; Centritto, M. An assessment of genetic diversity and drought tolerance in argan tree (Argania spinosa) populations: Potential for the development of improved drought tolerance. *Front. Plant Sci.* **2017**, *8*, 276. [CrossRef] [PubMed]
4. Reda Tazi, M.; Boukroute, A.; Berrichi, A.; Rharrabti, Y.; Kouddane, N. Growth of young argan tree seedlings (Argania spinosa L. Skeels) in northeast of morocco under controlled conditions at different NaCl concentrations. *J. Mater. Environ. Sci.* **2018**, *9*, 212–218.
5. Bani-Aameur, F.; Sipple-Michmerhuizen, J. Germination and seedling survival of Argan (*Argania spinosa*) under experimental saline conditions. *J. Arid. Environ.* **2001**, *49*, 533–540. [CrossRef]
6. Papoti, V.T.; Tsimidou, M.Z. Looking through the qualities of a fluorimetric assay for the total phenol content estimation in virgin olive oil, olive fruit or leaf polar extract. *Food Chem.* **2009**, *112*, 246–252. [CrossRef]
7. Bettaieb Rebey, I.; Bourgou, S.; Saidani Tounsi, M.; Fauconnier, M.-L.; Ksouri, R. Etude de la composition chimique et de l'activité antioxydante des différents extraits de la Lavande dentée (*Lavandula dentata*). *J. New Sci. Agri Biotech* **2017**, *39*, 2096–2105.
8. Bougandoura, B. Evaluation de l 'activité antioxydante des extraits aqueux et méthanolique de *Satureja calamintha* ssp. *Nepeta* (L.) Briq. *Nat. Technol.* **2013**, *9*, 14–19.
9. Chaouche, T.M.; Haddouchi, F.; Atik-bekara, F.; Ksouri, R.; Azzi, R.; Boucherit, Z.; Tefiani, C.; Larbat, R. Antioxidant, haemolytic activities and HPLC–DAD–ESI–MSn characterization of phenolic compounds from root bark of Juniperus oxycedrus subsp. oxycedrus. *Ind. Crop. Prod.* **2015**, *64*, 182–187. [CrossRef]
10. *ISO 5509:2009*; Animal and Vegetable Fats and Oils—Preparation of Methyl Esters of Fatty Acids. ISO: Geneva, Switzerland, 2009.
11. Chowdhury, N. *Advances in Agriculture Sciences*; AkiNik Publications: Delhi, India, 2020.
12. Charrouf, Z.; Guillaume, D. Argan oil: Occurrence, composition and impact on human health. *Eur. J. Lipid Sci. Technol.* **2008**, *110*, 632–636. [CrossRef]
13. Kouidri, M. Extraction et Caractérisation Physico-Chimique de l'Huile d'Argan Provenant d'Arbres Cultivés dans deux Régions de l'Algérie (Tindouf et Mostaganem). Doctoral Thesis, Université Hassiba Benbouali, Chlef, Algeria, 2008.
14. Mechqoq, H.; El Yaagoubi, M.; Momchilova, S.; Msanda, F.; El Aouad, N. Comparative study on yields and quality parameters of argan oils extracted by conventional and green extraction techniques. *Grain Oil Sci. Technol.* **2021**, *4*, 125–130. [CrossRef]
15. Sour, S. Effet Hypolipidémiant et Antioxydant de l'Huile d'Argania Spinosa. Doctoral Thesis, Université Abou Bekr Belkaïd, Tlecmen, Algeria, 2009.
16. Demnati, D.; Sánchez, S.; Pacheco, R.; Zahar, M. Comparative study of argan and olive fruits and oils. In Proceedings of the Premier Congrès International de l' Arganier, Agadir, Morocco, 15–17 December 2011; pp. 435–441.
17. Belahcen, O.; (Faculty of sciences Ben M'sik, University of Hassan II, Casablanca, Morocco); Harmel, H.; (Faculty of Sciences Ben M'sik, University of Hassan II, Casablanca, Morocco). Valorisation des produits et des sousproduits d'Arganier pour usage cosmétique. Personal communication, 2018. Unpublished work.
18. SNIMA. NORME MAROCAINE NM 08.5.090.pdf. 2003. Available online: https://www.imanor.gov.ma/Norme/nm-08-5-090/ (accessed on 13 November 2019).

environmental sciences proceedings

MDPI

Proceeding Paper

Electromagnetic Saline Water for Potato Growth and Water Relations †

Rawaa Akrimi [1,*], Hichem Hajlaoui [1], Mounir Denden [2] and Mahmoud Mhamdi [2]

1 Regional Center of Agricultural Research of Sidi Bouzid (CRRA) PB 357, Sidi Bouzid 9100, Tunisia; hajlaoui2001@yahoo.fr
2 Institut Supérieur Agronomique Chott Meriem, Sousse 4042, Tunisia; abidi_k@yahoo.fr (M.D.); mhamdimahmoud@yahoo.fr (M.M.)
* Correspondence: akrimirawaa2017@gmail.com

† Presented at the 2nd International Laayoune Forum on Biosaline Agriculture, 14–16 June 2022; Available online: https://lafoba2.sciforum.net/.

Abstract: Under water insecurity conditions, electromagnetic saline water may be experienced in irrigation. This study examined the impact of electromagnetic saline water on three potato varieties (Spunta, Bellini and Alaska). The trial includes three treatments; ground water (T1: 2.2 ms cm^{-1} EC), saline water (T2: 8.5 ms cm^{-1} EC) and saline water having undergone electromagnetic treatment (T3: 8.5 ms cm^{-1} EC) with Aqua-4D. The results revealed an improvement in yield with T3 compared to T2. Spunta and Alaska were more responsive to T3 than Bellini. The approving response of Alaska was associated with effective adjustment with proline, while Spunta was more efficient in water use.

Keywords: electromagnetic treatment; growth; potato; salinity; water status

check for updates

Citation: Akrimi, R.; Hajlaoui, H.; Denden, M.; Mhamdi, M. Electromagnetic Saline Water for Potato Growth and Water Relations. *Environ. Sci. Proc.* **2022**, *16*, 57. https://doi.org/10.3390/environsciproc2022016057

Academic Editor: Abdelaziz Hirich

Published: 16 June 2022

1. Introduction

Water scarcity is a worldwide problem for agricultural production. Tunisia is currently considered a food-deficit nation with an increasing trend in food imports due to lack of rainfall [1]. Additional brackish water supply can meet crops' water needs [2]. In this way, the valuation of saline water is gaining increasing attention in many countries suffering from salinization [2]. Efforts are now intended at identifying inexpensive desalination techniques that respect the environment and ensure crop productivity. Physical treatment of saline water is a recommended option. However, most of the considered methods have some drawbacks regarding energy consumption, costs, long reaction time for the treatment etc. [3]. One advantaged strategy, adopted in agricultural irrigation, is electromagnetic treatment [4]. Magnetic treatment consists of passing water through an electromagnetic softener, where a Lorentz force is exerted on each ion which is in the opposite direction to each other. The redirection of the particles increases the collisions between ions and therefore precipitates formation. Additionally, magnetic fields decrease salt hydration and enhance salt solubility, coagulation and crystallization [5].

It was found that electromagnetic saline water improved plant growth, water and ions uptake [6]. In this context, an attempt was made to test whether electromagnetic saline water can alleviate salinity impacts on potatoes. We supposed that electromagnetic saline water may enhance plant-water relations, growth and yield. The applicability of this technique may be supportive in case of dry conditions of the Tunisian spring season.

2. Material and Methods

2.1. Experimental Details

The present inestigation was carried out at the Regional Center of Agricultural Research of Sidi Bouzid (CRRA-Sidi Bouzid), Tunisia during March–June season of 2014–2015. The experimental layout had a factorial arrangement on the basis of a randomized complete block design with three replicates for each treatment. Potato varieties (Spunta, Bellini and Alaska) were irrigated with three irrigation treatments, T1 ground water (2.2 ms cm^{-1} electrical conductivity (EC)), T2 saline water (8.5 ms cm^{-1} EC) and T3 electromagnetic saline water (8.5 ms cm^{-1} EC). Electromagnetic treatment was done with an Aqua-4D® device which is composed of an electromagnet tube 60E (external diameter 65 mm, passage diameter 100 (DN 25) of 436 mm length and 60 L min^{-1} maximum flow, designed for transmitting the electromagnetic signals into the water and coupled to an electromagnet box (Command 60E Pro) intended to generate electromagnetic signals. The experimental area had a semi-arid climate with 200 mm mean precipitation and 1200 mm mean evapotranspiration. Crop water requirement (ETc) was estimated as reference evapotranspiration (ET0) multiplied by the potato crop coefficient (Kc).

2.2. Agronomic Parameters

Plant height (cm), fresh weight (FW) (g $plant^{-1}$) and dry weight (DW) (g $plant^{-1}$) were measured at the tuber initiation stage. The tuber weight per plant was determined at harvest.

2.3. Plant Water Status and Proline Determination

Leaf-relative water content (RWC) was calculated as follows:

$$RWC = \frac{FM - DM}{TM - DM} \times 100,$$

where FM is leaf fresh mass, DM is leaf dry mass determined after oven-drying at 75 °C for 24 h and TM is the turgid mass of the leaf, determined after the immersion of the fresh leaves in distilled water for 12 h. The leaf water potential (ψw) was determined before sunrise using a pressure chamber (Scholander model M-1000). Photosynthetic water use efficiency (PWUE) was calculated as the ratio between the amount of CO_2 fixed by photosynthesis (Pn) to the transpiration rate (E): PWUE = (Pn/E).

Proline was extracted from the leaves using the sulphosalicylic acid method. A solution of the obtained extract, glacial acetic acid and ninhydrin (1:1:1) was incubated at 100 °C for 1h and the absorbance was determined at 546 nm.

2.4. Statistical Analysis

The data were analyzed with SPSS 20 and the means were separated by the Duncan test ($p \leq 0.05$). The means were subjected to a three-way analysis of variance (ANOVA).

3. Results

3.1. Agronomic Traits

From Table 1, electromagnetic saline water (T3) increased the plant height (PH) of Spunta and Alaska by 34% and 41%, respectively. In addition, Alaska had the highest values of plant height. Hence, ANOVA results revealed that the PH was significantly affected by irrigation treatments and the variety factor. The T3 treatment enhanced the FW accumulation of Bellini and Alaska and the DW of Spunta and Alaska as compared to saline water (T2). The tuber yield varied significantly among the varieties and was increased with T3, but values do not exceed those obtained under ground water (T1).

Table 1. Changes in plant height (PH), fresh weight (FW), dry weight (DW) and yield of potato varieties (Spunta, Bellini and Alaska), for each irrigation treatment.

	Treatments	PH (cm)	FW (g Plant^{-1})	DW (g Plant^{-1})	Yield (g Plant^{-1})
Spunta	T1 (2.2 ms cm^{-1})	17.12 ± 0.95 ^{a}	868.36 ± 51 ^{a}	135.03 ± 11 ^{a}	1205.88 ± 41.86 ^{a}
	T2 (8.5 ms cm^{-1})	12.25 ± 1.34 ^{b}	772.36 ± 32 ^{b}	113.53 ± 8 ^{b}	918.72 ± 60.38 ^{c}
	T3 (8.5 ms cm^{-1})	16.50 ± 1.22 ^{a}	733.00 ± 40 ^{b}	131.16 ± 12 ^{a}	1016.97 ± 69.82 ^{b}
Bellini	T1 (2.2 ms cm^{-1})	18.75 ± 1.44 ^{a}	757.63 ± 51 ^{a}	141.2 ± 11 ^{a}	1088.00 ± 76.01 ^{a}
	T2 (8.5 ms cm^{-1})	9.52 ± 0.33 ^{b}	556.43 ± 32 ^{b}	100.13 ± 8 ^{b}	656.52 ± 75.41 ^{c}
	T3 (8.5 ms cm^{-1})	9.37 ± 1.18 ^{b}	482.30 ± 40 ^{c}	92.46 ± 12 ^{b}	865.50 ± 67.08 ^{b}
Alaska	T1 (2.2 ms cm^{-1})	22.50 ± 0.62 ^{a}	1136.13 ± 51 ^{a}	187.3 ± 11 ^{a}	1592.78 ± 92.62 ^{a}
	T2 (8.5 ms cm^{-1})	12.25 ± 0.72 ^{c}	675.66 ± 32 ^{c}	110.3 ± 8 ^{c}	914.24 ± 40.73 ^{c}
	T3 (8.5 ms cm^{-1})	17.37 ± 1.68 ^{b}	740.47 ± 40 ^{b}	144.16 ± 12 ^{b}	1380.08 ± 67.17 ^{b}
ANOVA	T	18.25 **	6.77 **	2.9 *	14.69 **
	V	52.25 **	1.7 ^{ns}	0.76 ^{ns}	29.36 **
	T × V	6.07 **	1.47 ^{ns}	0.54 ^{ns}	3.26 *

Different letters represent significant difference between treatments. T1: Ground water; T2: Saline water; T3: Electromagnetic saline water; T: Treatment; V: Variety. ns: not significant, * $p \leq 0.05$, ** $p \leq 0.01$.

3.2. Plant Water Status

The relative water content in potato leaves was in the order T1 > T3 > T2 (Table 2). The registered values of Ψw and PWUE increased with T3 in Spunta and Alaska. Alaska had the highest PWUE values, while Spunta had the highest RWC and Ψw. The RWC and Ψw were significantly influenced by water treatments.

Table 2. Changes in relative water content (RWC), water potential (Ψw) and photosynthetic water use efficiency (PWUE) of potato varieties (Spunta, Bellini and Alaska), for each irrigation treatment.

	Treatments	RWC (%)	Ψw (Bar)	PWUE
Spunta	T1 (2.2 ms cm^{-1})	88.37 ± 11.79 ^{a}	−1.37 ± 0.20 ^{a}	4.74 ± 0.20 ^{a}
	T2 (8.5 ms cm^{-1})	71.92 ± 5.54 ^{b}	−3.5 ± 0.40 ^{c}	4.29 ± 0.32 ^{c}
	T3 (8.5 ms cm^{-1})	88.20 ± 11.60 ^{a}	−2.15 ± 0.50 ^{b}	4.43 ± 0.36 ^{b}
Bellini	T1 (2.2 ms cm^{-1})	84.24 ± 22.04 ^{a}	−1.65 ± 0.14 ^{a}	4.43 ± 0.08 ^{a}
	T2 (8.5 ms cm^{-1})	71.83 ± 10.18 ^{c}	−3.25 ± 0.50 ^{b}	4.11 ± 0.31 ^{b}
	T3 (8.5 ms cm^{-1})	81.63 ± 23.01 ^{b}	−3.5 ± 0.65 ^{b}	4.05 ± 0.35 ^{b}
Alaska	T1 (2.2 ms cm^{-1})	84.82 ± 5.44 ^{a}	−2.25 ± 0.72 ^{a}	5.01 ± 0.34 ^{a}
	T2 (8.5 ms cm^{-1})	74.07 ± 4.09 ^{c}	−3 ± 0.50 ^{b}	4.17 ± 0.64 ^{c}
	T3 (8.5 ms cm^{-1})	82.84 ± 5.37 ^{b}	−2.75 ± 0.95 ^{a}	4.56 ± 0.58 ^{b}
ANOVA	T	29.78 **	13.40 **	0.18 ^{ns}
	V	1.74 ^{ns}	0.68^{ns}	1.14 ^{ns}
	T × V	0.96 ^{ns}	2.05 ^{ns}	0.31 ^{ns}

Different letters represent significant difference between treatments. T1: Ground water; T2: Saline water; T3: Electromagnetic saline water; T: Treatment; V: Variety. ns: not significant, ** $p \leq 0.01$.

3.3. Proline Content

The proline content in the Alaska leaves was set in the following decreasing trend T2 > T3 > T1 (Figure 1). However, there is no significant difference between T2 and T3 in Spunta and Bellini.

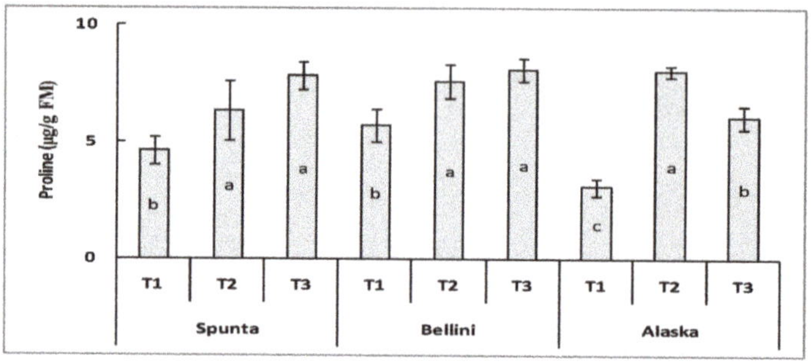

Figure 1. Changes in proline content of potato varieties under each irrigation treatment. Different letters indicate significant difference between treatments according to Duncan test ($p \leq 0.05$). T1: Ground water; T2: Saline water; T3: Electromagnetic saline water.

4. Discussion

Irrigation is an important determinant of potato development during the dry season of spring due to negligible rain. Moreover, the salinization of water resources makes the valuation of saline water very wanted. The present study showed that potato growth (plant height, FW and DW) and yield may be enhanced with electromagnetic saline water, depending on the assessed variety. These results agree with what [4,7]. The data of the present study may be due to the reduction of the soil sodium concentrations with electromagnetic water, by leaching it below the root zone [8]. Additionally, electromagnetic fields change the physical parameters of water (salt's solubility and electrolytic potential), which improved water and nutrient uptake, and therefore plant growth [7]. Spunta and Alaska were more responsive to electromagnetic saline water while Bellini plant height and DW were not affected by T3. This may be caused by high transpiration rate of Bellini, as demonstrates the low PWUE values of this variety. This is indicative of more susceptible water relations in Bellini. In fact, the RWC, Ψ_W and PWUE of Spunta and Alaska were higher than Bellini and were enhanced by T3. This result point to a stronger photosynthetic activity of Spunta and Alaska compared to Bellini [7]. The data also shows that the effective adjustment with proline may be a key criterion in maintaining growth under saline conditions of Alaska. Instead, the highest values of the Ψw and RWC of Spunta under T1 and T3 suggest that it was more efficient in water use.

5. Conclusions

Electromagnetic water improved the RWC and Ψw of Spunta and Alaska. Thus, plant height and fresh and dry biomass accumulation were enhanced. The yield of the three studied varieties was increased with T3 compared to untreated saline water. However, further research is required to ensure the safe use of electromagnetic saline water and to identify the exact mechanisms that direct plant response.

Author Contributions: R.A.; methodology, original draft preparation, H.H. and M.M.; editing, M.D.; supervision. All authors have read and agreed to the published version of the manuscript.

Funding: This research was funded by the Regional Center of Agricultural Research of Tunisia.

Institutional Review Board Statement: Not applicable.

Informed Consent Statement: Not applicable.

Data Availability Statement: The data that support the findings of this study are available from the corresponding author, upon reasonable request.

Acknowledgments: The authors would like to thank Planet Horizons Technologies SA Ecoparc de DavalA 9, 3960 Sierre, CH for supplying the Aqua-4D® 60E series device.

Conflicts of Interest: The authors declare no conflict of interest.

References

1. Sadok, W.; Schoppach, R.; Ghanem, M.E.; Zucca, C.; Sinclair, T.R. Wheat drought-tolerance to enhance food security in Tunisia, birthplace of the Arab Spring. *Eur. J. Agron.* **2019**, *107*, 1–9. [CrossRef]
2. Liu, Z.; Zhu, J.; Yang, X.; Wu, H.; Wei, Q.; Wei, H.; Zhang, H. Growth performance, organ level ionic relations and organic osmoregulation of *Elaeagnus angustifolia* in response to salt Stress. *PLoS ONE* **2018**, *13*, e0191552. [CrossRef] [PubMed]
3. Al-Raad, A.; Hanafiag, M.M.; Naje, A.S.; Ajeel, M.A.; Basheer, A.O.; Aljayashi, T.A.; Toriman, M.O. Treatment of Saline Water Using Electrocoagulation with Combined Electrical Connection of Electrodes. *Processes* **2019**, *7*, 242. [CrossRef]
4. Da Silva, J.A.T.; Dobránszki, J. Impact of magnetic water on plant growth. *Environ. Exp. Biol.* **2015**, *12*, 137–142.
5. Abedinpour, M.; Rohani, E. Effects of magnetized water application on soil and maize growth indices under different amounts of salt in the water. *J. Water Reuse Desalin.* **2017**, *7*, 319–325. [CrossRef]
6. Hilal, M.; Hilal, M. Application of magnetic technologies in desert agriculture. I-Seed germination and seedling emergence of some crops in a saline calcareous soil. *Egypt. J. Soil Sci.* **2000**, *40*, 413–422.
7. Akrimi, R.; Hajlaoui, H.; Rizzo, V.; Muratore, G.; Mhamdi, M. Agronomical traits, phenolic compounds and antioxidant activity in raw and cooked potato tubers growing under saline conditions. *J. Sci. Food Agric.* **2020**, *100*, 3719–3728. [CrossRef] [PubMed]
8. Zlotopolski, V. Magnetic treatment reduces water usage in irrigation without negatively impacting yield, photosynthesis and nutrient uptake in lettuce. *Int. J. Appl. Agric. Sci.* **2017**, *3*, 117–122. [CrossRef]

environmental sciences proceedings

MDPI

Proceeding Paper

Toward Better Preparedness of Mediterranean Rainfed Agricultural Systems to Future Climate-Change-Induced Water Stress: Study Case of Bouregreg Watershed (Morocco) †

Youssef Brouziyne [1,*], Aziz Abouabdillah [2], Lhoussaine Bouchaou [3,4], Oumaima Attar [4], Yassine Ez-zaouy [4], Lahcen Benaabidate [5] and Abdelghani Chehbouni [4,6]

1 International Water Management Institute (IWMI), MENA Office, Giza 12661, Egypt
2 Ecole Nationale d'Agriculture de Meknès, Meknès 50001, Morocco; aziz.abouabdillah@gmail.com
3 Laboratory of Applied Geology and Geo-Environment, Faculty of Sciences, Ibn Zohr University, BP/8106, Cité Dakhla, Agadir 1136, Morocco; l.bouchaou@uiz.ac.ma
4 Mohammed VI Polytechnic University (UM6P), International Water Research Institute, Benguerir 43150, Morocco; oumaima.attar@um6p.ma (O.A.); yassine.ez-zaouy@um6p.ma (Y.E.-z.); abdelghani.chehbouni@um6p.ma (A.C.)
5 Laboratory of Functional Ecology and Environmental Engineering, Faculty of Sciences and Technology, University of Sidi Mohamed Ben Abdellah, Fez 30000, Morocco; benaabidate@yahoo.fr
6 CESBIO, Université de Toulouse, CNRS, CNES, IRD, 31400 Toulouse, France
* Correspondence: youssef.brouziyne@gmail.com
† Presented at the 2nd International Laayoune Forum on Biosaline Agriculture, 14–16 June 2022; Available online: https://lafoba2.sciforum.net/.

Citation: Brouziyne, Y.; Abouabdillah, A.; Bouchaou, L.; Attar, O.; Ez-zaouy, Y.; Benaabidate, L.; Chehbouni, A. Toward Better Preparedness of Mediterranean Rainfed Agricultural Systems to Future Climate-Change-Induced Water Stress: Study Case of Bouregreg Watershed (Morocco). *Environ. Sci. Proc.* **2022**, *16*, 58. https://doi.org/10.3390/environsciproc2022016058

Academic Editor: Abdelaziz Hirich

Published: 16 June 2022

Abstract: Improving the preparedness of agricultural systems to future climate-change-induced phenomena, such as drought-induced water stress, and the predictive analysis of their vulnerability is crucial. In this study, a hybrid modeling approach based on the SWAT model was built to understand the response of major crops and streamflow in the Bouregreg catchment in Morocco to future droughts. During dry years, the simulation results showed a dramatic decrease in water resources availability (up to −40%) with uneven impacts across the study catchment area. Crop-wise, significant decreases in rainfed wheat productivity (up to −55%) were simulated during future extremely dry growing seasons.

Keywords: climate change; drought; water stress; rainfed agricultural system; crops yield; Bouregreg watershed

1. Introduction

At a global scale, the potential impact of climate change on major systems is an important question for decision makers, investors and farming communities. Water resources directly interfere with public health and safety, agriculture and food security, energy, industry, biodiversity and ecosystems and, thus, influence the socio-economic development of the nations. Water resources in Morocco are particularly vulnerable to human activities, such as increases in population, tourism, agricultural development and industrial growth being some examples of challenges faced by water resources in the country. In addition to these factors, the changing climate and all resulting phenomena (such as drought) represent a further threat that can have a significant impact on water resources in Morocco and all areas within a similar context. Farming activities are of crucial importance to the country, and are generally vulnerable to climate variability, which can affect the national economy in some situations. The objective of this study is to provide a representative case study of drought-induced water stress vulnerabilities and a catchment-scale impacts assessment in the future, with a focus on water and rainfed crop systems during dry years.

2. Material and Methods

2.1. Study Area

The study catchment area was Bouregreg, a 9656 km^2 watershed located in the northwestern part of Morocco; it is one of the most important watersheds in the country [1].

Essentially, there are different climatic zones in the study catchment area influenced by the ocean: subhumid in the extreme western part, semiarid in the central part and arid climate in the south-eastern [1]. The annual potential water resources of the Bouregreg watershed is 720 Million m^3, coming mostly from surface water. The stream network of the Bouregreg catchment area is organized into 3 main rivers: Korifla, Grou and Bouregreg rivers (Figure 1) [1,2]. In terms of land use, oak forests cover approximately 24% of the total area of the study watershed; farming activity is very important (olive trees, pulse crops, wheat in addition to minor parcels where some irrigated vegetables and grape crops are grown). Range and pasture lands are distributed across the watershed [1,3].

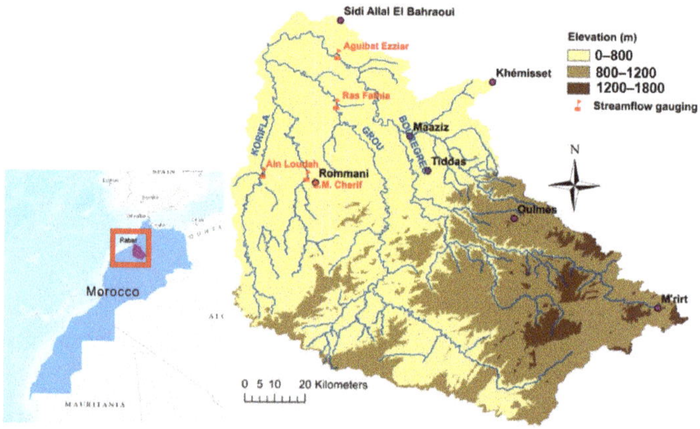

Figure 1. Location, topography and streamflow network of BW and the used gauging stations.

2.2. Models Presentation

For hydrological processes investigation, we used soil and water assessment tool (SWAT) model. SWAT is a hydrological model developed by the United States Department of Agriculture [4]. It simulates the cycle of water and sediments in large-scale watersheds and large basins with different soil types and management conditions; it also simulates the effect of agriculture practices, or any other anthropogenic actions that can affect water cycle [5,6]. Regarding drought analysis, interpolated rainfall outputs of SWAT over Bouregreg watershed were used to compute drought indicator such as the standardized precipitation index (SPI) along the study periods. In this study, SWAT model was run over the period of January 1990 to December 2005 (data collected from the local water department's measurements); the first two years were allocated to warm up process, and the period of January 1992 to December 2001 was dedicated to model calibration. Validation was performed over the period January 2002 to December 2005 (due to the limited amount of measured data). Following sensitivity analysis, streamflow and crops yields were examined during the calibration process. A maximal agreement between observed and predicted processes was the ultimate goal.

In order to assess future climate projections in the study watershed, dynamically downscaled output data from the Global Climate Model CNRM-CM5 were used [7]. Two representative concentration pathways (RCPs) were selected, RCP4.5 and RCP8.5. Baseline time series for both daily temperature (max and min) and rainfall data were from January 1990 to December 2010; for future datasets, the 2030–2050 period was collected from the outputs of the Global Climate Model CNRM-CM5. The SPI is a statistical monthly indicator that compares the cumulated precipitation during a period of a number of months with long-term cumulated rainfall distribution for the same location and accumulation period [8]. In this work, SPI-12 was calculated using the 2018 Version of SPI Generator application [9] of the National Drought Management Centre (University of Nebraska, USA) [8]. Once a drought event was determined with its start and end month, its duration and magnitude were then assigned. The duration of a drought event was equal to the number of months between its start (included) and end months (not included); the drought magnitudes were calculated as the sums of the SPI values that were always negative over consecutive months, and the severity as the maximum SPI value of each episode.

3. Results

3.1. Future Droughts Analysis

Figure 2 represents the distribution of monthly SPI-12 values over the study periods in the Bouregreg watershed.

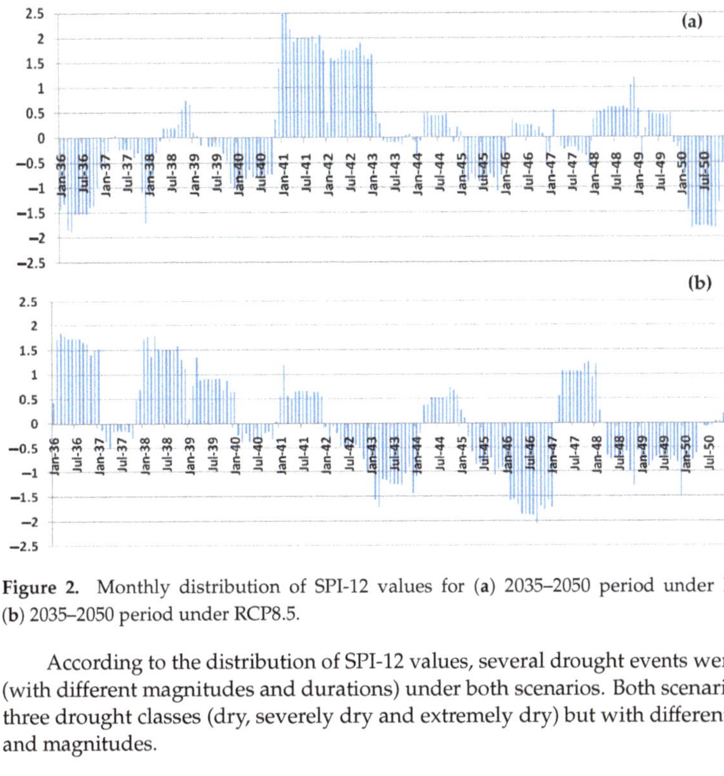

Figure 2. Monthly distribution of SPI-12 values for (**a**) 2035–2050 period under RCP4.5 and (**b**) 2035–2050 period under RCP8.5.

According to the distribution of SPI-12 values, several drought events were expected (with different magnitudes and durations) under both scenarios. Both scenarios showed three drought classes (dry, severely dry and extremely dry) but with different durations and magnitudes.

3.2. Drought Impact on Water Availability

During the 2035 to 2050 period, several decreases in the annual total water yield (TWYD) occurred and are shown in Figure 3. Under RCP4.5, the expected drought episode of January 2036 to April could affect water resources in the study watershed during 2036 and 2037 (67 mm and 57 mm, respectively). During the same simulation period under RCP8.5, the drought period of October 2045 till February 2047 could reduce the TWYD averages in the watershed to 30 mm and 33 mm in 2045 and 2046, respectively.

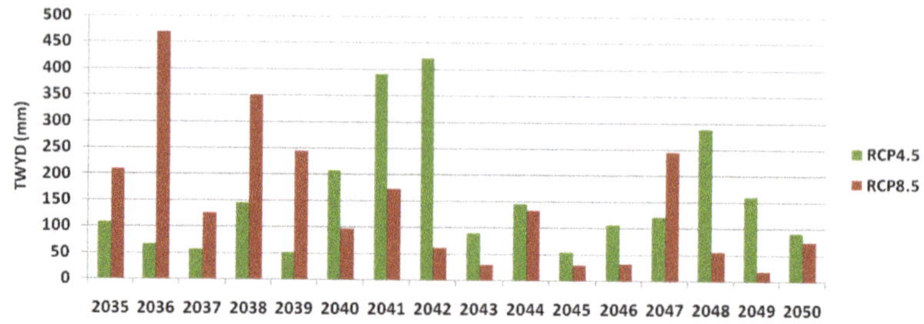

Figure 3. Distribution of annual TWYD averages under both RCPs over 2035–2050 period.

3.3. Future Crop Performance Analysis

Averages of wheat yields over the whole simulation periods were expected to decrease in most scenarios in comparison to the current one, but with a very high year-to-year variability (Figure 4).

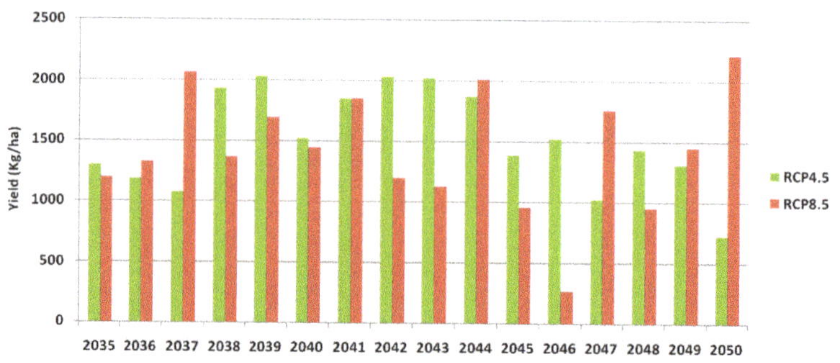

Figure 4. Annual yields averages of wheat crops (under both RCPs) over 2035–2050.

In Morocco, and more specifically in the Bouregreg watershed, wheat is sown in between autumn and winter, as most rain happens in wintertime, and temperatures are generally mild in this season. Early stress (excessive temperature and low soil moisture) (Figure 2) limits the tiller number (low ears per unit area) and later stress (after anthesis) can reduce the size of individual grains and their number [10].

4. Discussion

The projected droughts over the BW revealed that the watershed is likely to experience several droughts with different severity levels under both climate change scenarios. These results endorsed the findings of Gudmundsson and Seneviratne [11], since they reported that enhanced greenhouse forcing has contributed to increased drying in the Mediterranean region (including Northern Africa), and that this trend is likely to keep increasing over the century, especially under high levels of global warming.

5. Conclusions

The hydrology and agriculture of the Bouregreg watershed are presumed to experience significant impacts of climate-change-induced water stress; it is expected to have a high year-to-year water and crop yield variability due to several drought events with different severity levels. This context could lead to more challenging water resource situations in the watershed, mainly during the severe and long drought episodes.

Author Contributions: Conceptualization, Y.B.; methodology, Y.B., A.A., Y.E.-z.; writing—original draft, Y.B., O.A.; review and editing, L.B. (Lhoussaine Bouchaou); validation, A.C., L.B. (Lahcen Benaabidate) All authors have read and agreed to the published version of the manuscript.

Funding: This research received no external funding.

Institutional Review Board Statement: Not applicable.

Informed Consent Statement: Not applicable.

Data Availability Statement: Not applicable.

Conflicts of Interest: The authors declare no conflict of interest.

References

1. ABHBC. *Le Plan Directeur d'Amenagement Integré des Ressources en eau du Bassin Hydraulique du Bouregreg et de la Chaouia;* Agence du Bassin Hydraulique du Bouregreg et de la Chaouia: Benslimane, Morocco, 2011.
2. Belmatrik, S. *Dispositif de Prevision des Risques de Crues;* ABHBC: Benslimane, Morocco, 2010.
3. Rahdou, M. Modelisation Hydrologique du Bassin Versant de Bouregreg (S.S) et Evaluation de Taux d'Envasement du Barrage Sidi Mohamed Ben Abdellah (Maroc) par la Methode Bathymetrique. Master's Thesis, Faculty of Sciences and Techniques Fes, Fez, Morocco, 2010.
4. Arnold, J.G.; Srinivasan, R.; Muttiah, R.S.; Williams, J.R. Large area hydrologic modeling and assessment—PartI. *J. Am. Water Resour. Assoc.* **1998**, *34*, 73–89. [CrossRef]
5. Arnold, J.G.; Moriasi, D.N.; Gassman, P.; Abbaspour, K.C.; White, M.J.; Srinivasan, R.; Santhi, C.; Harmel, R.D.; Van Griensven, A.; Van Liew, M.W.; et al. SWAT: Model use, Calibration, and Validation. *Am. Soc. Agric. Biol. Eng.* **2012**, *55*, 1491–1508.
6. Moriasi, D.N.; Arnold, G.J.; VanLiew, W.M.; Bingner, L.R.; Harmel, D.R.; Veith, L.T. Model Evaluation Guidelines for Systematic Quantification of Accuracy in Watershed Simulations. *Trans. ASABE* **2007**, *50*, 885–900. [CrossRef]
7. Taylor, K.E.; Ronald, J.S.; Gerald, A.M. Overview of CMIP5 and the experiment design. *Bull. Am. Meteorol. Soc.* **2012**, *93*, 485–498. [CrossRef]
8. McKee, T.B.; Doesken, N.J.; Kleist, J. The Relationship of Drought Frequency and Duration to Time Scales. In Proceedings of the 8th Conference on Applied Climatology, Anheim, CA, USA, 17–22 January 1993; pp. 179–184.
9. NDMC. User Manual of: SPI Calculator. National Drought Mititgation Center—UNL. 2018. Available online: https://drought.unl.edu/monitoring/SPI/SPIProgram.aspx (accessed on 1 May 2022).
10. Brouziyne, Y.; Abouabdillah, A.; Hirich, A.; Zaaboul, R.; Benaabidate, L. Modeling sustainable adaptation strategies toward a climate-smart agriculture in a Mediterranean watershed under projected climate change. *Agric. Syst.* **2018**, *162*, 154–163. [CrossRef]
11. Gudmundsson, L.; Seneviratne, S.I. Anthropogenic climate change affects meteorological drought risk in Europe. *Environ. Res. Lett.* **2016**, *11*, 044005. [CrossRef]

environmental sciences proceedings

MDPI

Proceeding Paper

The Effect of Different Irrigation Water Salinity Levels on Nutrients Uptake, Biochemical Content and Growth Response of Blue Panicum, Quinoa and Silage Maize †

Assma Oumasst [1], Sara Azougay [1], Abdelaziz Mimouni [2] and Jamal Hallam [2,*]

[1] Science and Technology Department, Polydisciplinary Faculty of Taroudant, Ibn Zohr University, BP 271, Taroudant 83000, Morocco; assmaouma@gmail.com (A.O.); azou636@gmail.com (S.A.)

[2] Regional Center of Agricultural Research of Agadir, National Institute of Agricultural Research, Avenue Ennasr, BP 415 Rabat Principale, Rabat 10090, Morocco; abdelaziz.mimouni@inra.ma

* Correspondence: jamal.hallam@gmail.com or jamal.hallam@inra.ma

† Presented at the 2nd International Laayoune Forum on Biosaline Agriculture, 14–16 June 2022; Available online: https://lafoba2.sciforum.net/.

Abstract: This study aimed to investigate the functional response of *Panicum antidotale Retz, Chenopodium quinoa*, and *Zea mays L.* to different levels of saline irrigation water. Nutrient uptake, biochemical content, and growth response were evaluated to select potential crops best suited for biosaline agriculture. The results suggest that blue panicum is highly tolerant to saline water irrigation, followed by quinoa up to 10 dS m^{-1}, and then silage maize which is sensitive to saline conditions. The introduction of blue panicum as an alternative crop on salt-effected soils, such as the irrigated perimeter of Foum El Oued in Laâyoune in Morocco, would exhibit high performance better than traditional crops such as silage maize and therefore would improve the local farmers' income.

Keywords: salt-tolerant crops; mineral uptake; biosaline agriculture; climate change

Citation: Oumasst, A.; Azougay, S.; Mimouni, A.; Hallam, J. The Effect of Different Irrigation Water Salinity Levels on Nutrients Uptake, Biochemical Content and Growth Response of Blue Panicum, Quinoa and Silage Maize. *Environ. Sci. Proc.* **2022**, *16*, 60. https://doi.org/10.3390/environsciproc2022016060

Academic Editor: Abdelaziz Hirich

Published: 20 June 2022

1. Introduction

As part of the development of a sustainable solution to manage and mitigate the impact of salinity on the soils of irrigated areas in marginal lands, a larger Agroecological megaproject of INRA-Morocco's medium-term program was established. This study is part of that program and aims to select potential crops best suited for biosaline agriculture. Blue panicum and quinoa are classified as halophyte species, and silage maize is moderately sensitive to salt stress, but they differ in adaptation under salinity conditions. Testing their functional response to different levels of saline irrigation would contribute to the use of cultivated land suffering from salinity, which represents 20% of cultivated land in the world, and salt irrigated land [1].

2. Materials and Methods

The study was conducted on three plants: blue panicum (public type from Kuwait), quinoa (ICBA Q5), and silage maize (Dragma variety). The seeds were gathered from previous field trials at the INRA experimental station of Foum El Oued in Laâyoune. The experiment had a factorial design, with four levels of irrigation-water-salinity treatments (0.9 "tap water"; 3; 6 and 10 dS m^{-1}), and four replicates. The substrate used in the experiment was peat.

At the end of the experiment, the leaves, stems, and roots were harvested, and the roots were washed without substrate. The fresh plant material was weighed and then oven-dried at 80 °C until a constant weight. The physiological, biochemical, and mineral parameters studied during the experiment were the chlorophyll content, proline content, ionic content of leaves, and the amount of potassium (K^+), calcium (Ca^{2+}), and sodium

(Na^+) in the shoots and roots. The methodology adopted to measure these parameters is detailed in Oumasst et al. [2]. The normal distribution and homogeneity of variance between the plants were firstly tested by Shapiro–Wilk test and Barlett's test. The analysis of variance for different irrigation water treatments was performed using the SPSS program (IBM Corp. Released in 2011, Version 20, Armonk, NY, USA).

3. Results

3.1. Physiological and Biochemical Properties

Table 1 shows the mean of the stem height, aboveground and root dry biomass, chlorophyll a, chlorophyll b, carotenoid, and proline content of the blue panicum, quinoa, and silage maize of the tested treatments. The physiological and biochemical properties of blue panicum were not affected by the applied salinity treatments, except for the root dry matter biomass, which decreased by 39 % at 10 dS m^{-1} compared to the control. For quinoa, increasing irrigation-water-salinity only decreased plant growth and root dry matter biomass without significant effect on the other parameters compared to control. The increasing irrigation-water-salinity resulted in a significant decrease in silage maize growth in the aboveground and root dry matter biomass, chlorophyll a, chlorophyll b, and carotenoid compared to the control (T0). No effect of the applied salt concentrations on the proline content for all of the crop species was observed.

Table 1. Mean of stem height, aboveground and root dry biomass, chlorophyll a, chlorophyll b, carotenoid, and proline content of blue panicum, quinoa and silage maize by tested treatments. The symbols T0 (0.9 dS m^{-1}), T1 (3 dS m^{-1}), T2 (6 dS m^{-1}), and T3 (10 dS m^{-1}) refer to the four saline irrigation water treatments. Lower-case (a, b, c, d) show differences between treatments within each measured parameter for each plant species; Data with different lowercase are significantly different ($p < 0.05$). (Mean ± standard deviation; $n = 4$ for blue panic and silage maize, and $n = 3$ for quinoa).

		Stem Height (cm)	Aboveground Dry Biomass (g)	Root Dry Biomass (g)	Chlorophyll a (μg mL^{-1})	Chlorophyll b (μg mL^{-1})	Carotenoid (μg mL^{-1})	Proline (mg g^{-1})
Blue panicum	**T0**	99.25 ± 10.63 ab	11.71 ± 3.62 a	4.2 ± 1.24 ab	29.90 ± 2.69 a	18.06 ± 4.34 a	7.08 ± 0.86 a	0.41 ± 0.18 a
	T1	79.75 ± 5.25 a	14.34 ± 2.85 a	5.12 ± 0.88 b	27.10 ± 5.62 a	13.55 ± 4.71 a	6.93 ± 0.70 a	0.40 ± 0.14 a
	T2	107.25 ± 8.77 b	12.69 ± 0.91 a	3.3 ± 0.62 ab	29.72 ± 5.66 a	15.76 ± 2.20 a	9.19 ± 0.48 b	0.48 ± 0.14 a
	T3	98.75 ± 12.84 ab	10.67 ± 0.51 a	2.54 ± 0.71 a	26.06 ± 3.03 a	12.83 ± 4.26 a	8.00 ± 1.29 ab	0.58 ± 0.23 a
Quinoa	**T0**	85.53 ± 6.97 a	21.55 ± 1.27 a	2.08 ± 0.39 ab	21.62 ± 8.67 a	43.68 ± 20.7 a	2558.7 ± 1188 a	0.25 ± 0.11 a
	T1	61.96 ± 19.58 b	20.98 ± 0.89 a	2.34 ± 0.58 b	19.84 ± 3.66 a	28.47 ± 6.47 a	1762.24 ± 386 a	0.35 ± 0.01 a
	T2	60.16 ± 8.25 b	21.32 ± 1.12 a	1.75 ± 0.35 ab	17.81 ± 4.25 a	25.35 ± 5.75 a	1572.33 ± 361 a	0.35 ± 0.16 a
	T3	48.53 ± 7.39 c	20.11 ± 1.48 a	1.38 ± 0.24 a	14.33 ± 3.58 a	19.61 ± 4.33 a	1225.31 ± 278 a	0.37 ± 0.16 a
Silage maize	**T0**	90.45 ± 4.96 c	10.87 ± 0.79 c	4.8 ± 0.65 d	30.14 ± 2.89 c	29.11 ± 6.06 c	5.30 ± 1.80 ab	0.71 ± 0.17 a
	T1	85.35 ± 5.96 c	8.32 ± 0.57 b	3.622 ± 0.5 c	28.39 ± 3.37 c	18.51 ± 3.50 b	7.48 ± 0.73 b	0.62 ± 0.25 a
	T2	72.1 ± 4.09 b	8.7 ± 1.18 b	1.94 ± 0.57 b	16.61 ± 2.17 b	7.40 ± 1.77 a	3.09 ± 0.52 a	0.60 ± 0.08 a
	T3	56.95 ± 4.77 a	4.61 ± 0.82 a	0.64 ± 0.14 a	8.84 ± 2.83 a	6.64 ± 6.95 a	2.72 ± 1.63 a	0.33 ± 0.11 a

3.2. Plant Mineral Content

The means of the calcium Ca^{2+} (%), potassium K^+ (%) and sodium Na^+ (%) content in the leaves, stems, and roots of blue panicum, quinoa, and silage Maize for each treatment are shown in Figure 1. Except for the increasing calcium content in blue panicum, no significant differences were observed for potassium uptake in blue panicum and for calcium and potassium uptake in quinoa between treatments. At different parts of silage maize, Ca^{2+} and K^+ uptake decreased with increasing salinity levels to be significant for irrigation water with an electrical conductivity ≥ 6 dS m^{-1} relative to control. In contrast, the sodium content of leaves, stems and roots increased significantly with increasing water salinity for all three species.

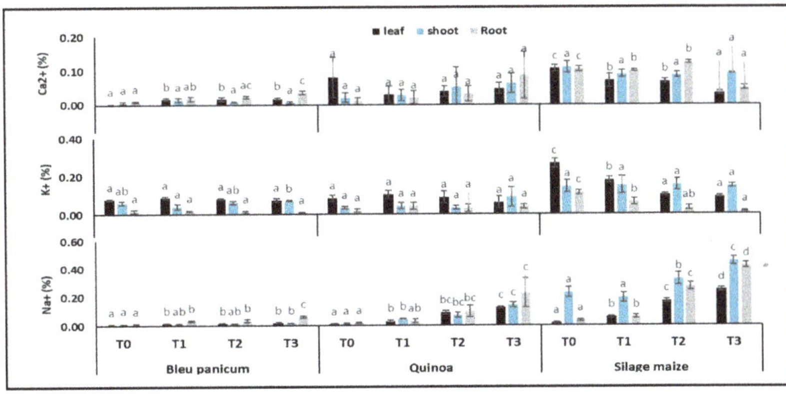

Figure 1. Mean content of calcium Ca^{2+} (%), potassium K^+ (%) and sodium Na^+ (%) in the leaves, stems, and roots of blue panicum, quinoa and silage maize for each treatment. The symbols T0 (0.9 dS m^{-1}), T1 (3 dS m^{-1}), T2 (6 dS m^{-1}), and T3 (10 dS m^{-1}) refer to the four saline irrigation water treatments. Lower-case (a, b, c, d) show differences between treatments for leaf, shoot and root within each tested plant species. Columns with the same letter over them are not significantly different ($p < 0.05$) (Mean ± standard deviation; $n = 3$).

4. Discussion

4.1. Physiological and Biochemical Properties

Blue panicum showed high tolerance to irrigation water salinity in terms of growth and production, as has been observed in many studies [3,4]. Furthermore, in terms of the biochemical properties, which did not affect the chlorophyll a, b, and proline leaf content and increased the carotenoid content only in the T2 (6 dS m^{-1}) treatment. Saline water irrigation in quinoa decreased the steam height and root dry biomass significantly to 10 dS m^{-1}. Similar results were obtained by Hirich et al. [5], who reported that quinoa is considered a facultative halophyte with a salt-tolerance threshold equal to 9 dS m^{-1}. The biochemical properties of quinoa were not affected by saline water irrigation. For silage maize, the physiological and biochemical properties decreased in all of the treatments at salt concentrations of > 6 dS m^{-1}, where the plants show stress symptoms. The decrease in chlorophyll levels in silage maize is due to the high enzymatic activity of chlorophyllase under high salinity conditions [6].

4.2. Plant Mineral Content

Our results indicate that if the cropped species are sensitive to high salinity conditions, the Na^+ content competes with the ions in the cell, which will lead to a decrease in their accumulation as the salinity increases. However, if the plant is tolerant to high or moderate saline conditions, the plant excludes the toxic ions Na^+ and Cl^- in order to maintain the uptake of K^+ and Ca^{2+}. The findings also show that the uptake of K^+ and Ca^{2+} in maize silage with a high sensitivity to saline conditions decreased with increasing salinity, compared to panicum blue and quinoa, where these ions remained stable due to their salt tolerance. This finding is consistent with the results reported by Roman et al. [7], which show that quinoa plants are able to maintain the uptake of potassium despite high Na^+ concentrations in the soil after extended salt stress. This is also consistent with the results of Shabala et al. [8], which indicated that the capacity of plants to exclude toxic ions such as Na^+ and Cl^- from transport systems is correlated to their salt tolerance. Silage maize showed a significantly higher Na^+ content than blue panicum and quinoa. This would be attributed to the ability differences of both crops to involve both mechanisms (expulsion and compartmentalization) of adaptation to salt stress [9].

Author Contributions: A.O., writing—original draft preparation; A.O. and S.A., methodology; J.H., supervision, review and editing; A.M., review. All authors have read and agreed to the published version of the manuscript.

Funding: This research received no external funding.

Institutional Review Board Statement: Not applicable.

Informed Consent Statement: Not applicable.

Data Availability Statement: Not applicable.

Acknowledgments: The authors thank the National Institute for Agronomic Research Morocco (CRRA-INRA) for supporting and providing research facilities and support.

Conflicts of Interest: The authors declare no conflict of interest.

References

1. Zaman, M.; Shahid, S.A.; Heng, L. *Guideline for Salinity Assessment, Mitigation and Adaptation Using Nuclear and Related Techniques*; Springer Nature: Cham, Switzerland, 2018; ISBN 3-319-96190-X.
2. Oumasst, A.; Azougay, S.; Taqarort, N.; Mimouni, A.; Hallam, J. Effets de La Salinité Sur l'absorption Des Nutriments, Les Paramètres Biochimiques et La Croissance Du Bleu Panicum (Panicum Antidotale Retz) et Du Maïs d'ensilage (Zea Mays L). *Afr. Mediterr. Agric. J.—Al Awamia* **2021**, 41–62. [CrossRef]
3. Koyro, H.-W.; Hussain, T.; Huchzermeyer, B.; Khan, M.A. Photosynthetic and Growth Responses of a Perennial Halophytic Grass Panicum Turgidum to Increasing NaCl Concentrations. *Environ. Exp. Bot.* **2013**, *91*, 22–29. [CrossRef]
4. Atia, A.; Debez, A.; Rabhi, M.; Barhoumi, Z.; Haouari, C.C.; Gouia, H.; Abdelly, C.; Smaoui, A. Salt Tolerance and Potential Uses for Saline Agriculture of Halophytes from the Poaceae. In *Sabkha Ecosystems*; Gul, B., Böer, B., Khan, M.A., Clüsener-Godt, M., Hameed, A., Eds.; Tasks for Vegetation Science; Springer International Publishing: Cham, Switzerland, 2019; Volume 49, pp. 223–237. ISBN 978-3-030-04416-9.
5. Hirich, A.; Choukr-Allah, R.; Jacobsen, S.-E. Deficit Irrigation and Organic Compost Improve Growth and Yield of Quinoa and Pea. *J. Agron. Crop Sci.* **2014**, *200*, 390–398. [CrossRef]
6. Santos, C.V. Regulation of Chlorophyll Biosynthesis and Degradation by Salt Stress in Sunflower Leaves. *Sci. Hortic.* **2004**, *103*, 93–99. [CrossRef]
7. Roman, V.J. *Salt Tolerance Strategies of the Ancient Andean Crop Quinoa*; Wageningen University: Wageningen, The Netherlands, 2021. [CrossRef]
8. Shabala, S. Learning from Halophytes: Physiological Basis and Strategies to Improve Abiotic Stress Tolerance in Crops. *Ann. Bot.* **2013**, *112*, 1209–1221. [CrossRef]
9. Munns, R.; Tester, M. Mechanisms of Salinity Tolerance. *Annu. Rev. Plant Biol.* **2008**, *59*, 651–681. [CrossRef]

environmental sciences proceedings

MDPI

Proceeding Paper

Will Roots Play a Decisive Role in Forage Sorghum Production under Salt Stress? †

Erick Amombo [1], Dennis S. Ashilenje [1], Karima Lazaar [1], Abdelaziz Hirich [1], Lamfeddal Kouisni [1], Krishna P. Devkota [1], Abdallah Oukarroum [2], Ghoulam Cherki [2,3], Mohamed El Gharous [4] and Abdelaziz Nilahyane [1,*]

1 African Sustainable Agriculture Research Institute (ASARI), Mohammed VI Polytechnic University (UM6P), Laâyoune 70000, Morocco; erick.amombo@um6p.ma (E.A.); dennis.ashilenje@um6p.ma (D.S.A.); karima.lazaar@um6p.ma (K.L.); abdelaziz.hirich@um6p.ma (A.H.); lamfeddal.kouisni@um6p.ma (L.K.); krishna.devkota@um6p.ma (K.P.D.)

2 AgroBioSciences Department (AgBS), Mohammed VI Polytechnic University (UM6P), Ben Guerir 43150, Morocco; abdallah.oukarroum@um6p.ma (A.O.); ghoulam.cherki@um6p.ma (G.C.)

3 Center of Agrobiotechnology and Bioengineering, Labelled Research Unit CNRST, Cadi Ayyad University (UCA), Marrakech 40000, Morocco

4 Agricultural Innovation and Technology Transfer Center (AITTC), Mohammed VI Polytechnic University (UM6P), Ben Guerir 43150, Morocco; mohamed.elgharous@um6p.ma

* Correspondence: abdelaziz.nilahyane@um6p.ma; Tel.: +212-666-945-933

† Presented at the 2nd International Laayoune Forum on Biosaline Agriculture, 14–16 June 2022; Available online: https://lafoba2.sciforum.net/.

check for updates

Citation: Amombo, E.; Ashilenje, D.S.; Lazaar, K.; Hirich, A.; Kouisni, L.; Devkota, K.P.; Oukarroum, A.; Cherki, G.; El Gharous, M.; Nilahyane, A. Will Roots Play a Decisive Role in Forage Sorghum Production under Salt Stress? *Environ. Sci. Proc.* **2022**, *16*, 61. https://doi.org/10.3390/environsciproc2022016061

Academic Editor: Redouane Choukr-Allah

Published: 20 June 2022

Abstract: Root developmental plasticity might provide breeders with the opportunity to generate crops with more resistant root system designs to abiotic pressures such as salt stress. The potential influence of diverse root designs in a panel of sorghum varieties' production and performance under salt stress was investigated in this study. Dry weight yield had a significant positive correlation with three root parameters: main root length, main root angle, and lateral root length. These root features have varying positive correlations with other salt tolerance indices. Interestingly, all root properties have a negative correlation with the electrolyte leakage. Except for the lateral root length, the negative correlations were significant in all other root features. This pattern holds true for other salt performance indices studied, such as total soluble sugar content, chlorophyll content, growth, and leaf area. Furthermore, there is an inter-variety variation in the contribution level of lateral root and main root length towards the total root length and this was found to influence yield and performance. These findings give a hint on the root architectural features that play a positive role in sorghum salt tolerance and improved forage yield.

Keywords: forage; root; salinity; sorghum; yield

1. Introduction

Recently, there is increasing interest in forage sorghum grown in desert areas. Some of the advantages cultivating this crop include high water use efficiency than other forage crops, higher heat and drought tolerance, versatility, and potentially high-quality nutrients for cows, sheep, and camels [1]. Compared to many forages, sorghum can produce more biomass and good feed quality if crop management practices are well optimized. In marginal and desertic areas, sorghum production is constrained by various abiotic stresses especially salt stress. Therefore, improved crop management and new varieties with increased tolerance to salinity can significantly increase their yields. However, the sorghum genera are very diverse genetically, phenotypically, and geographically and relatively few recognized species have been evaluated for potential forage selection.

Root growth and developmental flexibility may allow breeders to create crops with more resistant root system designs to abiotic stresses such as salt stress. However, this area remains underexplored in crop production systems. Thus, in this paper, we investigate the possible impact of diverse root characteristics on sorghum yield and performance under salt stress.

2. Materials and Methods

In this study, thirteen local sorghum varieties were used. The plants were cultivated in Laayoune, Morocco, under medium saline conditions (EC of 7884.5 S/cm) during the spring season of 2021 as part of an ongoing salt-tolerant forage germplasm phenotyping and genotyping project. Sampling for relative water content (RWC), chlorophyll content Chl), electrolyte leakage (EL), and total soluble carbohydrate (TSC) content was carried out during harvest. For root sampling, plants were removed, and roots scraped carefully to minimize damage at the soft dough stage when sorghum is apparently at the silage harvest maturity stage. The root data were processed for correlation analysis with other tested physiological parameters on SPSS (v.16, SPSS Inc., Chicago, IL, USA).

3. Results

Implications of Root Traits on Sorghum Performance and Dry Yield

Significant difference among varieties was observed for all root traits that were tested in sorghum (Table 1).

Table 1. Means of tested root parameters.

VARIETY	MRL [1]	MRV	MRA	LRN	LRL
1	61.56	11.82	12.74	77	74.74
2	73.91	26.70	13.63	112	144.37
3	114.30	24.09	23.40	248	355.34
4	57.27	13.42	4.24	95	92.37
5	72.37	15.38	17.21	193	134.32
6	133.20	9.731	6.00	104	161.33
7	48.05	26.49	1.79	63	70.68
8	95.16	23.84	10.94	124	177.64
9	60.85	12.67	7.84	108	158.32
10	60.46	18.18	10.60	142	140.86
11	29.36	5.80	18.43	77	112.93
12	18.84	21.28	123.00	129	289.18
13	121.10	31.15	33.99	247	301.49

[1] MRL—main root length; LRL—lateral root length; MRV—main root vector; MRA—main root angle; LRN—lateral root number.

To further analyze the possible interaction of various root features with yield and performance, a correlation analysis was carried out. Dry weight yield had a significant positive correlation with three root parameters: MRL (0.523), main root angle (MRA; 0.516), and lateral root length (LRL; 0.531). Among the performance indices, the lateral root number (LRN) was positively correlated with the RWC (0.568) while the correlations with Chl, leaf area, and height were insignificant despite being positive. All the tested root traits correlated negatively with EL. The negative correlations were significant except for LRL. However, for the non-root parameters, all the negative correlations were insignificant. Overall, the highest root–root correlations were observed between LRL and MRA, while the root–non-root parameters were LRN vs. RWC, LRL vs. RWC, height vs. dry yield, and LRL vs. dry yield. Among the non-root–non-root parameters, the highest positive correlation was between leaf length (LL) and leaf width (LW) as well as TSC and LW (Table 2).

Table 2. Pearson correlation coefficients of tested parameters.

	RWC	Height	EL	Chl	LL	LW	TSC	MRL	LRL	MRV	MRA	LRN	DW
RWC [1]	1												
Height	0.184	1											
EL	−0.049	−0.118	1										
Chl	0.083	0.400 *	−0.126	1									
LL	0.430 **	0.520 **	−0.056	0.360 *	1								
LW	0.388 *	0.387 *	−0.092	0.142	0.724 **	1							
TSC	0.186	0.255	−0.096	0.16	0.571 **	0.662 *	1						
MRL	0.016	0.063	−0.465 **	0.107	0.073	0.091	0.053	1					
LRL	0.439 *	0.042	−0.224	0.105	0.014	0.012	0.144	0.843 **	1				
MRV	0.071	0.102	−0.443 **	0.05	0.079	0.13	0.044	0.943 **	0.843 **	1			
MRA	0.101	0.036	−0.358 *	0.087	0.157	0.207	0.103	0.901 **	0.775 **	0.794 **	1		
LRN	0.568 **	0.023	−0.343 *	0.091	0.03	0.034	0.12	0.954 **	0.920 **	0.900 **	0.909 **	1	
DW	0.282	0.834 **	−0.532 *	0.301	0.654 **	0.417 *	0.389	0.523 **	0.531 **	0.239	0.516 **	0.351 *	1

[1] RWC—relative water content; EL—electrolyte leakage; Chl—chlorophyll content; LL—leaf length; LW—leaf width; TSC—total soluble carbohydrate; MRL—main root length; LRL—lateral root length; MRV—main root vector; MRA—main root angle; LRN—lateral root number; DW—dry weight. * Significant at 0.05. ** Significant at 0.01.

4. Discussion

The above-ground phenotypic traits measured in this study may be utilized as credible predictors of salt tolerance and forage yield. In terms of the physiological consequence of cellular water deficiency, the RWC, for example, is one of the most relevant markers of plant water status [2]. As a result, it can predict sorghum water status in terms of cellular hydration while accounting for the effects of both leaf water potential and osmotic adjustment. The preservation of considerably greater RWC in some sorghum varieties can be attributed to the resistance to salt-induced physiological drought, which is consistent with our earlier findings that high RWC was related to salt resistance [3]. Chlorophyll (a and b) functions as an intermediate in the transition of absorbed solar energy and its action in photosynthesis and organic compound production [4]. As a result, it can be utilized as a plant health indicator under stressful conditions. The ability of sorghum varieties to maintain relatively higher total Chl contents reflects their ability to maintain higher photosynthetic efficiency, organic compound accumulation, and growth under salt stress, as evidenced by relatively higher digestible sugar contents, leaf area, and plant height in most of the varieties. This is also evidenced by a strong positive correlation between Chl, height, RWC, and soluble carbohydrate contents. The EL reflects the stability index of cellular membranes. Here, we observe a negative correlation between RWC, Chl, height, and leaf area with EL, indicating that salinity could not only damage membranes but also limit growth and photosynthetic capacity. Additionally, a negative correlation with all root parameters indicates that EL can be a reliable phenotypic marker for salt sensitivity in large sorghum varieties.

Soluble carbohydrates not only serve as vital nutritive molecules, but they can also serve as significant osmolytes in plants, helping to promote osmotic adjustment [5]. A substantial correlation between soluble carbohydrates and RWC suggests that high soluble carbohydrate levels in sorghum had a favorable effect in water absorption and retention, which may boost photosynthesis and plant development. Roots play a vital role in plant response to stress and production. However, partly due to the difficulties in viewing them throughout the plant's life cycle, they have received less attention than other organs. In this case, we see structural variety in traits such as MRL, LRN, MRV, and MRA in a sorghum population and various correlations with salt performance indices. This demonstrates the presence of a significant amount of flexibility in sorghum root system architecture (RSA) during salt stress. The positive connection between LRN and RWC suggests that the design in sorghum varieties with the greatest LRN and the shortest MRL may be related to sorghum salt tolerance. As a result, they might be used to simulate sorghum with more efficient roots to improve tolerance and yield.

5. Conclusions

In this study, we see structural variety in root traits such as MRL, LRN, MRV, and MRA. This demonstrates the presence of a significant level of flexibility in sorghum RSA during salt stress. The positive connection between LRN and RWC suggests that the design in varieties with the greatest LRN and the shortest MRL may be related to sorghum salt tolerance and yield. As a result, they might be a target for developing and modeling sorghum with more efficient roots to improve tolerance and yield.

Author Contributions: Conceptualization, E.A., A.N., A.H. and A.O.; methodology, E.A., A.N. and K.P.D.; data collection, E.A., D.S.A., K.L. and A.N.; writing, editing, and reviewing, E.A., A.N., A.H., L.K., A.O., G.C. and M.E.G.; project administration, A.N.; funding acquisition, A.N. All authors have read and agreed to the published version of the manuscript.

Funding: This research is funded by OCP Phosboucraa Foundation, grant number FPB_SPA005_2020.

Institutional Review Board Statement: Not applicable.

Informed Consent Statement: Not applicable.

Acknowledgments: The authors acknowledge the OCP Phosboucraa Foundation for funding this project. Appreciation is extended to farmers for providing their farmlands to conduct the research trials and all field assistants.

Conflicts of Interest: The authors declare no conflict of interest.

References

1. Ramos, J.C.P.; Júnior, V.R.R.; Monção, F.P.; da Costa Parrela, R.A.; Caxito, A.M.; Cordeiro, M.W.S.; da Hora, F.F.; de Assis Pires, D.A. Effect of replacing forage sorghum silage with biomass sorghum silage in diets for F_1 Holstein × Zebu lactating cows. *Trop. Anim. Health Prod.* **2021**, *53*, 99. [CrossRef] [PubMed]
2. Sapes, G.; Sala, A. Relative water content consistently predicts drought mortality risk in seedling populations with different morphology, physiology and times to death. *Plant Cell Environ.* **2021**, *10*, 3322–3335. [CrossRef] [PubMed]
3. Amombo, E.; Li, X.; Wang, G.Y.; Fan, S.G.; Zhang, Y.K.; Fu, J.M. Screening of diverse tall fescue population for salinity tolerance based on SSR marker-physiological trait association. *Euphytica* **2018**, *214*, 220. [CrossRef]
4. Yang, Z.; Li, J.L.; Liu, L.N.; Xie, Q.; Sui, N. Photosynthetic regulation under salt stress and salt-tolerance mechanism of sweet sorghum. *Front. Plant Sci.* **2020**, *10*, 1722. [CrossRef] [PubMed]
5. Turner, N.C. Turgor maintenance by osmotic adjustment: 40 years of progress. *J. Exp. Bot.* **2018**, *69*, 3223–3233. [CrossRef] [PubMed]

environmental sciences proceedings

MDPI

Proceeding Paper

The Effect of Salt Stress on Proline Content in Maize (*Zea mays*) †

Shruti Nilesh Pingle *, Shruti Tanaji Suryawanshi, Kiran Ramesh Pawar and Sanjay N. Harke

Institute of Biosciences and Technology, Mahatma Gandhi Mission University (MGM), Aurangabad 431003, India; shrutits123@gmail.com (S.T.S.); kiranpawar510@gmail.com (K.R.P.); sanjayharke@gmail.com (S.N.H.)

* Correspondence: shrutipingle033@gmail.com

† Presented at the 2nd International Laayoune Forum on Biosaline Agriculture, 14–16 June 2022; Available online: https://lafoba2.sciforum.net/.

Abstract: In this study, the effect of applied NaCl on the growth and proline concentration of maize varieties, i.e., (*Zea mays* L.), Syngenta 7720, Syngenta 6668, eco-91, Syngenta 7710 and Advanta Pac 751 vertex 751, was investigated. The experiment was conducted in laboratory conditions using a completely randomized design with three replications. The soil used for the experiment was salinized by applying NaCI (common salt) solution at the rates of 50 mM, 100 mM, 150 mM, 200 mM and 250 mM every three days for the duration of 12 days. We found that, according to the data we have collected on germination percentage and proline content, maize varieties (Syngenta 7720 and eco-91) had a significant increase in proline content and a decrease in plant growth as the concentration increased.

Keywords: maize (*Zea mays* L.); salinity; proline

check for updates

Citation: Pingle, S.N.; Suryawanshi, S.T.; Pawar, K.R.; Harke, S.N. The Effect of Salt Stress on Proline Content in Maize (*Zea mays*). *Environ. Sci. Proc.* **2022**, *16*, 64. https://doi.org/10.3390/environsciproc2022016064

Academic Editor: Abdelaziz Hirich

Published: 23 June 2022

1. Introduction

Maize (*Zea mays* L.) plays an important role in human and animal nutrition in many developed and developing countries, [1]. It is a well-fit substitute crop for diversification of the rice-wheat system in Indo-Gangetic Plain (IGP). In India, the maize area has slowly expanded over the past few years, reaching 6.2 million ha (3.4% of the gross cropped area) in 1999/2000. In the country, it is grown under a wide range of climatic conditions, ranging from extreme semi-arid to sub-humid and humid. Traditionally, its growing area was more focused on Bihar, Madhya Pradesh, Rajasthan, and Uttar Pradesh, and to some extent in maize growing areas including Karnataka and Andhra Pradesh. It is also very popular in the low- and mid-hill areas of the western and northeastern regions [2].

Soil salinity is a key challenge in several regions of India, such as Gujarat, Uttar Pradesh, Maharashtra, West Bengal, Rajasthan and the coastal region of Odisha. Salt stress affects the growth and development of maize in several ways. However, the response of plants varies with the degree of stress at different crop growth stages. Short-term exposure of maize plants to salt stress influences plant growth to osmotic stress [3]. Salinity stress affects plant growth and productivity, water relation, photosynthesis and carbon assimilation, grain development, and filling. Salt resistance in some maize varieties is linked with higher potassium, low sodium and chloride fluxes, and cytoplasmic contents [4]. Photosynthesis is the most important process by which green plants convert solar energy into chemical energy in the form of organic compounds synthesized by the fixation of atmospheric carbon dioxide. Carbon fixation in maize is sensitive to salt stress [5]. Both stomatal and non-stomatal limitations and their combination are associated with salinity-induced reductions in maize photosynthesis [6]. Reduced stomatal conductance, impaired activities of carbon fixation enzymes, reduced photosynthetic pigments, and destruction of photosynthetic apparatus are among the key factors limiting carbon fixation capacity of maize plants under salt stress [5,6].

Proline is the most common endogenous osmolyte accumulated under various abiotic stresses including salinity [7,8]. When applied as an exogenous compound to crops, proline can improve salt tolerance [9]. For example, in salt-stressed maize, the foliar application of proline increased plant growth with a positive effect on yield characteristics [10]. The beneficial effects of exogenous proline application on salt stress tolerance have been the subject of several factors including crop growth stage and Na+/K+ ratio [11]. The role of proline in cell osmotic adjustment, membrane stabilization and detoxification of injurious factors in plants exposed to salt stress has been widely reported [12,13]. Thus, the objective of this study was to see the effect of salinity on proline.

2. Materials and Methods

The seeds of five maize varieties (Syngenta 7720, Syngenta 6668, eco-seeds eco-91, Syngenta 7710, Advanta Pac 751 vertex 751) were sown in a germination tray on coco peat. Six levels of salinity solutions, viz., 0, 50, 100, 150, 200 and 250 mM NaCl were used to evaluate the resistance of maize varieties. The seeds were treated with 5 mL of NaCl solution every 3 days for a total 12 days. Proline accumulation was determined by using the method described by Bates et al. [14]. A 0.2 g fresh plant sample of leaves was homogenized with 3% sulfosalicylic acid. The homogenate was centrifuged at 3000 RPM for 20 min. The supernatant was treated with 200 µL acetic acid and 200 µL ninhydrin, then boiled for 1 h in a hot water boiling bath. After boiling for 1 h, the solution was kept in ice water. Then, absorbance at 520 nm was determined by UV-Visible using a spectrophotometer. Germination percentage, rate of germination, number of leaves, shoot length, leaf color, days of maturity, root length, number of roots, weight of seed at time of sowing, germination stress tolerance index (GSI), promptness index (PI), root length stress index (RLST), shoot length of stress index (SLSI), and proline content were measured.

3. Results

In this experiment, the effect of salinity on proline content in five maize varieties was determined (Table 1). We found that as the salt concentration decreased, the rate of germination and germination percentage increased, and also, as the concentration level increased, the proline contained in maize also increased. Among the five evaluated varieties, two varieties, i.e., Syngenta 7720 and eco-91 (Figure 1a,b), had higher yields, mostly due to their increased level of proline concentration.

$$\mu\text{moles per gram tissue} = [(\mu\text{g proline/mL}) \times \text{mL toluene})/115.5\ \mu\text{g}/\mu\text{mole}]/[(\text{g sample})/5]. \quad (1)$$

(**a**) Syngenta 7720 (**b**) eco-seeds eco-91

Figure 1. Varieties that have significant increase in proline.

Table 1. (Proline content).

NaCl Concentration (mM)	Proline Content in 520 nm Solution				
	Advanta Pac 751 Vertex 751	**Syngenta 7720**	**Eco-91**	**Syngenta 6668**	**Syngenta 7710**
0 mM	0.6720	0.0616	1.4941	0.2799	0.1709
50 mM	0.7527	0.2111	1.9645	0.2345	0.6480
100 mM	0.8096	0.2799	1.3844	0.1064	0.0888
150 mM	1.0901	0.5573	1.5434	0.6576	0.0201
200 mM	0.6152	0.5689	1.6250	0.3982	0.8162
250 mM	0.4762	0.8432	1.7653	0.3039	0.5191

4. Discussion

Proline accumulation in salt-stressed plants is a primary defense response to maintain osmatic pressure in a cell, which is reported in many crops [15]. Proline content increases after an increase in NaCl concentration. We reviewed the effect, resistance mechanisms and management of salt stress in maize. Consistent with our findings, in 2010, Kaya et al. [16] reported that proline experiences salt stress. At 400 mM NaCl, sweetcorn leaves accumulated more than 600 μmol g^{-1} proline. There were significant increases in proline according to Farooq et al. [3]. Foliar application of proline increased plant growth with a positive effect on yield characteristics in maize. Proline improves resistance against salinity through cell osmotic adjustment, membrane stabilization, and detoxification of injured ions in plants exposed to salt stress [13].

5. Conclusions

In this study, the effects of applied NaCl on the growth and proline concentration of maize varieties were investigated. We have concluded that there was a significant increase in growth and proline concentration in maize varieties. With our study, we can help farmers who are suffering from soil salinity. We can use the method of proline accumulation, as described by Bates et al., on different types of crops.

Author Contributions: Conceptualization, S.N.P. and S.T.S.; methodology, S.N.P. and S.T.S.; investigation, S.N.P. and S.T.S.; resources, K.R.P. and S.N.H. data curation, S.N.P. and S.T.S.; writing —original draft preparation, S.N.P. and S.T.S.; writing—review and editing, S.N.P. and S.T.S.; visualization, S.N.P. and S.T.S.; supervision, K.R.P.; project administration, K.R.P.; funding acquisition, S.N.H. All authors have read and agreed to the published version of the manuscript.

Funding: MGM Institute of Biosciences and Technology, Aurangabad 431003, India.

Institutional Review Board Statement: Not applicable.

Informed Consent Statement: Not applicable.

Data Availability Statement: Not applicable.

Acknowledgments: We are thankful towards our college MGM institute of Biosciences and technology for providing us with all the necessary equipment.

Conflicts of Interest: The authors declare no conflict of interest.

References

1. Prasanna, B.M.; Vasal, S.K.; Kassahun, B.; Singh, N.N. Quality protein maize. *Curr. Sci.* **2001**, *81*, 1308–1319.
2. Joshi, P.K.; Singh, N.P.; Singh, N.N.; Gerpacio, R.V.; Pingali, P.L. *Maize in India Production Systems; Constraints and Research Priorities;* CIMMYT: Texcoco, Mexico, 2005; p. 1.
3. Farooq, M.; Hussain, M.; Wakeel, A.; Siddique, K.H.M. Salt stress in maize: Effects, resistance mechanisms, and management. A review. *Agron. Sustain. Dev.* **2015**, *35*, 461–481. [CrossRef]

4. Hajibagheri, M.A.; Harvey, D.M.R.; Flowers, J. Quantitative ion distribution within root cells of salt-sensitive and salt tolerant maize varieties. *New Phytol.* **1987**, *105*, 367–379. [CrossRef] [PubMed]
5. Omoto, E.; Taniguchi, M.; Miyake, H. Adaptation responses inC4 photosynthesis of maize under salinity. *J. Plant Physiol.* **2012**, *169*, 469–477. [CrossRef] [PubMed]
6. Gong, X.; Chao, L.; Zhou, M.; Hong, M.; Luo, L.; Wang, Y.; Jingwei, C.; Songjie, G.; Fashui, H. Oxidative damages of maize seedlings caused by exposure to a combination of potassium deficiency and salt stress. *Plant Soil* **2011**, *340*, 443–452. [CrossRef]
7. Szabados, L.; Savouré, A. Proline: A multifunctional amino acid. *Trends Plant Sci.* **2010**, *15*, 89–97. [CrossRef] [PubMed]
8. Slama, S.; Bouchereau, A.; Flowers, T.; Abdelly, C.; Savouré, A. Diversity, distribution and roles of osmoprotective compounds accumulated in halophytes under abiotic stress. *Ann. Bot.* **2015**, *115*, 433–447. [CrossRef] [PubMed]
9. Heuer, B. Role of proline in plant response to drought and salinity. In *Handbook of Plant and Crop Stress*, 3rd ed.; Pessarakli, A., Ed.; CRC Press: Boca Raton, FL, USA, 2010. [CrossRef]
10. Alam, R.; Das, D.; Islam, M.; Murata, Y.; Hoque, M. Exogenous proline enhances nutrient uptake and confers tolerance to salt stress in maize (*Zea mays* L.). *Progr. Agric.* **2016**, *27*, 409–417. [CrossRef]
11. Ashraf, M.F.M.R.; Foolad, M.R. Roles of glycine betaine and proline in improving plant abiotic stress resistance. *Environ. Exp. Bot.* **2007**, *59*, 206–216. [CrossRef]
12. Ashraf, M.; Foolad, M.R. Improving plant abiotic-stress resistance by exogenous application of osmoprotectants glycinebetaine and proline. *Environ. Exp. Bot.* **2007**, *59*, 206–216. [CrossRef]
13. Hare, P.D.; Cress, W.A.; Van Staden, J. Proline biosynthesis and degradation: A model system for elucidating stress related signal transduction. *J. Exp. Bot.* **1999**, *50*, 413–434. [CrossRef]
14. Bates, L.S.; Waldren, R.P.; Trare, L.D. Rapid determination of free proline for water stress studies. *Plant Soil* **1973**, *39*, 205–207. [CrossRef]
15. AI-Saady, N.A.; Khan, A.J.; Rajesh, L.; Esechie, H.A. Effect of salt stress on germination, proline Metabolism and chlorophyll content of Fenugreek (*Trignella foenum gracium* L.). *J. Plant Sci.* **2012**, *7*, 176–185. [CrossRef]
16. Kaya, C.; Tuna, A.L.; Okant, A.L. Effect of foliar applied kinetin and indole acetic acid on maize plants grown under saline conditions. *Turk. J. Agric. For.* **2010**, *34*, 529–538.

environmental sciences proceedings

MDPI

Proceeding Paper

Using Agronomic Parameters to Rate Quinoa (*Chenopodium quinoa* Willd.) Cultivars Response to Saline Irrigation under Field Conditions in Eastern Morocco †

Ilham Abidi [1,*], Abdelaziz Hirich [2], Didier Bazile [3,4], Hamid Mahyou [5,6], Fatima Gaboun [5,6] and Si Bennasseur Alaoui [1,*]

1 Crop Production, Protection and Plant Biotechnology, Hassan II Institute of Agronomy and Veterinary Medicine, Rabat 10112, Morocco
2 African Sustainable Agriculture Research Institute (ASARI), Mohammed VI Polytechnic University (UM6P), Laayoune 70000, Morocco; abdelaziz@gmail.com
3 CIRAD, UMR SENS, F-34398 Montpellier, France; didier.bazile@cirad.fr
4 SENS, Univ. Montpellier, CIRAD, F-34398 Montpellier, France
5 National Institute for Agricultural Research, Regional Agricultural Research Center, Oujda 60000, Morocco; hamid.mahyou@inra.ma (H.M.); fatima.gaboun@inra.ma (F.G.)
6 National Institute for Agricultural Research, Regional Agricultural Research Center, Rabat 10101, Morocco
* Correspondence: ilhamormvam@gmail.com (I.A.); b.alaoui@iav.ac.ma (S.B.A.)
† Presented at the 2nd International Laayoune Forum on Biosaline Agriculture, 14–16 June 2022; Available online: https://lafoba2.sciforum.net/.

Abstract: Salinity is becoming a serious threat to global food security, as it can significantly reduce crop yields and irreversibly damage soil fertility. Moreover, this problem is currently exacerbated by the impact of climate change, especially in drylands. Hence, introducing and adapting salinity-tolerant species, such as quinoa (*Chenopodium quinoa*, Willd.), could be among the ways to enhance the value of saline land, increasing its productivity and improving small farmers' income in rural areas. Quinoa, originally cultivated in the Andean region, has gained more attention throughout the Mediterranean region because it yields well even in marginal soils. It is also considered one of the world's healthiest foods, as its grains contain a balanced composition of minerals, vitamins, dietary fiber, fats, and high-quality, gluten-free proteins, with a balanced profile of all amino acids. In Morocco, quinoa was introduced in 2000, but its expansion is still limited to certain regions. In Eastern Morocco, for the first time, an experiment was carried out in 2019–2020 aiming to assess the response of five quinoa cultivars (INIA-420 Negra, Titicaca, Puno, ICBA-Q4 and ICBA-Q5) to saline irrigation. For this, we used two levels of water irrigation salinity: 1.50 $dS.m^{-1}$ as a no-salt control from Tagma's source in Tafoghalt village and 10.5 $dS.m^{-1}$ as salt treatment from local water drilling. Agronomic parameters, mainly dry matter, leaf area, grain yield and 1000-kernel weight, were measured to assess quinoa cultivars' responses to saline irrigation. Statistical analysis revealed that all investigated parameters were significantly affected by salinity, quinoa variety and their interaction ($p < 0.05$). Furthermore, significant differences in terms of salinity tolerance among the five quinoa cultivars were observed, with the highest (2.17 $t.ha^{-1}$) and lowest (0.33 $t.ha^{-1}$) yields recorded for ICBA-Q5 and INIA-420 Negra, respectively. However, the same varieties tested previously in Southern Morocco tolerated a higher level of salinity (12 $dS.m^{-1}$). We assume that other factors interfered with salinity and variety, such as the sowing date, which was relatively late and exposed the flowering and grain filling stages to high heat in May and June.

Keywords: quinoa cultivars; saline irrigation; tolerance; yield; Eastern Morocco

check for updates

Citation: Abidi, I.; Hirich, A.; Bazile, D.; Mahyou, H.; Gaboun, F.; Alaoui, S.B. Using Agronomic Parameters to Rate Quinoa (*Chenopodium quinoa* Willd.) Cultivars Response to Saline Irrigation under Field Conditions in Eastern Morocco. *Environ. Sci. Proc.* **2022**, *16*, 67. https://doi.org/10.3390/environsciproc2022016067

Academic Editor: Redouane Choukr-Allah

Published: 28 June 2022

1. Introduction

Nowadays, achieving food security for a growing population is a worldwide priority, especially in resource-poor and degraded marginal lands where agricultural productivity is still limited by continuous soil and water salinization [1]. Indeed, high salinity levels could substantially and irreversibly decrease biodiversity, as many crops are not able to grow in saline conditions. Salt-resilient crops such as quinoa could be considered as an alternative to current staple crops to sustain agricultural production, especially under saline irrigation [2,3].

Recently, quinoa (*Chenopodium quinoa*, Willd.), a facultative halophyte, has been drawing increasing interest worldwide given its high nutritional value and its capacity to enhance food security by tolerating environmental constraints such as frost [4], drought [5,6] and salinity [7–9]. In Morocco, quinoa was introduced in 2000, but its expansion is still limited to certain regions and few adapted genotypes. In Eastern Morocco, for the first time, a study was conducted in 2019–2020 aiming to investigate and assess the response of five quinoa cultivars to saline irrigation and to set up sustainable and suitable technical practices for local farmers in this rural area.

2. Materials and Methods

2.1. Experimental Site

This study was carried out on a farm located in the rural municipality of Boughriba in Berkane Province, 29.3 km from the Mediterranean Sea, towards the north east of Morocco (X = 769.014°; Y = 473.689°; Z = 526 m). The soil presented a silt loam texture with 1.45% organic matter, a pH of 7.3 and a CE of 0.8 $dS.m^{-1}$. The climate is semi-arid, with an irregular rainfall averaging 290 mm per year (average calculated from weather station data of Boughriba over a series of 30 years). During 2020, the total rainfall was 210 mm and the average temperature ranges from 10.7 to 26.1 °C, while the maximum and the minimum temperatures varied from 17.4 to 34 °C and from 3.9 to 20.3 °C, respectively (Figure A1).

2.2. Trial Setup

We laid out a field experiment in a split-plot design with three replicates, applying five quinoa cultivars in the main plots and two salinity levels of irrigation water in the subplots. The cultivars were INIA 420-Negra, Titicaca, Puno, ICBA-Q4, and ICBA-Q5, and have previously been tested in the Rhamna region in south Morocco. The irrigation water salinity levels were based on two available groundwater resources namely, Tagma's source in Tafoghalt village as a control treatment and water drilling as a salt treatment. The respective electrical conductivity EC values were approximately 1.5 $dS.m^{-1}$ and 10.5 $dS.m^{-1}$. We used drip irrigation with a flow rate of 2 $L.h^{-1}$. The sowing and harvesting dates, respectively, occurred on 13 March and 10 July. Regarding fertilization, all treatments received the same quantity of local manure before the sowing date. The surface of the unit plot was 12 m^2 (4 m × 3 m), and each consisted of seven rows with a 50 cm interline distance and a 30 cm interplant distance.

2.3. Sampling Procedure and Measured Parameters

During quinoa's growth cycle, one sample was taken in each phenological stage, mainly in the vegetative, flowering and grain filling stages, to measure agronomic parameters. These samplings occurred, respectively, on 50, 75 and 90 days after sowing. The monitored parameters were quinoa height, number of leaves, leaf area, and the fresh and dry weight of roots, leaves and the whole plant. The sampling area was 1 m^2 per replication, in which 9 to 10 plants were manually cut and transferred to the laboratory of the National Institute of Agronomical Research in Berkane. At harvest, kernels were dried in an oven at a temperature of 110 °C until their weight stabilized. Then, final dry matter, seed yield and 1000-kernel weight were assessed.

2.4. Statistical Analysis

Statistical analysis was performed using SAS 9.0 software. For a better interpretation, a two-way analysis of variance (ANOVA) was used to assess the effects of irrigation water salinity level, quinoa variety and their interaction on monitored parameters. When ANOVA was significant ($p < 0.05$), Student's test (LSD) was used for means comparison ($p \leq 0.05$).

3. Results and Discussion

3.1. Results

3.1.1. Agronomic Parameters during Quinoa Growth Cycle

All varieties achieved their growth cycle and developed differently depending on the salinity level of the irrigation water and variety. Analysis of variance (ANOVA) showed the significant effects of the two factors and their interaction on most measured parameters mainly in flowering and grain filling stages. Contrary, no significant effect was recorded on the leaf area in the last stage ($p < 0.05$) (Table 1). The assessed parameters were quinoa vegetative height (VH), root dry matter (RDM), leaf dry matter (LDM), total dry matter (TDM), leaf number (LN) and leaf area (LA).

Table 1. Variation of agronomic parameters during the quinoa growth cycle.

Parameters	VH (cm)			RDM (g/Plant)			LDM (g/Plant)			TDM (g/Plant)			LN			LA (cm^2)		
DAS	50	75	90	50	75	90	50	75	90	50	75	90	50	75	90	50	75	90
Salinity (S)	ns	***	***	*	***	***	**	***	ns	***	***	***	***	***	***	***	*	ns
Variety (V)	***	***	***	ns	***	***	*	***	***	***	***	***	***	***	***	***	*	ns
(S) × (V)	ns	***	***	**	***	***	ns	***	***	***	***	***	***	***	***	***	*	ns

ns: no significant, * $p < 0.05$, ** $p < 0.01$, *** $p < 0.001$.

3.1.2. Growth Parameters

The main results of the ANOVA for the investigated parameters are recorded in Table 2. At harvest time, the studied parameters were final dry matter, grain yield and 1000-kernel weight (1000 KW). The statistical analysis revealed that salinity, variety and their interaction affected significantly most parameters, except 1000 KW ($p < 0.05$).

Table 2. Analysis of variance (GLM procedure) of the effect of salinity, variety and their interaction on post harvesting parameters.

Factors	DF	Dry Matter (g/plt)	Kernel Yield (t/ha)	1000 Kernel Weight (g)
Salinity (S)	1	0.00 ***	0.00 ***	ns
Variety (V)	4	0.00 ***	0.00 ***	ns
Interaction (S × V)	4	0.00 ***	0.016 *	ns

ns = not significant, * denotes $p < 0.05$ and *** denotes $p < 0.001$.

3.1.3. Total Dry Matter and Grain Yield

Final dry matter (g/plant) and grain yield ($t.ha^{-1}$) were significantly affected by irrigation water salinity level (10.5 $dS.m^{-1}$) comparatively to the control (1.5 $dS.m^{-1}$) for all varieties ($p < 0.05$). Generally, a substantial decrease in dry matter and grain yield was recorded. However, this reduction depended on variety and was more noticeable for Titicaca (68%) and INIA-420 Negra (64%) in terms of dry matter and for Puno (48%) and Titicaca (46%) regarding grain yield ($p \leq 0.05$). Moreover, the average yield comparison showed that the best results were recorded for ICBA-Q5, with minimum loss (21%) when it was irrigated with saline water (10.5 $dS.m^{-1}$) ($p \leq 0.05$) (Table 3).

Table 3. Final dry matter and kernel yield according to water irrigation salinity level and quinoa variety.

CE ($dS.m^{-1}$)	Variety	Final Dry Matter (g/Plant)	Grain Yield (t/ha)
1.50	INIA 420-Negra	148.30 ± 4.97 **a**	0.55 ± 0.03 **c**
	Titicaca	104.23 ± 1.52 **b**	2.64 ± 0.43 **b**
	Puno	65.73 ± 2.93 **c**	3.55 ± 0.21 **a**
	ICBA-Q4	72.93 ± 2.60 **c**	2.61 ± 0.25 **b**
	ICBA-Q5	53.13 ± 2.04 **d**	2.74 ± 0.18 **ab**
10.50	INIA 420-Negra	52.83 ± 1.49 **a**	0.33 ± 0.01 **d**
	Titicaca	33.37 ± 0.93 **c**	1.45 ± 0.18 **c**
	Puno	49.70 ± 3.00 **a**	1.83 ± 0.06 **b**
	ICBA-Q4	40.30 ± 0.31 **b**	1.78 ± 0.06 **b**
	ICBA-Q5	35.67 ± 1.62 **bc**	2.17 ± 0.05 **a**

In each column and for each salinity level (EC), values followed by the same letters are statistically homogeneous ($p \leq 0.05$).

3.2. Discussion

According to this study, the investigated parameters during the quinoa life cycle were significantly affected by water irrigation salinity at 10.5 $dS.m^{-1}$ that caused losses in yield and dry matter at harvest, respectively, of 66% and 52%. Previous results obtained by Razzaghi et al. (2012) supported the fact that increasing salinity leads to quinoa yield and biomass reduction [10]. However, Hirich et al. (2014) noted that the same cultivars, Puno, Titicaca, ICBA-Q4 and ICBA-Q5, cultivated in South Morocco were able to grow successfully above 10.5 $dS.m^{-1}$ and their grain yield and dry matter were not significantly affected up to 17 $dS.m^{-1}$ [11]. Nonetheless, our study indicates that this result depends on quinoa varieties' response to salt stress. In fact, ICBA-Q5 recorded the highest grain yield (2.17 $t.ha^{-1}$) with the minimum loss (21%) under salt stress, while INIA-420 NEGRA showed the lowest grain yield whatever the salinity level. This Peruvian long-life cycle variety should be sown early to avoid the effect of high temperature mainly during the flowering and grain filling stages.

According to Figure A1 and the late sowing date (March 13), these critical stages coincide with the period of maximum temperature, reaching 30 °C. Thus, managing the sowing date is also essential to reduce the negative effect of heat stress during the flowering and grain filling stages, and therefore the eventual losses of grain yield and biomass [12]. We will repeat this study in the following seasons in order to confirm these results.

Author Contributions: Conceptualization, I.A. and S.B.A.; methodology, I.A.; software, I.A. and F.G.; validation, S.B.A., D.B., A.H. and H.M.; formal analysis, I.A.; investigation, I.A.; resources, I.A. and S.B.A.; data curation, I.A.; writing—original draft preparation, I.A.; writing—review and editing, I.A. and S.B.A. All authors have read and agreed to the published version of the manuscript.

Funding: This research received no external funding.

Data Availability Statement: Data collected from field measurements conducted on quinoa crops and soil and water analyses at the laboratory of the National Institute for Agronomic Research in Berkane.

Acknowledgments: This study was conducted on a private farm in the rural community of Boughriba in Berkane Province. Our acknowledgments go to the owner of the farm for his material support as well as the administration of the Boughriba community and Regional Agricultural Development Office (ORMVA) of Moulouya in Berkane for the facilitation of this study.

Conflicts of Interest: The authors declare no conflict of interest.

Appendix A

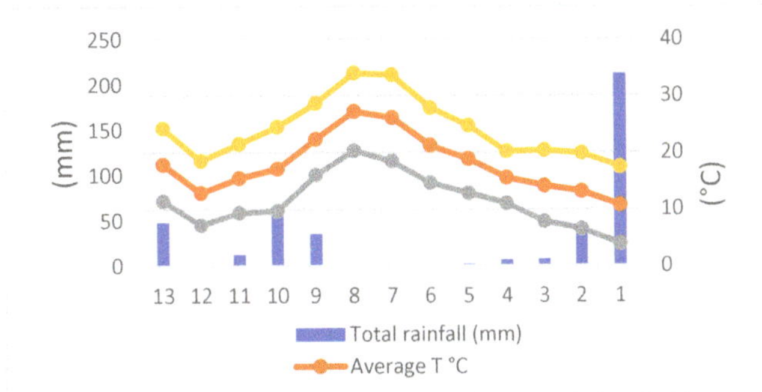

Figure A1. Climate parameters for the Berkane site (2020).

References

1. Mukhopadhyay, R.; Sarkar, B.; Jat, H.S.; Sharma, P.C.; Bolan, N.S. Soil salinity under climate change: Challenges for sustainable agriculture and food security. *J. Environ. Manag.* **2020**, *280*, 111736. [CrossRef] [PubMed]
2. Hussain, M.I.; Muscolo, A.; Ahmed, M.; Asghar, M.A.; Al-Dakheel, A.J. Agro-morphological, Yield and Quality Traits and Interrelationship with Yield Stability in Quinoa (*Chenopodium quinoa* Willd.) Genotypes under Saline Marginal Environment. *Plants* **2020**, *9*, 1763. [CrossRef] [PubMed]
3. Koyro, H.-W.; Eisa, S.S. Effect of salinity on composition, viability and germination of seeds of *Chenopodium quinoa* Willd. *Plant Soil* **2007**, *302*, 79–90. [CrossRef]
4. Jacobsen, S.E.; Monteros, C.; Christiansen, J.L.; Bravo, L.A.; Corcuera, L.J.; Mujica, A. Plant response of quinoa to frost at various phenological stages. *Eur. J. Agron.* **2005**, *22*, 131–139. [CrossRef]
5. Razzaghi, F.; Jacobsen, S.E.; Jensen, C.R.; Andersan, M.N. Ionic and photosynthetic homeostasis in quinoa challenged by sa-linity and drought-mechanisms of tolerance. *Func. Plant Biol.* **2014**, *42*, 136–148. [CrossRef] [PubMed]
6. Hirich, A.; Choukr-Allah, R.; Jacobsen, S.E.; El Youssfi, L.; El Homaria, H. Using deficit irrigation with treated wastewater in the production of quinoa (*Chenopodium quinoa* Willd.) in Morocco. *Rev. UDO Agric.* **2012**, *12*, 570–583.
7. Hirich, A.; Choukr-Allah, R.; Jelloul, A.; Jacobsen, S.E. Quinoa (*Chenopodium quinoa* Willd.) Seedling, Water Uptake and Yield Responses to Irrigation Water Salinity. *Acta Hort.* **2014**, *1054*, 145–152. [CrossRef]
8. Long, N.V. Effects of salinity stress on Growth and Yield of Quinoa. *Vietnam. J. Agric. Sci.* **2016**, *14*, 321–327.
9. Bouras, H.; Choukr-Allah, R.; Amouaouch, Y.; Bouaziz, A.; Devkota, K.P.; El Mouttaqi, A.; Bouazzama, B.; Hirich, A. How Does Quinoa (*Chenopodium quinoa* Willd.) Respond to Phosphorus Fertilization and Irrigation Water Salinity? *Plants* **2022**, *11*, 216. [CrossRef]
10. Koyro, H.W. Effect of salinity on growth, photosynthesis, water relations and solute composition of the potential cash crop halophyte *Plantago coronopus* (L.). *Environ. Exp. Bot.* **2006**, *56*, 136–146. [CrossRef]
11. Hirich, A.; Choukr-Allah, R.; Ezzaiar, R.; Shabbir, S.A.; Lyamani, A. Introduction of alternative crops as a solution to groundwater and soil salinization in the Laayoune area, South Morocco. *Euro-Mediterr. J. Environ. Integr.* **2021**, *6*, 1–13. [CrossRef]
12. Matías, J.; Rodríguez, M.J.; Cruz, V.; Calvo, P.; Reguera, M. Heat stress lowers yields, alters nutrient uptake and changes seed quality in quinoa grown under Mediterranean field conditions. *J. Agron. Crop Sci.* **2021**, *207*, 481–491. [CrossRef]

environmental sciences proceedings

MDPI

Proceeding Paper

Effect of Saline Irrigation Water on Growth and Productivity Growth of Sugar Beet (*Beta vulgaris* L.) under Nano Irrigation (Case of Moistube) †

Khadija Mouhdi [1], Abdeljabbar Attaoui [1], Brahim Bouizgarne [2], Mohamed Bakki [2], Khalid Azim [3] and Jamal Hallam [3,*]

[1] Department of Chemistry, Ben M'sik Faculty of Sciences, Hassan II University of Casablanca, B.P. 7955 Sidi Othmane, Casablanca 20000, Morocco; khadija.mouhdi1@gmail.com (K.M.); abdeljabbar.attaoui@gmail.com (A.A.)

[2] Team Microbial Biotechnology, Laboratory Plant Biotechnology, Faculty of Sciences of Agadir, Ibn Zohr University, B.P. 8106, Agadir 80000, Morocco; b.bouizgarne@uiz.ac.ma (B.B.); bakki@gmail.com (M.B.)

[3] CRRA of Agadir, National Institute of Agronomic Research, Avenue Ennasr, B.P. 415 Rabat Principale, Rabat 10090, Morocco; azim.khalid@yahoo.fr

* Correspondence: jamal.hallam@gmail.com or jamal.hallam@inra.ma

† Presented at the 2nd International Laayoune Forum on Biosaline Agriculture, 14–16 June 2022; Available online: https://lafoba2.sciforum.net.

Abstract: New technologies have been developed to maximize agricultural production with rational use of water resources, especially salt water. We conducted field experiments using either the Nano Irrigation system or the drip irrigation system on sandy loam soil to examine the response of sugar beet to different levels of saline irrigation water. Replicates (n = 4) of elementary plots, 4 m wide × 5 m long with two irrigation water salinity treatments (S1 = 1.6 dS m^{-1}, S2 = 6.3 dS m^{-1}), were established in a factorial design under Nano Irrigation. Soil chemical properties and morphological and physiological parameters of sugar beet were measured over two sampling periods. Irrigation with saline water resulted in proline accumulation in leaves and decreased chlorophyll content, leaf area, and root yield. The results suggest that irrigation water of 6 dS m^{-1} could be used to obtain an acceptable root biomass yield without significant short-term salinity issues in the cultivated soil.

Keywords: biosaline agriculture; salt stress; sugar beet; Nano Irrigation

Citation: Mouhdi, K.; Attaoui, A.; Bouizgarne, B.; Bakki, M.; Azim, K.; Hallam, J. Effect of Saline Irrigation Water on Growth and Productivity Growth of Sugar Beet (*Beta vulgaris* L.) under Nano Irrigation (Case of Moistube). *Environ. Sci. Proc.* **2022**, *16*, 72. https://doi.org/10.3390/environsciproc2022016072

Academic Editor: Abdelaziz Hirich

Published: 13 July 2022

1. Introduction

Sugar beet has a reputation among high added value crops for tolerating salt stress well. Although previous studies have investigated the effect of salinity on sugar beet [1–3], so far, no studies have been performed on the effects occurring under the Nano Irrigation system. Such studies are vital to fully understand how the sugar beet (a large, fleshy taproot that grows to be large) performs under a Nano Irrigation system and how it responds to the salinity of the water under the system. The "nano" system used in this study is a buried, porous tube irrigation system of relatively new technology (Moistube), and is composed of semi-permeable membranes whose pores are of the order of a nanometer [4]. This study follows two previous field experiments [5], that examined the water-saving abilities and performance of two alternative (fibrilroot) crops under Nano Irrigation. Our experience offers an objective vision of biosaline agriculture to better promote the growth of high added value crops by engaging accumulated knowledge and using technologies developed for the recovery of water and salty soils in arid and semi-arid regions.

2. Materials and Methods

Sugar beet (*B. vulgaris* L.) was cultivated under salt stress from 3 December 2020 to 25 May 2021. A factorial design was established to produce 11 elementary plots of 20 m^2, each with two salinity treatments (S1 = 1.6 dS m^{-1}, S2 = 6.35 dS m^{-1}) and distributed as follows: (i) Nano Irrigation with salinity S1 (NS1, $n = 4$), (ii) Nano Irrigation with salinity S2 (NS2, $n = 4$), drip irrigation with salinity S1 (GS1, $n = 1$), and (iii) drip irrigation with salinity S2 (GS2, $n = 2$). The soil and plant samples were collected after 2 months of treatment (flowering stage) and at the end of the growing cycle (harvest stage). Samples were analyzed at the Soil–Water–Plant laboratory at the Regional Center for Agronomic Research in Agadir.

3. Statistical Analyzes of Data

The data collected were analyzed using a general linear analysis of variance (ANOVA) model and SPSS software (IBM Corp. (Armonk, NY, USA), version 22) to estimate the statistical significance of the mean differences between the salinity treatments. The data were transformed to obtain homogeneity of variances, as necessary, and p-values < 0.05 were considered significant.

4. Results and Discussion

4.1. Changes in Soil Parameters

Figure 1 shows the means for soil electrical conductivity (a) and organic matter content (b) as a function of soil depth, the irrigation system used, and salinity treatments provided at the flowering and harvest stages. Similar organic matter content was reported for the different salinity treatments under both irrigation system treatments. The soil salinity increased considerably at the beginning of the experiment and then decreased, particularly at 5 cm soil depth.

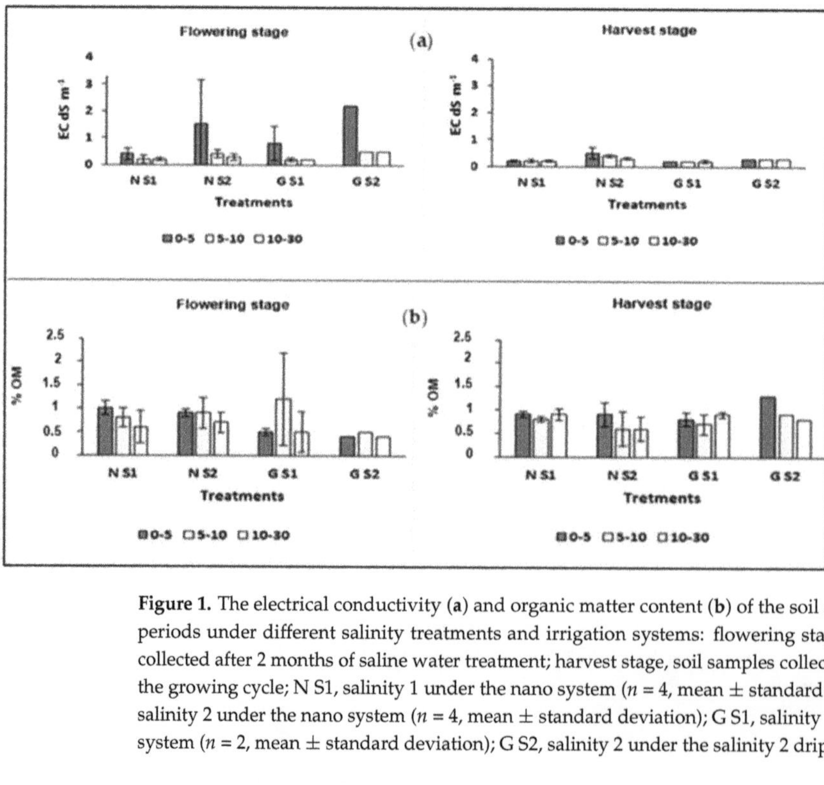

Figure 1. The electrical conductivity (**a**) and organic matter content (**b**) of the soil at two sampling periods under different salinity treatments and irrigation systems: flowering stage, soil samples collected after 2 months of saline water treatment; harvest stage, soil samples collected at the end of the growing cycle; N S1, salinity 1 under the nano system ($n = 4$, mean ± standard deviation); NS2, salinity 2 under the nano system ($n = 4$, mean ± standard deviation); G S1, salinity 1 under the drip system ($n = 2$, mean ± standard deviation); G S2, salinity 2 under the salinity 2 drip system ($n = 1$).

Organic matter and pH remained less influenced. The results of the current study are similar to those of other studies [6,7]. The studied soil has a buffering capacity; thus, the monitoring time is short, in part, to reveal differences in these [8]. Conversely, salt concentration is thought to be the result of high evaporation combined with water irrigation that is insufficient to allow the leaching of these salts into the depth of the soil.

4.2. Effect of Salinity on the Physiological Morphological and Productivity Parameters of the Plant

Figure 2 shows changes in proline content (a), photosynthetic pigment (b), leaf area (c), and root yield (d) in sugar beets at different development stages and as a function of salinity and the applied irrigation systems. The results show that irrigation with saline water resulted in the accumulation of proline in the leaves, and a decrease in chlorophyll content, leaf area, and root yield.

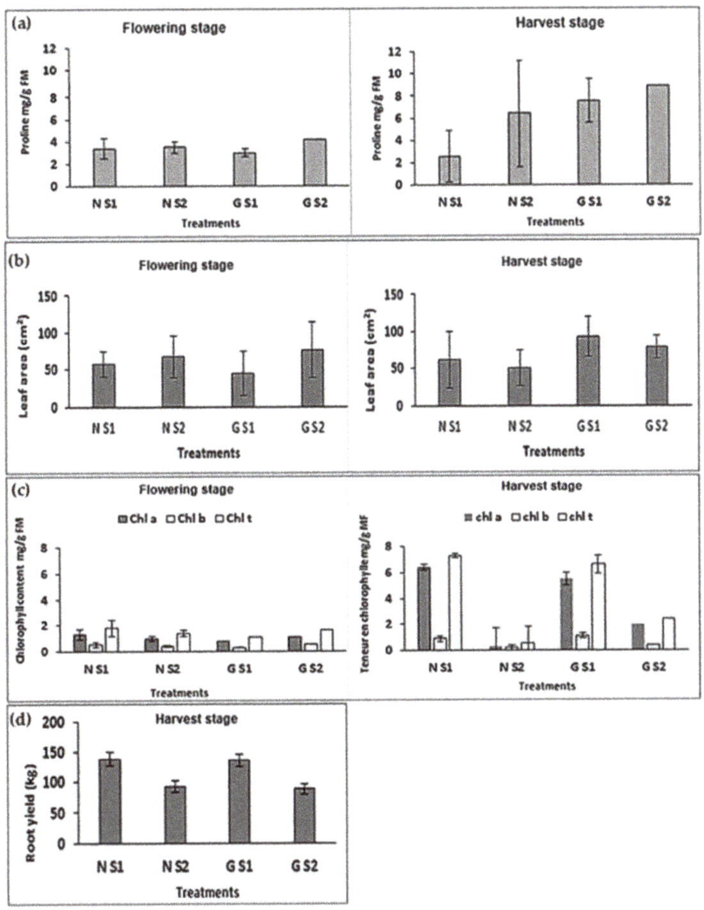

Figure 2. The average proline production (**a**), photosynthetic pigment (**b**), leaf area (**c**), and root yield (**d**) of sugar beets at two sampling periods under different salinity treatments and irrigation systems: flowering stage, leaf samples collected after 2 months of saline water treatment; harvest stage, leaf samples collected at the end of the crop cycle; N S1, salinity 1 under the nano system (n = 4, mean ± standard deviation); NS2, salinity 2 under the nano system (n = 4, mean ± standard deviation); G S1, salinity 1 under the drip system (n = 2, mean ± standard deviation); G S2, salinity 2 under the salinity 2 drip system (n = 1).

Irrigation water salinity showed significant proline accumulation, reflecting the effect of stress compared to control. Our results are consistent with previous studies on the response of plants to salt stress, which reported an increase in proline under salt stress conditions [9,10]. Similar effects would have resulted in a reduction in sugar beet growth in terms of root yield and leaf area. Our results also showed that sugar beets can grow well under the Nano Irrigation system without any significant compression effect on the buried tubes during root enlargement in the growth stages.

5. Conclusions

The results suggest that the Nano Irrigation system would be suitable for the cultivation of sugar beets, provided that the appropriate placement of the laterals—between the crop rows and at an appropriate depth—is chosen. Acceptable yields of root biomass can be obtained using irrigation water of 6 ds m^{-1} without having a significant impact on cultivated soils in the short term.

Author Contributions: K.M., writing—original draft preparation; K.M. and A.A., methodology; J.H., supervision, review and editing; K.A., B.B. and M.B., review. All authors have read and agreed to the published version of the manuscript.

Funding: This research received no external funding.

Institutional Review Board Statement: Not applicable.

Informed Consent Statement: Not applicable.

Data Availability Statement: Not applicable.

Conflicts of Interest: The authors declare no conflict of interest.

References

1. Khattak, R.A.; Jarrell, W.M. Effect of Saline Irrigation Waters on Soil Manganese Leaching and Bioavailability to Sugar Beet. *Soil Sci. Soc. Am. J. USA* **1989**, *53*, 142–146. [CrossRef]
2. Gzik, A. Accumulation of Proline and Pattern of α-Amino Acids in Sugar Beet Plants in Response to Osmotic, Water and Salt Stress. *Environ. Exp. Bot.* **1996**, *36*, 29–38. [CrossRef]
3. Moreno, F.; Cabrera, F.; Fernández-Boy, E.; Girón, I.F.; Fernández, J.E.; Bellido, B. Irrigation with Saline Water in the Reclaimed Marsh Soils of South-West Spain: Impact on Soil Properties and Cotton and Sugar Beet Crops. *Agric. Water Manag.* **2001**, *48*, 133–150. [CrossRef]
4. Dirwai, T.L.; Mabhaudhi, T.; Kanda, E.K.; Senzanje, A. Moistube Irrigation Technology Development, Adoption and Future Prospects: A Systematic Scoping Review. *Heliyon* **2021**, *7*, e06213. [CrossRef] [PubMed]
5. Hallam, J.; Brouziyne, Y.; Labbaci, A.; Lahlali, M.; Ait Aabd, N.; Elame, F. Assessing Water-Saving and Alternative Crops Performance under Nano Irrigation in Morocco (Case of Quinoa and Blue Panicum Crops under Moistube). *AFRIMED Agric. J. Al Awamia* **2021**, *133*, 116–138, *in press*.
6. Wang, R.; Kang, Y.; Wan, S.; Hu, W.; Liu, S.; Liu, S. Salt Distribution and the Growth of Cotton under Different Drip Irrigation Regimes in a Saline Area. *Agric. Water Manag.* **2011**, *100*, 58–69. [CrossRef]
7. Zhang, J.G.; Xu, X.W.; Lei, J.Q.; Li, S.Y.; Hill, R.L.; Zhao, Y. The Effects of Soil Salt Crusts on Soil Evaporation and Chemical Changes in Different Ages of Taklimakan Desert Shelterbelts. *J. Soil Sci. Plant Nutr.* **2013**, *13*, 1019–1028. [CrossRef]
8. Wang, X.; Tang, C.; Mahony, S.; Baldock, J.A.; Butterly, C.R. Factors Affecting the Measurement of Soil PH Buffer Capacity: Approaches to Optimize the Methods: Method to Measure Soil PH Buffer Capacity. *Eur. J. Soil Sci.* **2015**, *66*, 53–64. [CrossRef]
9. Abideen, Z.; Koyro, H.-W.; Huchzermeyer, B.; Ahmed, M.Z.; Gul, B.; Khan, M.A. Moderate Salinity Stimulates Growth and Photosynthesis of Phragmites Karka by Water Relations and Tissue Specific Ion Regulation. *Environ. Exp. Bot.* **2014**, *105*, 70–76. [CrossRef]
10. Ashraf, M.; Harris, P.J.C. Potential Biochemical Indicators of Salinity Tolerance in Plants. *Plant Sci.* **2004**, *166*, 3–16. [CrossRef]

environmental sciences proceedings

MDPI

Proceeding Paper

Salt Stress Effects on the Growth, Photosynthesis and Antioxidant Enzyme Activities in Maize (*Zea mays* L.) Cultivars †

Zayneb Kthiri *,‡, Mohamed Dhia Eddine Hammami ‡, Oumaima Marzougui ‡, Maissa Ben Jabeur ‡, Amal Aouadi, Chahine Karmous and Walid Hamada

Laboratory of Genetics and Cereals Breeding, National Institute of Agronomy of Tunisia, Carthage University, Tunis 1082, Tunisia; dhiahammami99@gmail.com (M.D.E.H.); marzougui.oumaima.esak@gmail.com (O.M.); maissa.benjabeur@hotmail.com (M.B.J.); amal.aouadi.esak@gmail.com (A.A.); karmouschahine@gmail.com (C.K.); w_hamada@yahoo.com (W.H.)

* Correspondence: zayneb.kthiri@gmail.com; Tel.: +216-53-556-610

† Presented at the 2nd International Laayoune Forum on Biosaline Agriculture, 14–16 June 2022; Available online: https://lafoba2.sciforum.net/.

‡ These authors contributed equally to this work.

Abstract: Salt stress is considered one of the most damaging abiotic stresses for maize productivity. In this study, two-hybrid varieties (Sancia and Agrister) were cultivated under 0, 2, 4, and 6 g NaCl L^{-1}. Under 6 g NaCl L^{-1}, shoot length, dry weight, and chlorophyll content were higher for Agrister compared to Sancia. Moreover, an increase in proline and H_2O_2 was recorded for Sancia at 2 and 4 g·L^{-1} and for Agrister at 6 g·L^{-1}. Laterally, Agrister enhances the antioxidant enzymes catalase and peroxidase, while Sancia decreased peroxidase by 25% compared to the control. In conclusion, Agrister seems to be more tolerant than Sancia.

Keywords: salt stress; maize; physiological traits; antioxidants enzymes

check for updates

Citation: Kthiri, Z.; Hammami, M.D.E.; Marzougui, O.; Jabeur, M.B.; Aouadi, A.; Karmous, C.; Hamada, W. Salt Stress Effects on the Growth, Photosynthesis and Antioxidant Enzyme Activities in Maize (*Zea mays* L.) Cultivars. *Environ. Sci. Proc.* **2022**, *16*, 73. https://doi.org/10.3390/environsciproc2022016073

Academic Editor: Abdelaziz Hirich

Published: 19 July 2022

1. Introduction

In Tunisia, around 11.6% of the total land area is affected by salinity [1]. The high levels of NaCl affect plant development by inducing osmotic stress, stomatal closure, ionic toxicity, and nutritional imbalance, leading to reduced photosynthesis and the inhibition of protein synthesis and metabolic enzymes [2,3]. The hyperosmotic and hyperionic stresses caused by the high salinity induce oxidative stress by generating reactive oxygen species (ROS) within plant cells [4]. To alleviate ROS damage, plants induce non-enzymatic (phenolics and flavonoids) and enzymatic (SOD, GPX, CAT, and APX) antioxidants [2]. After wheat and rice, maize (*Zea mays* L.) ranks third among the most important cereal crop globally. It has been reported that soil salinity is a significant threats to maize production [5]. Although maize is considered moderately salt-sensitive [6], screening tolerant cultivars should be one of the main aims to achieve sustainable agriculture. In this context, the present research is conducted to evaluate salt stress tolerance in two contrasting cultivars, Agrister and Sancia, as well as to identify the best salt-tolerant cultivar to be grown on cultivated fields with salt concerns.

2. Materials and Methods

2.1. Plant Cultivation and Processing

The seeds of two simple hybrid cultivars (Agrister and Sancia) of maize were obtained from the Limagrain agricultural seed Corporation (Verneuil, France). The seeds were sown in a total of 40. The pots contained a mixture of standard peat/perlite (2:1, *v*/*v*) and were grown in a greenhouse with a relative humidity of 60–80% and a photoperiod of 14 h/day. A complete random design with five replicates was used. Plants were subjected to three different levels of salt stress (2, 4, and 6 NaCl $g \cdot L^{-1}$) by uniformly irrigating with 0.5 L of each salt solution every 2 days; control plants were irrigated with 0.5 L of fresh water.

2.2. Measurements and Sampling

The measurements of physiological traits occurred after 2 weeks of salt stress imposition in the recent fully expanded leaves. The total chlorophyll content was measured as described by [7] and expressed as $g \cdot mL^{-1}$. The relative membrane permeability (RMP) was determined, based on electrolyte leakage of cells, as described by [8] and expressed in %. The content of total phenolic compounds was measured by the Folin–Ciocalteu method [9] and expressed as $mg \cdot g^{-1}$ FW. The content of hydrogen peroxide (H_2O_2) was measured according to [10] and expressed as mM. The guaiacol peroxidase was determined according to [11] and expressed as $mmol \cdot min^{-1} \cdot mg^{-1}$ P. The catalase activity was determined according to [12] and was expressed as $mmol.min^{-1} \cdot mg^{-1}$ P. The content of proline was measured according to [13] and expressed as $mg \cdot g^{-1}$ FW. After 3 weeks of salt stress imposition, shoots were sampled for length measurements, and then were dried at 70 °C for dry weight measurements.

2.3. Statistical Analyses

The effects of salinity and cultivar and their interaction on the measured traits were determined through a two-factor (salinity × cultivar) analysis of variance (ANOVA) with RStudio.

3. Results

3.1. Effect of Salt Stress on Physiological and Morphological Traits

The results show that the salinity significantly affected all measured parameters, such as shoot length, biomass and RMP ($p < 0.001$), and chlorophyll content ($p < 0.05$); however, the interaction (S × V) affected only the chlorophyll content and RMP ($p < 0.001$) (Table 1). The 6 g NaCl L^{-1} induces, in Sancia, a reduction in dry matter, shoot length, and chlorophyll content, and an increase in RMP. Unlikely, we recorded in Agrister a slight decrease in shoot length and an increase in chlorophyll content associated with a maintained dry matter.

Table 1. Physiological changes in the maize cultivars Agrister and Sancia under salt stress; analysis of shoot length, biomass, chlorophyll content, and relative membrane permeability.

Traits		Shoot Length (cm)		Shoot Dry Weight (g)		Chlorophyll Content (g/mL)		RMP (%)	
Cultivar		Agrister	Sancia	Agrister	Sancia	Agrister	Sancia	Agrister	Sancia
Salinity	$0\ g \cdot L^{-1}$	34.66 ± 0.24 a	33.58 ± 0.42 a	13.56 ± 0.98 a	10.96 ± 0.61 a	39.68 ± 0.45 b	39.44 ± 0.75 a	86.82 ± 0.89 d	83.84 ± 0.32 d
	$2\ g \cdot L^{-1}$	30.83 ± 0.47 b	28.67 ± 0.62 b	10.34 ± 0.13 b	9.13 ± 0.08 b	41.45 ± 0.03 ab	40.33 ± 0.58 a	88.51 ± 0.19 c	92.74 ± 1.18 c
	$4\ g \cdot L^{-1}$	29.83 ± 0.94 b	27.10 ± 0.29 c	10.00 ± 0.33 b	8.77 ± 0.26 b	41.67 ± 1.55 a	37.46 ± 0.38 b	94.38 ± 0.73 b	96.56 ± 0.65 b
	$6\ g \cdot L^{-1}$	28.08 ± 0.51 c	26.00 ± 0.49 d	9.18 ± 0.20 b	7.69 ± 0.13 c	42.29 ± 0.17 a	36.46 ± 0.74 b	97.14 ± 0.18 a	98.90 ± 0.55 a
ANOVA	df								
Salinity	1	168,515 ***		47,962 ***		8817 *		564,379 ***	
Cultivar	3	24,402 ***		15,999 ***		48,628 ***		10,055 **	
S × C	3	2118 n.s.		1951 n.s.		30,741 ***		41,812 ***	

The symbols indicate statistical significance (n.s., $p > 0.1$, ** $p < 0.01$; *** $p < 0.001$). The sum square values with statistical significance are shown (n.s.: non-significant, *: $p < 0.05$; **: $p < 0.01$; ***: $p < 0.001$).

3.2. Effect of Salt Stress on the Antioxidants Enzymes, H_2O_2, Proline, and Phenolic Compounds

The analysis of variance showed that all the traits were significantly affected by salinity, cultivar, and their interaction ($p < 0.001$). The 6 g NaCl L^{-1} stress induces a peak of H_2O_2 Sancia, but is maintained for Agrister. In addition, Agrister exhibited a higher phenolic content, peroxidases, and catalase activities than those of Sancia. Proline was high in Sancia at 2 and 4 $g \cdot L^{-1}$, while it was higher in Agrister at 6 $g \cdot L^{-1}$ (Figure 1).

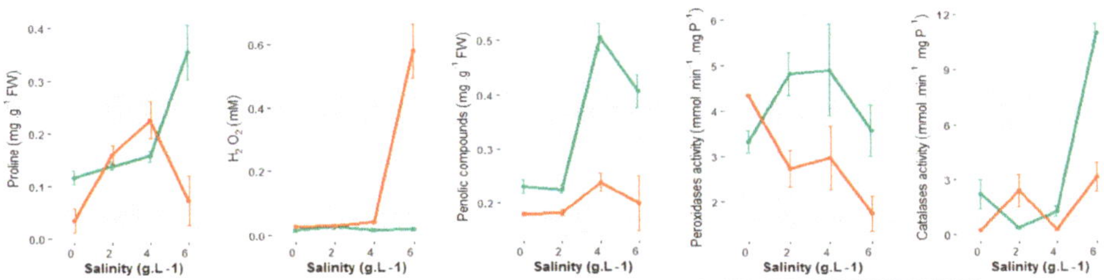

ANOVA	df	H_2O_2	Phenolic compounds	Catalase activity	Peroxidase activity	Proline
Salinity	1	0.33***	0.13***	159.66***	6.19***	0.07***
cultivar	3	0.133***	0.111***	28.711***	8.131***	0.027***
S×C	3	0.34***	0.05***	77.43***	9.68***	0.11***

Figure 1. The effect of salt stress (0, 2, 4, and 6 g·NaCl L^{-1}) on proline, H_2O_2, phenolic compounds, and peroxidase and catalase activities. The sum square values with statistical significance are shown (ns: non-significant, ***: $p < 0.001$).

4. Discussion

The applied salt stress affected Sancia development significantly compared to that of Agrister. This could be related to the accumulation of Na^+ in the roots, which reduces water absorption [14]. The reduction in total chlorophyll content is to be expected under stress conditions; its stability depends on membrane stability, which under saline conditions rarely remains intact [6]. Unlike Sancia, chlorophyll content and RMP were maintained in Agrister under the increasing salt stress, which shed light on its tolerance. Additionally, the higher level of proline in Agrister underlines the aptitude of the cultivar to minimize the salt effect by osmoregulation compared to Sancia. This is partially consistent with other studies on maize [3]. In response to salt stress, Agrister is able to mitigate the accumulation of H_2O_2 in plant cells by the biosynthesis of antioxidant metabolites (catalase, peroxidase, and phenolic compounds). Our results agree with those of [1,15], who suggested that CAT, peroxidases, and phenolic compounds were associated with salt stress tolerance. In conclusion, the hybrid Agrister was found to be more tolerant to salt stress than Sancia as indicated by the response of plant growth, chlorophyll content, and phenolic accumulation and their related antioxidant activity in the presence of NaCl.

Author Contributions: Z.K., O.M., M.B.J. and M.D.E.H. contributed equally to this work. Conceptualization, C.K.; methodology, Z.K. and A.A.; software, M.D.E.H.; data curation, O.M., writing—original draft preparation, Z.K. and M.B.J.; writing—review and editing, C.K., W.H. and M.B.J. All authors have read and agreed to the published version of the manuscript.

Funding: RA-LEARN has received funding from the European Union's Horizon 2020 research and innovation programme under grant agreement No. 811171.

Institutional Review Board Statement: Not applicable.

Informed Consent Statement: Not applicable.

Data Availability Statement: Data are available in a publicly accessible repository.

Acknowledgments: We thank the "Agroservices" company that provided us with the two hybrid cultivars.

Conflicts of Interest: The authors declare no conflict of interest.

References

1. Ben Khaled, A.; Hayek, T.; Mansour, E.; Hannachi, H.; Lachiheb, B.; Ferchichi, A. Evaluating salt tolerance of 14 barley accessions from southern Tunisia using multiple parameters. *J. Agric. Sci.* **2012**, *4*, 27–38. [CrossRef]
2. AbdElgawad, H.; Zinta, G.; Hegab, M.M.; Pandey, R.; Asard, H.; Abuelsoud, W. High salinity induces different oxidative stress and antioxidant responses in maize seedlings organs. *Front. Plant Sci.* **2016**, *7*, 276. [CrossRef] [PubMed]
3. Gao, Y.; Lu, Y.; Wu, M.; Liang, E.; Li, Y.; Zhang, D.; Yin, Z.; Ren, X.; Dai, Y.; Deng, D.; et al. Ability to Remove Na^+ and Retain K^+ Correlates with Salt Tolerance in Two Maize Inbred Lines Seedlings. *Front. Plant Sci.* **2016**, *7*, 1716. [CrossRef] [PubMed]

4. Ditta, A. Salt tolerance in cereals: Molecular mechanisms and applications. In *Molecular Stress Phsyiology of Plants*; Rout, G., Das, A., Eds.; Springer: New Delhi, India, 2013; pp. 133–154.
5. Huqe, M.A.S.; Haque, M.S.; Sagar, A.; Uddin, M.N.; Hossain, M.A.; Hossain, A.Z.; Rahman, M.M.; Wang, X.; Al-Ashkar, I.; Ueda, A.; et al. Characterization of Maize Hybrids (*Zea mays* L.) for Detecting Salt Tolerance Based on Morpho-Physiological Characteristics, Ion Accumulation and Genetic Variability at Early Vegetative Stage. *Plants* **2011**, *10*, 2549. [CrossRef] [PubMed]
6. Carpici, E.B.; Celik, N.; Bayram, G. The effects of salt stress on the growth, biochemical parameter and mineral element content of some maize (*Zea mays* L.) cultivars. *Afr. J. Biotechnol.* **2010**, *9*, 6937–6942.
7. Arnon, D.I. Copper enzymes in isolated chloroplasts polyphenol oxidase in Beta vulgaris. *Plant Physiol.* **1949**, *24*, 1–15. [CrossRef]
8. Filek, M.; Walas, S.; Mrowiec, H.; Rudolphy-Skórska, E.; Sieprawska, A.; Biesaga-Kościelniak, J. Membrane permeability and micro-and macroelement accumulation in spring wheat cultivars during the short-term effect of salinity-and PEG-induced water stress. *Acta Physiol. Plant.* **2012**, *34*, 985–995. [CrossRef]
9. Singleton, V.L.; Orthofer, R.; Lamuela-Raventós, R.M. Analysis of total phenols and other oxidation substrates and antioxidants by means of Folin-Ciocalteu reagent. In *Methods in Enzymology;* Academic Press: Cambridge, MA, USA, 1999; Volume 299, pp. 152–178.
10. Noreen, Z.; Ashraf, M. Assessment of variation in antioxidative defense system in salt treated pea (*Pisum sativum* L.) cultivars and its putative use as salinity tolerance markers. *J. Plant Physiol.* **2009**, *166*, 1764–1774. [CrossRef]
11. Egley, G.H.; Paul, R.N.; Vaughn, K.C.; Duke, S.O. Role of peroxidase in the development of water-impermeable seed coats in *Sida spinosa* L. *Planta* **1983**, *157*, 224–232. [CrossRef] [PubMed]
12. Beers, R.F.; Sizer, I.W. A spectrophotometric method for measuring the breakdown of hydrogen peroxide by catalase. *J. Biol. Chem.* **1952**, *195*, 133–140. [CrossRef]
13. Bates, L.S.; Waldren, R.P.; Teare, I.D. Rapid determination of free proline for water-stress studies. *Plant Soil* **1973**, *39*, 205–207. [CrossRef]
14. Parida, A.K.; Das, A.B. Salt tolerance and salinity effects on plants: A review. *Ecotoxicol. Environ. Saf.* **2005**, *60*, 324–349. [CrossRef] [PubMed]
15. Hichem, H.; Mounir, D. Differential responses of two maize (*Zea mays* L.) varieties to salt stress: Changes on polyphenols composition of foliage and oxidative damages. *Ind. Crops Prod.* **2009**, *30*, 144–151. [CrossRef]

environmental sciences proceedings

MDPI

Proceeding Paper

Growth of "Biquinho" Pepper Plants under Salt Stress in a Hydroponic System †

Alide Mitsue Watanabe Cova [1,*], Maria Augusta Amorim Bione [1], Vital Pedro da Silva Paz [2], Ubirani Oliveira Santos [2], Maria Raphaela Severo Rafael [1], Francisco José Nunes Modesto [1], Hans Raj Gheyi [2] and Tales Miler Soares [2]

1 Post Graduate Program in Agricultural Engineering, Federal University of Recôncavo da Bahia, Cruz das Almas 44380-000, Bahia, Brazil; gugabione@gmail.com (M.A.A.B.); mariaph1201@gmail.com (M.R.S.R.); fjmodesto@gmail.com (F.J.N.M.)

2 Center of Agricultural, Environmental and Biological Sciences, Federal University of Recôncavo da Bahia, Cruz das Almas 44380-000, Bahia, Brazil; vpspaz@gmail.com (V.P.d.S.P.); ubiraniufrb@gmail.com (U.O.S.); hgheyi@gmail.com (H.R.G.); talesmiler@gmail.com (T.M.S.)

* Correspondence: alidewatanabe@yahoo.com.br; Tel.: +55-75-99933-1088

† Presented at the 2nd International Laayoune Forum on Biosaline Agriculture, 14–16 June 2022, Available online: https://lafoba2.sciforum.net/.

Abstract: In the absence of matric potential, the hydroponic cultivation method of employing brackish water to prepare a nutrient solution permits satisfactory growth, even of non-leafy vegetables. The present study evaluated the growth variables, stem diameter, plant height, number of leaves, and the shoot dry mass in the different phenological stages of "Biquinho" pepper in response to the electrical conductivity of the nutrient solutions (ECsol), varying from 2.70 to 7.77 dS m^{-1} in a Nutrient Film Technique (NFT) hydroponic system. For the number of leaves, linear reductions of 3.89 and 9.29% (dS $m^{-1})^{-1}$ were observed at 10 and 30 days after transplanting (DAT), respectively. At 60 DAT, plants began to preserve their leaves up to ECsol of 4.60 dS m^{-1} and presented a linear decrease of 23.32% per unit increment above the salinity threshold. At 10 DAT, the plant height and stem diameter were not affected and only at 30 DAT were there significant differences due to ECsol, with a linear reduction of 3.98 and 5.27% (dS $m^{-1})^{-1}$, respectively. While at 60 DAT, the salinity response for these variables was represented by a plateau followed by a linear decrease, with salinity thresholds of 5.18 and 5.01 dS m^{-1}, and thereafter a relative decrease of 9.31 and 11.9% per unit increase in ECsol above the threshold, respectively, the values up to the salinity threshold being 0.87 m and 15.77 mm, respectively. The shoot dry mass up to 60 DAT reduced linearly with an increase in ECsol, but the plants under moderate salinity after acclimatization (90 DAT) surpassed the control treatment, and maximum dry mass accumulation was observed at an ECsol of 5.14 dS m^{-1}. The results reveal that in hydroponic cultivation of "Biquinho" pepper under salt stress, growth is attenuated in the reproductive phase.

Keywords: *Capsicum chinense* Jacq.; salt stress; phenological stage

check for updates

Citation: Cova, A.M.W.; Bione, M.A.A.; Paz, V.P.d.S.; Santos, U.O.; Rafael, M.R.S.; Modesto, F.J.N.; Gheyi, H.R.; Soares, T.M. Growth of "Biquinho" Pepper Plants under Salt Stress in a Hydroponic System. *Environ. Sci. Proc.* **2022**, *16*, 74. https://doi.org/10.3390/environsciproc2022016074

Academic Editor: Abdelaziz Hirich

Published: 29 July 2022

1. Introduction

Appreciated throughout the world since colonial times, peppers of the genus Capsicum are present in everyday life, being used fresh in the preparation of meals, the composition of salads and preserves, dry as a condiment/seasoning, and processed in the preparation of jams, sauces, and antipasti [1]. Among the cultivated species, *C. chinense* Jacq. is explored in several Brazilian regions [2]. The species *C. chinense* has high genetic variability for agronomic characteristics and capsaicin content, which is responsible for the pungency of peppers [3]. Among the genotypes, "Biquinho" is known for the length of the fruit apex, being one of the most consumed and cultivated in Brazil [3]. However, this crop can be affected by several abiotic stresses, including salinity [4].

Several studies have shown good results with the use of brackish water up to a certain level of salinity in the preparation of nutrient solution in a hydroponic system, including non-leafy species such as those of the genus *Capsicum* [5,6], *Abelmoschus esculentus* [7]. Therefore, the objective of the present study was to evaluate the growth of "Biquinho" pepper cultivated in the hydroponic system using brackish waters in the preparation of the nutrient solution.

2. Material and Methods

The experiment was conducted in a single-arch greenhouse, located at the Federal University of Recôncavo da Bahia, in the municipality of Cruz das Almas, Bahia, Brazil (12°40′19″ S, 39°06′23″ W, 220 m). The greenhouse is 7 m wide and 33 m long, with a ceiling height of 4 m, installed in the East–West direction, being protected on the sides by a black net (50% shade) and on the top by a 150-micron anti-UV plastic film. During the experiment, the minimum, maximum, and mean values of air temperature and relative humidity of the air were, respectively: 17.20, 35.70, and 23.88 °C; 66.46, 93.29, and 78.79%.

The experimental design adopted was randomized blocks, with seven levels of electrical conductivity of the nutrient solution (ECsol-2.70 (control), 3.64, 4.58, 5.28, 6.09, 6.90, and 7.77 dS m^{-1},), and 6 replicates. The nutrient solution (NS) adopted was recommended by Sonneveld and Straver [8]. The NS was prepared using municipal-supply water (ECw = 0.34 dS m^{-1}), and ECsol greater than 2.7 dS m^{-1} was obtained by dissolving NaCl. The hydroponic system used was the Nutrient Film Technique (NFT), with 42 experimental plots. Each plot was composed of five plants arranged in a PVC channel (0.075 m diameter and 6.0 m length). Each cultivation channel was independent, in respect of its hydraulic system, NS reservoir, and supply water tank. The spacing used was 0.83 m between plants and 0.70 m between rows. The NS circulated for 15 min at intervals of 15 min.

The plant material used in this study was pepper (*Capsicum chinense* Jacq.) of the varietal group "Biquinho", without pungency, using seeds of the Horticeres, sown in previously washed cubical cells of phenolic foam (0.02 × 0.02 × 0.02 m). Transplanting to the hydroponic system occurred 43 days after sowing (DAS). At 20 days after transplanting (DAT), a 50% shade net was installed at 3.5 m height to reduce radiation and avoid thermal stress in plants. The experiment was completed at 163 DAS, i.e., 120 DAT. The replenishment of the water consumed was carried out daily with supply water (ECw = 0.34 dS m^{-1}). The NS was substituted completely in all treatments at 28, 61, and 90 DAT.

The growth of pepper plants was evaluated by stem diameter (SD), plant height (PH), and the number of leaves (NL) 10, 30, and 60 DAT. All evaluations were performed in two pre-identified plants in the central part of each experimental plot. The shoot dry mass (SHDM) was evaluated 30, 60, 90, and 120 DAT after drying at 65 °C.

The data were subjected to analysis of variance (ANOVA) using the statistical SAS Program. When the effect by the F test was significant, the factor ECsol was subjected to linear and quadratic regression analysis to obtain the most adequate model representing the data. The models of Maas and Hoffman [9] and Plateau followed by exponential decay were also used. The Solver Microsoft Excel tool was used to analyze the parameters, aiming to minimize the sum of square deviations.

3. Results

For the NL, linear reductions of 3.89 and 9.29% (dS $m^{-1})^{-1}$ were observed at 10 and 30 DAT, respectively (Figure 1A,B). At 60 DAT, plants began to preserve their leaves up to ECsol of 4.60 dS m^{-1}, with a mean number of 1034.27 leaves per plant and a linear decrease of 23.32% per unit increment of ECsol above the salinity threshold (Figure 1C).

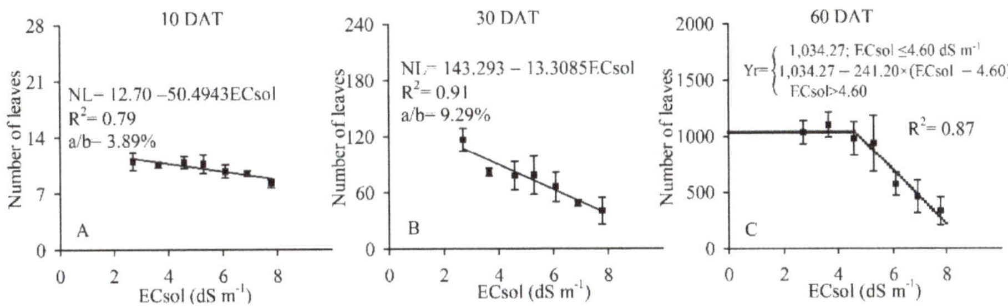

Figure 1. Mean number of leaves (NL) per plant at 10 (**A**), 30 (**B**), and 60 (**C**) days after transplanting (DAT) of "Biquinho" pepper cultivated in the hydroponic system as a function of the electrical conductivity of the nutrient solution salinity (ECsol).

At 10 DAT, mean PH and stem diameter were, respectively, 0.195 m and 7.85 mm for all treatments, and only at 30 DAT were there significant differences due to ECsol, with a linear reduction of 3.98 and 5.27% (dS m^{-1})$^{-1}$, respectively (Figure 2A,C). At 60 DAT, the variables PH and SD showed a plateau followed by a linear decrease, with salinity thresholds of 5.18 and 5.01 dS m^{-1}, and a relative decrease of 9.31 and 11.9% per unit increase in ECsol above the threshold, respectively. The values of these variables up to the salinity threshold were 0.87 m and 15.77 mm, respectively (Figure 2B,D).

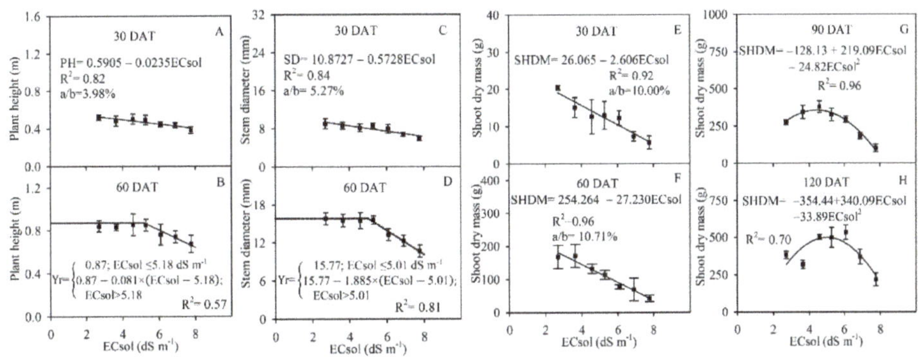

Figure 2. Mean plant height (**A**,**B**) and stem diameter (**C**,**D**) at 30 and 60 days after transplanting (DAT), respectively, and shoot dry mass per plant at 30 (**E**), 60 (**F**), 90 (**G**), and 120 DAT (**H**) of "Biquinho" pepper cultivated in the hydroponic system as a function of the electrical conductivity of the nutrient solution (ECsol).

The shoot dry mass (SHDM) accumulation of "Biquinho" pepper at 30 and 60 DAT decreased linearly by 10.00 and 10.71% per unit increase in ECsol (Figure 2E,F), respectively. Subsequently, at 90 and 120 DAT, the effect of ECsol on SHDM was represented by a quadratic equation, with maximum values no longer observed in the to control treatment (ECsol = 2.70 dS m^{-1}), but to intermediate salinity treatments: ECsol of 4.42 and 5.02 dS m^{-1}, with SHDM estimated at 355.12 and 498.67 g plant^{-1}, respectively (Figure 2G,H). Based on the equation of Figure 2H, SHDM values of the same magnitude (316.72 and 315.76 g plant^{-1}) were verified between the control treatment (ECsol = 2.70 dS m^{-1}) and ECsol of 7.34 dS m^{-1} at 120 DAT.

4. Discussion

The increase in salts in the NS reduced the growth variables NL, PH, SD, and SHDM. In the literature, there are several studies with different crops, including bell pepper, in the hydroponic system under salt stress, with a reduction in these growth variables [5,6,10] due to the reduction in water absorption, accumulation of toxic ions, and physiological effects, such as reduction in photosynthesis and stomatal conductance [6,10].

For basil crops under similar treatments and experimental conditions, the PH was not considered a good indicator of salinity effects compared to dry mass accumulation [11]. For basil and pepper, both shrub plants, the reduction in height tends to be a little intense; after a certain time, the plants start to mobilize more energy for the growth of the lateral branches, to the detriment of vertical growth, which tends to result in plants with heights that vary little, even under different levels of ECsol. These results corroborate results observed by Furtado et al. [5] in hydroponic bell pepper plants (*Capsicum annuum* L. "All Big") in which the absolute growth rate decreased from 0.096 to 0.03346 cm day^{-1} with a unit increase in salinity in the periods 60–75 and 75–90 DAT, respectively. According to Horticeres Sementes [12], the cultivar "Biquinho" used in the present study produces very uniform compact plants with a mean height of 0.60 m. Therefore, the plants of the present study grown under hydroponic conditions, even with high-salinity NS (ECsol = 7.7 dS m^{-1}), reached this mean standard of height (Figure 2B).

5. Conclusions

"Biquinho" pepper plants subjected to salinity soon after transplanting had, as consequences of the osmotic effect, restricted development of areal parts and the number of leaves, especially in the first weeks. The results reveal that in hydroponic cultivation of "Biquinho" pepper under salt stress, growth is attenuated in the reproductive phase.

Author Contributions: Conceptualization, M.A.A.B., V.P.d.S.P., H.R.G. and T.M.S.; methodology, M.A.A.B., U.O.S., M.R.S.R., F.J.N.M.; writing—review and editing A.M.W.C., H.R.G. and T.M.S.; supervision, project administration, and funding acquisition, V.P.d.S.P., H.R.G. and T.M.S. All authors have read and agreed to the published version of the manuscript.

Funding: Coordination for the Improvement of Higher Education Personnel—CAPES (001), the National Council for Scientific and Technological Development—CNPq (Universal Call 14/2011; Protocol 484645/2011-0) and Bahia State Research Support Foundation—FAPESB (Universal Call 14/2011; Protocol 484645/2011-0) and Federal University of Recôncavo da Bahia—UFRB (Protocol 1913/2013, Accord CNV 0075/2013).

Institutional Review Board Statement: Not applicable.

Informed Consent Statement: Not applicable.

Data Availability Statement: Not applicable.

Acknowledgments: Authors thank the National Institute of Science and Technology in Salinity (INCTSal) for the structural support over the years. support over the years.

Conflicts of Interest: The authors declare no conflict of interest.

References

1. Sanatombi, K.; Rajkumari, S. Effect of processing on quality of pepper: A review. *Food Rev. Intern.* **2020**, *36*, 626–643. [CrossRef]
2. Ribeiro, C.S.C.; Carvalho, S.I.C.; Heinrich, A.G.; Reifschneider, F.J.B. BRS Tui: A new Biquinho-type pepper cultivar released by Embrapa. *Hort. Bras.* **2018**, *36*, 295–297. [CrossRef]
3. Martinez, M.; Santos, C.P.; Verruma-Bernardi, M.R.; Carrilho, E.N.V.M.; Silva, P.P.M.; Spoto, M.H.F.; Ciarrocchi, I.R.; Sala, F.C. Agronomic, physical–chemical and sensory evaluation of pepper hybrids (*Capsicum chinense* Jacquin). *Sci. Hortic.* **2021**, *227*, 109819. [CrossRef]
4. Sá, F.V.S.; Souto, L.S.; Paiva, E.P.; Torres, S.B.; Oliveira, F.A. Initial development and tolerance of pepper species to salinity stress. *Rev. Caatinga* **2019**, *32*, 826–833. [CrossRef]
5. Furtado, G.F.; Cavalcante, A.R.; Chaves, L.H.G.; Santos Júnior, J.A.; Gheyi, H.R. Growth and production of hydroponic pepper under salt stress and plant density. *Am. J. Plant Sci.* **2017**, *8*, 2255–2267. [CrossRef]
6. Santos, A.C.; Silva, M.G.; Boechat, C.L.; Chagas, D.S.; Mendes, W.S. Brackish water: An option for producing hydroponic *Capsicum annuum* in laminar flows of mineral nutrients. *Rev. Colomb. Ciênc. Hortic.* **2018**, *12*, 147–155. [CrossRef]
7. Modesto, F.J.N.; Santos, M.Â.C.M.; Soares, T.M.; Santos, E.P.M. Crescimento, produção e consumo hídrico do quiabeiro submetido à salinidade em condições hidropônicas. *Irriga* **2019**, *24*, 86–97. [CrossRef]

8. Sonneveld, C.; Straver, N. *Nutrient Solutions for Vegetebles and Flowers Grown in Water or Substrates*, 10th ed.; Proefstation voor Tuinbouw onder Glas Te Naaldwijk: Naaldwijk, The Netherlands, 1994; 45p.
9. Maas, E.V.; Hoffman, G.J. Crop salt tolerance-current assessment. *J. Irrig. Drain. Div.* **1977**, *10*, 115–134. [CrossRef]
10. Cavalcante, A.R.; Santos Júnior, J.A.; Furtado, G.F.; Chaves, L.H.G. Gas exchanges and photochemical efficiency of hydroponic bell pepper under salinity and plant density. *Rev. Bras. Eng. Agríc. Ambient.* **2019**, *23*, 3–8. [CrossRef]
11. Bione, M.A.A.; Paz, V.P.S.; Silva, F.; Ribas, R.F.; Soares, T.M. Crescimento e produção de manjericão em sistema hidropônico NFT sob salinidade. *Rev. Bras. Eng. Agríc. Ambient.* **2014**, *18*, 1228–1234. [CrossRef]
12. HORTICERES SEMENTES. Pimenta Biquinho. Available online: http://www.horticeres.com.br/produtos/solanaceas/pimenta/pimenta-biquinho (accessed on 20 April 2017).

environmental sciences proceedings

MDPI

Proceeding Paper

Evaluation of New Cultivars of Plum (*Prunus salicina*) under Deficit Irrigation in a Semi-Arid Zone of Tunisia †

Monia Guizani [1,2,*] and Samira Maatallah [1,2]

1 Non-Conventional Water Valuation Research Laboratory (LR VENC), National Research Institute of Rural Engineering, Water and Forests (INRGREF), University of Carthage, Hedi EL Karray Street, El Menzah IV, Tunis 1004, Tunisia; samiraham2003@yahoo.fr

2 Regional Center for Agricultural Research, Institution of Research and Higher Education Agriculture (IRESA), Sidi Bouzid 9100, Tunisia

* Correspondence: guizanimonia2018@gmail.com; Tel.: +216-28329129

† Presented at the 2nd International Laayoune Forum on Biosaline Agriculture, 14–16 June 2022; Available online: https://lafoba2.sciforum.net/.

Abstract: The lack of precipitation influences the salinity of irrigation water. The experiment was conducted in a plum orchard in the Rgueb region. Two water regimes were applied. Stressed trees received 50% crop evapotranspiration (ETc) and the controls received 100% ETc. During the experimentation period, the phenological stages, flowering period and some pomological criteria were surveyed. The different parameters followed confirmed that the Black Star cultivar, with a shorter cycle, was the most resistant to water deficits and salinity. According to this study, it will be strategic to encourage the cultivation of the Black Star cultivar in the semi-arid region of Tunisia.

Keywords: Black Star; phonologic stage; plum fruit; salinity; water deficit

check for updates

Citation: Guizani, M.; Maatallah, S. Evaluation of New Cultivars of Plum (*Prunus salicina*) under Deficit Irrigation in a Semi-Arid Zone of Tunisia. *Environ. Sci. Proc.* **2022**, *16*, 2. https://doi.org/10.3390/environsciproc2022016002

Academic Editor: Abdelaziz Hirich

Published: 16 June 2022

1. Introduction

In Tunisia, in arid and semi-arid areas (such as Regueb, Tunisia), low annual rainfall and high evaporation rates affect the production of fruit trees, which require a water supply for orchards. Moreover, the excessive exploitation of water resources for irrigation accompanied by an uncontrolled use of fertilizers increases salinity. It is generally known that moderate water stress is a method that increases a plant's water use efficiency and improves fruit quality without affecting size, weight and even yield [1]. However, previous studies in mid-maturing Japanese plum have shown that water stress in a period close to harvest reduces fruit size and vegetative vigor [2]. In this context, the aim of this work is to evaluate the acclimatization of three plum cultivars (Black Diamond, Black Gold and Black Star) introduced in the Regueb region. The impact of climate and especially water deficit and high salinity on the development, production and quality of fruits is investigated.

2. Materials and Methods

2.1. Experimental Site and Plant Material

The orchard is located 4 km north of the town of Regueb. It covers an area of 4 ha with a dispositive in crisscross. It was installed in 2006. It contains three plum cultivars: Black Diamond (BD), Black Star (BS) and Black Gold (BG). The trees are irrigated by drip irrigation (4 L/h) with salinity of around 2.4 g EC (dS/m). The farmer applies an identical fertilization program for all the trees. These cultivars are characterized by the earliness of ripening of their fruits.

2.2. Water Treatment of the Plum Tree

Two water regimes were applied. For each cultivar, the trees considered as controls received between 12–30 L/d depending on seasons and phenological stages (100% ETc).

However, the treated trees received half of the irrigation water from the controls (50% ETc); this treatment can be considered as a moderate water stress (MWS) during the months of May–June–July.

2.3. Phenological Stages of Plum Trees

To define the phenological stage, we referred to the work published by [3] focusing on plum tree phenology. Each phenological stage was designated by a letter. The phenological stages of each cultivar were followed from budburst to fruiting.

2.4. Pomological Parameters of Plum Fruits

The fruits were harvested at the commercial ripening stage (full maturation stage). The extraction of sugars was performed according to the method described by [4] with some modifications. The carbohydrate content was determined according to [5]. Total acidity (TA) was determined according to [6]. The concentration of total phenols was measured by the method of [7]. Total flavonoids were measured according to [8].

2.5. Statistical Analysis

Statistical analysis (ANOVA and Duncan's test) was performed using SPSS statistical software (version 13 for Windows). For the comparison of the pomological parameters, the variables were subjected to a multivariate analysis using a general linear model including two fixed factors (cultivars and water treatment).

3. Results

3.1. Phenological Stages

The beginning of budburst takes place in February for BD and BG and March for BS (Table 1). The end of flowering and the appearance of fruits take place in May for the three varieties. The duration of the reproductive cycle from the bud swelling stage (B) to the young fruit stage (J) differed between cultivars. It was 68 days for the BD and BG cultivars; however, BS had the shortest development cycle (59 days). An intra-cyclic variability exists between the durations of the stages. All cultivars reach the fall of petals stage (I) together.

Table 1. Phenological stages of three plum cultivars grown in the Regueb region.

Varieties	Bud Swelling	First Bloom	Full Bloom	Post Bloom	Fruit Set	Young Fruit
Black Diamond	02/03	15/03	21/03	30/03	05/04	08/05
Black Gold	02/03	20/03	24/03	05/04	18/04	08/05
Black Star	13/03	20/03	30/03	10/04	18/04	08/05

3.2. Effect of Water Stress in the Pomological Parameters of Plum Fruits

The result of the variation of total sugar in the three plum cultivars studied under deficit irrigation strategy is shown in Table 2. We found that the level of total sugar contents in the plum fruit varied between 26.23 and 40.63 g/100 g FW. In addition, our results show that the sugar content in the stressed plum cultivars was higher than that in fruits harvested from the control trees BD and BS (37.90 and 40.63 g/100 g FW, respectively). However, deficit water irrigation did not affect the concentration of sugar in the fruits of Black Gold (Table 2).

The results shown in Table 2 show that, at the ripening stage and under full irrigation conditions, BS had the lowest level of acidity (3.65% malic acid), while BD fruits were the most acidic (3.96 % malic acid). In addition, MWS has a negative impact on the total acidity in all cultivars (Table 2). Our results show that the level of polyphenols in plum depends on the cultivars. Indeed, BG plums contain the highest levels (369.84 mg/100 g FW), followed by BS (263.52 mg/100 g FW). The lack of water leads to a slight improvement in the rate of these compounds in both BD and BS fruits. Furthermore, the results show a significant variation between the three cultivars concerning the level of flavonoids (Table 2). In fact, under control and deficit irrigation, BS plums were the richest in flavonoids with a concentration that reached 26 mg/100 g FW. The fruits of BD had the lowest concentration of flavonoids (did not exceed 16 mg/100 g FW).

Table 2. Variation of pomological parameters in the fruits of the three plum cultivars subjected (stressed) or not (control) to moderate water stress.

Varieties	Total Sugar (g/100 g FW)		Total Acidity (% Malic Acid)		Total Polyphenols (mg/100 g FW)		Flavonoids (mg/100 g FW)	
	Control	Stressed	Control	Stressed	Control	Stressed	Control	Stressed
Black Diamond	27.66 [cB]	37.90 [bA]	3.96 [aA]	3.75 [aB]	190.82 [cB]	219.96 [cA]	14.89 [cB]	16.01 [cA]
Black Gold	26.23 [bA]	27.10 [cA]	3.70 [bA]	3.54 [bB]	369.84 [aB]	375.57 [aA]	20.64 [bA]	21.71 [bA]
Black Star	30.53 [aB]	40.63 [aA]	3.65 [cA]	3.49 [cB]	263.52 [bB]	298.20 [bA]	26.23 [aA]	26.66 [aA]

Values are the means of three different samples (n = 3) ± standard deviation. The lowercase letters [a], [b] and [c] indicate significant differences ($p \leq 0.05$) among the three cultivars for each treatment separately. Capital letters [A] and [B] indicate significant differences ($p \leq 0.05$) for irrigation treatments.

4. Discussion

Phenology is one of the indicators used and accepted by many scientists to monitor climate change; it is an ideal manner to demonstrate the effects of global warming on the living world [9]. For all the cultivars studies, the flowering period extended from mid-March to the second week of April. However, in a previous study carried out on other plum cultivars grown in Romania, the flowering occurred from the end of March to the beginning of May [10]. In addition, the analysis of phenological data showed that the duration of flowering varied according to the variety, a characteristic that is influenced by climatic and genetic factors. In fact, the BS cultivar showed the longest flowering period (21 days). Moreover, according to [10], the longest flowering period is a positive characteristic of a cultivar's adaptation to unfavorable conditions. In addition to its impact on plant phenology, climatic conditions and irrigation strategies have a significant influence on fruit quality. In our study, the results show that the application of MWS can improve plum fruit quality. In fact, we found that water stress increased the concentration of total sugars, especially in BS fruit. According to [11], this increase is due to the overconcentration of sugars following the reduction in fruit size. Our also results show a decrease in the total acidity of plum fruit under MWS. Similar results were found for peach fruit [12]. In fact, according to [13], the decrease in total acidity can be explained by the decrease in malic and citric acid levels after their use in the processes of gluconeogenesis, fermentation and amino acids. Phenolic compounds are also major constituents of plum fruits [14]. There is a strong correlation between the richness of fruits in phenolic compounds and their antioxidant activity. Our study shows that moderate water stress increased the concentration of flavonoids and total polyphenols in plum fruits in the three cultivars studied. The same findings have been observed in other fruit species, such as the peach [15]. According to [16], the increase in these secondary metabolites is due, not only, to the reduction in fruit size under restrictive water conditions, but also to the fact that water stress stimulates the biosynthesis of phenolic compounds.

5. Conclusions

In conclusion, studies on the environmental factors' impact on phenology in fruit-tree species allow us to make decisions on suitable assortments for different culture zones, depending on local ecological conditions. According to the results obtained in this study, we found that the Black Star cultivar had the shortest phenological cycle, the longest flowering period and its fruit was improved by moderate water stress (high sugar and phenolic compound concentrations). Therefore, this cultivar may be appropriate for growing in the Regueb region (with a semi-arid climate).

Author Contributions: M.G. and S.M. wrote the first draft of the manuscript and revised it. All authors contributed to the analysis and discussion of the results. All authors have read and agreed to the published version of the manuscript.

Funding: This research was financially supported by the Tunisian Ministry of Higher Education and Scientific Research (Laboratoire LR VENC).

Institutional Review Board Statement: Not applicable.

Informed Consent Statement: Not applicable.

Data Availability Statement: Not applicable.

Acknowledgments: The authors thank all laboratory members of CRRA-Sidi Bouzid.

Conflicts of Interest: The authors declare that they have no conflict of interest.

References

1. Intrigliolo, D.S.; Castel, J.R. Effects of Irrigation on the Performance of Grapevine cv. Tempranillo in Requena, Spain. *Am. J. Enol. Vitic.* **2008**, *59*, 30–38.
2. Guizani, M.; Dabbou, S.; Maatallah, S.; Montevecchi, G.; Hajlaoui, H.; Rezig, M.; Helal, A.N.; Kilani-Jaziri, S. Physiological responses and fruit quality of four peach cultivars under sustained and cyclic deficit irrigation in center-west of Tunisia. *Agric. Water Manag.* **2019**, *217*, 81–97. [CrossRef]
3. Baggiolini, M. Stades Repères du Cerisier—Stades Repères du Prunier. Stades Repères de L'abricotier. Stades Repères du Pêcher. In *Guide Pratique de Défense des Cultures*; Acta Editions: Paris, France, 1980.
4. Kader, F.; Rovel, B.; Metche, M. Role of Invertase in Sugar Content in Highbush Blueberries (*Vaccinium corymbosum*, L.). *Food Sci. Technol.* **1993**, *26*, 593–595. [CrossRef]
5. DuBois, M.; Gilles, K.A.; Hamilton, J.K.; Rebers, P.A.; Smith, F. Colorimetric Method for Determination of Sugars and Related Substances. *Anal. Chem.* **1956**, *28*, 350–356. [CrossRef]
6. Tahi, H. Efficience D'utilisation de L'eau D'irrigation Chez la Tomate par la Technique de PRD (Partial Rootzone Dring) et Etude des Mécanismes Physiologiques et Biochimiques Impliqués. Ph.D. Thesis, Université Cadi Ayyad, Faculté des Sciences Semlalia-Marrakech, Marrakesh, Morocco, 2008; p. 153.
7. Montedoro, G.; Servili, M.; Baldioli, M. Simple and hydrolyzable phenolic compounds in virgin olive oil. Their extraction, separation, and quantitative and semiquantitative evaluation by HPLC. *J. Agric. Food Chem.* **1992**, *40*, 1571–1576. [CrossRef]
8. Zhishen, J.; Mengcheng, T.; Jianming, W. The determination of flavonoid contents in mulberry and their scavenging effects on superoxide radicals. *Food Chem.* **1999**, *64*, 555–559. [CrossRef]
9. Sparks, T.H.; Menzel, A. Observed changes in seasons: An Overview. *Int. J. Climatol.* **2002**, *22*, 1715–1725. [CrossRef]
10. Cosmulescu, S.; Baciu, A.; Cichi, M.; Gruia, M. The effect of climate changes on phenological phases in plum tree (*Prunus domestica* L.) in south-western Romania. *Hortic. Biol. Environ.* **2010**, *1*, 9–20.
11. Stefanelli, D.; Goodwin, I.; Jones, R. Minimal nitrogen and water use in horticulture: Effects on quality and content of selected nutrients. *Food Res. Int.* **2010**, *43*, 1833–1843. [CrossRef]
12. Genordet, M.; Lescourret, F.; Gomez, L.; Habib, R. Changes in fruit sugar concentrations in response to assimilate supply, metabolism and dilution: A modeling approach applied to peach fruit (*Prunus persica*). *Tree Physiol.* **2003**, *23*, 373–385.
13. Famiani, F.; Cultrera, N.G.M.; Battistelli, A.; Casulli, V.; Proietti, P.; Standadi, A.; Chen, Z.H.; Leegood, R.C.; Walker, R.P. Phosphoenolpyruvatecarboxykinase and its potential role in the catabolism of organic acids in the flesh of soft fruit during ripening. *J. Exp. Bot.* **2005**, *56*, 2959–2969. [CrossRef] [PubMed]
14. Walkowiak-Tomczak, D.; Reguła, J.; Łysiak, G. Physiochemical properties and antioxidant activity of selected plum cultivars fruit. *Acta Sci. Pol.* **2008**, *7*, 15–22.
15. Monet, R. *Le pêcher, Génétique et Physiologie*; INRA: Paris, France; Masson: Paris, France, 1983; p. 133.
16. Ojeda, H.; Ary, C.; Kraeva, E.; Carbonneau, A.; Deloire, A. Influence of pre and postveraison water deficit on synthesis and concentration of skin phenolic compunds during berry growth of *Vitisvinifera* L., cv Shiraz. *Am. Enol. Vitic.* **2002**, *53*, 261–267.

environmental sciences proceedings

Proceeding Paper

Selection of Salt-Stress-Tolerant Genotypes during Germination, Growth, and Development in Durum Wheat (*Triticum turgidum* subsp., *durum* Desf.) †

Fatima Henkrar [1,*], Kenza Stira [1] and Sripada Udupa [2,*]

1 Plant Biotechnology and Physiology Laboratory, Faculty of Sciences, Mohammed V University, Rabat 10000, Morocco; kenstira@gmail.com
2 International Center of Agricultural Research in the Dry Area (ICARDA), Rabat 10000, Morocco
* Correspondence: f.henkrar@um5r.ac.ma (F.H.); s.udupa@cgiar.org (S.U.)
† Presented at the 2nd International Laayoune Forum on Biosaline Agriculture, 14–16 June 2022; Available online: https://lafoba2.sciforum.net/.

Abstract: Salinity is a serious threat to agriculture, causing the inhibition of and alterations in germination and plant growth and development. Durum wheat is highly sensitive to salinity. Therefore, the main objective of this study was to establish a screening method of wheat genotypes, under saline conditions, at the germination and plant growth stages. Our results show a very significant effect of salt stress on the different parameters evaluated in durum wheat, for all treated genotypes. The tolerance screening test during growth and development was more effective than the germination test. The chlorophyll content allowed distinguishing tolerant from sensitive genotypes.

Keywords: salinity; *Triticum durum*; screening; tolerance; germination; growth stage; chlorophyll

Citation: Henkrar, F.; Stira, K.; Udupa, S. Selection of Salt-Stress-Tolerant Genotypes during Germination, Growth, and Development in Durum Wheat (*Triticum turgidum* subsp., *durum* Desf.). *Environ. Sci. Proc.* **2022**, *16*, 34. https://doi.org/10.3390/environsciproc2022016034

Academic Editor: Abdelaziz Hirich

Published: 16 June 2022

1. Introduction

Salinity is a major abiotic stress that affects and inhibits soil fertility and cereal growth and therefore reduces crop yield. The salt present in water irrigation or in the soil causes enormous damages that are manifested at several levels, in morphological aspects, and in physiological states [1]. Indeed, it causes a delay in germination and reduces the number of seedlings as well as the duration of their emergence, vegetative growth, and plant density, in addition to delaying the stages of growth and development (flowering, maturity, etc.). The effects of salinity vary according to the species, the intensity of the stress, and the duration of application [2]. Wheat is one of the most important cereal crops in the world, and it tends to be called the king of the grains. Wheat occupies a very important place in the market, whether at the international level, where the production reached almost 769.6 million tons in 2021 [3], or at the national level, with a total wheat production of 7.54 million tons (5.06 Mt of common wheat and 2.48 Mt of durum wheat) in 2020 [4]. However, wheat consumption in Morocco is still dependent on importation. Wheat crops are considered to be highly sensitive to salinity. It negatively affects the growth and development of this crop, leading to diminished grain yield and quality [5]. To improve breeding techniques, there is a need for an efficient screening method to select the correct and desired salt-tolerant varieties. Therefore, the main objective of this study was to compare the behavior of different durum wheat genotypes, under saline conditions, at both the germination and growth stages following [6] protocol. To achieve this objective, we (1) identified the median lethal concentration (LC50) of salt that can be effectively used to screen durum wheat plants during seed germination; (2) identified durum wheat genotypes that are tolerant to salinity during germination, growth, and development; and (3) evaluated the different biochemical and morphological changes in plants subjected to different salinity concentrations.

2. Materiel and Methods

2.1. Determination of the Median Lethal Salt Concentration (LC50) in Durum Wheat

To select an appropriate salt concentration for screening the tolerant genotypes of durum wheat during germination, we first investigated the effects of salt concentrations on germination rates. Nine saline solutions were created using 10 mL Hoagland solution in Petri dishes containing 0, 15, 30, 45, 60, 75, 90, or 105 mmol·L^{-1} of NaCl. A germination percentage was calculated using the following formula: GP= $(x/y) \times 100$, where x = number of seeds germinated after 7 days, and y = number of seeds per Petri dish (50).

2.2. Screen for Tolerance at the Growth and Development Stages

To assess salt tolerance in the growth and development stages, a total of 70 durum wheat genotypes were used for this experiment. The 70 genotypes including Simeto as a salt-tolerant genotype were planted separately in cone-tainers (Figure 1). When the third leaf was fully developed, 40 L of salt solution was added to the bottom of the container for the plants to take up. A total of 4 treatments with 3 replications were created including 3 treatments with 45, 90, and 135 mM of salt and 1 control without any salt. The plant height, number of dried leaves, and chlorophyll content (Chla and Chlb) were noted at the heading stage to determine the tolerant genotypes. The data from 70 durum wheat genotypes were analyzed using a GGE biplot.

Figure 1. Cone-tainer system installed in a net house.

3. Results

3.1. Determination of the Median Lethal Salt Concentration (LC50)

The number of germinated seeds decreased as the salinity concentrations increased (Table 1). For this reason, we used the concentrations at which 50% of seeds fail to germinate (LC50) to screen the tolerant genotype at the germination stage. The average LC50 value for the four genotypes was 45 mM (Figure 2). These results indicate that salt solutions used for screening germination tolerance should have concentrations higher than 45 mM. Furthermore, through this experiment, the screening of tolerant and sensitive genotypes can be determined at the germination stage. As shown in Figure 3, the genotype Icamor showed tolerance to salt at the germination stage. The germination percentage (GP) was still high even at the highest salt concentration, with 0.4 at 105 mM of salt.

Table 1. GP for four durum wheat genotypes under different concentrations of salt.

Concentration (mmol·L^{-1})	Karim	Kyperonda	Icamor	Nax1 DW	Mean
0	64.667	70.667	88.667	71.333	73.833
15	56.667	66.000	69.333	42.000	58.500
30	58.667	43.333	64.000	43.333	52.333
45	57.333	46.667	59.333	47.333	52.667
60	36.000	26.000	57.333	40.000	39.833
75	23.333	22.000	47.333	25.333	29.500

Table 1. *Cont.*

Concentration (mmol·L^{-1})	Karim	Kyperonda	Icamor	Nax1 DW	Mean
90	5.333	8.667	41.333	18.667	18.500
105	3.333	0.000	38.667	11.333	13.333
LSD	24.923	25.651	16.353	18.858	20.841

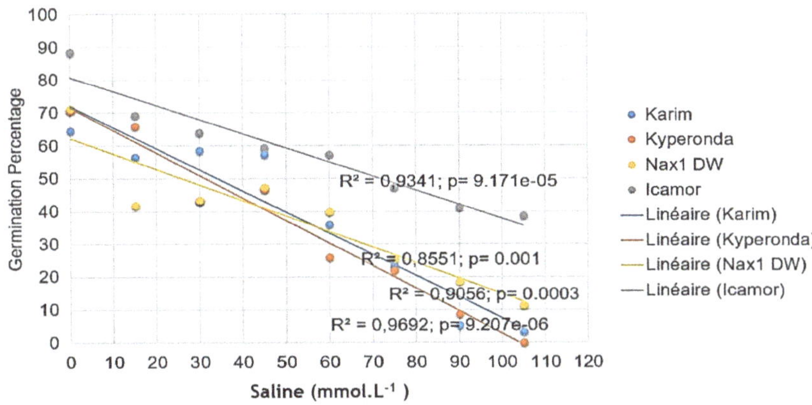

Figure 2. Regression analysis of the germination percentage for the four durum wheat genotypes under different concentrations of salt.

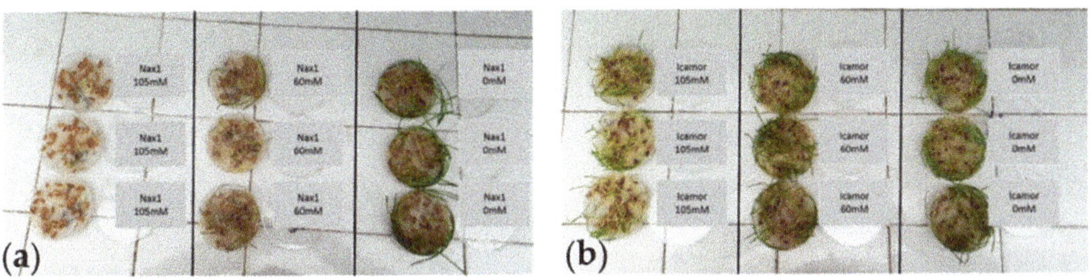

Figure 3. Germinated seeds of Icamor (**a**) and Nax1 (**b**) under salt concentrations of 0, 60, and 105 mM.

3.2. Screen for Tolerance at the Growth and Development Stages

The plants behaved differently under different salt concentrations. When the salt concentration was high, the leaf area was reduced, the number of yellow and dry leaves increased, the plants shortened, and their density decreased (Figure 4a). The concentration of 135 mM was the most discriminative salinity treatment. Therefore, the analysis was conducted using the 135 mM data. Figure 4b shows the "Rank genotypes" view of the biplot. The genotypes are ranked along the line with a single arrow. The tolerant genotypes that had a high chlorophyll content value under 135 mM salt stress are placed near "Chlorophyll", indicating that genotypes 46, 113, 103, 33, and 32 exhibited high chlorophyll content values with the lowest number of dried leaves. Similarly, for the dried leaves, genotype 158 registered the highest number of dried leaves with the lowest chlorophyll content value.

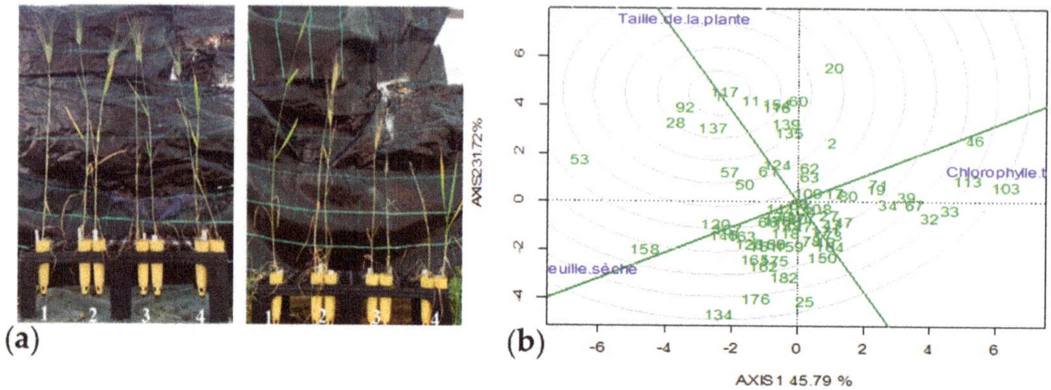

Figure 4. (**a**) Plant height of two genotypes, tolerant (left) and sensitive (right), under 4 treatments. Control (1), 45 mM (2), 90 mM (3), and 135 mM (4). (**b**) Ranking the 70 durum wheat genotypes in terms of chlorophyll content, number of dried leaves, and plant height under a 135 mM salt concentration.

The multiple comparison test (Duncan test) of the chlorophyll rate averages obtained by the plants developed under the 135 mM salt concentration revealed that among the 70 genotypes, 4 genotypes (46: Bidi-17; 103: Bidi-17-xSyrica; 113: Simeto; 32: Ajili) had a significantly high average chlorophyll content, and it seemed that these genotypes showed a tolerance to salinity (Table 2). Furthermore, our data reveal that the chlorophyll content in tolerant genotypes was higher than that in sensitive genotypes. The genotype Simeto was already reported by [7] as a tolerant genotype, which supports our results. This suggests that the chlorophyll content could be considered as an effective physiological trait for salt tolerance and could be used effectively in the breeding of tolerant durum wheat genotypes. However, the same phenomenon was not observed for the number of dried leaves and plant height.

Table 2. The highest chlorophyll content (Chla + b) at a 135 mM salt concentration using the Duncan test.

Genotype	Chlorophyll Content
46	13.659 ^{a}
103	12.021 ^{ba}
113	11.325 ^{ba}
32	10.748 ^{bac}

Different superscript letters in the chlorophyll content column indicate a statistically significant difference ($p < 0.05$) with n = 3.

4. Conclusions

Our results show a very significant effect of salt stress on the different parameters evaluated in durum wheat, for all treated genotypes. Indeed, salinity caused a decrease in germination, plant height, and chlorophyll content, and an increase in the number of dry leaves. The tolerance screening test during growth and development was more effective than the germination test. The chlorophyll content allowed distinguishing tolerant from sensitive genotypes; this result could be useful for salt tolerance screening tests. The results of this study would be a major asset in the selection and breeding improvement of durum wheat varieties. As future perspectives, it is important to apply this study to several levels of the wheat life cycle of stressed plants in the field, to use higher concentrations of salt, and to test several large populations of genotypes.

Author Contributions: F.H. performed data statistical analysis and wrote the manuscript, S.U. conceived, supervised the work, F.H. and K.S. carried out the experiments. All authors have read and agreed to the published version of the manuscript.

Funding: This work was financed by ICARDA.

Institutional Review Board Statement: Not applicable.

Informed Consent Statement: Not applicable.

Data Availability Statement: Not applicable.

Conflicts of Interest: The authors declare no conflict of interests.

References

1. Acosta-Motos, J.R.; Ortuño, M.F.; Bernal-Vicente, A.; Diaz-Vivancos, P.; Sanchez-Blanco, M.J.; Hernandez, J.A. Plant Responses to Salt Stress: Adaptive Mechanisms. *Agronomy* **2017**, *7*, 18. [CrossRef]
2. Gupta, B.; Huang, B. Mechanism of Salinity Tolerance in Plants: Physiological, Biochemical, and Molecular Characterization. *Int. J. Genom.* **2014**, *2014*, 701596. [CrossRef] [PubMed]
3. FAO. Available online: https://www.fao.org/faostat/en/#data/QCL (accessed on 27 January 2022).
4. USDA. Available online: https://www.fas.usda.gov/data/morocco-grain-and-feed-update-18 (accessed on 27 January 2022).
5. EL Sabagh, A.; Islam, M.S.; Skalicky, M.; Ali Raza, M.; Singh, K.; Anwar Hossain, M.; Hossain, A.; Mahboob, W.; Iqbal, M.A.; Ratnasekera, D.; et al. Salinity Stress in Wheat (*Triticum aestivum* L.) in the Changing Climate: Adaptation and Management Strategies. *Front. Agron.* **2021**, *3*, 661932. [CrossRef]
6. Bai, J.; Yan, W.; Wang, Y.; Yin, Q.; Liu, J.; Wight, C.; Ma, B. Screening Oat Genotypes for Tolerance to Salinity and Alkalinity. *Front. Plant Sci.* **2018**, *9*, 1302. [CrossRef] [PubMed]
7. Borrelli, G.M.; Ficco, D.B.M.; Giuzio, L.; Pompa, M.; Cattivelli, L.; Flagella, Z. Durum Wheat Salt Tolerance in Relation to Physiological, Yield and Quality Characters. *Cereal Res. Commun.* **2011**, *39*, 525–534. [CrossRef]

environmental sciences proceedings

MDPI

Proceeding Paper

Early Sowing of Quinoa to Enhance Water Use Efficiency and Yield under Arid Conditions in Morocco †

Nawal Taaime [1], Khalil El Mejahed [1], Mariam Moussafir [1], Rachid Bouabid [2], Abdallah Oukarroum [3], Redouane Choukr-Allah [1,4] and Mohamed El Gharous [1,*]

1 Agricultural Innovation and Technology Transfer Center, Mohammed VI Polytechnic University, Ben Guerir 43150, Morocco; nawal.taaime@um6p.ma (N.T.); khalil.elmejahed@um6p.ma (K.E.M.); mariam.moussafir@um6p.ma (M.M.); redouane.choukrallah@um6p.ma (R.C.-A.)
2 Department of Agronomy, National School of Agriculture, Meknes 50001, Morocco; rbouabid@enameknes.ac.ma
3 Agrobiosciences, Plant Stress Physiology Laboratory, Mohammed VI Polytechnic University, Ben Guerir 43150, Morocco; abdellah.oukarroum@um6p.ma
4 Department of Crop Production, Protection and Biotechnology, Hassan II Institute of Agronomy and Veterinary Medicine, Rabat 10101, Morocco
* Correspondence: mohamed.elgharous@um6p.ma

† Presented at the 2nd International Laayoune Forum on Biosaline Agriculture, 14–16 June 2022; Available online: https://lafoba2.sciforum.net/.

Abstract: Quinoa is a potential alternative crop for an adaptation strategy for salinization and climate change effects in dryland. A sowing date of February practiced in the Rehamna region, Morocco, resulted in stunted plants and low yields due to insufficient precipitations and high temperatures around the flowering stage. For this reason, a field experiment was conducted to investigate the effect of sowing date on quinoa water use efficiency, growth, and yield. The experiment was conducted at the UM6P experimental farm to evaluate five sowing dates for two short cycle quinoa cultivars. The results showed that the most early suitable sowing date of quinoa in the Rehamna region was December. Late sowing dates resulted in a significant decrease in WUE, growth and yield. The highest grain yield (0.84 t ha^{-1}) was obtained by ICBA-Q5 sown in December.

Keywords: sowing date; precipitation; temperature; water use efficiency; supplemental irrigation

Citation: Taaime, N.; El Mejahed, K.; Moussafir, M.; Bouabid, R.; Oukarroum, A.; Choukr-Allah, R.; El Gharous, M. Early Sowing of Quinoa to Enhance Water Use Efficiency and Yield under Arid Conditions in Morocco. *Environ. Sci. Proc.* **2022**, *16*, 66. https://doi.org/10.3390/environsciproc2022016066

Academic Editor: Abdelaziz Hirich

Published: 27 June 2022

Publisher's Note: MDPI stays neutral with regard to jurisdictional claims in published maps and institutional affiliations.

1. Introduction

Quinoa is used as alternative crop in the dryland of Morocco for soil salinity mitigation and drought tolerance. It is important to evaluate its performance after moving its cropping cycle backward. Planting date is one of the important agronomic practices for the success of the crop, and the optimal date depends on rainfall and its distribution, soil humidity and cultivar [1].

Rehamna is an arid region with annual precipitations rarely exceeding 170 mm. In addition, irrigated agriculture is threatened by water salinization that could reduce yields by up to 50% [2]. Quinoa tolerates better salinity and drought in comparison with cereals (wheat and barley) and has high nutritional and economic value, which makes it a strategic crop to increase farmers' income and improve food security [3]. Quinoa planted in February and March exhibited accelerated development and low yields due to high temperature and evapotranspiration associated with dry spells. Therefore, early planting of quinoa associated with supplemental irrigation, if needed, has increased rainwater use efficiency (WUE) and alleviated heat stress. In addition, the use of supplemental irrigation was reported to be a good strategy to secure and stabilize yield when rainfall is insufficient [4].

The only results published in Morocco so far are those carried out by [5] in Agadir, where quinoa sown in November and early December gave the highest grain yields with an average of 2.8 t ha^{-1}. Most previous research on quinoa had focused on evaluating the sowing date of quinoa long cycle cultivars with full irrigation. However, sowing quinoa short cycle cultivars with supplemental irrigation during sensitive stages was given little attention and needs to be more explored, especially under the arid climate of Morocco. Thus, the objective of the present study was to identify the most appropriate planting date for two commonly used quinoa short cycle cultivars (ICBA-Q5 and Titicaca) in Rehamna region, Morocco.

2. Material and Methods

A field experiment was carried out during the 2020–2021 cropping season at the experimental farm of UM6P. The soil is a sandy clay loam, having 1.86% of OM and pH of 8.28. Water used for supplemental irrigation has an Ec = 3.3 mS/cm. Two cultivars, ICBA-Q5 and Titicaca were used, and the planting dates were each month from 15 November to 15 March. The experimental design was a randomized complete blocks design with four replications. Full irrigation was applied to fulfill the crop's water requirement during the plant establishment, flowering, and seed filling stages. The quantity of irrigation water received by each treatment was around 140 mm and the quantity of rainfall received differed according to cultivars and sowing dates (Table 1).

Table 1. Amount of water received by each sowing date and cultivar.

	Rainfall (mm)+ Irrigation (mm) = Total Amount of Water Received (mm)				
	November	December	January	February	March
ICBA-Q5	77 + 140 = 217	108 + 142 = 250	76 + 139 = 215	65 + 144 = 209	8 + 148 = 156
Titicaca	119 + 142 = 261	108 + 141 = 249	78 + 140 = 218	65 + 142 = 207	9 + 145 = 154

Daily monitoring of plant development stages was carried out to determine the number of days from sowing to six true leaves, panicle emergence, flowering, and maturity. Before harvest, plant height was measured. At harvest, straw yield, grain yield, harvest index (HI), the weight of a thousand grains, and WUE were measured.

3. Results

3.1. Development Stages Length

Quinoa development stage lengths were significantly different for both sowing dates and cultivar (Table 2). Plant development was slow for the December planting date and more time was taken by the crop to reach the six true leaves, panicle emergence, flowering, and maturity, while late sowing (February and March) decreased this time significantly.

Table 2. Days from sowing to six true leaves, panicle emergence, flowering and maturity as affected by sowing dates and quinoa cultivars (mean ± SD, n =16).

	Days to Six True Leaves		Days to Panicle Emergence		Days to Flowering		Days to Maturity	
	ICBA-Q5	Titicaca	ICBA-Q5	Titicaca	ICBA-Q5	Titicaca	ICBA-Q5	Titicaca
November	(21.44 ± 0.23) d	(21.25 ± 0.2) d	(50.63 ± 0.39) e	(52.50 ± 0.30) d	(72.63 ± 0.39) c	(76.56 ± 0.31) b	101 ^{e}	119 ^{d}
December	(32.19 ± 0.22) a	(32.38 ± 0.19) a	(61.00 ± 0.28) b	(70.38 ± 0.39) a	(71.31 ± 0.40) d	(78.63 ± 0.39) a	112 ^{b}	125 ^{a}
January	(25.25 ± 0.12) c	(26.63 ± 0.13) b	(52.00 ± 0.49) d	(58.56 ± 0.42) c	(68.44 ± 0.51) e	(78.19 ± 0.44) a	103 ^{d}	127 ^{c}
February	(19.75 ± 0.18) e	(21.00 ± 0.16) d	(40.69 ± 0.49) h	(49.50 ± 0.49) f	(54.75 ± 0.53) h	(64.50 ± 0.40) f	98 ^{h}	111 ^{f}
March	(15.25 ± 0.12) g	(16.44 ± 0.16) f	(34.06 ± 0.37) i	(45.69 ± 0.49) g	(47.56 ± 0.51) i	(57.31 ± 0.40) g	83 ^{i}	109 ^{g}

Means followed by the same small letters are not significantly different at $p \leq 0.05$.

3.2. Plant Height

There was significant effect of sowing date and cultivar on the quinoa final height (Figure 1). However, no interaction between these two factors was recorded. Sowing in December gave the highest plant height for both cultivars, whereas February planting resulted in the lowest values.

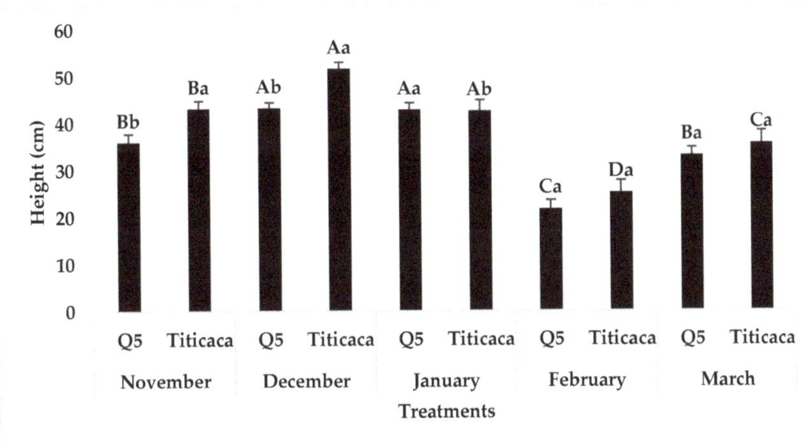

Figure 1. Final plant height of quinoa crop. For each cultivar, means followed by the same capital letters are not significantly different at $p \leq 0.05$. For each sowing date, means followed by the same small letters are not significantly different at $p \leq 0.05$.

3.3. Yield Components and WUE

December planting produced the highest grain yield (0.8 t ha^{-1}) for both cultivars (Table 3). Straw yield was significantly affected by both sowing date and cultivars. Planting in December, January and February gave the highest values (Table 3). The harvest index varied according to sowing dates and cultivars, and the highest values were recorded by early sowing dates (Table 3). The ICBA-Q5 cultivar had a higher 1000 grain weight than Titicaca cultivar.

Table 3. Yield components and panicle length of Titicaca and ICBA-Q5 for the different sowing date, Ben Guerir, Morocco (mean ± SD, n = 40).

	Grain Yield (t h^{-1})		Straw Yield (t h^{-1})		HI (%)		1000 Seed Weight (g)		WUE (kg mm^{-1} ha^{-1})	
	ICBA-Q5	Titicaca	ICBA-Q5	Titicaca	ICBA-Q5	Titicaca	ICBA-Q5	Titicaca	ICBA-Q5	Titicaca
November	(0.39 ± 0.05) ^{c}	(0.15 ± 0.01) ^{d}	(0.58 ± 0.07) ^{d}	(1.76 ± 0.08) ^{b}	(0.43 ± 0.04) ^{a}	(0.08 ± 0.01) ^{d}	(2.25 ± 0.04) ^{b}	(1.6 ± 0.05) ^{e}	(1.82 ± 0.23) ^{c}	(0.56 ± 0.06) ^{d}
December	(0.84 ± 0.05) ^{a}	(0.8 ± 0.04) ^{a}	(2.19 ± 0.14) ^{a}	(2.08 ± 0.17) ^{a}	(0.29 ± 0.02) ^{b}	(0.33 ± 0.04) ^{b}	(2.44 ± 0.05) ^{a}	(1.99 ± 0.03) ^{c}	(3.35 ± 0.2) ^{a}	(3.2 ± 0.16) ^{a}
January	(0.74 ± 0.05) ^{a}	(0.5 ± 0.05) ^{b}	(1.59 ± 0.12) ^{b}	(2.25 ± 0.12) ^{a}	(0.35 ± 0.03) ^{b}	(0.18 ± 0.02) ^{c}	(2.36 ± 0.04) ^{ab}	(1.76 ± 0.05) ^{d}	(3.42 ± 0.22) ^{a}	(2.34 ± 0.22) ^{b}
February	(0.22 ± 0.04) ^{d}	(0.18 ± 0.02) ^{d}	(2.14 ± 0.13) ^{a}	(2.22 ± 0.1) ^{a}	(0.11 ± 0.02) ^{d}	(0.08 ± 0.01) ^{d}	(1.69 ± 0.04) ^{de}	(1.4 ± 0.05) ^{f}	(1.05 ± 0.18) ^{d}	(0.86 ± 0.08) ^{d}
March	(0.08 ± 0.01) ^{d}	(0.09 ± 0.01) ^{d}	(0.96 ± 0.07) ^{c}	(0.77 ± 0.07) ^{cd}	(0.09 ± 0.01) ^{d}	(0.13 ± 0.01) ^{cd}	(1.28 ± 0.06) ^{f}	(1.04 ± 0.05) ^{g}	(0.53 ± 0.04) ^{d}	(0.61 ± 0.05) ^{d}

For each sowing date and cultivar combination, means followed by the same small letters are not significantly different at $p \leq 0.05$.

4. Discussion

Early sowing of quinoa short-cycle cultivars is a good strategy to secure good vegetative development and displace the growing cycle of the crop in the rainy season. A December planting date increased the number of days from sowing to maturity and enhanced plant height. However, late sowing decreased these parameters. Similar results were found in central Italy by [6], where high levels of temperature and photoperiod of late sowing decreased the number of days after emergence to panicle appearance and flowering. In other studies, late sowing also decreased the vegetative development of quinoa and resulted in stunted plants [5,7].

Sowing in December and January received the highest amount of rainwater and recorded the highest WUE. Low temperatures and a short photoperiod of this period decreased the evaporation, and the plant benefits better from water in the rainy season. Late sowing in February and March was associated with high evaporation, the rainwater was less effective for the plant use, and water losses were high. In addition, benefiting from rainwater during the rainy season helped in reducing the amount of salts applied in the irrigation water and thus contribute to mitigating the salinity effect in the arid and saline regions of Morocco.

In the present study, early planting dates (November, December, and January) placed the growing season of quinoa in the rainy season. However, it was observed that quinoa planted in November suffered from continuous low temperature and radiation from vegetative development to seed filling. This resulted in low straw and grain yield. A cool temperature is necessary to achieve good growth and a maximum yield of quinoa [8].

Author Contributions: Conceptualization, M.E.G. and K.E.M.; methodology, N.T.; software, N.T.; validation, M.E.G., K.E.M., and R.B.; formal analysis, N.T.; investigation, N.T. and M.M.; resources, M.E.G. and A.O.; data curation, M.E.G.; writing—original draft preparation, N.T.; writing—review and editing, M.E.G., K.E.M., R.B., R.C.-A. and A.O.; visualization, N.T. and M.M.; supervision, M.E.G. and K.E.M.; project administration, M.E.G. All authors have read and agreed to the published version of the manuscript.

Funding: This research received no external funding.

Institutional Review Board Statement: Not applicable.

Informed Consent Statement: Not applicable.

Data Availability Statement: Not applicable.

Conflicts of Interest: The authors declare no conflict of interest.

References

1. FAO. Quinoa: An ancient crop to contribute to world food security. In *Regional Office for Latin America and the Caribbean*; Bojanic, A., Ed.; Food and Agriculture Organization: Rome, Italy, 2011; p. 63.
2. Taleb, H. Water management in Morocco. In *Management of Intentional and Accidental Water Pollution*; Dura, G., Kambourova, V., Simeonova, F., Eds.; Springer: Dordrecht, The Netherlands, 2006; pp. 177–180.
3. Hirich, A.; Rafik, S.; Rahmani, M.; Fetouab, A.; Azaykou, F.; Filali, K.; Ahmadzai, H.; Jnaoui, Y.; Soulaimani, A.; Moussafir, M.; et al. Development of Quinoa Value Chain to Improve Food and Nutritional Security in Rural Communities in Rehamna, Morocco: Lessons Learned and Perspectives. *Plants* **2021**, *10*, 301. [CrossRef]
4. Fox, P.; Rockström, J. Supplemental irrigation for dry-spell mitigation of rainfed agriculture in the Sahel. *Agric. Water Manag.* **2003**, *61*, 29–50. [CrossRef]
5. Hirich, A.; Choukr-Allah, R.; Jacobsen, S.-E. Quinoa in Morocco—Effect of Sowing Dates on Development and Yield. *J. Agron. Crop Sci.* **2014**, *200*, 371–377. [CrossRef]
6. Casini, P. Seed Yield of Two New Quinoa (Chenopodium Quinoa Willd.) Breeding Lines as Affected by Sowing Date in Central Italy. *Acta Agric. Slov.* **2019**, *113*, 51. [CrossRef]
7. Garcia, M.; Raes, D.; Jacobsen, S.-E. Evapotransporation analysis and irrigation requirements of quinoa (*Chenopodium quinoa*) in the Bolivian highlands. *Agric. Water Manag.* **2003**, *60*, 119–134. [CrossRef]
8. Galwey, N.W. Exploited plants quinoa, grain amaranth. Crop yield: Physiology and processes. *Agron. J.* **1989**, *85*, 120–128.

environmental sciences proceedings

MDPI

Proceeding Paper

Development of Salt-Tolerant Alfalfa Clones by In Vitro Culture †

Claudia Ruta [1,*], Giuseppe De Mastro [1], Sandrine Flajoulot [2] and Jean-Christophe Caillet [3]

1 Department of Agricultural and Environmental Science, University of Bari Aldo Moro, Via Amendola 16/a, 70125 Bari, Italy; giuseppe.demastro@uniba.it
2 Cérience, RN 147-4 Av de la CEE-La cour, d'Henon, 86170 Cissé, France; flajoulot@gmail.com
3 GIE GRASS, 1 allée de la sapinière, La Litière, 86600 Saint Sauvant, France; caillet@gmail.com
* Correspondence: claudia.ruta@uniba.it
† Presented at the 2nd International Laayoune Forum on Biosaline Agriculture, 14–16 June 2022; Available online: https://lafoba2.sciforum.net/.

Abstract: In this research, in vitro culture was applied to evaluate five alfalfa lines for salt tolerance and to select and multiply the most resistant clones. The pregerminated seeds were tested on a semisolid culture medium enriched with four different NaCl concentrations. Saline stress response was estimated by evaluating the survival capacity, growth, sensitivity indexes and electrolyte leakage. The most salt-tolerant plantlets for each line were multiplied to clone true-to-type plants. At the same time, the selected clones were evaluated again in vitro on middle salt stress conditions to confirm the salt tolerance. In conclusion, in vitro culture allowed the rapid selection of alfalfa genotypes to develop salt-tolerant clones.

Keywords: salinity test; in vitro selection; NaCl; multiplied clones

check for updates

Citation: Ruta, C.; De Mastro, G.; Flajoulot, S.; Caillet, J.-C. Development of Salt-Tolerant Alfalfa Clones by In Vitro Culture. *Environ. Sci. Proc.* **2022**, *16*, 71. https://doi.org/10.3390/environsciproc2022016071

Academic Editor: Abdelaziz Hirich

Published: 13 July 2022

1. Introduction

Alfalfa is an important forage crop that exhibits moderate sensitivity to salt levels in irrigation water and soil. Since alfalfa is polyploid and highly heterozygous due to cross-pollination, populations vary widely in their salt tolerance; furthermore, there is enormous variability within populations. Therefore, this genetic variation can be exploited through screening by allowing the selection of plants with superior performance when exposed to such stress.

In vitro tissue culture can be applied to large and diverse populations for screening vigorous salt-tolerant individuals, uncovering desired phenotypes in a short time, limited space and low cost while other factors (nutrients, lighting, temperature) are kept constant and optimally controlled [1,2].

The objective of this study was the evaluation of five alfalfa lines for salt tolerance and the selection and multiplication of the most resistant clones.

2. Materials and Methods

2.1. In Vitro Screening Selection for Salt Tolerance of Five Alfalfa Lines

The experimental trials were carried out on five alfalfa lines (A, B, C, D, E) of the Jouffray-Drillaud Company (now become Cérience)-RN 147-4 Av de la CEE-La cour d'Hénon CISSE, FR 86170, thanks to a research agreement with the Department of Agricultural and Environmental Science (DAES)-University of Bari Aldo Moro. The seeds were surface sterilised (70% ethanol for 1 min followed by NaOCl (active chlorine 14–15%) for 20 min) two times and then rinsed with deionised and sterile water. Then they were placed in sterile conditions on wet filter paper in Petri dishes and germinated for five days at 20 °C in a growth chamber. Germinated seedlings of each line were transferred on a growth medium (basal medium- BM) [3] enriched with sucrose (20 g L^{-1}) and different concentrations of sodium chloride (0, 100, 150, 200 mM NaCl). The values corresponding to electrical conductivity (EC) were about 6, 15, 20 and 25 mS respectively, measured by Model 120 Microprocessor Conductivity Meter. Seedlings were grown on the selective medium for ten weeks in a growth chamber at 23 °C, with a photoperiod of 16 h and under a light intensity of 50 μE s^{-1} m^{-2}. Then on 20 samples per line and treatment survival capacity (%), shoot and root length (g), shoot number node, fresh and dry weight of shoot and root system were evaluated. The salt influence on the plant growth was expressed through sensitivity index calculated for height, node number and fresh weight of shoots and length, number and fresh weight of roots, according to the formula: IS = (Ps − Pt)/Pt × 100, in which Ps is the value of each parameter of the salt-stressed plants and Pt is the value of the same parameter in non-stressed plants [4]. Moreover, Electrolyte leakage was determined on the fresh leaves according to Lutts et al. [5], using the formula: EL = (Lt/L0) × 100, where Lt and L0 are the values of conductivity after and before the release of electrolytes in the solution, respectively, measured by conductimeter WTW (LF330 model-Weilheim, Germany).

2.2. In Vitro Establishment of Clones from Salt Tolerant Genotype

The salt-tolerant selected genotypes were micropropagated using BM enriched with sucrose (20 g L^{-1}) and 6- Benzilaminopurine (BAP) 0.5 mg L^{-1} to induce shoot proliferation and the same medium hormone free for the rooting. Each micropropagated clone, obtained from the corresponding selected genotype, identified as being salt stress tolerant in in vitro screening, was transferred in Jiffy pots of 8 cm^2 containing a commercial peat mixture soil (organic carbon 46%, organic nitrogen 1–2%, organic matter 80%) and mixed with perlite at a 2:1 (*v*/*v*) ratio. Acclimatisation took place in a climatic greenhouse at 18–25 °C with mist, reducing the humidity level from 85–90% to 50–60% over 20 days in natural daylight. After four weeks, the plants were trimmed and managed in natural climatic conditions before shipping to the Jouffray-Drillaud Company.

2.3. Evaluation of Salt Tolerance of Selected and Multiplied Clones

To evaluate the salt tolerance of selected and multiplied clones, plant samples were grown for 30 days on BM added with 0.75 mM NaCl or on the salt-free control medium, according to the method developed by Dziadczyk et al. [6]. The increase of shoot fresh weight, expressed as relative growth rate [RGR: (ln Fwfinal − ln FWinitial) × 100/total days of growth (where total days of growth are 30 and FW is the fresh weight of the plants)], and proline content were measured in both experimental conditions. Proline content was determined following the protocol described by Bates et al. [7].

3. Results and Discussion

3.1. Evaluation of the Response of Five Alfalfa Varieties to Salt Stress

Salt presence in the growth media interfered with both physiological and biochemical processes even if a different establishment of the lines to the in vitro condition was shown. Indeed, regarding the survival capacity of the in vitro seedlings measured after 10 weeks of growth on the salt media, LC and LD seemed the lines more sensitive about survival, although they were also the same ones that showed a reduced survival on the control medium (Figure 1). Instead, LB showed the best performance on the salt media. Considering the total values, LB showed the highest survival (72.7%), while the others raised values between 57.8 % (LD) and 47.37 % (LC).

The morphological parameters evaluated showed an obvious reduction of the growth of the plantlets, depending from the salt stress. It is known that slower growth is an adaptive feature for plant survival under stress. Because the behaviors are also linked to the genotypes that had different responses in term of enhancement, for standardising the data, IS was used to discuss the effect of salt stress on seedlings of five alfalfa lines. In Table 1 the indexes of salt sensitivity (IS) estimated, based on the total fresh biomass, are reported.

These data showed LA and LD the most sensitive to salt stress at all the concentrations of NaCl, while LB and LC were the less sensitive.

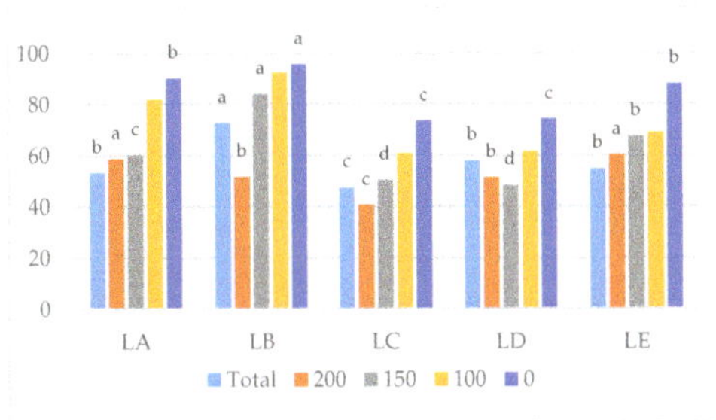

Figure 1. Survival of alfalfa five lines measured after 10 weeks on culture medium enriched with 0, 100, 150 and 200 mM NaCl. The histograms represent the average of 20 replicates. The different letters show the statistical differences among the varieties for each salt treatment. (Test SNK-$p \leq 0.01$).

Table 1. Indexes of salt sensitivity (IS) estimated based on the total fresh biomass of five alfalfa lines grown for 10 weeks on a basal medium enriched with different salt concentrations. Data represent the average of 20 replicates.

NaCl (mM)	Alfalfa Lines				
	LA	LB	LC	LD	LE
200	−64.66 a	−55.77 b	−54.48 b	−69.09 a	−65.33 a
150	−52.12 b	−36.54 c	−33.29 c	−61.82 a	−38.49 c
100	−50.86 a	−32.69 b	−29.77 b	−49.09 a	−31.93 b

The different letters show the statistical differences among the varieties for each salt treatment. (Test SNK-$p \leq 0.01$).

To better understand how the salt stress caused the growth reduction and the tissues and organs more involved, IS of shoot height, node number and fresh weight and IS of root length, number and fresh weight were compared among the lines. To simplify the comparison, IS values were divided into three classes: 0 to −33% (low stress); >−33% to −66% (moderate stress); >−66% to −100% (high stress) and represented respectively with *; **; *** (Table 2). One of the strongest effect of salt stress on the growth of all five lines was on the elongation of shoots and roots. At 200 and 150 mM NaCl concentrations, only LC responded with moderate stress to the shoot height, while the other lines resulted highly stressed. A comparison between the IS weight of shoots and roots showed a higher sensitivity to salt stress of the roots than the shoots that showed IS representing moderate or low classes of stress. Therefore, the most sensitive organ appeared to be the root. Moreover, a simple consideration on the frequencies of the high class stress, for all the sensitive indexes considered and independently from the concentration of NaCl, confirmed LA and LD the most sensitive, followed by LE, LB and LC.

The effect of salt stress on membrane permeability, a physiological measurement often used to differentiate the genotypes that can grow at different salinity levels, was estimated according to the electrolyte leakage (EL). Figure 2 showed that the salt stress caused a high increase of EL in all the tested lines. At severe stress, LB was the least affected (83.7%) and LC and LD were more affected by this treatment (about 92%). The results of electrolyte leakage confirmed the high sensitivity to salt stress of LA and LD lines, and indicated as most tolerant LB, the line that already showed the best performance for the survival on the salt media and for the low IS.

Table 2. Effect of salt stress, expressed as classes of IS values, on the different morphological parameters of five alfalfa lines grown for 10 weeks on a basal medium enriched with different salt concentrations.

		LA			LB			LC			LD			LE		
		200	150	100	200	150	100	200	150	100	200	150	100	200	150	100
Shoot	Height	***	***	***	***	***	**	**	**	*	***	***	**	***	***	**
	Nodes	**	**	*	**	**	*	**	**	*	***	***	***	**	**	*
	Fresh weight	*	*	*	**	*	*	**	*	*	**	**	*	**	*	*
Root	Length	***	***	***	***	**	*	**	**	*	***	***	***	***	**	**
	Number	***	***	**	*	*	*	**	**	*	**	*	**	**	**	*
	Fresh weight	***	***	***	***	**	**	**	**	**	***	***	***	***	***	*

*: 0 to −33% (low stress); **: >−33% to −66% (moderate stress); ***: >−66% to −100% (high stress).

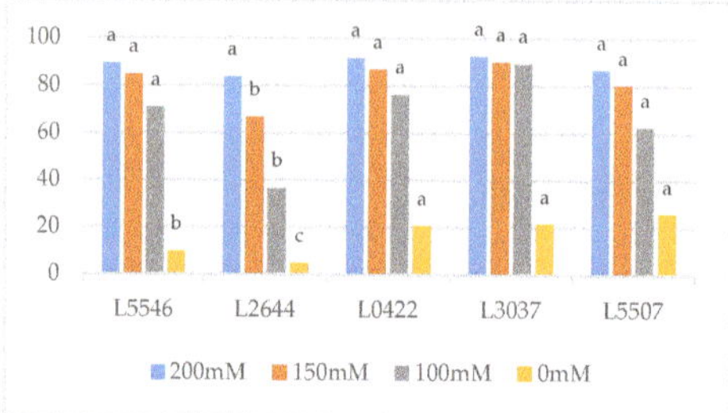

Figure 2. Electrolyte leakage of five alfalfa lines measured after 10 weeks on culture medium enriched with 0, 100, 150 and 200 mM. The histograms represent the average of 20 replicates. The different letters show the statistical differences among the varieties for each salt treatment. (Test SNK-$p \leq 0.01$).

3.2. Micropropagation of the Salt Tolerant Plants

All the lines showed a good response to the micropropagation technique, obtaining 4/5 shoots for each subculture starting from each selected shoot. The clones raised a high percentage of rooting. The rooted clones were transferred to acclimatisation in the climatic greenhouse and then hardened in natural climate conditions before sending to Jouffray-Drillaud Company.

3.3. Salt Tolerance of Selected Clones

The results of RGR showed an excellent capacity of the selected salt tolerant clones to grow both on mild salt stress (75 mM NaCl) and on control (0 mM NaCl) (data not reported). Taking into consideration that the rapid accumulation of proline in vegetal tissues as response to salt stress has been associated with the ability of proline to act as an osmolyte to adjust the plant under drought/saline condition, the greater proline accumulation in the leaves of all alfalfa selected clones grown on mild salt stress in comparison of the control proline content (Table 3), confirmed the effectiveness of this protocol.

Table 3. Proline content of salt tolerant selected clones of five alfalfa lines measured after 30 days of growth on culture medium enriched with 0 and 75 mM NaCl. Data represent the average of 20 replicates.

NaCl (mM)	Proline Content (μmol g^{-1} FW)				
	LA	LB	LC	LD	LE
0	32.1 b	27.3 b	40.1 b	32.2 b	33.6 b
75	50.0 a	52.4 a	49.7 a	50.0 a	50.5 a

The different letters show the statistical differences among the salt treatments for each line. (Test SNK-$p \leq 0.01$).

Author Contributions: Conceptualization, C.R. and G.D.M.; methodology, C.R.; validation, C.R., G.D.M., S.F. and J.-C.C.; investigation, C.R.; resources, S.F. and J.-C.C.; data curation, C.R. and G.D.M.; writing—original draft preparation, C.R.; writing—review and editing, C.R., G.D.M., S.F. and J.-C.C.; supervision, G.D.M.; project administration, G.D.M.; funding acquisition, C.R. and G.D.M. All authors have read and agreed to the published version of the manuscript.

Funding: This research was funded by the Jouffray-Drillaud Company (now become Cérience,)-RN 147-4 Av de la CEE-La cour, d'Hénon 86170 CISSE, France.

Informed Consent Statement: Not applicable.

Conflicts of Interest: The author declare no conflict of interest.

References

1. Arzani, A. Improving salinity tolerance in crop plants: A biotechnological view. *In Vitro Cell. Develop. Biol.-Plant* **2008**, *44*, 373–383. [CrossRef]
2. Arab, M.M.; Yadollahi, A.; Eftekhari, M.; Ahmadi, H.; Akbari, M.; Khorami, S.S. Modeling and optimizing a new culture medium for in vitro rooting of G× N15 Prunus rootstock using artificial neural network-genetic algorithm. *Sci. Rep.* **2018**, *8*, 9977. [CrossRef] [PubMed]
3. Ruta, C.; De Mastro, G.; Ancona, S.; Tagarelli, A.; De Cillis, F.; Benelli, C.; Lambardi, M. Large-Scale Plant Production of *Lycium barbarum* L. by Liquid Culture in Temporary Immersion System and Possible Application to the Synthesis of Bioactive Substance. *Plants* **2020**, *9*, 844. [CrossRef]
4. Hamrouni, L.; Abdallah, F.B.; Abdelly, C.; Ghorbel, A. La culture in vitro: Un moyen rapide et efficace pour sélectionner des génotypes de vigne tolérant la salinité. *Comptes Rendus Biol.* **2008**, *331*, 152–163. [CrossRef] [PubMed]
5. Lutts, S.; Kinet, J.M.; Boharmunt, J. Effect of various salt and mannitol on ion and proline accumulation in relation to osmotic adjustment in rice callus cultures. *J. Plant Physiol.* **1996**, *149*, 186–195. [CrossRef]
6. Dziadczyk, P.; Bolibok, H.; Tyrka, M.; Hortyñski, J.A. In vitro selection of strawberry (Fragaria × ananassa Duch.) clones tolerant to salt stress. *Euphytica* **2003**, *132*, 49–55. [CrossRef]
7. Bates, L.S.; Waldren, R.P.; Teare, I.D. Rapid determination of free proline for water stress studies. *Plant Soil* **1973**, *39*, 205–207. [CrossRef]

environmental sciences proceedings

MDPI

Proceeding Paper

Identification of QTLs for Morpho-Physiological Traits under Saline Stress in *Indica* MAGIC Rice Population †

Bayragondlu M. Lokeshkumar [1], Saraswathipura L. Krishnamurthy [1,*], Suman Rathor [1], Arvinder S. Warraich [1], Satyendra Yadav [1], Parbodh C. Sharma [1,*] and Rakesh Kumar Singh [2]

1 Division of Crop Improvement, Central Soil Salinity Research Institute, Indian Council of Agricultural Research (ICAR), Karnal 132001, India; lokeshbm8050@gmail.com (B.M.L.); suman.rathore08@gmail.com (S.R.); arvindersinghwarraich@gmail.com (A.S.W.); ysatyendra791@gmail.com (S.Y.)

2 Crop Diversification and Genetics, International Center for Biosaline Agriculture, Dubai 14660, United Arab Emirates; r.singh@biosaline.org.ae

* Correspondence: krishnagene@gmail.com (S.L.K.); pcsharma.knl@gmail.com (P.C.S.)

† Presented at the 2nd International Laayoune Forum on Biosaline Agriculture, 14–16 June 2022; Available online: https://lafoba2.sciforum.net/.

Abstract: Rice is sensitive to salinity at both the seedling and reproductive stages, and it reduces the grain yield from 50 to 100%. In this study, 27 SNP markers significantly associated with 15 QTLs were identified. Three major QTLs were associated with shoot length (*Sal_SL 7.1*),shoot sodium content (*Sal_Na 1.1*) and shoot magnesium content (*Sal_Mg 2.1*). Five QTLs for root length (*Sal_RL 1.1, Sal_RL 3.1, Sal_RL 6.1 Sal_RL 8.1* and *Sal_RL 12.1*), shoot K^+/Na^+ homeostasis (*Sal_K/Na 1.1, Sal_K/Na 4.1, Sal_K/Na 5.1, Sal_K/Na 7.1* and *Sal_K/Na 10.1*) and single QTLs for shoot potassium (*Sal_K 6.1*) and calcium content (*Sal_Ca 5.1*) were also detected. QTL *Sal_K/Na 1.1* was found responsible for the ionic ratio associated with the *Saltol* region, and *Sal_K/Na 10.1* was associated with the gene *OSCA1;4*,which is a hyperosmolality-gated calcium-permeable channel that acts as an osmosensor under salt stress condition. A candidate gene haplotype analysis revealed ten significant genes, *LOC_Os06g03940, LOC_Os10g42820, LOC_Os07g36230, LOC_Os02g06410, LOC_Os06g48610, LOC_Os12g12950, LOC_Os03g12050, LOC_Os08g02690, LOC_Os07g47560*, and *LOC_Os05g08840*, responsible for abiotic stress tolerance. The identified potential candidate genes can be used for functional characterization to understand the complex mechanism of salinity tolerance in rice.

Keywords: rice; salinity; SNP; QTL; GWAS; haplotype; MAGIC; ion-homeostasis

check for updates

Citation: Lokeshkumar, B.M.; Krishnamurthy, S.L.; Rathor, S.; Warraich, A.S.; Yadav, S.; Sharma, P.C.; Singh, R.K. Identification of QTLs for Morpho-Physiological Traits under Saline Stress in *Indica* MAGIC Rice Population. *Environ. Sci. Proc.* **2022**, *16*, 29. https://doi.org/10.3390/environsciproc2022016029

Academic Editor: Abdelaziz Hirich

Published: 16 June 2022

1. Introduction

Rice is one of the most important cereal grains, which feed more than half of the world population. However, it is under constant pressure due to biotic and abiotic stresses, which seriously affect its production. Among these stresses, soil salinity is one major constraint, which hinders rice production around the globe. The Global Map of Salt-Affected Soils has estimated more than 833 million hectares of land as salt affected [1]. Salinity affects 50% of the world's irrigated land, and it is estimated to be rising at a much faster pace [2]. Under saline stress, rice seedlings accumulated high Na^+, which interferes with the cellular machinery of a plant cell. Furthermore, it disturbs the water potential of the plant, which leads to low water uptake [3,4]. Seedling stage salinity tolerance is a crucial stage in rice's life cycle, as it helps the plant to survive the early adverse affects of salinity. Moreover, seedling stage salinity facilitates early plant establishment, which ensures good vegetative growth at the reproductive stage [5]. To identify the genes responsible for salinity tolerance, several QTL mapping studies have been conducted in the past [6,7]. Major *Saltol*, QTL mapped on chromosome 1, was successfully transferred to a wide array of high yielding genotypes through a series of marker-assisted breeding programs [8–10].

In the present study, we employed 27,041 filtered SNPs for identification of novel QTLs for two morphological and five physiological traits using MAGIC rice lines developed from the inter-crossing of eight advanced founder lines.

2. Material and Methods

The plant material consisted of 391 rice MAGIC lines developed from the inter-crossing of eight founder lines at the International Rice Research Institute in the Philippines [11]. These lines were evaluated in hydroponic condition at ICAR-CSSRI, Karnal, using Yoshida nutrient solution. The EC of the solution was raised to ~10 dS/m after 14 DAS, and the anticipated level of salinity was sustained for the next 14 days. Genotyping of the founder and magic lines was done via genotyping by sequencing (GBS) using the Illumina HiSeq method. To identify QTLs for morph-physiological traits, the R-based GAPIT program was employed using the Bayesian-information and Linkage-disequilibrium Iteratively Nested Keyway (BLINK) model. Finally, the marker-associated SNPs with an LOD threshold > 4 were considered to be significant SNPs for traits of interest. Candidate gene haplotype analysis for trait-associated QTLs was conducted using Candi-HAP V2,and violin plots were visualized in ggplot2 program in R.

3. Results and Discussion

QTL mapping for salinity tolerance is an effective way to locate robust genomic regions controlling complex traits. In the present study,15 QTLs were identified for salt tolerance with LOD >4 for 7morph-physiological traits (Table 1).

A maximum of five QTLs were recorded for root length (Sal_RL 1.1, Sal_RL 3.1, Sal_RL 6.1 Sal_RL 8.1 and Sal_RL 12.1) and shoot K^+/Na^+ homeostasis (Sal_K/Na 1.1, Sal_K/Na 4.1, Sal_K/Na 5.1, Sal_K/Na 7.1 and Sal_K/Na 10.1). The major QTLs, Sal_Mg 2.1 for Mg^{2+} content, Sal_Na 1.1 for Na^+ content and Sal_SL 7.1 for shoot length, were identified; these explained about 23.74%, 14.14% and 9.43% of the phenotypic variation, respectively. A peak marker at genomic location 11703845 of newly identified QTL Sal_K/Na 1.1was located in previously reported Saltol QTL region [8]. Around 15 genes were found to overlap marker-associated SNPs with diverse functions, such as transcription factor, kinase activity membrane protein with regulatory role in enzyme activity and some other useful functions. The other QTL Sal_Na 1.1 was also found near the Saltol QTL region; the peak of this QTL, lying in the genomic region of LOC_Os01g02750, belongs to the LRR-kinase family protein, and it is a putative candidate gene for cold tolerance [12].Candidate gene haplotype analysis revealed ten significant genes, LOC_Os06g03940, LOC_Os10g42820, LOC_Os07g36230, LOC_Os02g06410, LOC_Os06g48610, LOC_Os12g12950, LOC_Os03g12050, LOC_Os08g02690, LOC_Os07g47560 andLOC_Os05g08840, as having a key role in abiotic stress tolerance. From these, the genes associated with physiological traits were presented in Figure 1. The peak SNP marker of Sal_Mg 2.1 lay in the genomic region of the CBS domain-containing membrane protein (CDCPs) gene (LOC_Os02g06410 (OsCBSX9)) and has a significant role in regulation of the thioredoxin system and stress response/tolerance in rice [13]. The peak marker of Sal_RL 6.1 contains the CCT family protein gene LOC_Os06g48610, which regulates salt tolerance through the abscisic acid-dependent signalling pathway with pleiotropic effects on morphological traits [14]. The peak SNP of Sal_RL_8.1 is associated with the MA3 domain-containing protein coding gene (LOC_Os08g02690). In Arabidopsis, the MA3 domain-containing protein (ECIP1) plays a key role in regulation of salt stress response by interacting with the central membrane protein of ethylene signalling (EIN2) [15].A single SNP peak for the K^+ content was associated with the gene LOC_Os06g03940, which encodes for the defence response spastin protein in rice under stress [16]. A group of SNPs associated with the shoot Ca^{2+} content lay in the genomic region of the Hsp70 domain-containing gene (LOC_Os05g08840) play important role in housekeeping functions under heat stress [17,18]. A significant candidate gene haplotype for LOC_Os10g42820, found to be associated with K^+/Na^+, belongs to the DUF221 domain-containing gene family [19]. Early-responsiveness to the dehydration protein gene LOC_Os10g42820 (OSCA1;4), responsible for salt and drought tolerance, was detected on QTL Sal_K/Na 10.1. The gene OSCA1;4 is a hyperosmolality-gated calcium-permeable channel that activates an inward current after receiving an osmotic signal exerted by salt stress [20]. Further, a number of uncharacterized genes exhibiting significant haplotypes associated with different traits were also detected in our study. The identified QTLs and respective candidate genes can be further investigated to confirm their role in salt tolerance in rice.

Table 1. Associated QTLs with SNP position and probable candidate genes for salinity tolerance in MAGIC population.

Sr. No	Trait	QTL	Chromosome	Position	*p* Value	LOD	Locus ID	Gene Annotation
1	Shoot length	*Sal_SL 7.1*	7	28443952	2.096×10^{-6}	5.68	*LOC_Os07g47560*	Expressed protein
2		*Sal_RL 1.1*	1	32773208	2.647×10^{-6}	5.58	*LOC_Os01g56790*	Expressed protein
3		*Sal_RL 3.1*	3	6323650	1.206×10^{-5}	4.92	*LOC_Os03g12050*	Expressed protein
4	Root length	*Sal_RL 6.1*	6	29408939	8.851×10^{-8}	7.05	*LOC_Os06g48610*	CCT motif protein
5		*Sal_RL 8.1*	8	1122340	6.223×10^{-5}	4.21	*LOC_Os08g02690*	MA3 domain
6		*Sal_RL 12.1*	12	7176380	2.351×10^{-6}	5.63	*LOC_Os12g12950*	Expressed protein
7	Na^+	*Sal_Na 1.1*	1	954124	1.509×10^{-5}	4.82	*LOC_Os01g02750*	LRk-type protein
8	K^+	*Sal_K 6.1*	6	1590002	1.548×10^{-5}	4.81	*LOC_Os06g03940*	Spastin, putative
9			5	4870881	9.084×10^{-5}	4.04		
10			5	4877897	9.084×10^{-5}	4.04		
11	Ca^{2+}	*Sal_Ca 5.1*	5	4878113	9.084×10^{-5}	4.04	*LOC_Os05g08840*	DnaK family protein
12			5	4878119	9.084×10^{-5}	4.04		
13			5	4878122	9.084×10^{-5}	4.04		
14	Mg^{2+}	*Sal_Mg 2.1*	2	3204128	1.407×10^{-6}	5.85	*LOC_Os02g06410*	CBS domain
15		*Sal_K/Na 1.1*	1	11703845	2.617×10^{-6}	5.58	*LOC_Os01g20950*	Sulfotransferase domain
16			4	24365577	9.658×10^{-5}	4.02		
17			4	24387802	9.658×10^{-5}	4.02		
18			4	24408803	9.658×10^{-5}	4.02		
19			4	24415765	4.248×10^{-5}	4.37		
20		*Sal_K/Na 4.1*	4	24454895	8.873×10^{-5}	4.05		
21	K^+/Na^+		4	24466526	8.873×10^{-5}	4.05	*LOC_Os04g41229*	bHLH domain
22			4	24471326	8.873×10^{-5}	4.05		
23			4	24471339	8.873×10^{-5}	4.05		
24			4	24497553	8.873×10^{-5}	4.05		
25		*Sal_K/Na5.1*	5	1480306	4.36×10^{-6}	5.36	*LOC_Os05g03500*	Expressed protein
26		*Sal_K/Na 7.1*	7	21655201	5.235×10^{-6}	5.28	*LOC_Os07g36230*	BTB domain
27		*Sal_K/Na 10.1*	10	23089194	3.406×10^{-6}	5.47	*LOC_Os10g42820*	Early-responsive dehydration protein

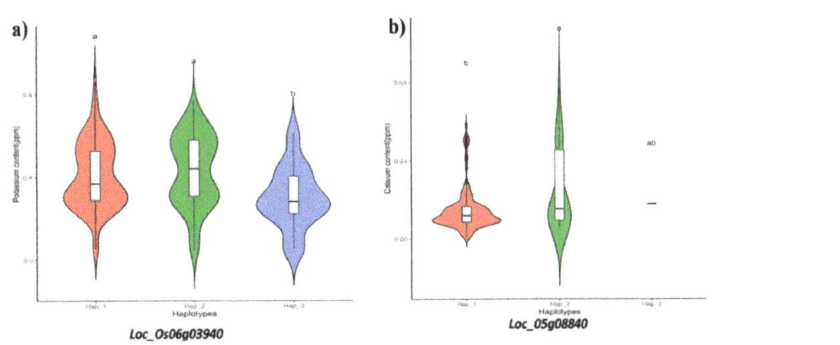

Figure 1. *Cont.*

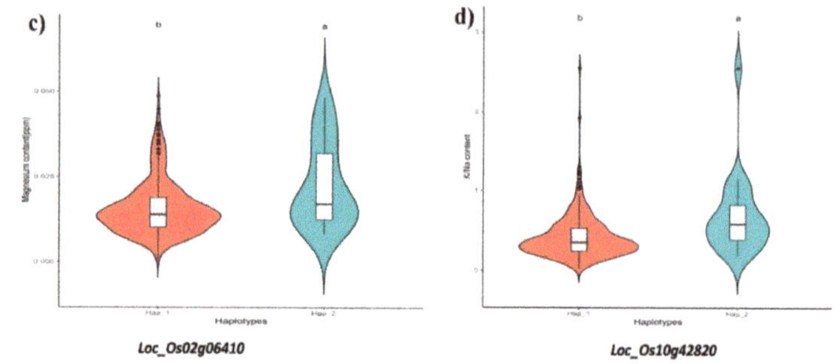

Figure 1. Violin plots indicating the significant haplotypes of candidate genes for (**a**) K^+ content, (**b**) Ca^{2+} content, (**c**) Mg^{2+} content and (**d**) K^+/Na^+(x-axis=haplotypes; y-axis= trait value). Each box plot with in the violin plot represents minimum, lower quartile, median, upper quartile, and maximum values. Different letters above the violin plots indicate statistically significant differences for the respective haplotypes, at a significance level of $p < 0.05$ (Duncan test).

4. Conclusions

In the present study, we identify three major QTLs associated with shoot length, shoot sodium content and shoot magnesium content. The QTL for Sal_K/Na 1.1 for the ionic ratio, lying within the Saltol region, and the osmosensor gene OSCA1;4, responsible for hyperosmolality-gated calcium-permeable channel, were detected. In our study, we found significant haplotypes for ten potential candidate genes that play key role in abiotic stress tolerance. The identified SNP markers linked to salt tolerant QTLs and candidate genes with significant allelic variation are potential prospects for marker-assisted breeding for salinity tolerance in rice.

Author Contributions: Conceptualization, R.K.S. and S.L.K.; methodology, S.L.K. and R.K.S.; software, B.M.L. and A.S.W.; validation, S.L.K., P.C.S. and R.K.S.; formal analysis, B.M.L. and A.S.W.; investigation, S.R. and S.Y.; resources, S.L.K. and P.C.S.; data curation, B.M.L., S.Y.; writing—original draft preparation, A.S.W. and S.Y.; writing—review and editing, B.M.L. and S.R.; visualization, B.M.L.; supervision, S.L.K.; project administration, S.L.K. All authors have read and agreed to the published version of the manuscript.

Funding: This research was funded by ICAR-New Delhi and IRRI-Phillipines.

Institutional Review Board Statement: Not applicable.

Informed Consent Statement: Not applicable.

Data Availability Statement: Not applicable.

Acknowledgments: The authors are thankful to the Director of ICAR-CSSRI for support to carry out the proposed research work.

Conflicts of Interest: The authors declare no conflict of interest.

References

1. FAO Director-General. Global map of salt-affected soils. In Proceedings of the World Map of Salt-Affected Soils, Rome, Italy, 20–22 October 2021.
2. Singh, R.K.; Kota, S.; Flowers, T.J. Salt tolerance in rice: Seedling and reproductive stage QTL mapping come of age. *Theor. Appl. Genet.* **2021**, *134*, 3495–3533. [CrossRef] [PubMed]
3. Warraich, A.S.; Krishnamurthy, S.L.; Sooch, B.S.; Vinaykumar, N.M.; Dushyanthkumar, B.M.; Bose, J.; Sharma, P.C. Rice GWAS reveals key genomic regions essential for salinity tolerance at reproductive stage. *Acta Physiol. Plant.* **2020**, *42*, 134. [CrossRef]
4. Reddy, N.B.L.; Kim, B.K.; Yoon, I.S.; Kim, K.H.; Kwon, T.R. Salt Tolerance in Rice: Focus on Mechanisms and Approaches. *Rice Sci.* **2017**, *24*, 123–144. [CrossRef]

5. Babu, N.N.; Krishnan, S.G.; Vinod, K.K.; Krishnamurthy, S.L.; Singh, V.K.; Singh, M.P.; Singh, R.; Ellur, R.K.; Rai, V.; Bollinedi, H.; et al. Marker Aided Incorporation of Saltol, a Major QTL Associated with Seedling Stage Salt Tolerance, into Oryza sativa 'Pusa Basmati 1121'. *Front. Plant Sci.* **2017**, *8*, 41. [CrossRef] [PubMed]
6. Prakash, N.R.; Lokeshkumar, B.M.; Rathore, S.; Warraich, A.S.; Yadav, S.; Vinaykumar, N.M.; Dushyanthkumar, B.M.; Krishnamurthy, S.L.; Sharma, P.C. Meta-analysis and validation of genomic loci governing seedling and reproductive stage salinity tolerance in rice. *Physiol. Plant.* **2022**, *174*, e13629. [CrossRef] [PubMed]
7. Mazumder, A.; Rohilla, M.; Bisht, D.; Krishnamurthy, S.L.; Barman, M.; Sarma, R.; Sharma, T.; Mondal, T. Identification and mapping of quantitative trait loci (QTL) and epistatic QTL for salinity tolerance at seedling stage in traditional aromatic short grain rice landrace Kolajoha (Oryza sativa L.) of Assam. *India. Euphytica* **2020**, *216*, 75. [CrossRef]
8. Gregorio, G.B. Tagging Salinity Tolerance Genes in Rice Using Amplified Fragment Length Polymorphism (AFLP). Ph.D. Thesis, University of the Philippines Los Banõs, Laguna, Philippines, 1997.
9. Krishnamurthy, S.L.; Pundir, P.; Warraich, A.S.; Rathor, S.; Lokeshkumar, B.M.; Singh, N.K.; Sharma, P.C. Introgressed Saltol QTL Lines Improves the Salinity Tolerance in Rice at Seedling Stage. *Front. Plant Sci.* **2020**, *11*, 833. [CrossRef] [PubMed]
10. Yadav, A.K.; Kumar, A.; Grover, N.; Ellur, R.K.; Krishnan, S.G.; Bollinedi, H.; Bhowmick, P.K.; Vinod, K.K.; Nagarajan, M.; Krishnamurthy, S.L.; et al. Marker aided introgression of 'Saltol', a major QTL for seedling stage salinity tolerance into an elite Basmati rice variety 'Pusa Basmati 1509'. *Sci. Rep.* **2020**, *10*, 13877. [CrossRef] [PubMed]
11. Bandillo, N.; Raghavan, C.; Muyco, P.A.; Sevilla, M.A.; Lobina, I.T.; Dilla-Ermita, C.J.; Tung, C.W.; McCouch, S.; Thomson, M.; Mauleon, R.; et al. Multi-parent advanced generation inter-cross (MAGIC) populations in rice: Progress and potential for genetics research and breeding. *Rice* **2013**, *6*, 11. [CrossRef] [PubMed]
12. Chai, C.; Shankar, R.; Jain, M. Genome-wide discovery of DNA polymorphisms by whole genome sequencing differentiates weedy and cultivated rice. *Sci. Rep.* **2018**, *8*, 14218. [CrossRef] [PubMed]
13. Kumar, R.; Subba, A.; Kaur, C.; Ariyadasa, T.U.; Sharan, A.; Pareek, A.; Sopory, S.K.; Singla-Pareek, S.L. OsCBSCBSPB4 is a two cystathionine-β-synthase domain-containing protein from rice that functions in abiotic stress tolerance. *Curr. Genom.* **2017**, *19*, 50–59. [CrossRef] [PubMed]
14. Li, Y.; Xu, M. CCT family genes in cereal crops: A current overview. *Crop J.* **2017**, *5*, 449–458. [CrossRef]
15. Lei, G.; Shen, M.; Li, Z.G.; Zhang, B.; Duan, K.X.; Wang, N.; Cao, Y.R.; Zhang, W.K.; Ma, B.; Ling, H.Q.; et al. EIN2 regulates salt stress response and interacts with a MA3 domain-containing protein ECIP1 in Arabidopsis. *Plant Cell Environ.* **2011**, *34*, 1678–1692. [CrossRef] [PubMed]
16. Fekih, R.; Tamiru, M.; Kanzaki, H.; Abe, A.; Yoshida, K.; Kanzaki, E.; Saitoh, H.; Takagi, H.; Natsume, S.; Undan, J.R.; et al. The rice (Oryza sativa L.) lesion mimic resembling, which encodes an AAA-type ATPase, is implicated in defence response. *Mol. Genet. Genom.* **2015**, *290*, 611–622. [CrossRef] [PubMed]
17. Hartl, F.; Bracher, A.; Hayer-Hartl, M. Molecular chaperones in protein folding and proteostasis. *Nature* **2011**, *475*, 324–332. [CrossRef] [PubMed]
18. Wang, Y.; Huang, M.; Gao, P.; Chen, H.; Zheng, Y.; Yang, C.; Yang, Z.; Sun, Q. Expression of heat shock protein (HSP) genes and antioxidant enzyme genes in hybrid rice II YOU 838 during heat stress. *Aust. J. Crop Sci.* **2021**, *4*, 38–43. [CrossRef]
19. Ganie, S.A.; Pani, D.R.; Mondal, T.K. Genome-wide analysis of DUF221 domain-containing gene family in Oryza species and identification of its salinity stress-responsive members in rice. *PLoS ONE* **2017**, *12*, e0182469. [CrossRef] [PubMed]
20. Zhai, Y.; Wen, Z.; Han, Y.; Wang, F.; Xi, C.; Liu, J.; Gao, P.; Zhao, H.; Wang, Y.; Wang, Y.; et al. Heterogeneous expression of plasma-membrane-localised OsOSCA1.4 complements osmotic sensing based on hyperosmolality and salt stress in *Arabidopsis osca1* mutant. *Cell Calcium* **2020**, *91*, 102261. [CrossRef] [PubMed]

environmental sciences proceedings

MDPI

Proceeding Paper

Development of Salt-Tolerant Rice Varieties to Enhancing Productivity in Salt-Affected Environments †

Saraswathipura L. Krishnamurthy [1,*], Byragondlu M. Lokeshkumar [1], Suman Rathor [1], Arvinder S. Warraich [1], Satyendra Yadav [1], Raj K. Gautam [1,2], Rakesh Kumar Singh [1,3] and Parbodh C. Sharma [1,*]

1 ICAR—Central Soil Salinity Research Institute, Karnal 132001, India; lokeshbm8050@gmail.com (B.M.L.); suman.rathore08@gmail.com (S.R.); arvindersinghwarraich@gmail.com (A.S.W.); ysatyendra791@gmail.com (S.Y.); rgautampb@gmail.com (R.K.G.); r.singh@biosaline.org.ae (R.K.S.)
2 ICAR—National Bureau of Plant Genetic Resources, New Delhi 110012, India
3 International Centre for Bio Saline Agriculture, Dubai 122002, United Arab Emirates
* Correspondence: krishnagene@gmail.com (S.L.K.); pcsharma.knl@gmail.com (P.C.S.)
† Presented at the 2nd International Laayoune Forum on Biosaline Agriculture, 14–16 June 2022; Available online: https://lafoba2.sciforum.net/.

Abstract: Among abiotic stresses, salt stress is the most complicated problem posing a major challenge for maintaining world food supplies as well as food security as it covers 1125 m ha globally and 6.73 m ha in India. It is very essential to increase rice productivity in salt-affected soils for food security and sustainability in salt-affected environments. The pass port data of 9000 rice Germplasm has been established for 30 traits and a mini core of 1500 lines has been developed. Approximately 20,000 rice lines have been screened for salinity and sodicity for both seedling and reproductive stages. The highly tolerant rice lines are being used in breeding programme to enhance the genetic diversity. The six QTLs in chromosomes 1, 2, 8, 9, 10 has been identified for the Na/K ratio, spikelet fertility and grain yield for the reproductive stage salt tolerance. The *Saltol* QTL has been transferred to Indian mega rice varieties namely, Pusa44 and Sarjoo52, to improve the salinity tolerance at the seedling stage. Similarly, the *qSSISFHS8.1* (QTL for spikelet fertility) QTL is being transferred into mega rice varieties namely, PR114, Pusa44 and Sarjoo52, to improve the salinity tolerance at the reproductive stage. The marker assisted selection accelerates the breeding activities to develop the salt-tolerant varieties as well as transfer the QTLs to HVYs through marker assisted back cross breeding. The Central Soil Salinity Research Institute (CSSRI) has developed 13 salt-tolerant rice varieties, namely CSR10, CSR13, CSR23, CSR27, Basmati CSR30, CSR36, CSR43, CSR46, CSR49, CSR52, CSR56, CSR60 and CSR76, through conventional breeding approaches to increase the productivity of salt-affected soils for resource poor farmers. The popularity of these salt-tolerant rice varieties can be ascertained from the facts that large quantity of breeder seeds (1056 quintals) and truthfully labelled seeds (3126 quintals) has been produced and sold during last 22 years (2000–2021), more demand from the farmers and large area app. One point two million hectares is covered by these salt-tolerant rice varieties every year. Numerous salt-tolerant rice lines are being developed at CSSRI to cope with salt-affected soils and enable sustainable agriculture under salt-affected soils.

Keywords: rice; salinity; sodicity

Citation: Krishnamurthy, S.L.; Lokeshkumar, B.M.; Rathor, S.; Warraich, A.S.; Yadav, S.; Gautam, R.K.; Singh, R.K.; Sharma, P.C. Development of Salt-Tolerant Rice Varieties to Enhancing Productivity in Salt-Affected Environments. *Environ. Sci. Proc.* **2022**, *13*, 30. https://doi.org/doi:10.3390/environsciproc2022016030

Academic Editor: Abdelaziz Hirich

Published: 13 June 2022

1. Introduction

Rice is one of the important staple food crops and is consumed by approximately 50% of the global population as their main source of energy. The world population is increasing and will reach 9.5 billion by year 2050 and world food production will need to increase by 70% [1]. Therefore, there is a need to increase the productivity of rice by overcoming the lower crop productivity caused by various biotic and abiotic stresses. Among the abiotic stresses, salinity and sodicty are the most important environmental factors hampering crop productivity as they cause ionic disequilibrium and disrupt the metabolic activities of the plants [2]. Salinity and sodicity stress is highly influenced by environmental fluctuations such as rising temperature and relative humidity [3]. Salt stress affects all the stages of in plants, but early seedling and reproductive stages are the most sensitive in rice [4,5]. Salinity stress reduces root length, shoot length at the seedling stage and tiller number, number of spikes, panicle length and spikelet fertility at the reproductive stage [6,7]. Sodicity also affects crop productivity. Many workers have reported the effect of sodic soils on growth and development in rice [8–10]. Collection of the rice germplasm from salt-affected soils and other locations and their evaluation is crucial in breeding programmes. After the green revolution, breeding programs were adapted to develop saline tolerant genotypes at IRRI [11]. Extensive research has been redirected towards the advancement of salt resistance at the seedling stage of rice, but little effort has been made to recognize QTLs related to the reproductive stage salinity [12–19]. QTL mapping is generally based solely on the phenotypic performance under stress; therefore, the utilization of comparative performance of various genotypes under stress and non-stress conditions will be useful in identifying and mapping the QTLs for the development of stress-resistant varieties [20]. Introgression of QTLs into high-yielding mega varieties is an essential step in molecular breeding programmes once the identification of QTLs and fine mapping is completed. Conventional breeding has been very successful in developing salt-tolerant varieties of rice. The dissemination of salt-tolerant rice varieties to farmer's fields of salt-affected locations is a crucial step to enhance the productivity of salt-affected soils.

2. Methods

Plant Materials

A total of 9000 rice Germplasms were collected and evaluated for 30 different qualitative and quantitative traits and a mini core of 1500 lines was developed from the 9000 rice Germplasm. Approximately 20,000 rice lines were screened for salinity and sodicity for both seedling and reproductive stages during 2020–2022 [7,21,22] to identify salt-tolerant lines. Different mapping populations, namely GWAS panel [18,23] Bi-parental mapping populations of CSR11/MI48 and CSR11/MI48 [17]; CSR10/PS5 [16], CSR27/MI48 [24]; and MAGIC population, were used to identify the QTLs governing salt tolerance. Different segregating populations were evaluated in salt stress conditions to select good performing recombinants. We developed and released the 13 salt-tolerant rice varieties for different zones of the country after evaluation of AICRP rice. Breeder seed and truthfully label seed of these salt-tolerant rice varieties were produced and sold to different seed producing agencies and farmers.

3. Results and Discussion

3.1. Collection and Evaluation of Rice Germplasms

We collected the 9000 rice Germplasms from the National Bureau of Plant Genetic Resources (NBPGR), New Delhi. We evaluated them for 30 qualitative and quantitative traits. We developed a mini core collection of 1500 rice Germplasms and evaluated them in salt stress conditions. We were able to identify 85 tolerant rice lines that are being used in breeding programmes to enhance the salt tolerance. We also evaluated 7500 rice lines from 2012–2021 for saline stress of EC ~ 10.00 dS/m at the seedling stage along with the sensitivity check IR29 and tolerance check FL478. We identified 241 tolerant rice lines. We are using these lines in our breeding programme to diversify the genetic base and enhance salinity tolerance.

3.2. Identification and Introgression QTLs for Salt Tolerance

We evaluated 391 MAGIC lines in saline stress of EC ~ 10.00 dS/m and sodic stress of pH of 9.7 at the seedling stage. We used GBS data of 391 MAGIC lines and identified QTLs for salinity and sodicity tolerance. One QTL was identified on chromosome 1 that had already been reported. Two novel QTLs tolerant to sodicity were mapped on chromosomes

6 and 11. An aluminium activated malate gene transporter, which is related to sodic stress, was localised on chromosome 6. Another gene, Os11g0565400 (Ring finger protein), was found on chromosome 11. These two genes could be used for the enhancement of sodicity tolerance in rice. We identified marker trait associations (MTA) for physiological traits at the reproductive stage. We identified 28 marker trait associations (MTA) on different chromosomes' physiological traits. In another study, we also identified two significant marker trait associations (MTA) on chromosome 1 for Na/K homeostasis. The biparental mapping population (CSR11/MI48) was used to identify the QTLs for sodicity tolerance at the reproductive stage. We phenotyped the CSR11/MI48 RIL population in normal (pH ~7.5) and sodic (pH ~9.5) stress across three seasons for grain yield and its contributing traits and genotyped using 50K SNP. We identified a major QTL qSSI6.2 on chromosome 6 for grain yield at sodic stress with a phenotypic variation of 32%. We introgressed Saltol QTL into two mega rice varieties, namely Pusa44 and Sarjoo52, through marker-assisted back crossbreeding. We used FL478 as donor parents and developed 10 near-isogenic lines (NILs) in the background of pusa44 and 8 near-isogenic lines (NILs) in the background of Sarjoo52. These NILs were tolerant up to a salinity of EC ~ 10.00 dS/m at the seedling stage. A major QTL qSSISFH8.1 for spikelet fertility was identified and fine mapped for salinity tolerance at the reproductive stage on chromosome 8 in CSR27. Introgressing of qSSISFH 8.1 into three mega rice varieties namely, Pusa44, PR114 and Sarjoo52 through marker-assisted back crossbreeding is underway. We developed BC_3F_2 populations in the Pusa44, PR114 and Sarjoo52 background. We also followed the stringent phenotyping to select the near isogenic lines (NILs).

3.3. *Development of Salt-Tolerant Rice Varieties*

Recently, we developed six salt-tolerant rice varieties namely, CSR46, CSR49, CSR52, CSR56, CSR60 and CSR76, through conventional breeding approaches to increase the productivity of salt-affected soils for resource poor farmers. Recently, we developed six salt-tolerant rice varieties. CSR46 is recommended for sodic soils of Uttar Pradesh. CSR46 is tolerant to a sodicity pH of 9.9 and salinity of EC ~ 8.00 dS/m. It is a medium duration rice variety with long slender grain. It has a yield potential of 7.0 t/ha in normal and 4.0 t/ha in salt stress situations. Rice variety CSR56 tolerates a sodicity pH of 9.9. It is a medium duration rice variety with long bold grain. It has a yield potential of 7.5 t/ha in normal and 4.1 t/ha in salt-stress situations. It is recommended for sodic soils of Uttar Pradesh and Haryana. CSR60 is tolerant to a sodicity pH of 9.9. It is medium duration rice variety with long slender grain. It has yield potentials of 7.5 t/ha in normal and 4.3 t/ha in salt stress situation. It is recommended for sodic soils of Uttar Pradesh and Pondicherry. CSR76 is tolerant to a sodicity pH of 9.9. It is medium duration rice variety with long slender grain. It has yield potentials of 7.5 t/ha in normal and 4.5 t/ha in salt stress situation. It is recommended for sodic soils of Uttar Pradesh. Many salt-tolerant rice varieties have been released for commercial cultivation in India. We have produced 1056 and 3126 quintals of breeder and truthfully labelled seeds of salt-tolerant rice varieties, respectively, during last 22 years (2000–2021). These salt-tolerant rice varieties covered an area of approximately 1.02 m ha every year.

4. Conclusions

These salt tolerant rice varieties cover about 1.6 m ha of salt-affected land and are estimated to earn USD 4732 million annually to the national exchequer. These salt tolerant varieties have been adopted by land reclamation corporations, benefiting 430,000 famers. With the introduction of these salt-tolerant varieties, about 0.6 million tons of gypsum has been saved, which would have cost USD 250,000. The salt-tolerant rice varieties facilitated the management of salt-affected soils and sustainable agriculture under salt-affected soils.

Author Contributions: Conceptualization, S.L.K. and P.C.S.; methodology, S.L.K., R.K.G. and R.K.S.; software, B.M.L., S.R., S.Y. and A.S.W.; validation, S.L.K., P.C.S. and R.K.S.; formal analysis, B.M.L., S.R., S.Y. and A.S.W.; investigation, S.L.K., R.K.S., R.K.G. and P.C.S.; resources, S.L.K. and P.C.S.; data curation, B.M.L. and S.Y.; writing—original draft preparation, S.L.K., B.M.L., S.R., A.S.W. and S.Y.; writing—review and editing, S.L.K., B.M.L. and S.R.; visualization, B.M.L.; supervision, S.L.K. and P.C.S.; project administration, S.L.K. and P.C.S. All authors have read and agreed to the published version of the manuscript.

Funding: Authors thanks to Direct CSSRI, Karnal and DST for funding for the project code—CRG/2020/003078. Authors thanks to ICAR for funding for the project Incentivizing in Agriculture Research, From QTL to Variety" (Grant No. BT/PR/14544/AGR/02/745/2010) through Department of Biotechnology, Government of India.

Informed Consent Statement: Not applicable.

Data Availability Statement: Not applicable.

Acknowledgments: Authors thank to Director, CSSRI for support to carry out the experiments.

Conflicts of Interest: The authors declare no conflict of interest.

References

1. FAO; IFAD; WFP. The State of Food Insecurity in the World. In *The Multiple Dimensions of Food Security*; FAO: Rome, Italy, 2013.
2. Munns, R.; Tester, M. Mechanisms of salinity tolerance. *Annu. Rev. Plant Biol.* **2008**, *59*, 651–681. [CrossRef] [PubMed]
3. Tack, J.; Singh, R.K.; Nalley, L.L.; Viraktamath, B.C.; Krishnamurthy, S.L.; Lyman, N.; Jagadish, K.S.V. High vapor pressure deficit drives salt-stress induced rice yield losses in India. *Glob. Change Biol.* **2015**, *21*, 1668–1678. [CrossRef] [PubMed]
4. Singh, R.K.; Redoña, E.D.; Refuerzo, L. Varietal improvement for abiotic stress tolerance in crop plants: Special reference to salinity in rice. In *Abiotic Stress Adaptation in Plants: Physio-Logical, Molecular and Genomic Foundation*; Pareek, A., Sopory, S.K., Bohnert, H.J., Eds.; Springer: New York, NY, USA, 2010; pp. 387–415.
5. Singh, R.; Flowers, T. Physiology and molecular biology of the effects of salinity on rice. In *Handbook of Plant and Crop Stress*; Pessarakli, M., Ed.; CRC Press: Boca Raton, FL, USA, 2010; pp. 899–939.
6. Ali, S.; Gautam, R.K.; Mahajan, R.; Krishnamurthy, S.L.; Sharma, S.K.; Singh, R.K.; Ismail, A.M. Stress indices and selectable traits in SALTOL QTL introgressed rice genotypes for reproductive stage tolerance to Sodicity and salinity stresses. *Field Crops Res.* **2013**, *154*, 65–73. [CrossRef]
7. Krishnamurthy, S.L.; Gautam, R.K.; Sharma, P.C.; Sharma, D.K. Effect of different salt stresses on agro-morphological traits and utilization of salt stress indices for reproductive stage salt tolerance in rice. *Field Crops Res.* **2016**, *190*, 26–33. [CrossRef]
8. Krishnamurthy, S.L.; Sharma, S.K.; Kumar, V.; Tiwari, S.; Singh, N.K. Analysis of genomic region spanning Saltol using SSR markers in rice genotypes showing differential seedlings stage salt tolerance. *J. Plant Biochem. Biot.* **2015**, *25*, 331–336. [CrossRef]
9. Singh, Y.P.; Singh, D.; Sharma, S.K.; Krishnamurthy, S.L. Evaluation of rice genotypes for yield, physiological and biological traits in sodic soil. *J. Soil Salin. Water Qual.* **2013**, *5*, 40–49.
10. Krishnamurthy, S.L.; Sharma, S.K.; Gautam, R.K.; Kumar, V. Path and Association analysis and Stress indices for salinity tolerance traits in promising rice (*Oryza sativa* L.) genotypes. *Cereal Res. Commun.* **2014**, *42*, 474–483. [CrossRef]
11. Gregorio, G.B. Tagging Salinity Tolerance Genes in Rice Using Amplified Fragment Length Polymorphism (AFLP). Ph.D. Thesis, University of the Philippines, Los Baños, Philippines, 1997.
12. Zeng, L.; Shannon, M.C.; Grieve, C.M. Evaluation of salt tolerance in rice genotypes by multiple agronomic parameters. *Euphytica* **2002**, *127*, 235–245. [CrossRef]
13. Mohammadi, R.; Mendioro, M.S.; Diaz, G.Q.; Gregorio, G.B.; Singh, R.K. Mapping quantitative trait loci associated to yield and yield components under reproductive stage salinity stress in rice (*Oryza sativa* L.). *J. Genet.* **2013**, *92*, 433–443. [CrossRef] [PubMed]
14. Bimpong, I.K.; Manneh, B.; Diop, B.; Ghislain, K.; Sow, A.; Amoah, N.K.A.; Gregorio, G.B. New quantitative trait loci for enhancing adaptation to salinity in rice from Hasawi, a Saudi landrace into three African cultivars at the reproductive stage. *Euphytica* **2014**, *200*, 45–60. [CrossRef]
15. Hossain, H.; Rahman, M.A.; Alam, M.S.; Singh, R.K. Mapping of quantitative trait loci associated with reproductive- stage salt tolerance in rice. *J. Agron. Crop Sci.* **2015**, *201*, 17–31. [CrossRef]
16. Pundir, P.; Devi, A.; Krishnamurthy, S.L.; Sharma, P.C.; Vinaykumar, N.M. QTLs in salt rice variety CSR10 reveals salinity tolerance at reproductive stage. *Acta Physiol. Plant.* **2021**, *43*, 35. [CrossRef]
17. Tiwari, S.; Krishnamurthy, S.L.; Kumar, V.; Singh, B.; Rao, A.R.; Amitha Mithra, S.V.S.; Rai, V.; Singh, A.K.; Singh, N.K. Mapping QTLs for salt tolerance in rice (*Oryza sativa* L.) by bulked segregant analysis of recombinant inbred lines using 50K SNP chip. *PLoS ONE*. **2015**, *11*, e0153610. [CrossRef] [PubMed]
18. Kumar, V.; Singh, A.; Mithra, A.S.V.; Krishnamurthy, S.L.; Parida, S.K.; Jain, S.; Mohapatra, T. Genome-wide association mapping of salinity tolerance in rice (*Oryza sativa*). *DNA Res.* **2015**, *22*, 133–145. [CrossRef] [PubMed]

19. Warraich, A.S.; Krishnamurthy, S.L.; Sooch, B.S.; Vinaykumar, N.M.; Dushyanthkumar, B.M.; Bose, J.; Sharma, P.C. Rice GWAS reveals key genomic regions essential for salinity tolerance at reproductive stage. *Acta Physiol. Plant.* **2020**, *42*, 134. [CrossRef]
20. Raman, A.; Verulkar, S.; Mandal, N.; Variar, M.; Shukla, V.; Dwivedi, J.; Singh, B.; Singh, O.; Swain, P.; Mall, A.; et al. Drought yield index to select high yielding rice lines under different drought stress severities. *Rice* **2012**, *5*, 31. [CrossRef] [PubMed]
21. Sharma, S.K.; Jogendra, S.; Chauhan, M.S.; Krishnamurthy, S.L. Multivariate Analysis of Phenotypic Diversity of Rice germplasm in North-West India. *Indian J. Agric. Sci.* **2014**, *84*, 120–124.
22. Krishnamurthy, S.L.; Sharma, P.C.; Sharma, D.K.; Ravikiran, K.T.; Singh, Y.P.; Mishra, V.K.; Burman, D.; Maji, B.; Mandal, S.; Sarangi, S.K.; et al. Identification of mega-environments and rice genotypes for general and specific adaptation to saline and alkaline stresses in India. *Sci. Rep.* **2017**, *7*, 7968. [CrossRef] [PubMed]
23. Warraich, A.S.; Krishnamurthy, S.L.; Sooch, B.S.; Lokeshkumar, B.M.; Vinaykumar, N.M.; Dushyanthkumar, B.M.; Rai, V.; Singh, N.K.; Sharma, P.C. Population structure and association studies for reproductive stage salinity tolerance in rice (*Oryza sativa* L.). *Indian J. Exp. Biol.* **2021**, *59*, 328–338.
24. Pandit, A.; Rai, V.; Bal, S.; Sinha, S.; Kumar, V.; Chauhan, M.; Singh, N.K. Combining QTL mapping and transcriptome profiling of bulked RILs for identification of functional polymorphism for salt tolerance genes in rice (*Oryza sativa* L.). *Mol. Genet. Genom.* **2010**, *284*, 121–136. [CrossRef] [PubMed]

environmental sciences proceedings

MDPI

Proceeding Paper

Dissection of Genomic Regions for Ion Homeostasis under Sodic Salt Stress in MAGIC Rice Population †

Suman Rathor [1], Saraswathipura L. Krishnamurthy [1,*], Bayragondlu M. Lokeshkumar [1], Arvinder S. Warraich [1], Satyendra Yadav [1], Parbodh C. Sharma [1,*] and Rakesh Kumar Singh [2]

1 ICAR—Central Soil Salinity Research Institute, Karnal 132001, India; suman.rathore08@gmail.com (S.R.); lokeshbm8050@gmail.com (B.M.L.); arvindersinghwarraich@gmail.com (A.S.W.); ysatyendra791@gmail.com (S.Y.)

2 International Center for Biosaline Agriculture, Dubai 122002, United Arab Emirates; r.singh@biosaline.org.ae

* Correspondence: krishnagene@gmail.com (S.L.K.); pcsharma.knl@gmail.com (P.C.S.)

† Presented at the 2nd International Laayoune Forum on Biosaline Agriculture, 14–16 June 2022; Available online: https://lafoba2.sciforum.net/.

Abstract: Salt tolerance mechanisms are regulated by balance in cell ionic concentrations such as K^+, Na^+, H^+, Ca^{2+}, and Mg^{2+}. In this study, we examined major QTLs for the traits K^+/Na^+ homeostasis, shoot magnesium content (Mg^{2+}), shoot calcium content (Ca^{2+}), and shoot length. The QTLs for K^+/Na^+ homeostasis *Sod K/Na.1* are associated with three candidate genes: *LOC_Os02g48290, LOC_Os02g48340, and LOC_Os02g48350*, and *Sod_Ca.1* is associated with the gene *LOC_Os08g15020*. Three significant candidate gene haplotypes for shoot length, *Sod_SL.1* (*LOC_Os10g36690*), sodium content *Sod_Na.1* (*LOC_Os01g41770*), and magnesium content *Sod_Mg.1* (*LOC_Os10g31040*) were identified. The identified candidate genes encode dehydration response proteins, leucine rich repeat proteins, citrate transporter proteins, and diacylglycerol O-acyltransferase (DGATs), and play a key role in salt and abiotic stress tolerance. The identified novel QTLs and potential candidate genes could be used for functional characterization to help further supplement our understanding of the genetic makeup of sodicity stress tolerance in rice.

Keywords: GWAS; MAGIC; GBS; sodicity; rice; aluminium tolerance; QTL; salt stress

Citation: Rathor, S.; Krishnamurthy, S.L.; Lokeshkumar, B.M.; Warraich, A.S.; Yadav, S.; Sharma, P.C.; Singh, R.K. Dissection of Genomic Regions for Ion Homeostasis under Sodic Salt Stress in MAGIC Rice Population. *Environ. Sci. Proc.* **2022**, *16*, 39. https://doi.org/10.3390/environsciproc2022016039

Academic Editor: Abdelaziz Hirich

Published: 16 June 2022

1. Introduction

Rice is an important staple food for more than 3.5 billion people in the world and is cultivated in 114 countries [1]. Exploding population with urbanization are not only threats to food security, but also affect global climate change by increasing temperature and decreasing cultivable lands. The effect of climate change on the agriculture sector is creating massive emerging problems such as biotic and abiotic stresses. Among the abiotic stresses, salt stress (salinity and sodicity) is the most important environmental factor hampering crop productivity. More than 6% of the world's (900 Mha) soil is facing intrusion by salt [1]. During salt stress, crops experience ion imbalances, ion toxicity, and reduced water potential, which affects the normal plant metabolism and crop yield [2]. Many researchers have reported the effect of sodic soils on growth and development in rice [3]. Excess Na^+ accumulation creates ionic stress in the aerial parts of plants since Na^+ interferes with plant physiology, including imbalances in the homeostasis of other ions such as K^+, Ca^{2+}, and Mg^{2+}. Hence, high cytosolic K^+/Na^+ ratios become a key salt tolerance trait. Several QTLs were identified for salinity tolerance in rice [4,5]. A major QTL, *Saltol*, was mapped on chromosome 1 and was introgressed into mega rice varieties [6–8]. However, less information is available on the molecular basis for sodicity tolerance. In this study, we intended to find the effect of sodicity on the *indica* MAGIC population and to identify the novel QTLs for sodicity tolerance in rice at the seedling stage.

2. Materials and Methods

The plant material consisted of 391 rice MAGIC lines developed from an intercrossing of eight founder lines at the International Rice Research Institute, Philippines [9]. These lines were evaluated under control and alkaline condition at ICAR-CSSRI, Karnal. The seeds of rice lines were sown in soil contained within a tray. The desired pH of the soil (pH ~9.7–9.8) was created on 14 DAS using sodium bicarbonate ($NaHCO_3$) and sodium carbonate (Na_2CO_3) solution. The data on the morphological traits, shoot length, and tissue samples were taken to measure physiological parameters such as shoot Na^+, K^+, Ca^{2+}, and Mg^{2+} content (ppm) 14 days after stress, and the K^+/Na^+ ratio was calculated. Genotyping of founder and MAGIC lines was done by sequencing using the Illumina Hi Seq method. The processed credible 27,041 SNP sites were then used for marker-trait association studies (MTA).

MTAs were identified using Bayesian-information and Linkage-disequilibrium Iteratively Nested Keyway (BLINK) implemented using the genomic association and prediction integrated tool (GAPIT). The SNP positions of genes associated with traits of interest were considered as candidate genes. The haplotype analyses for identified candidate genes were conducted by CandiHap V2 (https://github.com/xukaili/CandiHap, accessed on 20 December 2020). Visualization of the violin plot was done using the ggplot2 package in R programme.

3. Results and Discussion

Genotype by sequencing (GBS) data of 391 *indica* MAGIC rice lines were used for association for morph-physiological traits such as shoot length, shoot Na^+, K^+ Ca^{2+}, Mg^{2+}, and tissue K^+/Na^+ ratio. The eight QTLs distributed on chromosomes 1, 2, 4, 8, and 10 that were responsible for sodicity tolerances at the seedling stage are presented in Table 1.

The QTL *Sod_SL.1* was responsible for shoot length exhibiting 9.01% phenotypic variance. The peak SNP (Position—19621750 on Chromosome 10) of *Sod_SL.1* QTL was lying in the genic region of *LOC_Os10g36690* and encoded a dehydration response protein to have a significant role under salt stress [10]. The given gene is characterized as a member of the dehydration responsive element-binding (DREB) super family of genes, and over expression of the *OsDREB1F* gene is confirmed to enhance salt tolerance in rice [11]. Three QTLs *Sod_Na.1*, *Sod_Na.2*, and *Sod_Na.3* for shoot sodium content were displaying 3.46, 6.22, and 5.23% phenotypic variance, respectively. The peak SNPs position-23642384 on Chromosome 1 was lying in the genic region of *LOC_Os01g41770* encoding leucine-rich repeat protein. The role of the leucine-rich repeat receptor-like kinases gene (*OsSTLK*) in response to salt tolerance was reported in rice [11,12]. Over expression of *OsSTLK* exhibited reduced malondialdehyde (MDA) content, electrolyte leakage, and reactive oxygen species (ROS) under salt stress conditions. The only QTL (*Sod_K.1*) was detected with an LOD of 3.43 for shoot potassium content lying in the genic region of *LOC_Os02g32814*. The effects of both sodicity and aluminium became evident at pH~9.0 and waning at pH > 9.2 [12,13]. Hence, the role of genes responsible for metal toxicity under sodic condition cannot be ignored. QTLs for shoot calcium content (*Sod_Ca.1*) and magnesium content (*Sod_Mg.1*) exhibited 9.8 and 20.65 phenotypic variance. The peak SNP responsible for calcium content present on the gene *LOC_Os08g15020* encoding the MYB family transcription factor plays a key regulatory role under drought and salinity [14]. Further, an independent study has also functionally characterized the positive role of the MYB gene under salt stress in rice [15]. The gene in this QTL is a probable MYB, which has the capacity to stimulate plant growth through calcium signaling under salt stress [16]. The peak SNP associated with QTL *Sod_Mg.1* was lying in the genic region of *LOC_Os10g31040*, responsible for citrate transporter protein. In the QTL *Sod_K/Na.1* responsible for K^+/Na^+ homeostasis, we observed six peak SNP markers tracked between the 289 kb region from 29.313 to 29.602 Mb. The peak SNP positions in *Sod_K/Na.1* were lying in the regions of *LOC_Os02g48290*, *LOC_Os02g48340*, and *LOC_Os02g48350* (*OsDGAT*), which encodes thioredoxin reductase, RNA recognition motif containing (RRM) protein and diacylglycerol O-acyltransferase

(DGATs), respectively. The role of thioredoxin reductase in seedling development, photosynthetic metabolism, and plant growth in response to varying light conditions and starch degradation in guard and mesophyll cells under osmotic stress was reported earlier [17–19]. RNA binding proteins (RBP) are able to increase the yeast Na^{+}-tolerance, *Beta vulgaris*. Salt-tolerant (BvSATO) genes BvSATO1, BvSATO2, BvSATO4, and BvSATO6 were RRM containing proteins involved in RNA metabolism, developmental processes, and played an important role in salt tolerance [20]. The co-expression of *OsTCP19* and *LOC_Os02g48350 (OsDGAT)* regulating the triacylglycerol biosynthesis by modulatingABI4-mediated pathways under salt and drought stress conditions was reported in rice [21]. The haplotype analyses for candidate genes linked to seedling stage sodicity tolerance suggested that three genes (*LOC_Os10g36690, LOC_Os01g41770*, and *LOC_Os10g31040*) show significant haplotypes associated with QTLs Sod_SL.1, Sod_Na.1, and Sod_Mg.1, respectively (Figure 1). Identified QTLs associated with sodicity tolerant and respective candidate genes responsible for abiotic stress can be further reinvestigated to confirm their role in salt tolerance in rice.

Table 1. Associated QTLs with SNP position and probable candidate genes for sodicity tolerance in MAGIC population.

Sl. No	Trait	QTL	Chromosome	Position	*p* Value	LOD	Locus ID	Gene Annotation
1	Shoot length	*Sod_SL.1*	10	19621750	1.50×10^{-5}	4.82	*LOC_Os10g36690*	Dehydration response related protein, putative, expressed
2	Na	*Sod_Na.1*	1	23642384	1.54×10^{-5}	4.81	*LOC_Os01g41770*	Leucine rich repeat protein, putative, expressed
3		*Sod_Na.2*	4	11638572	7.35×10^{-5}	4.13	*LOC_Os04g20749*	Expressed protein
4		*Sod_Na.3*	8	4773918	2.38×10^{-6}	5.62		
5	K	*Sod_K.1*	2	19479355	0.000375	3.43	*LOC_Os02g32814*	Heavy metal-associated domain containing protein, expressed
6	Ca	*Sod_Ca.1*	8	9069848	1.11×10^{-5}	4.95	*LOC_Os08g15020*	MYB family transcription factor, putative, expressed
7	Mg	*Sod_Mg.1*	10	16228910	3.41×10^{-7}	6.47	*LOC_Os10g31040*	Citrate transporter protein, putative, expressed
8	K/Na	*Sod_K/Na.1*	2	29313685	1.60×10^{-8}	7.80		
9			2	29550035	2.60×10^{-5}	4.58	*LOC_Os02g48290*	Thioredoxinreductase 2, putative, expressed
10			2	29596887	1.22×10^{-5}	4.91	*LOC_Os02g48340*	RNA recognition motif containing protein, putative, expressed
11			2	29602570	1.22×10^{-5}	4.91	*LOC_Os02g48350*	Diacylglycerol O-acyltransferase, putative, expressed
12			2	29602596	1.22×10^{-5}	4.91		
13			2	29602600	1.22×10^{-5}	4.91		

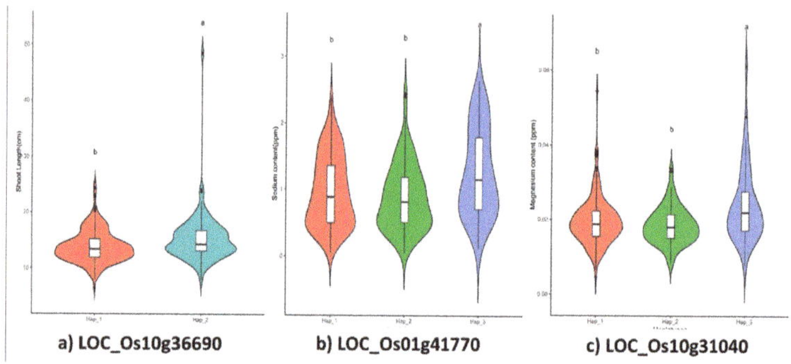

Figure 1. Violin plot indicating the significant haplotypes for candidate genes associated with sodic tolerance QTL; (**a**) shoot length; (**b**) shoot sodium content; (**c**) magnesium content among *indica* MAGIC rice lines (X-axis = Haplotypes; Y-axis = Trait value). Each box plot with in the violin plot represents minimum, lower quartile, median, upper quartile, and maximum values. Different letters above the violin plots indicate statistically significant differences for the respective haplotypes, at a significance level of $p < 0.05$ (Duncan test).

4. Conclusions

In the present study, we identified thirteen SNPs associated with eight QTL regions responsible for sodicity tolerance. We detected major QTLs for the traits K^+/Na^+ homeostasis, shoot magnesium content (Mg^{2+}), shoot calcium (Ca^{2+}), and shoot length, explaining about 32.02%, 20.65%, 9.83%, and 9.01% phenotypic variance, respectively. We found three significant candidate gene haplotypes associated with the QTLs for shoot length (LOC_Os10g36690), shoot sodium content (LOC_Os01g41770), and shoot magnesium content *Sod_Mg.1* (LOC_Os10g31040). The identified QTL regions and candidate genes play significant roles in sodic stress tolerance, which can be further investigated.

Author Contributions: Conceptualization, S.L.K., P.C.S. and R.K.S.; methodology, S.R., B.M.L., A.S.W. and S.Y.; software, S.R., B.M.L., A.S.W. and S.Y.; validation, S.R., B.M.L., A.S.W. and S.Y. formal analysis, S.R., B.M.L., A.S.W. and S.Y.; investigation, S.R., B.M.L., A.S.W. and S.Y.; resources, S.R., B.M.L., A.S.W. and S.Y.; data curation, S.R., B.M.L., A.S.W. and S.Y.; writing—original draft preparation, S.R., B.M.L. and A.S.W.; writing—review and editing, S.L.K., P.C.S. and R.K.S.; visualization, S.L.K., P.C.S. and R.K.S.; supervision, S.L.K., P.C.S. and R.K.S.; project administration S.L.K., P.C.S. and R.K.S.; funding acquisition, S.L.K., P.C.S. and R.K.S. All authors have read and agreed to the published version of the manuscript.

Funding: The authors are thankful to ICAR-New Delhi, and IRRI-Phillipines for funding this research project.

Institutional Review Board Statement: Not applicable.

Informed Consent Statement: Not applicable.

Data Availability Statement: Not applicable.

Acknowledgments: The authors thanks to Director, Central Soil Salinity Research Institute for completing the research work.

Conflicts of Interest: The authors declare no conflict of interest.

References

1. FAO. Extent of Salt Affected Soils. 2014. Available online: http://www.fao.org/soils-portal/soil-management/management-of-someproblem-soils/salt-affected-soils/more-information-on-saltaffected-soils/en/ (accessed on 3 December 2014).
2. Krishnamurthy, S.L.; Gautam, R.; Sharma, P.; Sharma, D. Effect of different salt stresses on agro-morphological traits and utilisation of salt stress indices for reproductive stage salt tolerance in rice. *Field Crop. Res.* **2016**, *190*, 26–33. [CrossRef]

3. Krishnamurthy, S.; Sharma, P.; Sharma, D.; Ravikiran, K.; Singh, Y.; Mishra, V.; Burman, D.; Maji, B.; Mandal, S.; Sarangi, S.K.; et al. Identification of mega-environments and rice genotypes for general and specific adaptation to saline and alkaline stresses in India. *Sci. Rep.* **2017**, *7*, 7968. [CrossRef] [PubMed]
4. Warraich, A.; Krishnamurthy, S.; Sooch, B.; Vinaykumar, N.; Dushyanthkumar, B.; Bose, J.; Sharma, P.C. Rice GWAS reveals key genomic regions essential for salinity tolerance at reproductive stage. *Acta Physiol. Plant.* **2020**, *42*, 134. [CrossRef]
5. Prakash, N.R.; Lokeshkumar, B.M.; Rathor, S.; Warraich, A.S.; Yadav, S.; Vinaykumar, N.M.; Dushynthkumar, B.M.; Krishnamurthy, S.L.; Sharma, P.C. Meta-analysis and validation of genomic loci governing seedling and reproductive stage salinity tolerance in rice. *Physiol. Plant.* **2022**, *18*, e13629. [CrossRef] [PubMed]
6. Gregorio, G.B. Tagging Salinity Tolerance Genes in Rice Using Amplified Fragment Length Polymorphism (AFLP). Ph.D. Thesis, University of the Philippines Los Banõs, Laguna, Philippines, 1997.
7. Babu, N.; Krishnan, S.; Vinod, K.; Krishnamurthy, S.; Singh, V.; Singh, M.; Singh, R.; Ellur, R.K.; Rai, V.; Bollinedi, H.; et al. Marker Aided Incorporation of Saltol, a Major QTL Associated with Seedling Stage Salt Tolerance, into Oryzasativa 'Pusa Basmati 1121'. *Front. Plant Sci.* **2017**, *8*, 41. [CrossRef] [PubMed]
8. Krishnamurthy, S.L.; Pundir, P.; Warraich, A.S.; Rathor, S.; Lokeshkumar, B.M.; Singh, N.K.; Sharma, P.C. Introgressed Saltol QTL Lines Improves the Salinity Tolerance in Rice at Seedling Stage. *Front. Plant Sci.* **2020**, *11*, 833. [CrossRef] [PubMed]
9. Bandillo, N.; Raghavan, C.; Muyco, P.; Sevilla, M.; Lobina, I.; Dilla-Ermita, C.; Tung, C.; McCouch, S.; Thomson, M.; Mauleon, R.; et al. Multi-parent advanced generation inter-cross (MAGIC) populations in rice: Progress and potential for genetics research and breeding. *Rice* **2013**, *6*, 11. [CrossRef] [PubMed]
10. Shavrukov, Y. Salt stress or salt shock: Which genes are we studying? *J. Exp. Bot.* **2012**, *64*, 119–127. [CrossRef] [PubMed]
11. Wang, Q.; Guan, Y.; Wu, Y.; Chen, H.; Chen, F.; Chu, C. Overexpression of a rice OsDREB1F gene increases salt, drought, and low temperature tolerance in both *Arabidopsis* and rice. *Plant Mol. Biol.* **2008**, *67*, 589–602. [CrossRef] [PubMed]
12. Lin, F.; Li, S.; Wang, K.; Tian, H.; Gao, J.; Zhao, Q.; Du, C. A leucine-rich repeat receptor-like kinase, OsSTLK, modulates salt tolerance in rice. *Plant Sci.* **2020**, *296*, 110465. [CrossRef] [PubMed]
13. Brautigan, D.; Rengasamy, P.; Chittleborough, D. Aluminium speciation and phytotoxicity in alkaline soils. *Plant Soil* **2012**, *360*, 187–196. [CrossRef]
14. Tang, Y.; Bao, X.; Zhi, Y.; Wu, Q.; Guo, Y.; Yin, X.; Zeng, L.; Li, J.; Zhang, J.; He, W.; et al. Overexpression of a MYB Family Gene, OsMYB6, Increases Drought and Salinity Stress Tolerance in Transgenic Rice. *Front. Plant Sci.* **2019**, *10*, 168. [CrossRef] [PubMed]
15. Yang, A.; Dai, X.; Zhang, W.H. A R2R3-type MYB gene, OsMYB2, is involved in salt, cold, and dehydration tolerance in rice. *J. Exp. Bot.* **2012**, *63*, 2541–2556. [CrossRef] [PubMed]
16. Fasani, E.; DalCorso, G.; Costa, A.; Zenoni, S.; Furini, A. The *Arabidopsis thaliana* transcription factor MYB59 regulates calcium signalling during plant growth and stress response. *Plant Mol. Biol.* **2019**, *99*, 517–534. [CrossRef]
17. Ojeda, V.; Pérez-Ruiz, J.M.; González, M.; Nájera, V.A.; Sahrawy, M.; Serrato, A.J.; Geigenberger, P.; Cejudo, F.J. NADPH ThioredoxinReductase C and Thioredoxins Act Concertedly in Seedling Development. *Plant Physiol.* **2017**, *174*, 1436–1448. [CrossRef] [PubMed]
18. Valerio, C.; Costa, A.; Marri, L.; Issakidis-Bourguet, E.; Pupillo, P.; Trost, P.; Sparla, F. Thioredoxin-regulated beta-amylase (BAM1) triggers diurnal starch degradation in guard cells, and in mesophyll cells under osmotic stress. *J. Exp. Bot.* **2011**, *62*, 545–555. [CrossRef]
19. Thormählen, I.; Meitzel, T.; Groysman, J.; Öchsner, A.B.; von Roepenack-Lahaye, E.; Naranjo, B.; Cejudo, F.J.; Geigenberger, P. Thioredoxin f1 and NADPH-Dependent ThioredoxinReductase C Have Overlapping Functions in Regulating Photosynthetic Metabolism and Plant Growth in Response to Varying Light Conditions. *Plant Physiol.* **2015**, *169*, 1766–1786. [CrossRef] [PubMed]
20. Rosa Téllez, S.; Kanhonou, R.; CastelloteBellés, C.; Serrano, R.; Alepuz, P.; Ros, R. RNA-Binding Proteins as Targets to Improve Salt Stress Tolerance in Crops. *Agronomy* **2020**, *10*, 250. [CrossRef]
21. Mukhopadhyay, P.; Tyagi, A. *OsTCP19* influences developmental and abiotic stress signaling by modulatingABI4-mediated pathways. *Sci. Rep.* **2015**, *5*, 9998. [CrossRef] [PubMed]

environmental sciences proceedings

MDPI

Proceeding Paper

Phenotypic Evaluation of Recombinant Inbred Lines for Sodicity Tolerance at Reproductive Stage in Rice †

Ashish Nain [1,2], Saraswathipura L. Krishnamurthy [1,*], Parbodh C. Sharma [1,*], Bayragondlu M. Lokeshkumar [1], Mukesh Kumar [2] and Arvinder S. Warraich [1]

1 Division of Crop Improvement, ICAR—Central Soil Salinity Research Institute, Karnal 132001, India; a.s.nain001@gmail.com (A.N.); lokeshbm8050@gmail.com (B.M.L.); arvindersinghwarraich@gmail.com (A.S.W.)

2 Department of Genetics and Plant Breeding, CCS Haryana Agricultural University, Hisar 122004, India; mukeshhau@yahoo.com

* Correspondence: krishnagene@gmail.com (S.L.K.); pcsharma.knl@gmail.com (P.C.S.)

† Presented at the 2nd International Laayoune Forum on Biosaline Agriculture, 14–16 June 2022; Available online: https://lafoba2.sciforum.net/.

Abstract: Salt stress is the most widespread soil problem in rice-growing countries, and it causes significant reductions in rice production worldwide. Identifying QTLs for sodicity tolerance at the reproductive stage is crucial to enhance the grain yield in sodic stress ecology. From this perspective, we developed recombinant inbred lines (RILs) from MTU 1001/Kalarata. A set of 176 recombinant inbred lines (RILs) was evaluated, along with the tolerant parent Kalarata and the sensitive parent MTU 1001, in a controlled microplot with a sodicity of pH ~ 9.5 at ICAR - Central Soil Salinity Research Institute (CSSRI), Karnal. The lines RIL34 (13.1 g/plant), RIL83 (12.7 g/plant), RIL40 (11 g/plant), RIL41 (10.2 g/plant), and RIL124 (10.1 g/plant) were top yielders. The yield-contributing traits, namely, plant height (cm), panicle length (cm), total tiller, productive tiller, biological weight (g/plant), and spikelet fertility (%), were highly affected in sodic stress conditions. The tolerant lines RIL34, RIL83, RIL40, RIL41, and RIL124 could be used for breeding programs and further studies to dissect the molecular and physiological mechanisms of reproductive stage sodic stress tolerance in rice.

Keywords: sodicity tolerance; RILs; rice; reproductive stage

Citation: Nain, A.; Krishnamurthy, S.L.; Sharma, P.C.; Lokeshkumar, B.M.; Kumar, M.; Warraich, A.S. Phenotypic Evaluation of Recombinant Inbred Lines for Sodicity Tolerance at Reproductive Stage in Rice. *Environ. Sci. Proc.* **2022**, *16*, 47. https://doi.org/10.3390/environsciproc2022016047

Academic Editor: Abdelaziz Hirich

Published: 16 June 2022

1. Introduction

Rice is a diploid (2n = 2x = 24) cereal crop, which serves as a primary staple food for more than 3.5 billion people in the world, and it is cultivated in more than 114 countries. In developing countries, an exploding population with urbanization and a decrease in cultivable lands are threatening food security. Therefore, achieving food security is of global concern. To meet the food demand, we must increase global rice production by at least 70% to feed the projected number of 9.6 billion people by 2050 [1]; however, at that time, India will have a population size of around 162 million, which will require about 136 mt of rice [2]. Recently, the agricultural sector has been facing various emerging problems of biotic and abiotic stresses. Abiotic stresses, such as drought, temperature, and salinity, seriously affect rice yield to a great extent. Among these, salt stress (both salinity and alkalinity) is the most predominant soil problem in rice-cultivating countries in the world, and it is a serious threat to increasing rice production worldwide [3]. Salt-affected soil has been reported in more than 100 countries, and it is estimated that, out of around one billion hectares of land worldwide, around 580 mha is facing the problem of alkalinity [4]. In India, the total salt-affected area has been estimated to be 6.73 million hectares, of which 2.97 million hectares is saline, whereas 3.77 million hectares is affected by alkaline soils. Rice is a salt-sensitive crop, and exposure to salt stress has serious negative effects on its vegetative growth and grain yield. The effects of salt stress on the morphological, physiological, and biochemical traits of rice have been reported. Salt stress has been found to significantly reduce grain yield by affecting tillering, spikelet filling, plant height, biomass production, and 1000 grain weight [5–7]. Therefore, the development of salt tolerance during the reproductive stage is the only option left for the efficient management of salt-affected soils. Alkalinity tolerance in rice is a complex trait that makes the screening of phenotypes under field conditions very difficult. Furthermore, to improve alkalinity tolerance in rice, it is very important to find sufficient genetic variation and to formulate suitable screening techniques that are reliable and able to identify alkaline-tolerant genotypes [8]. Grain yield is a complex trait, and it is influenced by various genetic and environmental factors. To improve grain yield, an evaluation of breeding lines is the most important step to select the superior genotype. From this perspective, the present investigation was carried out to evaluate an RIL population and to study the association between grain yield and important yield-contributing traits in rice under alkaline conditions.

2. Materials and Methods

A population of 176 recombinant inbred lines (RILs, F_7 generation) derived from a cross between MTU1001 (alkaline sensitive) and Kalarata (alkaline tolerant) was evaluated for alkalinity tolerance at the reproductive stage in alkaline soil microplots (6 $\times$ 3 m) at a facility at ICAR-CSSRI, Karnal. A complete set of the RIL population, along with parents, was raised in normal field conditions in a nursery with the recommended package of practices. Thirty-day-old seedlings were transplanted into alkaline microplots (pH ~ 9.5) in a randomized complete block design with two replications. Before transplanting plant material, pH2 of microplots was measured by dissolving one part of soil in two parts of distilled water. Each genotype was planted in a single row of 3.0 m length following a row x plant spacing of 20 $\times$ 15 cm. One month after transplanting, at the maximum tillering stage, alkaline stress was imposed by irrigating alkaline water created using sodium bicarbonate. Subsequently, the pH of the microplots was recorded weekly and maintained at around pH~9.5 throughout the cropping season up to maturity. Morphological data were recorded as per the standard evaluation system of rice [9]. The data were recorded from 5 randomly labeled plants from each RIL, parent, and check. Morphological characteristics were days to 50% flowering, salt injury score, plant height (cm), panicle length (cm), total tillers per plant, productive tiller per plant, spikelet fertility (%), 1000 grain weight (g), biomass/plant (g), grain yield/plant (g), and harvest index (%). The salt injury score was recorded based on visual symptoms following the IRRI-modified standard evaluation system (SES) for rice, with a score of "1" meaning highly tolerant and a score of "9" meaning highly sensitive [9]. The phenotypic data were used to estimate ANOVA, histograms, and correlations among yield-contributing traits. All the analyses and diagrammatic visualization of data were carried out using the variability package of R program.

3. Results and Discussion

Analysis of variance (ANOVA) was carried out for all morphological parameters, and it was found that the genotypic mean sum of square was highly significant, indicating variation within genotypes. There were prominent differences in the responses of rice plants to alkaline stress at the reproductive stage. The alkaline injury score was recorded after forty days of stress, and it was found that 67, 73, 30, and 12 genotypes exhibited tolerant, moderate tolerant, sensitive, and highly sensitive reactions with salt injury scores of 3, 5, 7, and 9, respectively, in response to stress (Figure 1). Sensitive RILs scoring 9 died without producing any grain yield. The classification was carried out based on their phenotypic performances in all measurements, including plant height, panicle length, tiller number, and biological surveys, of which the alkaline injury score and survivability were conclusive parameters. Grain yield per plant (g) was the trait most affected by alkalinity. The reductions in plant growth, and grain yield and its contributing traits that take place under alkaline stress occur through osmotic effects, which reduce a plant's ability to absorb water and cause reduced growth [10,11].

Figure 1. Response of RIL population in alkaline stress conditions.

At the reproductive stage, RIL 19 (145.5 cm) had the highest plant height, and the lowest value was found in RIL 165 (31.43 cm); the plant height ranged from 31.43 to 145.5 cm. The highest panicle length was recorded in RIL 32 (26.6 cm), while no panicle was produced by highly sensitive RILs. The total number of tillers was recorded to be the highest in RIL 170 (11) and the lowest in RIL 9 (2) however, the highest productive tillers per plant was observed in RIL 170 (10), and no productive tillers were found in RIL 132, RIL 162, RIL 163, RIL164, RIL 165, or RIL 169. The highest spikelet fertility (%) was observed in RIL 68 (95.57%), while the lowest spikelet fertility (%) was observed in RIL 161 (8.51). The maximum biomass per plant (g) was observed in RIL 57 (34.5 g), while the minimum biomass was observed in RIL 162 (3.33 g). The range of grain yield per plant (g/plant) was 0 in RILs with salt injury score 9 to 13.1 g for RIL 34 in alkaline stress. The lines RIL34 (13.1 g/plant), RIL83 (12.7 g/plant), RIL40 (11 g/plant), RIL41 (10.2 g/plant), and RIL124 (10.1 g/plant) showed the significantly highest grain yields amongst all the RILs evaluated in alkaline stress conditions. The top yielding recombinant inbred lines showed the highest biomass production, spikelet fertility, test weight, and productive tillers. The RILs recorded with the highest performance under sodic stress are the most likely to perform well in saline stress as reported by [7].

These results are in agreement with the findings of [10,12] in rice. Histograms of the distribution of frequency of the most affected traits, i.e., spikelet fertility (%) and grain yield per plant (g), are depicted in Figure 2.

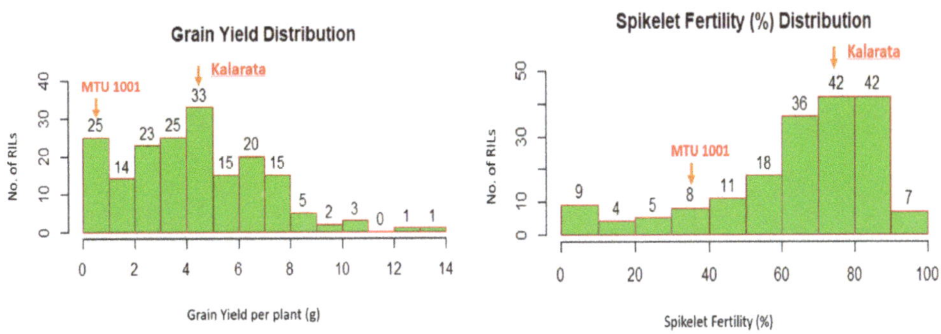

Figure 2. Histograms of distribution of RIL population for grain yield per plant (g) and spikelet fertility (%).

A positive correlation between two traits permits simultaneous improvement in both the traits while confining selection to any one of the associated traits [10]. Grain yield per plant showed a significant positive association with days to 50% flowering (0.24), plant height (0.73), panicle length (0.64), total tillers per plant (0.36), productive tillers per plant

(0.63), test weight (0.6), spikelet fertility (0.7), biological yield per plant (0.87), and harvest index (0.75), while it showed a significant negative correlation with the alkaline injury score (0.84) (Figure 3). Similar findings were previously reported by [10,13].

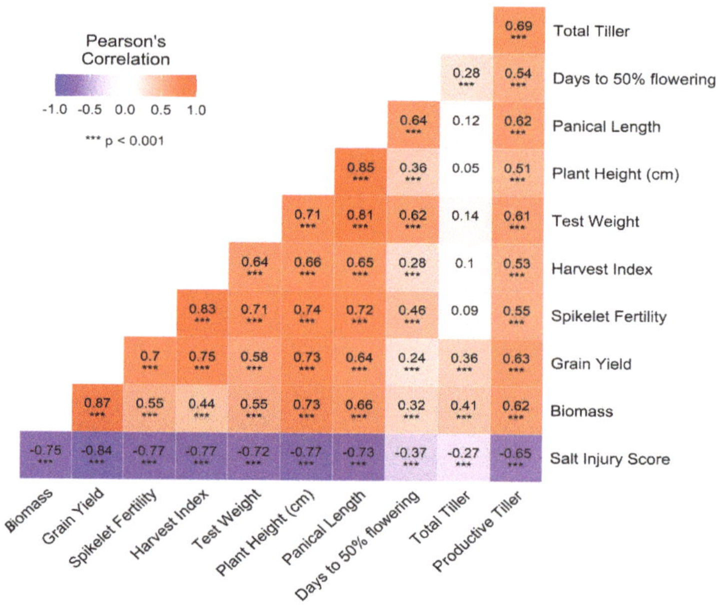

Figure 3. Correlations among morphological traits under alkaline conditions.

4. Conclusions

In this study, positive correlations were found between grain yield and its contributing traits, which may be useful as selection criteria for the improvement of grain yield in alkaline stress conditions. The tolerant lines RIL34, RIL83, RIL40, RIL41, and RIL124 could be used for breeding programs and further studies of the molecular and physiological mechanisms of sodic stress tolerance in rice at the reproductive stage. The findings of this study may help to simplify the breeding of salinity tolerance in rice in order to adapt to climate change.

Author Contributions: Conceptualization, S.L.K. and P.C.S.; methodology, A.N., S.L.K., B.M.L. and A.S.W.; software, A.N. and A.S.W.; validation, S.L.K. and P.C.S..; formal analysis, S.L.K. and B.M.L.; resources, P.C.S. and S.L.K.; data curation, A.N., S.L.K. and B.M.L.; writing—original draft preparation, A.N.; writing—review and editing, S.L.K., P.C.S., B.M.L. and M.K.; visualization, S.L.K., B.M.L. and M.K.; supervision, S.L.K. and P.C.S. All authors have read and agreed to the published version of the manuscript.

Funding: This research received no external funding.

Informed Consent Statement: Not applicable.

Acknowledgments: The authors thank the Director of the Central Soil Salinity Research Institute for completing the research work.

Conflicts of Interest: The authors declare no conflict of interest.

References

1. Godfray, H.C.; Beddington, J.R.; Crute, I.R.; Haddad, L.; Lawrence, D.; Muir, J.F.; Pretty, J.; Robinson, S.; Thomas, S.M.; Toulmin, C. Food security: The challenge of feeding 9 billion people. *Science* **2010**, *327*, 812–818. [CrossRef] [PubMed]
2. *IIRR, 2015;* Vision 2050. ICAR-Indian Institute of Rice Research: Hyderabad, India, 2015; p. 2.
3. Hoang, T.; Tran, T.; Nguyen, T.; Williams, B.; Wurm, P.; Bellairs, S.; Mundree, S. Improvement of salinity stress tolerance in rice: Challenges and opportunities. *Agronomy* **2016**, *6*, 54. [CrossRef]
4. FAO and ITPS. *Status of the World's Soil Resources (SWSR)—Main Report;* Food and Agriculture Organization of the United Nations and Intergovernmental Technical Panel on Soils: Rome, Italy, 2015.
5. Morales, S.G.; Trejo-Téllez, L.I.; Gómez Merino, F.C.; Caldana, C.; Espinosa-Victoria, D.; Herrera Cabrera, B.E. Growth, photosynthetic activity, and potassium and sodium concentration in rice plants under salt stress. *Acta Scientiarum. Agron.* **2012**, *34*, 317–324. [CrossRef]
6. De Leon, T.B.; Linscombe, S.; Gregorio, G.; Subudhi, P.K. Genetic variation in southern USA rice genotypes for seedling salinity tolerance. *Front. Plant Sci.* **2015**, *6*, 374. [CrossRef]
7. Krishnamurthy, S.L.; Gautam, R.K.; Sharma, P.C.; Sharma, D.K. Effect of different salt stresses on agro-morphological traits and utilisation of salt stress indices for reproductive stage salt tolerance in Rice. *Field Crops Res.* **2016**, *190*, 26–33. [CrossRef]
8. Kranto, S.; Chankaew, S.; Monkham, T. Theerakulpisut, P.; Sanitchon, J. Evaluation for salt tolerance in rice using multiple screening methods. *J. Agric. Sci. Technol.* **2016**, *18*, 1921–1931.
9. *IRRI (2014);* Standard Evaluation System for Rice. 5th ed. International Rice Research Institute: Los Banos, The Philippines.
10. Krishnamurthy, S.L.; Sharma, S.K.; Gautam, R.K.; Kumar, V. Path and association analysis and stress indices for salinity tolerance traits in promising rice (*Oryza sativa* L.) Genotypes. *Cereal Res. Commun.* **2014**, *42*, 474–483. [CrossRef]
11. Wang, H.; Wu, Z.; Chen, Y.; Yang, C.; Shi, D. Effects of salt and alkali stresses on growth and ion balance in rice (*Oryza sativa* L.). *Plant Soil Environ.* **2011**, *57*, 286–294. [CrossRef]
12. Chakraborty, S.; Das, P.K.; Guha, B.; Sarmah, K.K.; Barman, B. Quantitative genetic analysis for yield and yield components in Boro Rice (*Oryza sativa* L.). *Not. Sci. Biol.* **2010**, *2*, 117–120. [CrossRef]
13. Sanghera, G.S.; Kashyap, S.C. Genetic parameters and selection indices in F3 progenies of hill rice genotypes. *Not. Sci. Biol.* **2012**, *4*, 110–114. [CrossRef]

environmental sciences proceedings

MDPI

Proceeding Paper

Identifying Rice Genotypes Suitable for Aerobic Direct-Seeded Conditions †

Alisha Goyal [1,2], Parbodh C. Sharma [1,*], Renu Munjal [2], Saraswathipura L. Krishnamurthy [1,*], Hanuman S. Jat [3], Bayragondlu M. Lokeshkumar [1] and Virender Kumar [4]

1 Divison of Crop Improvement, ICAR—Central Soil Salinity Research Institute, Karnal 132001, India; alishagoel9222@gmail.com (A.G.); lokeshbm8050@gmail.com (B.M.L.)
2 Department of Botany and Plant Physiology, CCS Haryana Agricultural University, Hisar 125004, India; munjalrenu66@gmail.com
3 Divison of Soil and Crop Management, ICAR—Central Soil Salinity Research Institute, Karnal 132001, India; hsjat_agron@yahoo.com
4 Sustainable Impact Platform, IRRI, HQ, Los Baños 4030, Philippines; virender.kumar@irri.org
* Correspondence: pcsharma.knl@gmail.com (P.C.S.); krishnagene@gmail.com (S.L.K.)
† Presented at the 2nd International Laayoune Forum on Biosaline Agriculture, 14–16 June 2022; Available online: https://lafoba2.sciforum.net/.

Abstract: Direct-seeded rice (DSR) requires less labor and less water. The low input cost of DSR makes it a better alternative than the conventional transplanted system for rice. DSR helps to reduce production risk, facilitates crop production, and limits irrigation water use. In the present study, a total of 44 rice genotypes were evaluated in a randomized block design (RBD) with three replications over two seasons (2020 and 2021) under DSR conditions at the Central Soil Salinity Research Institute (CSSRI), Karnal. The average grain yield ranged from 1114 kg/ha (CSR 62) to 5198 kg/ha (CSR MAGIC-167), biomass ranged from 6670 kg/ha (CSR 52) to 14,744 kg/ha (CSR MAGIC-117), plant height ranged from 67 cm (CSR 52) to 113 cm (CSR 47), panicle length ranged from 19 cm (CSR 53) to 30 cm (CSR 66), and total tillers ranged from to 7 (CSR MAGIC-117) to 13 (CSR 2748-4441-193). Out of 44 genotypes, maximum grain yield was observed in genotype CSR MAGIC-167 (5198 kg/ha) followed by CSR 58 (5117 kg/ha), CSR 49 (5014 kg/ha), and CSR RIL-06-178 (4904 kg/ha). The best performing genotypes, namely CSR MAGIC-167, CSR 58, CSR 49, and CSR RIL-06-178, should be further evaluated in larger and multilocation trails under DSR situations, and stably performing lines could be released as commercial varieties of DSR.

Keywords: rice; DSR; aerobic; CSR

Citation: Goyal, A.; Sharma, P.C.; Munjal, R.; Krishnamurthy, S.L.; Jat, H.S.; Lokeshkumar, B.M.; Kumar, V. Identifying Rice Genotypes Suitable for Aerobic Direct-Seeded Conditions. *Environ. Sci. Proc.* **2022**, *16*, 68. https://doi.org/10.3390/environsciproc2022016068

Academic Editor: Abdelaziz Hirich

Published: 1 July 2022

1. Introduction

Rice (*Oryza sativa* L.) is a major food crop that plays a crucial role in national food security. It serves as a staple food for more than half of the world's population [1]. Hence, it is very important to increase rice production to meet the exorbitant growth in global population [2,3]. In India, rice is grown in an area of 43.79 million hectares with a production of 116.42 million tonnes [4]. Currently, the dominant method of rice planting is manual transplanting cultivation. Transplanted rice (TPR) is a traditional method that constitutes labor-intensive activities with low productivity [5] and the use of a vast quantity of irrigation water [6]. A shortage in labor and water lead to delayed transplanting which results in a reduction in rice yield and, due to this, a delay in wheat sowing occurs, which also further reduces wheat yield [7].

Recently, many farmers in the tropical region have shifted their cropping method from transplanting to direct-seeded rice (DSR). Direct-seeded rice is the most feasible alternative rice production technique because it not only saves labor requirements but also preserves natural resources, particularly underground water. It is a beneficial and profitable rice production system for current as well as future agricultural scenarios [8]. In Asia, DSR occupied a total of 21% of rice plantation areas at the beginning of the 21st century [9]. DSR refers to the process of establishing the crop from seeds sown directly in the field as compared to transplanting seedlings from the nursery [10]. Many studies have shown that 25% of labor inputs could be saved in DSR compared to the transplanting method [11]. DSR is endowed with multiple benefits, such as using water and labor more efficiently as well as reducing production costs [11]. It is faster and easier to plant, provides higher economic returns, and emits less methane than transplanting rice cultivation [12]. DSR also reduces production risk, facilitates crop production, and shortens crop growth duration, i.e., early maturity. Furthermore, weed infestations, which are a major constraint in DSR, can be overcome through the use of smart agricultural practices and strategies [13]. Although, DSR is very beneficial for cultivators, it still faces a lot of challenges, one of which is the limited accessibility of suitable cultivars [12,14]. In DSR cultivation, only those cultivars are used that have already been developed for transplanting rice. Consequently, this causes variations in grain yield, poor crop establishment, and high weed infestation [15]. So, the main objective of this study was to identify a suitable genotype for aerobic direct-seeded rice based on growth and yield parameters.

2. Materials and Methods

A field experiment was conducted to evaluate the performance of different rice genotypes in the DSR system in response to yield and yield-contributing parameters in an RBD design. A total of 44 rice genotypes were screened by direct-seeded rice at the Central Soil Salinity Research Institute (CSSRI), Karnal, during the two seasons of Kharif 2020 and Kharif 2021, in reclaimed sodic soils. The machinery used for the DSR was a Turbo Happy Seeder. Rice seeding depth ranged from 1 cm to 2 cm. Irrigation was given at intervals of 5 to 6 days depending on the rainfall during the crop growing period. Data were collected at the time of maturity. Yield and yield attributes such as plant height (cm), total tillers, panicle length (cm), yield (kg/ha), and biomass (Kg/ha) were recorded. The collected data were statistically analyzed using analysis of variance by STAR software (STAR 2.0.1) and correlation analysis was performed using the corrplot package in the R program.

3. Results and Discussion

Analysis of variance was performed for all characters concerning genotypes, year, and year–cultivar interaction for the direct-seeded rice, and the results are given in Table 1. Highly significant differences were observed among all parameters in the genotypes. The growth and yield parameters of the direct-seeded rice varied significantly between genotypes. Out of 44 genotypes, maximum grain yield was recorded in genotype CSR MAGIC-167 (5198 kg/ha) with optimal biomass (11,758 kg/ha), plant height (91 cm), panicle length (23 cm), and total tillers (9), while minimum grain yield was observed in CSR 62 (1114 kg/ha) with smaller plant height (85 cm), panicle length (21 cm), total tillers (9 No), and biomass (10,349 kg/ha).

Table 1. Analysis of variance for yield, biomass, plant height, tillers, and panicle length of rice genotypes with the direct seeding method.

Source	DF	Yield	Biomass	Height	Tillers	Panicle Length
Genotype	43	8,124,772.54 **	14,051,913.10 **	670.98 **	9.06 **	29.85 **
Year	1	141,567.34	55,559,133.40 **	491.90 **	131.10 **	7.51 *
Genotype: Year	43	951,238.66 *	6,190,075.88 **	64.02 **	7.30 **	4.21 **
Pooled Error	132	597,957.95	2,153,000.63	18.29	3.87	1.45

*, ** = Significant at 5% and 1% level, respectively.

Grain yield ranged from 1114 kg/ha to 5198 kg/ha (Figure 1). The highest grain yield was observed in CSR MAGIC-167 (5198 kg/ha) followed by CSR 58 (5117 kg/ha) and CSR 49 (5014 kg/ha), while the lowest grain yield was recorded in CSR 62 (1114 kg/ha) followed by CSR27SM-132 (1174 kg/ha) and CSR 52 (1202 kg/ha). Maximum biomass was observed in genotype YET 72 (13,386 kg/ha) followed by CSR 53 (13,702 kg/ha) and CSR MAGIC-117 (14,744 kg/ha), while minimum biomass was observed in CSR 52 (6670 kg/ha) followed by CSR-2748-4441-193 (7620 kg/ha) and CSR-2748-140 (7834 kg/ha). Biomass was ranged from 6670 kg/ha to 13,386 kg/ha. These results conform with the findings of [16], who evaluated different Indian rice varieties for grain yield and quality attributes in direct-seeded and transplanted rice production systems.

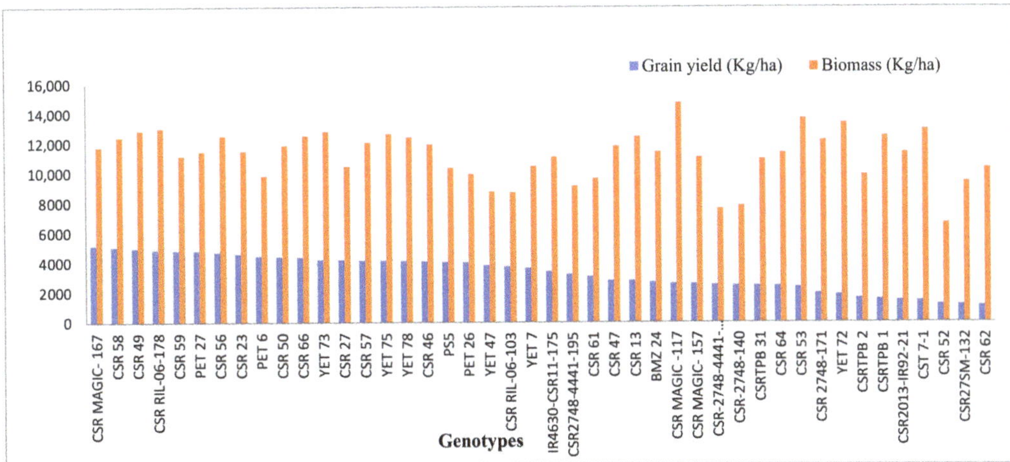

Figure 1. Grain yield and Biomass of rice genotypes in direct seeding method.

A positive association of grain yield was observed with its component traits. Grain yield showed a significant and positive correlation with tillers (0.51), biomass (0.22), plant height (0.41), and panicle length (0.52). A negative but non-significant correlation was observed between biomass and tillers (−0.06). Similar findings were reported by [17,18].

4. Conclusions

Based on the study findings, it is concluded that CSR MAGIC-167, CSR 58, CSR 49, and CSR RIL-06-178 were best performing genotypes. These lines should be evaluated in large trails in a farmer's field with multiplication trails of AICRP to identify the real potential of these genotypes. They could then be commercially released as optimal rice varieties for DSR situations. These genotypes could be utilized in subsequent breeding programs and could lay the foundation for genetic analysis.

Author Contributions: Conceptualization, P.C.S., S.L.K. and H.S.J.; methodology, A.G., P.C.S., S.L.K., V.K. and B.M.L.; software, A.G. and B.M.L.; validation, A.G. and B.M.L.; formal analysis, S.L.K. and B.M.L.; resources, P.C.S., H.S.J., V.K. and S.L.K.; data curation, A.G., P.C.S., S.L.K. and B.M.L.; writing—original draft preparation, A.G.; writing—review and editing, P.C.S., S.L.K., B.M.L., H.S.J., V.K. and R.M.; visualization, P.C.S., S.L.K., B.M.L., H.S.J. and R.M.; supervision, P.C.S. and S.L.K.; project administration, P.C.S., S.L.K., H.S.J.; All authors have read and agreed to the published version of the manuscript.

Funding: The authors acknowledge the funding from Direct-Seeded Rice Consortium, Philippines and ICAR–IRRI Collaboration.

Informed Consent Statement: Not applicable.

Acknowledgments: All authors acknowledge The Director, Central Soil Salinity Research Institute for completing the research work andDirect—Seeded Rice Consortium for funding.

Conflicts of Interest: The authors declare no conflict of interest.

References

1. Wang, W.X.; Jie, D.U.; Zhou, Y.Z.; Zeng, Y.J.; Tan, X.M.; Pan, X.H.; Shi, Q.H.; Wu, Z.M.; Zeng, Y.H. Effects of different mechanical direct seeding methods on grain yield and lodging resistance of early indica rice in South China. *J. Integr. Agric.* **2021**, *20*, 1204–1215. [CrossRef]
2. Chen, S.; Liu, S.; Zheng, X.; Yin, M.; Chu, G.; Xu, C.; Yan, J.; Chen, L.; Wang, D.; Zhang, X. Effect of various crop rotations on rice yield and nitrogen use efficiency in paddy–upland systems in southeastern China. *Crop J.* **2018**, *6*, 576–588. [CrossRef]
3. Li, S.; Jiang, H.; Wang, J.; Wang, Y.; Pan, S.; Tian, H.; Duan, M.; Wang, S.; Tang, X.; Mo, Z. Responses of plant growth, physiological, gas exchange parameters of super and non-super rice to rhizosphere temperature at the tillering stage. *Sci. Rep.* **2019**, *9*, 10618. [CrossRef] [PubMed]
4. Ministry of Agricultural and Farmers Welfare. *Pocket Book of Agricultural Statistics 2019*; Ministry of Agricultural and Farmers Welfare: New Delhi, India, 2019.
5. Bhushan, L.; Ladha, J.K.; Gupta, R.K.; Singh, S.; Tirol-Padre, A.; Saharawat, Y.S.; Gathala, M.; Pathak, H. Saving of water and labor in a rice-wheat system with no-tillage and direct seeding technologies. *Agron. J.* **2007**, *99*, 1288–1296. [CrossRef]
6. Pandey, S.; Velasco, L. Trends in crop establishment methods in Asia and research issues. *Rice Is Life Sci. Perspect. 21st Century* **2005**, 178–181.
7. Kakraliya, S.K.; Singh, L.; Yadav, A.K.; Jat, H.; Kakraliya, S.; Choudhary, K.; Datta, A. Smart package and practices for direct seeded rice (DSR). *Innov. Farming* **2016**, *1*, 49–55.
8. Quilloy, F.A.; Labaco, B.; Dixit, S.; Casal, C., Jr. Crop Establishment in Direct-Seeded Rice: Traits, Physiology, and Genetics. *Rice Improv.* **2021**, 171–202. [CrossRef]
9. Bhandaria, S.; Khanala, S.; Dhakalb, S. Adoption of direct seeded rice (dsr) over puddled-transplanted rice (TPR) for resource conservation and increasing wheat yield. *Rev. Food Agric.* **2020**, *1*, 44–51. [CrossRef]
10. Xu, L.; Li, X.; Wang, X.; Xiong, D.; Wang, F. Comparing the grain yields of direct-seeded and transplanted rice: A meta-analysis. *Agronomy* **2019**, *9*, 767. [CrossRef]
11. Liu, H.; Hussain, S.; Zheng, M.; Peng, S.; Huang, J.; Cui, K.; Nie, L. Dry direct-seeded rice as an alternative to transplanted-flooded rice in Central China. *Agron. Sustain. Dev.* **2015**, *35*, 285–294. [CrossRef]
12. Susilawati, H.L.; Setyanto, P.; Kartikawati, R.; Sutriadi, M.T. The opportunity of direct seeding to mitigate greenhouse gas emission from paddy rice field. *In IOP Conf. Ser. Earth Environ. Sci.* **2019**, *393*, 12042. [CrossRef]
13. Rao, A.N.; Johnson, D.E.; Sivaprasad, B.; Ladha, J.K.; Mortimer, A.M. Weed management in direct-seeded rice. *Adv. Agron.* **2007**, *93*, 153–255. [CrossRef]
14. Joshi, E.; Kumar, D.; Lal, B.; Nepalia, V.; Gautam, P.; Vyas, A.K. Management of direct seeded rice for enhanced resource-use efficiency. *Plant Knowl. J.* **2013**, *2*, 119–134.
15. Sandhu, N.; Yadav, S.; Kumar Singh, V.; Kumar, A. Effective Crop Management and Modern Breeding Strategies to Ensure Higher Crop Productivity under Direct Seeded Rice Cultivation System: A Review. *Agronomy* **2021**, *11*, 1264. [CrossRef]
16. Khokhar, J.S.; Sarial, A.K. Evaluation of Indian Rice Varieties under Direct Seeded and Transplanted Rice Production Systems for Grain Yield and Quality Traits. *Int. J. Pure Appl. Biosci.* **2018**, *6*, 1107–1115. [CrossRef]
17. Yaqoob, M.; Anjum, R.; Hussain, M.; Shah, M.J. Genetic diversity analysis and character association in some Chinese hybrid rice under dry conditions. *Pak. J. Agric. Res.* **2012**, *25*, 249–256.
18. Eradasappa, E.; Nadarajan, N.; Ganapathy, K.N.; Shanthala, J.; Satish, R.G. Correlation and path analysis for yield and its attributing traits in rice (*Oryza sativa* L.). *Crop Res. Hisar.* **2007**, *34*, 156.

environmental sciences proceedings

MDPI

Proceeding Paper

Seed-Enhancement Technologies Promote Direct Seeding and Overcoming Biotic and Abiotic Barriers in Degraded Dryland Ecosystem †

Heba Jarrar * and Ali El-Keblawy

Department of Applied Biology, College of Sciences, University of Sharjah, Sharjah P.O. Box 27272, United Arab Emirates; akeblawy@sharjah.ac.ae

* Correspondence: hebajr85@yahoo.com; Tel.: +971-528908502

† Presented at the 2nd International Laayoune Forum on Biosaline Agriculture, 14–16 June 2022; Available online: https://lafoba2.sciforum.net/.

Abstract: Restoration programs face several challenges in degraded drylands and desert environments, such as high temperatures, soil salinity, water scarcity, loose soils with low water-holding capacity, and poor fertility. Seed-enhancement technologies (SETs) are proposed to improve soil physical and chemical properties, improve seed germination and seedling recruitment, and promote plant growth. SETs improve seed, seedling, and adult plant growth through three main approaches: adding materials to seed coats (seed coating), removing barriers of seed coats (seed scarification), and physiologically altering metabolites through priming (seed priming). These three main approaches (categories) are further divided into several other subcategories. This review aims to define the general categorization of SETs, adopt the proper SETs for arid lands, and indicate the benefits of SETs to overcome the biotic and abiotic barriers in agricultural systems and the ecological restoration of degraded dryland ecosystems. In general, integrating different seed-enhancement technologies (SETs) for rehabilitating degraded lands with a mixture of seeds from various species is recommended, especially since some technologies tend to be species-specific.

Keywords: degraded dryland rehabilitation; seed-enhancement technologies; sustainable restoration; seed coating; seed priming; seed scarification; biotic and abiotic stresses

Citation: Jarrar, H.; El-Keblawy, A. Seed-Enhancement Technologies Promote Direct Seeding and Overcoming Biotic and Abiotic Barriers in Degraded Dryland Ecosystem. *Environ. Sci. Proc.* **2022**, *16*, 1. https://doi.org/10.3390/environsciproc2022016001

Academic Editor: Abdelaziz Hirich

Published: 16 June 2022

1. Introduction

Seed-enhancement technologies (SETs) contribute to dryland restoration, specifically by overcoming the biotic and abiotic barriers to efficient seed employment, promoting precision agriculture practices, using cutting-edge technologies to succeed in planting seeds, and facing degradation effects in arid land while minimizing inputs, mitigating environmental impacts, and paving the path for more sustainable plant enhancement [1].

2. SET Categorization and Sub-Categorization

SETs introduce treatments that physically, biologically, and chemically manipulate and apply materials to the seed to enhance germination and seedling emergence, alleviate the effect of both abiotic and biotic stress, and accelerate early seedling growth. SETs' capability to promote direct seeding and alleviate biotic and abiotic challenges is achieved through five different applications: (i) modify seed physical attributes that could break physical seed dormancy and adjust seed size and shape for precise seed delivery; (ii) enhance the seed's physiological status and help to overcome stress tolerance during germination and even later in seedling and adult plant growth; (iii) alleviate seed storability concerns through the capability of storing seeds after treatment; (iv) offer seed protection against pathogens, pests, and seed predators; and (v) improve the seed microenvironment by promoting the seed's zone via water retention, supporting the adsorption of nutrients and biomaterials, resolving physical and chemical soil inconveniences, and providing a proper microhabitat for germination and enhance seedling establishment. Different SET categories and subcategories are illustrated in Figure 1.

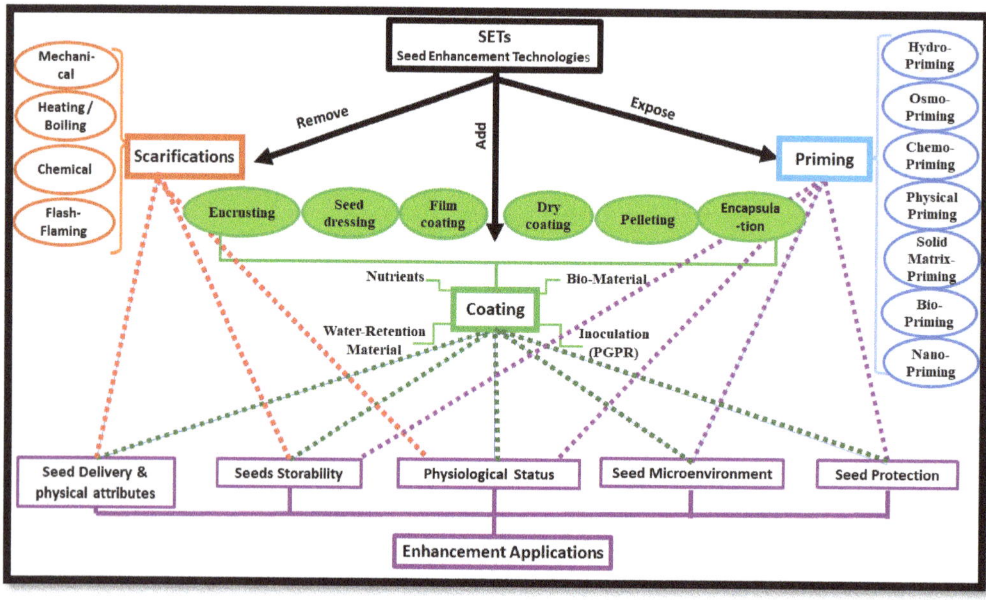

Figure 1. Three main concepts of seed-enhancement technologies (SET: removal, adding, and priming). Each concept is further divided into other SET categories. The main five enhancement applications are illustrated, and the roles of each of the three concepts in terms of seed-enhancement aspects are shown.

2.1. Modifying Seed Coat (Seed Coating): The Addition Concept

The addition of exterior materials to the seed can modify seeds' physical attributes and facilitate their delivery process, enhance their physiological status and microenvironment, protect seeds in soils, and prolong seed storability. This can be achieved with different techniques and constituents under various subcategories that have been recognized, such as seed encapsulation, seed pelleting, encrusting techniques, etc. [1].

2.2. Modifying Seed Coat (Seed Scarification): The Removal Concept

Scarification comprises the techniques that lead to eliminating the outer part of the seed to achieve one (or more) enhancement applications to overcome degraded- and arid-land-restoration barriers. Additionally, it enhances the physical attributes of seeds that affect the delivery process and break seed dormancy. On the other hand, scarification SET shows an effect on seed storability and seed dormancy [2]. Scarification can be recognized through several approaches, including chemical, mechanical, heating, flaming, and flash flaming.

2.3. Modifying Seed Physiology: Seed Priming

Seed priming is a fast, cost-effective technology for treating seeds with mild doses of natural and synthetic materials that allow partial seed imbibition to start the germination metabolic processes without actual germination. There are several subcategories of seed priming, including (i) water (hydropriming); (ii) osmotic solution (osmopriming); (iii) inorganic and organic chemicals (chemical priming) that include halo (salt), hormonal, and biostimulants priming; (iv) beneficial microbes (biopriming); (v) nanoparticles (nano priming); (vi) solid matrices (matrix priming); and (vii) physical priming, such as using low or high temperatures (thermopriming) via radiation or a magnetic field.

3. SETs for Overcoming Challenges Affecting Seeding Success

Direct seeding is the main technique used for large-scale restoration/rehabilitation of degraded lands. This technique faces several challenges in dryland soils. SETs can enhance the success of the seedling establishment and alleviate biotic and abiotic challenges by affecting one or more of the main five applications illustrated in Figure 1;

(1) SETs can modify seeds' physical attributes, such as seed size, shape, and flowability, improving seed delivery. For example, in pelleting technique, outer materials are added to the seed through coating [3], enhancing seeds' shape, weight, and bulk density, enabling more control over precision seeding, and eventually improving plant establishment [4]. Additionally, with scarification, seeds become more homogeneous and easier to manipulate in the restoration process, particularly by enhancing seeds' sowing distribution requirements (i.e., shape, weight, flowability, bulk density, etc.) [5].

(2) SETs enhance seed physiological status via breaking seed dormancy and enhancing stress tolerance during germination and even later in seedling and adult plant growth. These technologies can alter the physiological status of the seed through imbibition methods, including priming, coating with biomaterials and inoculants [5], and breaking seed dormancy through processes such as chemical and mechanical scarification [4].

(3) SETs improve seed storability, which concerns the capability of storing seeds after treatment. The advantage of priming over other SETs, such as seed encrusting and pelleting, is the ability of the former to improve seed storability [6]. On the other hand, applying scarification may indirectly reduce seed volume, representing significant savings in space for banking more native seeds. Seeds can be stored under the appropriate conditions (temperature, light, humidity) to optimize shelf life and seed storability [7].

(4) SETs can protect seeds against pathogens, pests, and seed predators through coating as well as priming SETs by imbuing the seed with protection agents, including fungicides, insecticides, and predator deterrents/repellents, to improve plant establishment [3].

(5) SETs also improve the seed microenvironment by promoting the seed's zone via (i) increasing water availability by water-retention materials and surfactants, (ii) nutrients, (iii) microorganisms, (iv) hormones, and (v) beneficial biomolecules to seedlings, paving the way for more sustainable and effective delivery of growth factors for seeds and seedlings [5].

4. Seed-Enhancement Technologies for Dryland Peculiarities

The success of SETs, specifically in drylands, is mainly correlated to their efficiency, sustainability (long-term effect), and overall beneficial impact on enviro-socio-economic aspects. More sustainable SETs mean longer-time enhancement throughout different plant-growth stages. For example, SETs that increase soil microbial activity cause improvement in rhizosphere conditions and/or enhance seedling survival, boosting the restoration of non-fertile soils of the deserts. Moreover, SETs incorporating organic nutrient sources, water-retention materials, biomaterials, and beneficial soil microorganisms support more sustainable improvement in seed emergence and seedling survival under the harsh climatic conditions of deserts [6].

SETs effectively help plants to overcome several abiotic and edaphic stresses in drylands, particularly soil salinity, drought stress, soil alkalinity, and heat. For example, several SETs, such as hydropriming, osmo-priming, halo-priming, chemo-priming, nano-priming, radiation-priming (laser and UV), and magnet-priming, help plants overcome salinity stress during germination and even later in seedling and adult plant growth [8]. Moreover, drought stress has been mitigated using hydro-priming, chemo-priming, and nano-priming [9], while overcoming heat stress was achieved by thermo-priming.

The addition concept (coating) enhances the seed microenvironment. For instance, a coating with water retention materials increases seeds' water-holding capacity in dryland, maximizing the benefits of precipitation in the root zone and mitigating drought stress. In addition, adding materials such as salicylic acid helps in enhancing seedling growth and survival under drought conditions [6].

Author Contributions: Conceptualization, A.E.-K.; writing—original draft preparation, H.J., A.E.-K.; writing—review and editing, H.J., A.E.-K.; All authors have read and agreed to the published version of the manuscript.

Funding: This research received no external funding.

Institutional Review Board Statement: Not applicable.

Informed Consent Statement: Not applicable.

Data Availability Statement: Not applicable.

Conflicts of Interest: The authors declare no conflict of interest.

References

1. Brown, V.S.; Erickson, T.E.; Merritt, D.J.; Madsen, M.D.; Hobbs, R.J.; Ritchie, A.L. A Global Review of Seed Enhancement Technology Use to Inform Improved Applications in Restoration. *Sci. Total Environ.* **2021**, *798*, 149096. [CrossRef] [PubMed]
2. Visscher, A.M.; Yeo, M.; Gomez Barreiro, P.; Stuppy, W.; Latorre Frances, A.; di Sacco, A.; Seal, C.E.; Pritchard, H.W. Dry Heat Exposure Increases Hydrogen Peroxide Levels and Breaks Physiological Seed Coat-Imposed Dormancy in *Mesembryanthemum crystallinum* (Aizoaceae) Seeds. *Environ. Exp. Bot.* **2018**, *155*, 272–280. [CrossRef]
3. Maity, A.; Dunna, V.; Singh, S.K.; Gupta, C.K. Layered Pelleting of Seed with Nutrient Enriched Soil Enhances Seed Germination in Dinanath Grass (*Pennisetum pedicellatum*). *Range Manag. Agrofor.* **2017**, *38*, 70–75.
4. Berto, B.; Ritchie, A.L.; Erickson, T.E. Seed-Enhancement Combinations Improve Germination and Handling in Two Dominant Native Grass Species. *Restor. Ecol.* **2021**, *29*, e13275. [CrossRef]
5. Zvinavashe, A.T.; Lim, E.; Sun, H.; Marelli, B. A Bioinspired Approach to Engineer Seed Microenvironment to Boost Germination and Mitigate Soil Salinity. *Proc. Natl. Acad. Sci. USA* **2019**, *116*, 25555–25561. [CrossRef] [PubMed]
6. Pedrinii, S.; Stevens, J.C.; Dixont, K.W. Seed Encrusting with Salicylic Acid: A Novel Approach to Improve Establishment of Grass Species in Ecological Restoration. *PLoS ONE* **2021**, *16*, e0242035. [CrossRef] [PubMed]
7. Pedrini, S.; Lewandrowski, W.; Stevens, J.C.; Dixon, K.W. Optimising Seed Processing Techniques to Improve Germination and Sowability of Native Grasses for Ecological Restoration. *Plant Biol.* **2019**, *21*, 415–424. [CrossRef] [PubMed]
8. Migahid, M.M.; Elghobashy, R.M.; Bidak, L.M.; Amin, A.W. Priming of *Silybum marianum* (L.) Gaertn Seeds with H_2O_2 and Magnetic Field Ameliorates Seawater Stress. *Heliyon* **2019**, *5*, e01886. [CrossRef] [PubMed]
9. Rahimi, D.; Kartoolinejad, D.; Nourmohammadi, K.; Naghdi, R. Increasing Drought Resistance of *Alnus subcordata* C.A. Mey. Seeds Using a Nano Priming Technique with Multi-Walled Carbon Nanotubes. *J. For. Sci.* **2016**, *62*, 269–278. [CrossRef]

environmental sciences proceedings

MDPI

Proceeding Paper

Setting Up of a New Local and Ecological Substrate for Tomato Soil-Less Cultivation to Cope with Saline Soils †

Mohamed Majdoubi [1], Soumia El Malahi [1,2,*], Redouane Choukr-Allah [1] and Mimoun Mokhtari [1]

1 Department of Horticulture, Agronomic and Veterinary Institute Hassan II, 80010 Agadir, Morocco; mohamedmajdoubi3@gmail.com (M.M.); redouane53@yahoo.fr (R.C.-A.); mokhtari.mi@gmail.com (M.M.)
2 Higher School of Technology, University of Cadi Aayad, 43000 El Kelâa des Sraghna, Morocco
* Correspondence: s.elmalahi@uca.ma
† Presented at the 2nd International Laayoune Forum on Biosaline Agriculture, 14–16 June 2022; Available online: https://lafoba2.sciforum.net/.

Abstract: Soilless culture is one of the alternative techniques to cope with saline soil. The objective of this study is to identify a new local substrate based on cactus fiber. Three mixtures of cactus fiber and coconut fiber substrates (M0, M1, M2, M3, and C) at a ratio of (1:0, 3:1, 1:1, 1:3, and 0:1), respectively, were evaluated on round tomato cultivation under greenhouse. The mixtures had an organic matter content that exceeded 82%, a C/N ratio between 41.3% and 46%. The mixtures did not cause any phytotoxicity in lettuce and bean. The economic study showed that the use of M2 substrate was the most profitable.

Keywords: cactus fiber; coconut fiber; salinity; soilless culture; substrate; tomato

Citation: Majdoubi, M.; El Malahi, S.; Choukr-Allah, R.; Mokhtari, M. Setting Up of a New Local and Ecological Substrate for Tomato Soil-Less Cultivation to Cope with Saline Soils. *Environ. Sci. Proc.* **2022**, *16*, 6. https://doi.org/10.3390/environsciproc2022016006

Academic Editor: Abdelaziz Hirich

Published: 16 June 2022

1. Introduction

Global agriculture faces the challenge of ensuring optimal productivity while respecting the environment and preserving natural resources [1]. Soilless farming offers an alternative to problems associated with agricultural intensification such as soil fatigue, salinization, and soil diseases [2]. However, this cropping system requires a high level of technical expertise and must be associated with high value-added crops in order to make a return on an initial heavy investment. In Morocco, this production system is growing rapidly, particularly in the Souss-Massa region, which faces water scarcity problems [3]. Imported organic substrates such as coconut fibers are expensive in terms of foreign exchange outflow, especially since the quantities imported in potting soil and the areas of soilless crops are constantly increasing over the years [4]. On the other hand, mineral substrates such as sand, despite its good stability of properties that offers the possibility of use over several years [5], remain subject to problems such as low water retention capacity, risks of contamination by pathogens or the presence of limestone that disrupts the pH, and substrate compaction that leads to root asphyxia [6]. It is then necessary to look for local substrates that can replace imported substrate and that have valid physicochemical, mechanical, sanitary, and economic properties. In the present study, we will study the evolution of physico-chemical parameters of substrate mixtures based on cactus fiber and coconut fiber as well as their impact on the productivity and profitability of a round tomato variety in the Souss-Massa region (Morocco).

2. Materials and Methods

2.1. Preparation of the Cactus Fiber Substrate

The dried cactus paddles were sorted, crushed, and then sieved in two times; the first in a 10 mm × 10 mm sieve and then with another 2 mm × 2 mm sieve to have a granulometry between 2 and 10 mm.

2.2. Physico-Chemical Analysis of the Three Mixtures

The study consisted in determining the physical properties (water retention capacity, apparent density (Da), real density, and porosity), chemical properties (organic matter content (OM), pH, electrical conductivity (EC) at 20 °C, cation exchange capacity (C.E.C in meq/100 g of dry matter), the ratio carbon/nitrogen: C/N, the mineral nitrogen NH_4^+ and NO_3^-, and the assimilable phosphorus ($H_2PO_4^-$); the potassium K, calcium Ca, and the trace elements in mg/liter (Fe, Zn, B and Mo)) and sanitary of the substrates.

2.3. Growth Attributes Assessment and Fruit Quality and Folaire Analysis

Plants growth was monitored by measuring: Root length, Stem height length, Leaf area. Moreover, the mineral elements were analyzed from leaves. Total nitrogen was analyzed by the Kjeldhal method, total phosphorus by colorimetry by complexing with molybdate and ammonium vanadate, and organic carbon by the loss on ignition method at 600 °C and K+, Na+, Mg++, and Ca++ ions by spectrophotometry. An analysis of fruit quality was also made, i.e., firmness, acidity rate, ripeness index, and juice rate according to the studied mixtures.

2.4. Phytotoxicity Study of the Different Mixtures Used

The test was carried out in vitro to analyze the effect of each aqueous extract of the mixture of cactus fiber and coconut fiber (1:0; 3:1; 1:1; 1:3; 0:1) on the germination of the seeds of the lettuce (*Lactuca sativa*) variety SUCRINE; distilled water was used as the control. The experiment layout used is a completely randomized device with five replicates. Each treatment consists of 20 seeds.

2.5. Economic Study of New Substrates Based on Cactus Fibers and Coconut Fibers for Greenhouse Tomatoes

In order to demonstrate that the use of cactus fibers as a growing substrate for greenhouse tomato can be advantageous on a commercial scale. The trial was conducted in a Canary Islands-type greenhouse on an area of 80 m^2. The containers filled with the different substrate mixtures were 1 m long, 20 cm wide and 20 cm deep, offering a volume of 13.33 L/plant. The planting density was 2 plants/m^2 (i.e., 20,000 plants/ha). For the determination of production costs, the fixed charges related to the structure were determined: greenhouse and irrigation system. On the other hand, the variable charges related to the consumption of water, fertilizers, and phytosanitary products were taken in account. On the other side, the marketable yield after eliminating the sorting differences was calculated.

3. Results and Discussion

3.1. Physico-Chemical Characteristics of the Mixtures Based on Cactus and Coconut Fibers

The physico-chemical characteristics of the mixtures of substrate of cactus fiber and coconut fiber for which its ratio are, respectively, (M0, M1, M2, M3, and C) at a rate of (1:0, 3:1, 1:1, 1:3, and 0:1) are represented in Table 1 below.

According to [7], the C.E.C of an ideal substrate is between 10 to 30 meq/100 g dry weight, which coincides with the results of the analysis of the substrates studied and which are between 13.33 and 26.9 meq/100 g. In terms of water retention, the results obtained show that the addition of coconut fiber has generated an increase in water retention capacity. Concerning total porosity, the substrates meet the considered standard of total porosity of the culture substrates ($Pt \geq 50\%$). It should be remembered that in soilless culture, the physical characteristics (granulometry and porosity) of the culture substrate are considered among the decisive factors of the morphological quality of the plants. They act directly on all the root functions of the plants, in particular on the absorption of water, mineral elements, and respiration [8].

Table 1. Physico-chemical characteristics of the substrates mixtures (M) and the Control (C).

	M0	M1	M2	M3	C
% OM	82	85	90	92	96
Cation exchange capacity CEC (meq/100 g of dry weight)	13.33	14.50	16.66	20.80	26.90
Water retention (mL/L)	225	240	260	330	350
Bulk density Da (g/cm^3)	0.30	0.25	0.23	0.20	0.18
Real density Dr (g/cm^3)	1.67	1.60	1.57	1.55	1.55
Total porosity in % of volume	82	84	85	87	90
pH at 23 °C	8.4	7.9	7.4	7.1	6.6
EC (mS/cm)	3.32	3.15	2.92	2.08	0.40
C/N	43.5	41.3	43.48	46.00	52.00

It is considered that ECs between 0.6 mS/cm and 1 mS/cm are suitable for most of the plants. Between 1 and 4 mS/cm, most plants experience a significant decrease in growth and development. Above 4 mS/cm, only halophytes can tolerate these conditions [9]. The EC measurements of the different substrates, except for the control substrate, revealed that the electrical conductivity of the substrates belong to a "saline" class. This salinity higher than 2 mS/cm of the substrates is related to their high minerals content (Table 2). A washing of these substrates before planting eliminates the excess of salts. A good fertilizing irrigation management reduces the risks of this salinity.

Table 2. The content of mineral elements in the different substrates.

Substrat	NH_4^+	NO_3^-	Phosphorus	Potassium	Ca	Mg	Fe	Mn	Cu	Zn
M0	12.60	42.00	710	180	0.19	0.007	0.071	0.004	0.0017	0.0013
M1	28.00	38.50	700	176	0.19	0.004	0.063	0.004	0.0017	0.0014
M2	57.75	38.75	340	80	0.15	0.004	0.042	0.003	0.0018	0.0013
M3	56.00	38.60	90	40	0.11	0.004	0.014	0.001	0.0018	0.0090
C	58.00	37.00	80	30	0.10	0.003	0.010	0.001	0.0015	0.0080

3.2. Evaluation of the Phyto-Toxic Power of the Different Mixtures Used

Statistical analysis showed that there was no significant difference between the germination rates by substrate. In 1987, Mustin reports that the germination rate for a non-phyto-toxic substrate is higher than 50% [10]. The values obtained in this trial show that our substrates are not toxic for the germination of lettuce seeds.

3.3. Economic Return of the Substrate Mixtures Based on Cactus Fiber and Coconut Fiber on the Cultivation of Round Tomato under Greenhouse

After calculating the profitability in an open system, the use of the substrate (M2) appeared to be more economical beneficial with a profit margin of 224,782.5 Dh/ha against 179,187.5 Dh/ha for substrate (M1), 223,180 Dh/ha for the control substrate (C), and 197,015 Dh/ha for substrate (M3). The conclusions will be different, however, if we place the trial in the case of a closed-circuit soilless production system because the substrate (M1) is less efficient in terms of water consumption.

4. Conclusions

Substrates (T) and (M2) offered the best commercial yields of 9.66 kg/m^2 and 9.52 kg/m^2 respectively. In terms of water and fertilizer consumption, substrates (T) and (M2) were more efficient than (M3) and (M1). Concerning mineral nutrition and water consumption/plant, the analysis showed that all substrates were within the standard norms except for a slight water deficiency in the case of substrate (M1) due to its low water retention capacity. Physiologically, there were not many differences between the treatments. Economically, in the case of an open system that does not allow for recovery or recycling of drainage solutions, and under the conditions of the trial, the use of the substrate (M2) is profitable since it provides profit margins of about 224,782.5 Dh/ha.

Author Contributions: Conceptualization, M.M. (Mohamed Majdoubi); methodology, M.M. (Mimoun Mokhtari) and M.M. (Mohamed Majdoubi); software, S.E.M.; validation, R.C.-A. and M.M. (Mimoun Mokhtari); formal analysis, M.M. (Mohamed Majdoubi) and S.E.M.; resources, S.E.M.; data curation, M.M. (Mohamed Majdoubi) and S.E.M.; writing—original draft preparation, S.E.M.; writing—review and editing, M.M. (Mimoun Mokhtari) and R.C.-A.; supervision, M.M. (Mimoun Mokhtari) and R.C.-A.; project administration, M.M. (Mimoun Mokhtari) and R.C.-A. All authors have read and agreed to the published version of the manuscript.

Funding: This research received no external funding.

Institutional Review Board Statement: Not applicable.

Informed Consent Statement: Not applicable.

Data Availability Statement: Not applicable.

Conflicts of Interest: The authors declare no conflict of interest.

References

1. Axelos, M.A.V.; Bamière, L.; Colin, F.; Dourmad, J.Y.; Duru, M.; Gillot, S.; Kurek, B.; Mathias, J.D.; Méry, J.; O'Donohue, M.; et al. *Réflexion Prospective Interdisciplinaire Bioéconomie*; INRAE: Paris, France, 2020; p. 70.
2. Muller, A.; Ferré, M.; Engel, S.; Gattinger, A.; Holzkämper, A.; Huber, R.; Müller, M.; Six, J. Can soil-less crop production be a sustainable option for soil conservation and future agriculture? *Land Use Policy* **2017**, *69*, 102–105. [CrossRef]
3. Choukr-Allah, R.; Nghira, A.; Hirich, A.; Bouchaou, L. Water Resources Master Plan for Sustainable Development of the Souss-Massa River Basin. In *The Souss-Massa River Basin, Morocco;* Choukr-Allah, R., Ragab, R., Bouchaou, L., Barceló, D., Eds.; Springer International Publishing: Berlin/Heidelberg, Germany, 2016; Volume 53, pp. 1–26.
4. Guidi, T. Selection of Substrate Amendements Improving Hydrological Properties of Extensive Green Roofs in Japan. Master's Thesis, Université de Liège, Liège, Belgium, 2019.
5. Gonzalez-Julian, J. Processing of MAX phases: From synthesis to applications. *J. Am. Ceram. Soc.* **2021**, *104*, 659–690. [CrossRef]
6. Slukovskaya, M.V.; Ivanova, L.A.; Kremenetskaya, I.P.; Gorbacheva, T.T.; Drogobuzhskaya, S.V.; Lashchuk, V.V.; Markovskaya, E.F. Rehabilitation of Industrial Barren in Arctic Region Using Mining Wastes. *Open Ecol. J.* **2018**, *11*, 1–13. [CrossRef]
7. Özenç, D.B. Growth and Transpiration of Tomato Seedlings Grown in Hazelnut Husk Compost Under Water-Deficit Stress. *Compost. Sci. Util.* **2008**, *16*, 125–131. [CrossRef]
8. Landis, T.D.; Jacobs, D.F.; Wilkinson, K.M.; Luna, T. Growing media. *Contain. Tree Nurs. Man.* **1990**, *2*, 41–85.
9. Mathieu, C.; Pieltain, F.; Jean, A.; Chossat, J.C.; Valentin, C. *Physical Analysis of Soils: Selected Methods*; Lavoisier TEC & DOC: Paris, France, 1998; p. 275.
10. Mustin, M. *Le Compost: La Gestion de la Matière Organique*; Francois Dubusc: Paris, France, 1987; p. 954.

environmental sciences proceedings

MDPI

Proceeding Paper

Effect of Magnetic Treatment of Irrigation Water on a Greenhouse Tomato Crop under Salinity Conditions †

Moulay Hicham El Omari [1,*], Samba Sene [1], Jamaa Zim [2], Soumia El Malahi [3], Mounia Ennami [1,4] and Houda Taimourya [1,4]

1 Department of Horticulture, Agronomic and Veterinary Institute Hassan II, Agadir 80000, Morocco; senesamba59@gmail.com (S.S.); ennami.mounia@yahoo.fr (M.E.); h.taimourya@iav.ac.ma (H.T.)
2 Department Plant Protection, Agronomic and Veterinary Institute Hassan II, Agadir 80000, Morocco; j.zim@iav.ac.ma
3 Faculty of Science, Ibn Zohr University, Agadir 80000, Morocco; soumia.elmalahi98@gmail.com
4 Department of Production, Protection and Plant Biotechnology, Agronomic and Veterinary Institute Hassan II, Rabat 10090, Morocco
* Correspondence: helomari3@gmail.com; Tel.: +212-661-191-831
† Presented at the 2nd International Laayoune Forum on Biosaline Agriculture, 14–16 June 2022; Available online: https://lafoba2.sciforum.net/.

Abstract: The use of magnetic technology in irrigation is increasing significantly in the world as it has been established in several research studies that its application on irrigation water can improves crop productivity and even alleviate plant salt stress. The aim of this study was to investigate the effect of magnetic treatment of saline water on greenhouse tomatoes in Morocco. The preliminary results showed significant decrease in soil conductivity measured at different depths. The soil irrigated with magnetically treated water had a high-water retention capacity. A slight increase in the yield with plants exhibiting an early flowering was recorded, and fruit quality was also improved in terms of weight and juice content.

Keywords: irrigation water; magnetic treatment; water treatment; salinity; tomato; greenhouse

Citation: El Omari, M.H.; Sene, S.; Zim, J.; El Malahi, S.; Ennami, M.; Taimourya, H. Effect of Magnetic Treatment of Irrigation Water on a Greenhouse Tomato Crop under Salinity Conditions. *Environ. Sci. Proc.* **2022**, *16*, 31. https://doi.org/10.3390/environsciproc2022016031

Academic Editor: Abdelaziz Hirich

Published: 16 June 2022

1. Introduction

Some studies report that magnetic treatment of water improves water quality by breaking water clusters into smaller water clusters [1,2] and reducing the surface water tension [3] that affects the water's hydrophobicity, and some reports also mention reducing water conductivity [3,4].

Also, it has been found that magnetic treated water had an impact on some properties of the soil such as decrease of soil NaCl content, which causes salinity, in some case with magnetic treated water irrigation [5,6]; increase in soil nutrient content and microbial activity [7]. Some trials with saline water have given satisfactory results both on the vegetative development of plant and on the yield [8–10].

The aim of this study is to determine the effects of magnetic treatment of saline irrigation water on plants and soils in order to investigate the possibility of using saline water for agriculture by magnetic treatment.

2. Materials & Methods

2.1. Experimental Site

The study was conducted at the ASNI domains (NL 30°19′1″; WL 9°30′11″) located in the rural commune of Inchaden, circle of Belfaa, Province of Chtouka-Ait Baha (Souss-Massa Region) from July to October 2021. The soil (loamy soil) and water mineral analysis results are reported in the tables below (Tables 1 and 2).

Table 1. Water irrigation analysis.

pH	EC	HCO_3	Cl^-	SO_4^{2-}	NO_3^-	Ca^{2+}	Mg^{2+}	Na^+	K^+	B
	dS/m 25°	meq/L	meq/L	meq/L	meq/L	meq/L	meq/L	meq/L	meq/L	mg/L
7.35	2.02	3.80	11.7	1.07	0.38	3.95	5.53	9.44	0.07	0.21

Table 2. Soil mineral analysis.

pH	EC	H_2PO_4	Cl^-	SO_4^{2-}	NO_3^-	NH_4^+	Ca^{2+}	Mg^{2+}	Na^+	K^+	B
	dS/m 25°	mg/L	meq/L	meq/L	meq/L	meq/L	meq/L	meq/L	meq/L	meq/L	mg/L
6.47	4.52	27.9	10.8	11.5	9.26	0.22	10.2	7.70	16.3	5.34	0.97

2.2. Treatment Device

The magnetic treatment of water was done with magnet device Alva5 from the company AlvaTech Water Revolution. The Alva5 device is presented as an irrigation pipe (50 cm diameter and 81 cm length) with a control box.

The irrigation water was magnetically treated by passing once through this device. In the following document, the treated water with the Alvatech device will be noted T1

2.3. Experimental Design

The experimental setup chosen for this study is a complete randomized block design consisting of 6 blocks of 5 m width (i.e., two ridges) and in each block there are two replicates for the different treatments (T0: untreated water the control and T1) represented by the experimental units. The experimental units are as wide as the blocks and 6m long. In total we have 12 repetitions for each treatment.

2.4. Data Recorded

The data recorded during this study can be divided into three main groups. The first is the parameters related to plant growth and development of the plant. The second is the parameters related to yield and fruit quality, followed by soil parameters. The flowering tomato plant index was calculated to estimate early plant maturity. The first metric is the percentage of flowers opened through the truss and the second is the percentage of flowers set through the truss.

2.5. Statistical Analysis

The data collection and organization were made with EXCEL spreadsheet. The analysis of variance (ANOVA) was performed with the MINITAB 16 software. Tukey's test was adopted to compare the means between different treatments, at 5% significance level.

3. Results

3.1. Yield and Fruit Quality

In this part, three parameters were studied. The first is the earliness of the plants, it is based essentially on the flowering and setting of the truss. Statistical tests of the calculated values show that there is a highly significant difference between treatments for the %Open flower Truss N°1 and %Open flower Truss N°2 index, but not for the two others Table 3 below summarizes the precocity index.

Table 3. Precocity index. The means followed by the same letter are not statistically different at $p < 5\%$ according to the Tukey test.

	%Open Flower Truss N°1	%Set Flower Truss N°1	%Open Flower Truss N°2	%Set Flower Truss N°2
T0	10.72 ± 8.23 (a)	88.53 ± 8.57	20.56 ± 9.00 (a)	28.21 ± 24.29
T1	3.54 ± 3.84 (b)	94.03 ± 7.39	31.59 ± 7.99 (b)	29.61 ± 10.11

In the same perspective of evaluating earliness, the distances between trusses were also measured, from the base to the fourth truss. However, the statistical analysis of these data indicates that there is no significant difference between the two treatments. Same results were obtained with the yield even if the treated plot produced 10% more than the control. The statistical tests of the data of the analysis of the quality of the fruits obtained indicate that there is a significant difference between the average weight of the fruits, the firmness of the fruits and the juice content of the fruits.

3.2. Soil Parameters

The installation of capacitive sensors allowed us to monitor the soil moisture over several depths. The measurements were taken every 15 min and stored in a cloud file on the website http://cloud.yobeen.com/ (accessed on 1 August 2021). Thus, the graphs above represent the evolution of soil moisture at 10 cm depth from 4–11 October 2021 and from 11–19 October 2021. The Figure 1 shows clearly a difference between the evolution of the moisture in the soil. The soil irrigated with treated water hold more water and the decrease of the soil moisture is slower.

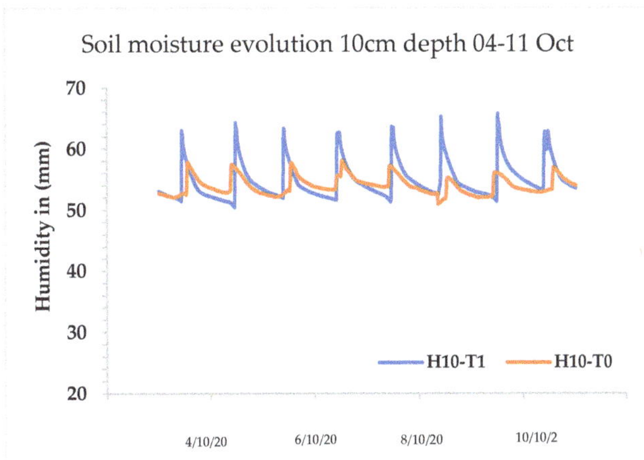

Figure 1. Evolution of soil moisture at 10 cm depth (a) From 4 to 11 October.

The electric conductivity ($EC_{1/5}$: (dS/m) and pH of the different layers were measured from representative samples of the plots. The measured values show that at all horizons, the electrical conductivities of the plot irrigated with magnetically treated water are lower than those of the control, Table 4 bellow summarizes the variation of conductivity and pH.

Table 4. Soil Conductivity and pH.

	Layer1 0–20 cm		Layer2 20–40 cm		Layer3 40–60 cm	
	EC (dS/m)	pH	EC (dS/m)	pH	EC (dS/m)	pH
T0	5.5	7.94	4.1	7.84	4.5	7.89
T1	3.0	7.85	2.9	7.94	4.1	7.83

4. Discussion

The results obtained in soil conductivity are partly due to the leaching capacity of magnetically treated water. In addition, it prevents the build-up of soil salinity [6] because the treated water has a greater affinity to bind with soil particles and therefore percolates with more difficultly [3] solubilizing by the way more mineral elements. This same phenomenon could also explain the evolution of moisture at the soil level.

Since the plant has better water and mineral status, it's more likely to produce more with better quality fruit than stressed plant. Similar results as ours concerning yield and fruit quality were obtained in other studies [7–10].

5. Conclusions

Magnet processing technology as a means to utilize resources affected by salinity for production. As we demonstrated in our study, magnetically treated water was able to reduce soil conductivity by nearly half and keep it within a range that is favorable for plant development. On the other hand, irrigation with magnetically treated water improves the moisture status of the soil by retaining more water. Under these conditions, even if the irrigation water is salty, it creates an ideal environment for plants to grow.

Author Contributions: Conceptualization M.H.E.O., H.T. and S.S.; methodology, M.H.E.O. and H.T.; software, M.E.; validation, M.H.E.O. and H.T.; investigation, S.S.; data curation, S.S., M.E. and H.T.; writing—original draft preparation, S.S., S.E.M. and J.Z.; writing—review and editing, M.H.E.O., S.S., J.Z. and H.T. All authors have read and agreed to the published version of the manuscript.

Funding: This research received no external funding.

Institutional Review Board Statement: Not applicable.

Informed Consent Statement: Not applicable.

Data Availability Statement: Not applicable.

Acknowledgments: We'd also like to express our gratitude to COMATECH SARL who provided us with the device and helped us find the experimental site, ''Les Domaines ASNI'' we thank in the meantime, to conduct the experiment.

Conflicts of Interest: The authors declare no conflict of interest.

References

1. Toledo, E.J.; Ramalho, T.C.; Magriotis, Z.M. Influence of magnetic field on physical–chemical properties of the liquid water: Insights from experimental and theoretical models. *J. Mol. Struct.* **2008**, *888*, 409–415. [CrossRef]
2. Su, N.; Wu, C. Effect of magnetic field treated water on mortar and concrete containing fly ash. *Cem. Concr. Compos.* **2003**, *25*, 681–688. [CrossRef]
3. Surendran, U.; Sandeep, O.; Joseph, E.J. The impacts of magnetic treatment of irrigation water on plant, water and soil characteristics. *Agric. Water Manag.* **2016**, *178*, 21–29.
4. Ashraf, M.W. Magnetic treatment of irrigation water and its effect on water salinity. In Proceedings of the 2 nd International Conference on Food and Agricultural Sciences, Auckland, New Zealand, 12 November 2014; Volume 77, pp. 1–5.
5. Hachicha, M.; Kahlaoui, B.; Khamassi, N.; Misle, E.; Jouzdan, O. Effect of electromagnetic treatment of saline water on soil and crops. *J. Saudi Soc. Agric. Sci.* **2018**, *17*, 154–162. [CrossRef]
6. Zlotopolski, V. The impact of magnetic water treatment on salt distribution in a large unsaturated soil column. *Int. Soil Water Conserv. Res.* **2017**, *5*, 253–257. [CrossRef]
7. Cui, H.; Liu, X.; Jing, R.; Zhang, M.; Wang, L.; Zheng, L.; Kong, L.; Wang, H.; Ma, F. Irrigation with Magnetized Water Affects the Soil Microenvironment and Fruit Quality of Eggplants in a Covered Vegetable Production System in Shouguang City, China. *J. Soil Sci. Plant Nutr.* **2020**, *20*, 2684–2697. [CrossRef]
8. Taimourya, H.; Oussible, M.; Baamal, L.; EL Harif, A.; Zaid, E.; Abdelkarim, G.; Abdelaziz, S. Magnetic treatment of culture medium enhance growth and minerals uptake of strawberry (Fragaria × ananassa Duch.) and tomato (Solanum lycopersicum) in Fe deficiency conditions. *Int. J. Sci. Eng. Res.* **2017**, *8*, 1414–1436.
9. Samarah, N.H.; Hani, B.; Makhadmeh, I.M. Effect of Magnetic Treatment of Water or Seeds on Germination and Productivity of Tomato Plants under Salinity Stress. *Horticulturae* **2021**, *7*, 220. [CrossRef]
10. Akrimi, R.; Hajlaoui, H.; Batelli, G.; Ruggiero, A.; Badri, M.; Grillo, S.; Mhamdi, M. Electromagnetic water enhanced metabolism and agro-physiological responses of potato (*Solanum tuberosum* L.) under saline conditions. *J. Agron. Crop Sci.* **2021**, *207*, 44–58. [CrossRef]

environmental sciences proceedings

MDPI

Proceeding Paper

Integrated Agronomic Practices to Enhance Forage Productivity and Quality of Blue Panicum (*Panicum antidotale* Retz.) under Saline and Arid Conditions in the South of Morocco †

Ayoub El Mouttaqi *, Ihssane Mnaouer, Abdelaziz Nilahyane, Dennis Ashilenje, Erick Amombo, Mohamed Belcaid, Mohamed Ibourki, Karima Lazaar, Krishna Prasad Devkota, Lamfeddal Kouisni and Abdelaziz Hirich

Biosaline Agriculture Research Program, African Sustainable Agriculture Research Institute (ASARI), Mohammed VI Polytechnic University (UM6P), Laayoune 70000, Morocco; ihssane.mnaouer@um6p.ma (I.M.); abdelaziz.nilahyane@um6p.ma (A.N.); dennis.ashilenje@um6p.ma (D.A.); erick.amombo@um6p.ma (E.A.); med.belcaid80@gmail.com (M.B.); mohamed.ibourki@um6p.ma (M.I.); karima.lazaar@um6p.ma (K.L.); krishna.devkota@um6p.ma (K.P.D.); lamfeddal.kouisni@um6p.ma (L.K.); abdelaziz.hirich@um6p.ma (A.H.)

* Correspondence: ayoub.elmouttaqi@um6p.ma; Tel.: +212-675673127

† Presented at the 2nd International Laayoune Forum on Biosaline Agriculture, 14–16 June 2022; Available online: https://lafoba2.sciforum.net/.

Abstract: Soil salinity and drought are the two major challenges agriculture is facing in marginal environments, such as desert areas, which limit the growth and productivity of traditional cultivated crop species. In the Sahara Desert of the southern region of Morocco, livestock is the major agricultural activity, and forage supply is severely affecting livestock production. Blue panicum (*Panicum antidotale* Retz.) is an alternative salt-tolerant crop recently introduced to Morocco's Sahara to enhance forage availability for livestock. The aim of this study was to increase the productivity and nutritional quality of blue panicum through the use of integrated agronomic practices, such as planting methods (bed or flat), crop establishment (direct seeding, transplantation, and rhizome propagation), organic and/or mineral amendments, and grass–legumes intercropping. The preliminary results showed that in the highly saline conditions, the combined application of organic amendments, seedling transplanting, and bed planting method gave the highest production (56 t/ha/year) of fresh biomass, which is higher than 50% compared to the control. This study suggests that in the southern region of Morocco and the Sahara Desert, the adoption of the blue panicum with integrated agronomic practices sustains livestock production.

Keywords: blue panicum; salinity; cropping practices; yield; quality

Citation: El Mouttaqi, A.; Mnaouer, I.; Nilahyane, A.; Ashilenje, D.; Amombo, E.; Belcaid, M.; Ibourki, M.; Lazaar, K.; Devkota, K.P.; Kouisni, L.; et al. Integrated Agronomic Practices to Enhance Forage Productivity and Quality of Blue Panicum (*Panicum antidotale* Retz.) under Saline and Arid Conditions in the South of Morocco. *Environ. Sci. Proc.* **2022**, *16*, 50. https://doi.org/10.3390/environsciproc2022016050

Academic Editor: Redouane Choukr-Allah

Published: 16 June 2022

1. Introduction

Increasing soil salinity and declining irrigation quality are global challenges, especially in semi-arid, arid, and desert climatic conditions. Worldwide, 37% of the salt-affected soils are located in the arid and desert regions [1]. In Morocco, most of the southern and eastern territories are salt affected, which covers approximately 67% of the total area of the country. These regions are characterized by harsh environmental conditions manifested by water scarcity, soil and groundwater salinity, high temperature variability, low or no rainfall, high wind speed, poor soil quality, and low soil organic carbon content. The impact of climate change has further worsened the situation, leading to an increased accumulation of salt in the topsoil. All these factors are affecting forage productivity in the regions.

Morocco's Sahara region is mainly used as rangeland for extensive livestock production of dromedary (camel), goats, and sheep. In the past, a few rehabilitation initiatives were launched to improve forage crop's productivity, but they had little success due to the vastness of the territory with major challenges [2]. To meet the increasing forage demand, the introduction and adoption of alternative crops could be a judicious choice [3].

Blue panicum (*Panicum antidotale* Retz.) is a newly introduced grass in the Sahara region, and the initial results show promising performance in terms of productivity and quality compared to traditional forage crops [4]. Additionally, earlier studies recommended the application of organic amendments to improve agricultural production in salt-affected lands [5]. This study aims at forage productivity enhancement of blue panicum through the adoption of integrated agronomic practices in contrasting environments in the southern region of Morocco.

2. Materials and Methods

2.1. Experimental Sites

On-farm experiments were conducted in five contrasting locations of the southern region of Morocco (Table 1). The experimental sites had the maximum temperature ranging from 30 to 40 °C, minimum temperature 8–11 °C, and annual rainfall of 26–53 mm.

Table 1. Climate and soil characteristics of the experimental sites/platform.

Site	Coordinates		Climate Data					Soil Characteristics		
	Longitude	Latitude	Altitude (m)	T_{max} (°C)	T_{min} (°C)	Rainfall (mm)	Wind Speed (Km.h^{-1})	Texture	ECe (dS/m)	pH
Tadkhast	13°12′28″ W	26°58′34″ N	76	34	9	31	19.1	Sandy loam	0.60	8.56
Boujdour	14°02′35″ W	25°36′37″ N	168	36	10	27	18.9	Sand	0.40	9.04
Bir Anzarane	14°35′24″ W	23°57′40″ N	204	40	8	26	15.5	Loamy sand	1.35	9.03
Tarfaya	12°51′30″ W	27°52′21″ N	17	30	11	53	20.5	Sandy loam	2.70	8.98
Es-smara	12°07′22″ W	26°32′43″ N	313	38	9	29	16.7	Sandy loam	10.05	8.24

2.2. Crop Management and Experimental Design

Four agronomic practices were compared in this study: (a) planting methods (bed or flat), (b) crop establishment (direct seeding, transplantation, and rhizome propagation), (c) organic and/or mineral amendments, and (d) grass–legumes intercropping. Experiments were conducted for the first year during 2020–2021 cropping season and implemented in a randomized complete block design with four replications in each site/platform. Plant material consisted of a public variety of blue panicum, while Siriver variety of alfalfa (*Medicago sativa*) and ILRI 15077 sesbania (*Sesbania sesban*) accession were used for the intercropping trial. Before plantation, the soil was plowed, and the crop was established. Amendments were incorporated at a level of 30 and 5 t/ha for farmyard manure and commercial compost, respectively, while NPK (10-30-10) fertilizer was applied at a rate of 100 kg/ha. Irrigation water was supplied using a drip irrigation system with 50 cm spacing. A leaching fraction was considered regarding the site's characteristics. Irrigation water quality parameters for each site are shown in Table 2. Experimental units were harvested manually at 40 days cutting intervals. The yield is for the cumulative harvest results for the entire year.

Table 2. Irrigation water quality in different platforms.

Site	EC (dS/m)	pH	Cation's Concentration (ppm)				Anion's Concentration (ppm)		
			Na^+	K^+	Mg^{2+}	Ca^{2+}	Cl^-	$SO_4{}^{2-}$	$HCO_3{}^-$
Tadkhast	3.85	7.07	25.47	25.47	64.32	167.04	802.10	174.12	646.60
Boujdour	6.20	6.97	27.16	27.16	51.52	279.19	599.66	2373.54	122.00
Bir Anzarane	6.64	6.94	1306.30	49.19	173.03	283.46	2655.46	360.12	212.28
Tarfaya	8.67	7.29	1306.30	49.19	173.03	283.46	2655.46	360.12	212.28
Es-smara	12.40	7.16	1163.33	32.64	242.15	421.70	2567.80	855.09	585.60

2.3. Plant and Soil Analysis

Several agro-morphological parameters were monitored during the crop growing period and at harvest. The major parameters measured include plant height, number of tillers, number of leaves, panicle length, plant fresh weight, and fresh biomass. Harvested plants were oven dried at 60 °C for three days to determine the dry matter. Dry biomass for each treatment was ground to investigate the nutritional quality of the forage. Fresh leaves sampling was conducted to assess chlorophyll and proline content. Soil samples were collected after harvest to monitor soil salinity and organic carbon stock.

3. Results

Figure 1 represents the effect of agronomic practices on yield under different salinity levels. Preliminary results showed no significant difference between practices at a low salinity level. However, the bed plantation improved the fresh biomass production by 90% compared to the control (direct seeding or transplantation) under a medium salinity range. The use of organic amendment doubled blue panicum's annual fresh biomass production at high salinity conditions to reach 62 T·ha^{-1}.

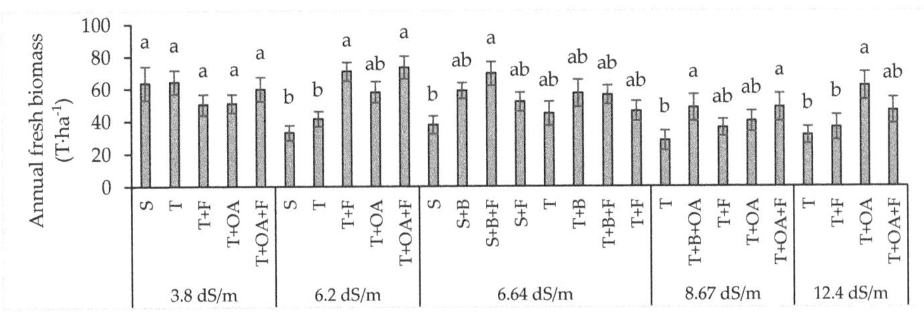

Figure 1. Annual fresh biomass of blue panicum regarding several cropping practices and salinity. (S: Seeding/T: Transplantation/OA: Organic amendment/F: Fertilization/B: Bed plantation). Error bars indicate the standard deviation. a and b present Tuckey's test at $p = 0.05$.

Results of the intercropping trial indicate that the performance of blue panicum was better under the intercropping system with legumes compared to sole crop. However, annual fresh biomass production was increased by 46% (to reach 58 T·ha^{-1}) with 4 alfalfa:1 blue panicum planting ratio. Furthermore, the land equivalent ratio was more than 3 in combination with alfalfa and approximately two times better than in combination with sesbania (Figure 2).

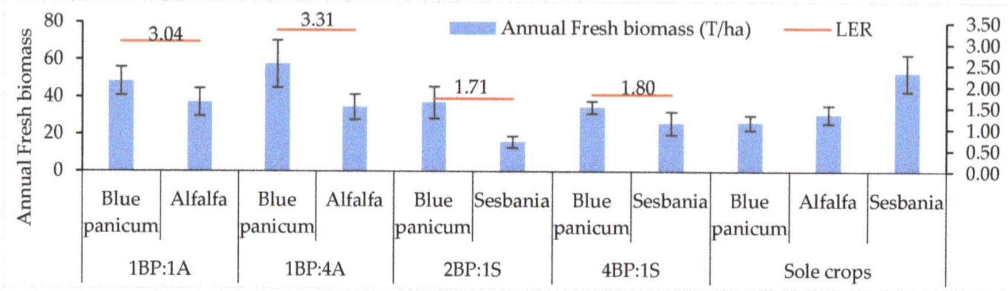

Figure 2. Fresh biomass production and land equivalent ratio of blue panicum as affected by intercropping combination. Error bars indicate the standard deviation.

4. Discussion

Preliminary results of the experiment indicate a significant positive effect of different agronomic practices on the productivity of blue panicum except for low salinity conditions. However, the combined effect of organic amendments, transplantation, and bed planting method increased annual biomass production by more than 50% under high salinity and drought conditions. Our result is consistent with the results by Cuevas et al. [5]. They also found amendments and bed planting methods improved crop yield by 50% in saline soil conditions. Similarly, consistent with Al-Shareef et al. [6], intercropping of blue panicum with alfalfa has a significant advantage over sole cropping.

Author Contributions: Conceptualization, A.H.; methodology, A.H., A.N., A.E.M. and I.M.; software, A.E.M.; validation, A.H., A.N. and K.P.D.; formal analysis, A.E.M., D.A. and E.A.; investigation, A.E.M., D.A., E.A., M.B., M.I., K.L. and I.M.; resources, A.H.; data curation, A.H.; writing—original draft preparation, A.H., A.E.M. and I.M.; writing—review and editing, A.H., A.N., L.K. and K.P.D.; visualization, A.E.M.; supervision, A.H.; project administration, A.H.; funding acquisition, A.H. All authors have read and agreed to the published version of the manuscript.

Funding: This research was funded by OCP Phosboucraa Foundation within the project titled "Scaling up the cultivation of alternative crops in salt-affected farms in southern provinces of Morocco (N° AS03)".

Institutional Review Board Statement: Not applicable.

Informed Consent Statement: Not applicable.

Data Availability Statement: Not applicable.

Acknowledgments: The authors gratefully appreciate the efforts of the farmers to provide us with farmlands to carry out the trials.

Conflicts of Interest: The authors declare no conflict of interest.

References

1. FAO. *The State of the World's Land and Water Resources for Food and Agriculture—Systems at Breaking Point*; Synthesis Report; FAO: Rome, Italy, 2021. [CrossRef]
2. Aït Houssa, A.; Drissi, S.; Asehraou, A.; Asfers, A.; Oubaki, L.; Chraibi, H. Changements climatiques au Maroc: Quels systèmes de culture et quelles biotechnologies pour s'y adapter? *Rev. Mar. Sci. Agron. Vét.* **2017**, *5*, 210–221.
3. Ismail, S.; Rao, N.K.; Dagar, J.C. Identification, Evaluation, and Domestication of Alternative Crops for Saline Environments. In *Research Developments in Saline Agriculture*; Dagar, J., Yadav, R., Sharma, P., Eds.; Springer: Singapore, 2019. [CrossRef]
4. Hirich, A.; Choukr-Allah, R.; Ezzaiar, R.; Shahid, A.S.; Lyamani, A. Introduction of alternative crops as a solution to groundwater and soil salinization in the Laayoune area, South Morocco. *Euro-Mediterr. J. Environ. Integr.* **2020**, *6*, 52. [CrossRef]
5. Cuevas, J.; Daliakopoulos, I.N.; del Moral, F.; Hueso, J.J.; Tsanis, I.K. A Review of Soil-Improving Cropping Systems for Soil Salinization. *Agronomy* **2019**, *9*, 295. [CrossRef]
6. Al-Shareef, A.R.; Ismail, S.M.; El-Nakhlawy, F.S. Blue panic-alfalfa combination as affected by irrigation water regimes and forage mixing ratio under subsurface drip irrigation in arid regions. *Grassl. Sci.* **2018**, *64*, 234–244. [CrossRef]

environmental sciences proceedings

MDPI

Proceeding Paper

Farmer's Participatory Genetic and Agronomic Approaches for Higher Rice Productivity in Sodicity Stress †

Yash Pal Singh [1,*] and Rakesh Kumar Singh [2]

1 Regional Research Station, Central Soil Salinity Research Institute, Indian Council of Agricultural Research (ICAR), Lucknow 226002, India
2 International Center for Biosaline Agriculture, Al Ruwayyah 2, Academic City, Dubai P.O. Box 14660, United Arab Emirates; r.singh@biosaline.org.ae
* Correspondence: ypsingh.agro@gmail.com; Tel.: +91-7309563010
† Presented at the 2nd International Laayoune Forum on Biosaline Agriculture, 14–16 June 2022; Available online: https://lafoba2.sciforum.net/.

Abstract: In salt affected soils, neither physical nor chemical remediation methods are cost-effective for saline/sodic soil reclamation. Salt-affected lands are estimated at about 955 million ha worldwide, afflicting over 6% of the world's total arable land, of which sodic soils constitute 581 million ha. Regaining the agricultural potential and enhancing the productivity and profitability of rice in sodic soils, it necessitates the development of advanced technologies for the sustainable reclamation of these soils, suitable salt tolerant varieties (STVs) through farmer's participatory varietal selection (FPVS) approaches and their matching management practices (Mmp).The results from the study showed that combining Mmp with STV resulted in 35% higher yields over traditional variety (TV) with farmer's management practices and proved cost effective nutrient management approach to maximize the productivity and profitability of rice in sodic soils.

Keywords: cost effective reclamation technologies; matching management practices; *Oryza sativa*; farmer's participatory varietal selection; salt tolerant varieties; crop productivity

Citation: Singh, Y.P.; Singh, R.K. Farmer's Participatory Genetic and Agronomic Approaches for Higher Rice Productivity in Sodicity Stress. *Environ. Sci. Proc.* **2022**, *16*, 52. https://doi.org/10.3390/environsciproc2022016052

Academic Editor: Abdelaziz Hirich

Published: 16 June 2022

1. Introduction

The presence of salts in the soil is an alarming threat to agricultural productivity and sustainability. According to the FAO Land and Nutrition Management Service [1], over 6% of the world's land is affected by salinity, which accounts 955 million ha in 100 countries, of which sodic/solonetz soils constitute 581 million ha. Chemical amendments like gypsum ($CaSO_4 \cdot 2H_2O$) have been used most extensively for the reclamation of sodic soils [2] but is a costly affair because of its requirement in large quantity (12–16 tha^{-1}) and high market price (USD 60 t^{-1}). Therefore, it is imperative to develop cost effective farmers' participatory sodic soil reclamation technologies.

Rice is an important crop grown in salt-affected soils, but its yield is much lower because of its high sensitivity to salinity and sodicity at early seedling stage [3], high mortality and poor crop establishment [4]. Increasing and sustaining yields in these soils will require the development of high yielding STVs and their integration with cost effective crop and nutrient management practices in farmers' participatory mode. In this paper we limit the discussion on developing advanced cost effective reclamation technologies, the development of farmers' preferred high yielding salt tolerant variety of rice through the farmers' participatory varietal selection (FPVS) approach and improved agronomic management practices to ensure better crop establishment and higher productivity of STV of rice.

2. Materials and Methods

2.1. Development of Advanced Cost-Effective Reclamation Technologies

Four times replicated field experiments laid out in a split plot design with four gypsum levels (control, 15% GR, 25% GR and 50% GR) and two varieties 'CSR 13 (STV)' and 'Pant 4 (TV)' were conducted for three years in highly sodic soil (pH 10.5, EC 1.43 dSm^{-1}, ESP 89) at the ICAR-Central Soil Salinity Research Institute (CSSRI), Regional Research Station, Lucknow, Uttar Pradesh, India (26°47′58″ N and 80°46′24″ E). As per treatments, gypsum was incorporated once throughout the study in surface soil (0–15 cm). Thirty day old seedlings of both the varieties were transplanted at 20 × 15 cm row to row and plant to plant spacing. After three years of study, soil samples were collected and analyzed to monitor the improvement in soil properties. All the observation related to crop growth, yield attributing characters and yields were recorded and analyzed statistically using the statistical package MSTAT-C.

2.2. Development of High Yielding Salt Tolerant Variety through Farmer's Participatory Varietal Selection Approach

A set of 126 geographically and genetically diverse rice genotypes includes local genotypes and advanced salt tolerant genotypes were screened through researchers' managed on-station and on-farm trials. Based on sodicity tolerance and high yield potential, 18 genotypes were selected for FPVS. After 3 years of FPVS studies, conducted at 17 sodic environments, 6 genotypes were selected and further evaluated, taking into account traits desired by men and women farmers. Preference scores for each genotype were calculated. The genotype 'CSR89IR-8' consistently ranked first based on both farmers' preference ranking and grain yield and later released as variety 'CSR 43' (Table 1).

Table 1. Combined effect of gypsum levels and varieties on grain yield (t/ha^{-1}) of rice.

Gypsum Levels (% G.R.)	1st Year		2nd Year		3rd Year	
	CSR 13	Pant 4	CSR 13	Pant 4	CSR 13	Pant 4
0	0.32	0.07	0.55	0.36	0.63	0.42
15	2.73	1.57	3.01	2.77	3.28	2.81
25	4.09	3.48	4.38	3.76	4.52	4.11
50	4.36	4.05	4.79	4.36	4.96	4.75
Sem (±)	0.09		0.03		0.08	
CD (p = 0.05)	0.27		0.12		0.28	

2.3. Development of Management Practices

2.3.1. Optimizing Number of Seedling/$Hill^{-1}$ and Spacing

Field experiments with three replications consisted of two rates of seedlings/$hill^{-1}$, T1 (2 seedlings) and T2 (4 seedlings), and three spacing S1: 15 × 15 cm, S2: 15 × 20 cm, S3: 20 × 20 cm was conducted at the experimental farm of ICAR-CSSRI, Regional Research Station, Lucknow, Uttar Pradesh, India, on a sodic soil (pH 9.2, EC 0.61 dSm^{-1}). Thirty-day-old seedlings were transplanted with recommended doses of fertilizer.

2.3.2. Optimizing Nitrogen Requirement

A field experiment with three replicates and six N treatments ((N $kgha^{-1}$) N1: 0, N2: 100, N3: 125, N4: 150, N5: 175 and N6: 200) was conducted at ICAR-CSSRI, Research farm, Shivri, Lucknow. Thirty-day-old seedlings were transplanted with 4 seedlings/$hill^{-1}$ spaced at 15 × 20 cm. Full P_2O_5, K_2O and $ZnSO_4$ (60–40–25 kg N–P_2O_5–K_2O–$ZnSO_4{\cdot}7H_2O$ ha^{-1}) and half of N were applied as basal, and the remaining N was applied in equal splits at 30 and 60 days after transplanting. All the relevant data were collected and analyzed (ANOVA) using WINDOSTAT.

3. Results and Discussion

3.1. Cost Effective Reclamation Technology

Salt tolerant variety 'CSR 13' significantly excelled 'Pant 4' in grain yield at gypsum levels from 15 to 50% G.R. However, the magnitude of the combined effect of the reduced dose of gypsum and salt tolerant variety of rice was reflected as 17% increase in grain yield over the tradition variety. It indicated that the gypsum, at 25% G.R. with the sodicity-tolerant variety, could save about 43% of the total initial expenditure for the reclamation of sodic soils over the recommended dose of gypsum (50% G.R.) with non-sodicity tolerant traditional high yielding varieties. These observations indicate the saving of gypsum to the tune of 25% by using salt tolerant variety in sodic soils (Table 1). The application of a reduced dose of gypsum and cultivating salt-tolerant varieties saved 50% of gypsum without any significant yield loss [5].

3.2. Development of Salt Tolerant Variety through FPVS

After intensive evaluation and selection within the 126 genotypes, only 6 genotypes were selected (Table 2). Farmers ranked CSR-89IR-8, CSR36 and NDR359 as their first, second and third preferred genotypes, respectively, with positive correlations between preference scores of male and female farmers (r = 0.91**) and between farmers and researchers (r = 0.74*). Genotype 'CSR-89IR-8' ranked first in all desired traits. Farmers selected CSR-89IR-8 as their preferred genotype because of several traits, including its good taste, aroma, color, non-cohesiveness when cooked and higher grain yield. The FPVS program conducted in sodic soils [6] has confirmed the relevance and utility of participatory approaches in these less favorable environments. The screening of a large number of genotypes helped to narrow the number of genotypes suitable for sodic soils. Farmers' preferences revealed that they generally preferred high yielding, salt tolerant, short duration and lodging-resistant varieties of rice. Based on farmers' rankings, genotype 'CSR-89IR-8' fulfils all the preferred traits. These attributes together culminated in the release of this genotype as CSR 43 [6].

Table 2. FPVS process for developing salt tolerant variety. The arrow sign is indicating that out of three outstanding performing genotypes only one is selected to develop as a variety.

Year	Process	Number of Genotypes Planted in Sodic Soil	Number of Genotypes Selected	Number of Trials Sites in Sodic Soil
1st	Screening of Genotypes	126 ➞	18	2
2nd	To conduct PVS	18 ➞	12	5
3rd	To conduct PVS	12 ➞	9	6
4th	To conduct PVS	9 ➞	6	6
5th	Final selection	6 ➞	2	6
6th	Up-scaling	2 + 1 (farmer variety) ➞	1 selected and released as variety "CSR43"	32

3.3. Improved Management Practices

3.3.1. Optimizing Number of Seedlings/Hill^{-1} and Spacing

Productive tillers/hill^{-1}, grain and straw yields with four seedlings/hill^{-1} were, respectively, 15.9%, 22.7% and 30.3% higher than those with two seedlings/hill^{-1}. Spacing had a significant ($p < 0.05$) effect on floret fertility and grain yield. Using 15 × 20 cm spacing resulted in about 4–7% more spikeletspanicle^{-1} over both 15 × 15 and 20 × 20 cm spacing though these differences were not significant. Using 15 × 20 cm spacing resulted in 11% higher grain yield and 16% higher straw yield over the values obtained with 20 × 20 cm spacing. Grain yield was significantly enhanced by the transplanting of four seedlings/hill^{-1} at a 15 × 20 cm spacing. Higher number of seedlings/hill^{-1} resulted in significantly higher dry matter production and more productive tillers/hill^{-1}, whereas medium spacing enhanced floret fertility resulting in higher grain yield over other treatments [7].

3.3.2. Optimizing Nitrogen Requirement for Salt Tolerant Variety

The application of 200 kg Nha^{-1} produced 70%, 53.8%, 21.5%, 9.7% and 3.5% more productive tillers/hill^{-1} than the application of 0, 100, 125, 150 and 175 kg Nha^{-1}, respectively. Number of spikeletspanicle^{-1} and 1000 grain weight increased significantly by increasing N to 200 and 175 kgha^{-1}. Applying 100 kg Nha^{-1} resulted in 125% increase in grain yield over control. Grain yield was statistically similar when 175, 200 and 150 kg Nha^{-1} were used but these yields were significantly higher than when 0, 100 and 125 kg of N were applied. A similar pattern was also observed for straw yield. Nitrogen application is necessary to maintain plant growth and enhance grain yield. Similar beneficial effects of N application in similar environments were reported by Singh et al. [7]. The grain yield of STV responded positively to N application up to 169 kgha^{-1}; however, gross margin and BCR were highest with 150 kg Nha^{-1}. This N rate can, therefore, be recommended for STV in sodic soils.

4. Conclusions

The sodic soils treatment with a reduced dose of gypsum (25% G.R.) and growing a salt-tolerant variety of rice can save 25% of gypsum and proved to be highly economical and sustainable technology for increasing rice productivity. The salt-tolerant variety produced about 0.5 tha^{-1} additional grain yields over farmers' current varieties across locations. Suggested recommendations include the transplanting of four seedlings/hill^{-1} at a spacing of 15 × 20 cm and the use of 150–60–40–25 kg $N–P_2O_5–K_2O–ZnSO_4·7H_2O$ ha^{-1} in the field.

Author Contributions: Y.P.S.: Conceptualization, field experimentation, data collection, compilation, editing, analysis, writing original draft; R.K.S.: Reviewing and editing the draft. All authors have read and agreed to the published version of the manuscript.

Funding: This research was funded by UPCAR, Lucknow (UP), India, during 2014–2017.

Institutional Review Board Statement: Not applicable.

Informed Consent Statement: Not applicable.

Data Availability Statement: The data are available at the affiliated institute and not made public.

Acknowledgments: Authors acknowledge the facilities provided by Director, ICAR-CSSRI, Karnal, India, and the farmers for participation.

Conflicts of Interest: The authors declare no conflict of interest.

References

1. FAO. Land and Plant Nutrition Management Service. 2008. Available online: htpp://www.fao.org/ag/agl/agll/spush (accessed on 8 September 2016).
2. Oster, J.D.; Jayewardene, N.S. Agricultural management of sodic soils. In *Sodic Soils: Distribution, Properties, Management and Environmental Consequences*; Sumner, M.E., Naidu, R., Eds.; Oxford University Press: New York, NY, USA, 1998; pp. 125–147.
3. Aslam, M.; Qureshi, R.H.; Ahmad, N. Mechanism of salinity tolerance in rice (*Oryza sativa* L.). In *Towards the Rational Use of High Salinity Tolerant Plants*; Leith, H., Massoum, A.A., Eds.; Kluwer Academic Publisher: Amsterdam, The Netherlands, 1993; pp. 135–138.
4. Ismail, A.M.; Heuer, S.; Thomson, M.J.; Wissuwa, M. Genetic and genomic approaches to develop rice germplasm for problem soils. *Plant Mol. Biol.* **2007**, *65*, 547–570. [CrossRef] [PubMed]
5. Singh, Y.P.; Singh, R.; Sharma, D.K. Combined effect of reduced dose of gypsum and salt tolerant varieties of rice and wheat on rice-wheat cropping system in sodic soils. *Indian J. Agric. Sci.* **2009**, *54*, 24–28.

6. Singh, Y.P.; Nayak, A.K.; Sharma, D.K.; Gautam, R.K.; Singh, R.K.; Singh, R.; Mishra, V.K.; Paris, T.; Ismail, A.M. Farmers participatory varietal selection: A sustainable approach for the 21st century. *Agroecol. Sustain. Food Syst.* **2014**, *38*, 427–444. [CrossRef]
7. Singh, Y.P.; Mishra, V.K.; Singh, S.; Sharma, D.K.; Singh, D.; Singh, U.S.; Singh, R.K.; Haefele, S.M.; Ismail, A.M. Productivity of sodic soils can be enhanced through the use of salt tolerant rice varieties and proper agronomic practices. *Field Crops Res.* **2016**, *190*, 82–90. [CrossRef] [PubMed]

environmental sciences proceedings

MDPI

Proceeding Paper

Blue Panicum-Alfalfa Mixture Buffers against Effects of Soil Salinity on Forage Productivity †

Dennis S. Ashilenje, Erick Amombo, Ayoub El Mouttaqi, Ihssane Mnaouer, Mohamed Ibourki, Abdelaziz Hirich, Krishna P. Devkota and Abdelaziz Nilahyane *

African Sustainable Agriculture Research Institute (ASARI), Mohammed VI Polytechnic University (UM6P), Laâyoune 70000, Morocco; dennis.ashilenje@um6p.ma (D.S.A.); erick.amombo@um6p.ma (E.A.); ayoub.elmouttaqi@um6p.ma (A.E.M.); ihssane.mnaouer@um6p.ma (I.M.); mohamed.ibourki@um6p.ma (M.I.); abdelaziz.hirich@um6p.ma (A.H.); krishna.devkota@um6p.ma (K.P.D.)

* Correspondence: abdelaziz.nilahyane@um6p.ma; Tel.: +212-666945933

† Presented at the 2nd International Laayoune Forum on Biosaline Agriculture, 14–16 June 2022; Available online: https://lafoba2.sciforum.net/.

Abstract: Soil salinity limits forage productivity in ~1125 million ha of particularly arid lands across continents. Grass-legume mixtures are known to enhance soil nutrient cycling and forage productivity in non-saline conditions. We tested if these benefits can be replicated in the saline soils ($EC_e > 14$) of Laayoune-Morocco during the year 2021, based on a randomized complete block design (RCBD) experiment with 50–50 and 30–70% blue panicum grass (*Panicum antidotales* Reitz), alfalfa (*Medicago sativa* L.) mixtures compared to their monocrops. There was no significant variation ($p = 0.36$) in forage accumulation across treatment, but the blue panicum grass and alfalfa mixture had slightly greater forage yield compared to each in monocrop (13% and 5% Mg ha^{-1}). Blue panicum grass proportionate dry matter exceeded alfalfa in mixtures (62 and 54%). During establishment, blue panicum grass buffers against effects of high salinity on forage productivity in mixtures with alfalfa.

Keywords: alfalfa; blue panicum; crop mixture; forage productivity; soil salinity

check for updates

Citation: Ashilenje, D.S.; Amombo, E.; El Mouttaqi, A.; Mnaouer, I.; Ibourki, M.; Hirich, A.; Devkota, K.P.; Nilahyane, A. Blue Panicum-Alfalfa Mixture Buffers against Effects of Soil Salinity on Forage Productivity. *Environ. Sci. Proc.* **2022**, *16*, 70. https://doi.org/10.3390/environsciproc2022016070

Academic Editor: Redouane Choukr-Allah

Published: 4 July 2022

1. Introduction

Soil salinity is a major problem limiting crop production in arid and semi-arid lands across the globe including Morocco where, an estimated 3.6 million ha is affected [1]. This problem is triggered by intrusion of saline water from the Atlantic Ocean and persistent use of saline irrigation water, which deposits toxic sodium (Na^+) ions in the soil. Therefore, soil EC levels are moderate (4 ds m^{-1}) to extreme (18 ds m^{-1}), which adversely affects forage crops. Recent reports indicate that alfalfa suffers yield losses of up to 26% due to salt stress in Morocco [2,3]. Sodium toxicity suppresses seed germination [4] and impairs plant osmotic potential, photosynthesis and cell elongation, ultimately suppressing forage yields [5]. Salt tolerant crops, including blue panicum grass, can maintain net photosynthesis, water potential and dry matter accumulation hence mitigate against increasing salinity [6]. Nevertheless, ref [7] have found blue panicum grass yields to an average maximum of 15 tons ha^{-1} in saline conditions of southern Morocco, compared to the potential 60 Mg ha^{-1} [8]. This study explores measures to alleviate effects of soil salinity by leveraging adaptive features of alfalfa and blue panicum grass in mixtures.

Perennial grass-legume mixtures are renowned for their complementarity in carbon (C) sequestration, nitrogen (N) acquisition, weed suppression and overyielding monocrops [9–11]. In southern Morocco, where salinity compounds temporal variation of temperatures, yields of alfalfa and blue panicum monocrops slump because of their C_3 and C_4 photosynthetic mechanisms that cannot withstand extremely hot summers and cold winters, respectively. Consequently, this research explores mechanisms of blue panicum grass-alfalfa mixtures to stabilize yields across seasons, when persistently exposed to saline growing conditions.

We sought to understand if a grass-legume mixture complementarity can buffer against soil salinity and enhance forage productivity compared to monocrops. Hence the objective of this study was to evaluate correlations between forage yield and soil salinity and to test for species complementarity in overall forage productivity. We hypothesized that the halophytic nature of blue panicum grass can alleviate salinity and complement alfalfa growth and hence increase forage productivity of their mixtures unlike monocrops.

2. Materials and Methods

RCBD experiments were established in two sites in Foum El Oued-Laayoune with moderate and high soil salinity in April 2021. The study had four treatments, including blue panicum grass and alfalfa monocrops and their combination in 50–50 and 30–70% seeding proportions in four replicates. Crops were established by adjusting seeding rates up, to account for pure live seed estimates of germination and purity shortfalls from 100%. Soil salinity regimes were maintained by irrigation using moderate and highly saline water in drip irrigation according to crop evapotranspiration. Forage was harvested at approximately 10% flowering stage of alfalfa and oven dried at 60 °C for 72 h to determine forage productivity ha^{-1} along with soil EC. The standard deviation of forage dry matter across different harvests of each treatment were divided by the corresponding means, to determine temporal stability of forage production. Data were subjected to ANOVA (stat package of R) to determine site × treatment, site or treatment effects on forage productivity, temporal stability and soil salinity. Mean separation was conducted using Fisher's protected LSD.

3. Results

3.1. Initial Soil and Water Physical and Chemical Properties

Initial soil properties indicated that both sites had sandy loam soil texture, see Table 1. There was greater dominance of sand in medium (68%) compared to highly saline soils (60%) coinciding with soil EC_e values of 7.9 and 52.5 dS m^{-1} and water EC of 8 and 17 dS m^{-1}.

Table 1. Initial soil texture and soil and water salinity in sites of medium and high salinity in Laayoune-Morocco during April 2021.

Soil Characteristic	Salinity	
	Medium	**High**
Textural class	Sandy loam	Sandy loam
Ratio of sand, loam and clay (%)	68:20:12	60:26:14
Soil EC_e (dS m^{-1})	7.8	52.5
Water EC (dS m^{-1})	8	17

3.2. Forage Acumulation

There was no significant (p = 0.3) site × treatment interaction, but each of these factors singly ($p < 0.05$) influenced accumulated forage dry biomass. The 50–50 mixture of blue panicum grass and alfalfa accumulated the greatest amount of forage (10.4 Mg ha^{-1}), slightly exceeding that of alfalfa monocrop (10.2 Mg ha^{-1}) and the 30–70 m mixture (9.6 Mg ha^{-1}); however, it surpassed that of blue panicum grass monocrop (8.1 Mg ha^{-1}) based on averages across sites (Table 2). The cumulative forage biomass from medium salinity (12.4 Mg ha^{-1}) was almost double the mean yield from high salinity 6.7 Mg ha^{-1}. Cumulative forage dry biomass across treatments ranged between 10.2 and 13.6 Mg ha^{-1} in moderate salinity and 6 to 7.2 Mg ha^{-1} in high salinity.

Table 2. Cumulative forage dry biomass of blue panicum (BP) and alfalfa (Alf) monocrops and their 50–50 and 30–70% mixtures in medium and high salinity from June to December 2021.

Treatment	Cumulative Forage Dry Matter		
	Medium Salinity	High Salinity	Mean
		Mg ha^{-1}	
100BP	10.2a	6.0a	8.1b
100Alf	13.6a	6.8a	10.2a
50BP50Alf	13.6a	7.2a	10.4a
30BP70Alf	12.4a	6.8a	9.6ab
Mean	12.4A	6.7B	

Value within column followed by similar letters in lower case and means across salinity followed by similar letters in upper case are not significantly different (LSD, $p > 0.05$).

3.3. Temporal Stability

There was no significant ($p = 0.3$) site × treatment interaction, but their main effects against dry matter temporal stability were significant ($p < 0.05$). Blue panicum had the highest mean temporal stability across sites, comparable to alfalfa monocrop and the 50–50 mixture (0.36), but greater than that of the 30–70 blue panicum alfalfa mixture (0.34) (Table 3). Crops had greater mean temporal stability in medium (0.44) compared to high salinity (0.35). Blue panicum grass dry biomass dominated in mixtures established in high soil salinity (62 and 54%). In medium salinity, blue panicum was more abundant (71%) than alfalfa in the 50–50 mixture, unlike the 30–70 mixture where alfalfa dominated (61%).

Table 3. Temporal stability of 50–50 and 30–70% blue panicum grass (BP), alfalfa (Alf) mixtures and monocrop dry matter in medium and high salinity from June to December 2021.

Treatment	Temporal Stability		
	Medium Salinity	High Salinity	Mean
100BP	0.60a	0.45a	0.53a
100Alf	0.37a	0.35a	0.36ab
50BP50Alf	0.35a	0.38a	0.36ab
30BP70Alf	0.45a	0.20a	0.34b
Mean	0.44A	0.35B	

Value within column followed by similar letters in lower case and means across salinity followed by similar letters in upper case are not significantly different (LSD, $p > 0.05$).

4. Discussion

The inability of blue panicum grass to alleviate soil salinity contrary to our hypothesis might be explained by inadequate establishment of blue panicum grass roots to perform phytoremediation. Grasses alleviate salinity through root mechanisms of degrading calcite and releasing Ca^{2+} which replace Na^{+} from the soil solution [12] for exclusion into plant roots and shoots. Across sites, the 50–50% blue panicum grass alfalfa mixture slightly enhanced yields compared to monocrops. Substantial complementary advantages of mixtures usually manifest in years following establishment, particularly attributed to enhanced deposit of crop residues and turnover of soil N and C [13]. Apparently, dominance and stability of blue panicum grass shoot biomass might buffer the productivity of mixtures despite persistently saline conditions. In non-saline and warm conditions, blue panicum grass is expected to thrive via the C_4 photosynthetic machinery compared to the C_3 photosynthesis of alfalfa [14]. Temporal stability of biomass affirmed blue panicum grass tolerance to salinity, usually associated with its ability to maintain favorable potassium (K)/Na ratio, photosynthesis and proline mediated osmotic balance [15].

5. Conclusions

During establishment, blue panicum grass does not alleviate soil salinity, but its biomass stability and dominance buffers productivity of its mixture with alfalfa. Apparently, complementary benefits of grass–legume mixtures to increase forage productivity might extend to conditions of moderate to high soil salinity. Equitable blue panicum grass and alfalfa forage biomass persistence against increasing salinity can help reduce the pitfalls in forage yields due to salt injury and varying temperatures in arid conditions.

Author Contributions: Conceptualization, D.S.A., A.N., K.P.D. and A.H.; methodology, D.S.A., A.N.; data collection, E.A., A.E.M., M.I. and I.M.; project administration, A.N.; funding acquisition, A.N. All authors have read and agreed to the published version of the manuscript.

Funding: This research is funded by OCP Phosboucraa Foundation, grant number FPB_SPA002_2020.

Institutional Review Board Statement: Not applicable.

Informed Consent Statement: Not applicable.

Data Availability Statement: Not applicable.

Acknowledgments: The authors thank the OCP Phosboucraa Foundation for funding this project, farmers who providing their farmlands for research trials, Fadoua Karouach and Abdelaziz Adadi for helping to establish and manage the field trials, and Driss Zouine for transport logistics.

Conflicts of Interest: The authors declare no conflict of interest.

References

1. Antipolis, S. Les menaces sur les sols dans les pays mediterraneens. *Plan Bleu* **2003**, *80*, 70.
2. Farissi, M.; Faghire, M.; Bargaz, A.; Bouzgaren, A.; Makoudi, B.; Sentenac, H.; Ghoulam, C. Growth nutrients concentrations, and enzymes involved in plants nutrition of alfalfa populations under saline conditions. *J. Agric. Sci. Technol.* **2014**, *16*, 301–314.
3. Mouradi, M.; Latrach, L.; Mohamed, F.; Bouzgarne, A.; Ghoulam, C. Impact of the salt stress on agronomic potential of Moroccan population of alfalfa (*Medicago sativa* L.) under the field conditions of Marrakesh. *Appl. J. Environ. Eng.* **2018**, *4*, 309–320.
4. Patel, P.R.; Kajal, S.S.; Patel, V.R.; Patel, V.J.; Khristi, S.M. Impact of salt stress on nutrient uptake and growth of cowpea. *Braz. J. Plant Physiol.* **2010**, *22*, 43–48. [CrossRef]
5. Patel, N.T.; Pandey, A.N. Effect of soil salinity on growth, water status and nutrient accumulation in seedlings of Sueda nudiflora (Chenopodiaceae). *An. Biol.* **2009**, *31*, 61–70.
6. Hussain, M.I.; Farouq, M.; Muscolo, A.; Rehman, A. Crop diversification and saline water irrigation as potential strategies to save fresh water resources and reclamation of marginal soils—A review. *Environ. Sci. Pollut. Res.* **2020**, *27*, 28695–28729. [CrossRef] [PubMed]
7. El Mouttaqi, A.; Mnaoeur, I.; Nilahyane, A.; Belcaid, M.; Kouisni, L.; Hirich, A. Effect of cutting time interval on agronomic performance and forage quality of blue panicum (*Panicum antidotale* Retz) grown under arid and saline conditions in Southern Morocco. In Proceedings of the Doctoral Day, UM6P Interdisciplinary Doctoral Day, Mohammed VI Polytechnic University, Bengeurir, Morocco, 28 June 2021.
8. Ali, H.; Gul, B.; Adnan, M.Y.; Ahmed, M.Z.; Aziz, I.; Gulzar, S.; Ansari, R.; Khan, M.A. NPK mediated improvements in biomass production, photosynthesis and Na+ regulation in *Panicum antidotale* under saline conditions. *Pak. J. Bot.* **2016**, *46*, 1975–1979. Available online: https://www.pakbs.org/pjbot/ (accessed on 18 August 2021).
9. Dhakal, D. Forage Grass-Legume Mixtures for Improved Forage Yield, Forage Quality, Stand Persistence, Soil Properties and Economics. Ph.D. Thesis, University of Wyoming, Laramie, WY, USA, 2015.
10. Adjesiwor, A.T.; Islam, M.A.; Zheljazkov, V.D.; Ritten, J.P.; Garcia y Garcia, A. Grass-legume seed mass ratios and nitrogen rates affect forage accumulation, nutritive value, and profitability. *Crop Sci.* **2017**, *57*, 2852–2864. [CrossRef]
11. Ashilenje, D.S. Plant Community Structure and Nitrogen Dynamics Affect Productivity and Environment of Meadow Bromegrass—Legume Cropping Systems. Ph.D. Thesis, University of Wyoming, Laramie, WY, USA, 2018.
12. Quadir, M.; Oster, J.D.; Schubert, S.; Noble, A.D.; Sahrawat, K.L. Phytoremediation of Saline and Saline-Sodic Soils. *Adv. Agron.* **2007**, *96*, 197–247.
13. Cong, W.; van Ruijven, J.; Mommer, L.; De Deyn, G.B.; Berendse, F.; Hoffland, E. Plant species richness promotes soil carbon and nitrogen stocks in grasslands without legumes. *J. Ecol.* **2014**, *102*, 1163–1170. [CrossRef]
14. Wiggans, D.R.; Singer, J.W.; Moore, K.J.; Lamkey, K.R. Response of continuous maize with stover removal to living mulches. *Agron. J.* **2012**, *104*, 917–925. [CrossRef]
15. Ahmed, M.S.A.; Ashraf, S.A.; Ali, Q. Soil salinity as a selection pressure is a key determinant for the evaluation of salt tolerance in Blue Panicgrass (*Panicum antidotale* Retz). *Flora Morphol. Distrib. Funct. Ecol. Plants* **2010**, *205*, 37–45. [CrossRef]

environmental sciences proceedings

MDPI

Proceeding Paper

Sustainable Irrigation and Abiotic Tolerant Crops in South Italy within TRUSTFARM Project †

Cataldo Pulvento [1,*], Osama Ahmed [2,3], Mohamed Houssemeddine Sellami [4,5], Antonella Lavini [4] and Giuseppe De Mastro [1]

1 Department of Agricultural and Environmental Science (DISAAT), University of Bari, 70121 Bari, Italy; giuseppe.demastro@uniba.it
2 Department of Agricultural Markets, Leibniz Institute of Agricultural Development in Transition Economies (IAMO), 06120 Halle (Saale), Germany; naser@iamo.de
3 Agricultural Economics Department, Faculty of Agriculture, Cairo University, Giza 12613, Egypt
4 Institute for Mediterranean Agricultural and Forestry Systems, National Research Council of Italy (CNR-ISAFOM), 80055 Portici, Italy; mohamed.sellami@isafom.cnr.it (M.H.S.); antonella.lavini@cnr.it (A.L.)
5 Interdepartmental Research Centre on the "Earth Critical Zone", University of Naples Federico II, 80055 Portici, Italy
* Correspondence: cataldo.pulvento@uniba.it
† Presented at the 2nd International Laayoune Forum on Biosaline Agriculture, 14–16 June 2022; Available online: https://lafoba2.sciforum.net/.

Abstract: Today, irrigated agriculture is even more influenced by climate change with consequent negative effects on food security. The Mediterranean area is most affected by climate change, leading to greater exposure to uncertainty and production risks. In these environments, water stress, rainfall variability, and soil salinization have been accentuated. Improving crop productivity by minimizing such effects is possible through intelligent climate farming practices (CSFP). Towards resilient and sustainable integrated agro-ecosystems through appropriate climate-smart farming practices (TRUSTFATM) is a project funded by the European Union's Horizon 2020 research and innovation program with the aim to design integrated agro-ecosystems by conserving natural resources and using the principles of the circular economy for developing climate-resilient production systems in Egypt, Morocco, Italy, France, and Senegal. The Department of Agricultural and Environmental Science (DISAAT) of the University of Bari is responsible for coordinating the activities (starting in 2022) related to the introduction of new crop varieties and management of water and efficient irrigation systems, such as deficit irrigation, use of marginal quality (saline water) irrigation water, and introduction of abiotic stress-tolerant crops.

Keywords: salinity; climate change; tolerant crops

Citation: Pulvento, C.; Ahmed, O.; Sellami, M.H.; Lavini, A.; De Mastro, G. Sustainable Irrigation and Abiotic Tolerant Crops in South Italy within TRUSTFARM Project. *Environ. Sci. Proc.* **2022**, *16*, 8. https://doi.org/10.3390/environsciproc2022016008

Academic Editor: Abdelaziz Hirich

Published: 16 June 2022

1. Introduction

Globally, climate change is one of the biggest challenges. It is causing a significant change in the average values of meteorological elements, such as precipitation and temperature, and rainfall, while negatively affecting agriculture, such as causing biotic (increased insect, pest, and disease populations, uplifting weed growth, declining beneficial soil microbes, and threatening pollinators) and abiotic stresses, occurrence of severe drought or flood, extremes of temperature, salinity, and alkalinity, etc. [1]. The Mediterranean region is most affected by the impact of climate change, leading to greater exposure to uncertainty and production risks. In these environments, water stress, rainfall variability, and soil salinization have been accentuated. Extreme climatic events enhance the risk of desertification in Southern Italy, where extreme weather conditions are increasing tornadoes, and sudden hailstorms, alternating with persistent drought periods. Southern Italy has recorded a decrease in rainfall of up to 20%, with an increase in the number of highly intensive precipitation events [2]. The reduction in rainfall and the increase in precipitation intensity without a doubt affects the total availability of water resources in the soil and the extension of agricultural areas, resulting in higher runoff, soil erosion, less accumulation of water in the reservoirs, and reduced availability of water for irrigation purposes. Furthermore, a warmer climate and drought lead to an increase in evapotranspiration demand from crops.

Under these conditions, it is urgent to improve the productivity of the agricultural system through intelligent climate farming practices (CSFP) using the principle of soil and water conservation when introducing new crops that tolerate drought and salinity stress into these systems and managing them.

The adverse effect of climate change on crops could be mitigated, by the introduction of adapted cultivars with adjusted planting date, and changing fertilizer and irrigation practices [3,4]. It has been demonstrated that cultivar adaptation is the most effective, followed by irrigation.

An increasing number of research projects have been funded in recent years, with the aim to innovate traditional agricultural systems and to improve their resilience to climate change. A recent one is the TRUSTFATM project "Towards resilient and sustainable integrated agro-ecosystems through appropriate climate-smart farming practices", developed with the aim to design integrated agro-ecosystems by conserving natural resources and using the principles of the circular economy in order to develop more climate-resilient production systems in Egypt, Morocco, Italy, France, and Senegal. The project was funded by the European Union's Horizon 2020 research and innovation program. The main aim of the TRUSTFARM project is to design integrated crop–livestock agro-ecosystems in order to improve productivity and make production systems more resilient to climate change. The project includes a series of activities aimed at the construction of resilient cropping systems; in Italy, in particular, field trials will be carried out aimed at identifying new varieties resistant to abiotic stresses and sustainable irrigation practices that will make high yields even in changing climatic conditions.

2. Field Trials

Biannual experimental field trials starting from 2022 to evaluate the production performance of climate-resilient crops and best irrigation practices will be carried out for the Italian case study.

2.1. *Evaluation of Climate Resilient Crops*

The effect of environmental stresses on crop physiology is highly variable. The response that each plant gives to single specific stress, such as drought, depends on numerous variables (time of onset, severity, and duration). In addition, plants employ different environmental stress response strategies, such as avoidance of stress, stress tolerance, and escape from stress [3].

Furthermore, the growing negative impact of climate change on a given environment increases the probability of plants being exposed to several stresses simultaneously, such as drought and salinity; in this case, the response of the plant to a combination of stress cannot be explained by the sum of the single effects, but it is a unique and different response.

In marginal environments where the production of traditional crops is compromised, attempts have been made in recent years to introduce new resistant crops. Extremophiles have evolved and reproduced in very marginal habitats, such as highly saline environments, which can also be prone to drought and heat, floods, sodium, and alkalinity [5].

There are several promising species that have interesting characteristics both from the point of view of resistance to abiotic stress and from the point of view of the quality characteristics of the seeds produced.

Examples are quinoa (*Chenopodium quinoa* Willd.) and grain amaranth (*Amaranthus* spp.) (Figure 1), of which the cultivation in recent years has expanded to several countries out of its area of origin due to increasing interest, market development, research, and promotion [6]. Thanks to their high-quality protein content, these crops are considered promising candidates for enhancing high-quality plant protein food production in the world. Quinoa and amaranth are important for the high tolerance to abiotic stresses and for their nutritious characteristics [7].

Figure 1. Quinoa (**a**) and amaranth (**b**).

2.2. *Deficit Irrigation Practices*

Improvement of irrigation management is one of the main options in Mediterranean cropping systems to increase the efficiency of water use [8]. Besides improving agricultural practices under dryland farming, water saving is important to sustain high productivity under more arid conditions in the future.

Many trials have been carried out in the past years aimed towards finding the best solution to manage irrigation with saline water.

Sprinkler irrigation is not suitable because salts may cause injury to leave tissues, while subsurface and drip irrigation appear to be ideal methods. In particular, drip irrigation reduces the water use resulting in less salt deposited in the soil layers, avoiding leaf burn; furthermore, in drip irrigation, the high frequency of irrigation interventions prevents the soil from drying out, thus preventing the salts from concentrating in the soil and ensuring their leaching below the root system [9].

Among the sustainable irrigation strategies, supplementary irrigation consists in the application of small quantities of irrigation water to crops that are normally grown in dry conditions [10]. A very interesting approach is related to deficit irrigation, which consists of the use of irrigation volumes lower than the optimal ones with water savings of up to 50/70% of the quantity used for full irrigation and with the same levels of yields. Deficit irrigation could also be applied only during drought-sensitive growth phases characteristic of each plant species.

Author Contributions: Conceptualization, C.P.; methodology, C.P.; writing—original draft preparation, C.P. and O.A.; writing—review and editing, O.A., G.D.M., M.H.S., A.L. and C.P.; visualization, C.P.; supervision, C.P. and G.D.M. All authors have read and agreed to the published version of the manuscript.

Funding: This research was funded by the European Union's Horizon 2020 Research and Innovation Program (TRUSTFARM project).

Institutional Review Board Statement: Not applicable.

Informed Consent Statement: Not applicable.

Data Availability Statement: Not applicable.

Conflicts of Interest: The authors declare no conflict of interest.

References

1. Teshome, D.T.; Zharare, G.E.; Naidoo, S. The threat of the combined effect of biotic and abiotic stress factors in forestry under a changing climate. *Front. Plant Sci.* **2020**, *11*, 601009. [CrossRef] [PubMed]
2. Brunetti, M.; Caloiero, T.; Coscarelli, R.; Gullà, G.; Nanni, T.; Simolo, C. Precipitation variability and change in the Calabria region (Italy) from a high resolution daily dataset. *Int. J. Climatol.* **2012**, *32*, 57–73. [CrossRef]

3. Kissoudis, C.; Van De Wiel, C.; Visser, R.G.; Van Der Linden, G. Future-proof crops: Challenges and strategies for climate resilience improvement. *Curr. Opin. Plant Biol.* **2016**, *30*, 47–56. [CrossRef] [PubMed]
4. Raza, A.; Razzaq, A.; Mehmood, S.S.; Zou, X.; Zhang, X.; Lv, Y.; Xu, J. Impact of climate change on crops adaptation and strategies to tackle its outcome: A review. *Plants* **2019**, *8*, 34. [CrossRef] [PubMed]
5. Hamed, K.B.; Ellouzi, H.; Talbi, O.Z.; Hessini, K.; Slama, I.; Ghnaya, T.; Bosch, S.M.; Savour, A.; Abdelly, C. Physiological response of halophytes to multiple stresses. *Funct. Plant Biol.* **2013**, *40*, 883–896. [CrossRef] [PubMed]
6. Sellami, M.H.; Pulvento, C.; Lavini, A. Agronomic practices and performances of quinoa under field conditions: A systematic review. *Plants* **2021**, *10*, 72. [CrossRef] [PubMed]
7. Pulvento, C.; Sellami, M.H.; Lavini, A. Yield and quality of Amaranthus hypochondriacus grain amaranth under drought and salinity at various phenological stages in southern Italy. *J. Sci. Food Agric.* **2021**. *Online ahead of print*. [CrossRef] [PubMed]
8. Jacobsen, S.E.; Jensen, C.R.; Liu, F. Improving crop production in the arid Mediterranean climate. *Field Crops Res.* **2012**, *128*, 34–47. [CrossRef]
9. Dasberg, S.; Or, D. Practical applications of drip irrigation. In *Drip Irrigation*; Springer: Berlin/Heidelberg, Germany, 1999; pp. 125–138.
10. Oweis, T.; Pala, M.; Ryan, J. Stabilizing rainfed wheat yields with supplemental irrigation and nitrogen in a Mediterranean climate. *Agron. J.* **1998**, *90*, 672–681. [CrossRef]

environmental sciences proceedings

MDPI

Proceeding Paper

Desert Truffles and Truffles in Morocco: Biodiversity of Promising Fungi to Combat Desertification †

Fatima Henkrar [1,*], Chaimae Meyad [2], Sanae Sabaa [2] and Lahsen Khabar [2,*]

[1] Plant Biotechnology and Physiology Laboratory, Faculty of Sciences, Mohammed V University-Rabat, Rabat 10 000, Morocco

[2] Botanical, Mycology and Environment Laboratory, Faculty of Sciences, Mohammed V University-Rabat, Rabat 10 000, Morocco; chaimae.meyad@um5r.ac.ma (C.M.); sanae.sabaa@um5s.net.ma (S.S.)

* Correspondence: f.henkrar@um5r.ac.ma (F.H.); l.khabar@um5r.ac.ma (L.K.)

† Presented at the 2nd International Laayoune Forum on Biosaline Agriculture, 14–16 June 2022; Available online: https://lafoba2.sciforum.net/.

Abstract: The desert truffle (*Terfezia*) and other truffles (*Tuber*) have a significant economic and ecological value and are considered as important fungi in Morocco. Desert truffles are important to combat desertification and enhance soil fertility. As these fungi form a mycorrhizal symbiosis with several specific desert shrubs, it protects the soil from degradation and assists plant growth in the semi- arid and desert areas. The aim of this short paper is to present the different species of desert truffle that exist in Morocco and identify their macroscopic and microscopic characteristics and their host plants as well as their areas of distribution. There are strong analogies between the species found in Morocco with those previously discovered in other countries. *Terfezia arenaria*, *T. leptoderma* and *Delastria rosea* were mainly available in Mamora forest and *T. boudieri* in Oualidia. Otherwise, the species *Picoa juniperi*, *Terfezia claveryi*, *Tirmania pinoyi* and *T. nivea* were present in the Oriental regions of Morocco.

Keywords: desert truffles; truffles; morphology; microscopy

Citation: Henkrar, F.; Meyad, C.; Sabaa, S.; Khabar, L. Desert Truffles and Truffles in Morocco: Biodiversity of Promising Fungi to Combat Desertification. *Environ. Sci. Proc.* **2022**, *16*, 35. https://doi.org/10.3390/environsciproc2022016035

Academic Editor: Abdelaziz Hirich

Published: 16 June 2022

1. Introduction

In Morocco, truffles and desert truffles (Terfess) are important fungi with economic and ecological value. These are edible species and are considered important seasonal and local trade in the country. Generally, Moroccan truffles grow in the sand and are harvested in the vicinity of herbaceous plants such as *Helianthemum*, *Cistus*, or *Pinus* tree in a mycorrhizal association [1–4]. The most common method of identification for harvesting is by the "mark" produced by the fungi (the soil is often swollen and cracked on the surface at the base of the host plant) [4].

Terfess has a considerable ecological interest, especially in arid and semi-arid areas. Its cultivation using biotechnological methods, in particular the controlled mycorrhization with their native host or non-host plants, would be interesting to utilize infertile lands, conserve these fungal species, and promote the social and economic activities in the region [5]. Moreover, the host plants of these fungi are xerothermophile species and their planting can help to preserve the land against desertification and degradation. The fungi act as extensions of the plant root and are capable of capturing water where the plant cannot. The aim of the study was to characterize the different species of desert truffles "Terfess" and other Moroccan truffles (*Tuber*), to identify the areas for harvest as well as areas of their distribution.

2. Materials and Methods

A survey was carried out to collect truffles from several regions of Morocco, with different climatic and edaphic conditions. The morphological observations of the collected samples were realized visually or using a stereomicroscope. The microscopic studies were performed using crushed ascocarp fragments in water. The scanning electron microscopy studies were preceded through two methods: progressive dehydration with acetone (25%, 50%, 75%, 95% and 100%) and acetolysis according to the Erdtman method [6].

3. Results

The distinction of truffle species was classically based on both the ascocarps morphology and the ascosporal ornamentation. Eight desert-truffle species were identified. Three species were collected from Mamora Forest, located in the Rabat region (34°15′52″ N, 6°39′27″ W). The first species was *Delasria rosea* (=*Terfezia rosea*), locally called 'Bitter Terfess of Taida' and collected under the pine (*Pinus pinaster* var. atlantica and *Pinus halepensis*). The ascocarps were sub-globose or turbinate, more or less bumpy, 3 to 5 cm in diameter. The ascocarp color was white to pinkish or white blackened. Each elongated asci (140–176 × 56–64 µm) contained two to four globose ascospores, orange to yellow and ornamented with an alveoli network surmounted by short spines (Figure 1a). The second species was *Terfezia arenaria*, the most popular edible mushroom in this region, commonly called 'Pink Terfess of Mamora'. It was harvested on acid soil, in a semi-arid climate. *Terfezia arenaria* established mycorrhizal symbiosis with *Helianthemum guttatum*, and it was detected by the 'mark' method. The ascocarps examined were from 2 to 10 cm in diameter and weighed between 4 and 200 g with variable shape (sub-globose, cordiform, bumpy). The asci were ovoid or sub-globose (80–96 × 71–79 µm). They contained eight ascospores, spherical (22–26 µm diameter) and covered with truncated cylindrical warts characteristic of the species (Figure 1b). *Terfezia leptoderma* was collected on acid soil under *Helianthemum guttatum*. The ascocarps were 2 to 5 cm in diameter, generally globose ovoid or pyriform and humped. Asci were octospores, globose, and sessile at maturity; they measured 54–70 × 62–80 µm. The ascospores were black at maturity, spherical and 20–24 µm in diameter. They were covered with spines truncated at the ends (Figure 1c).

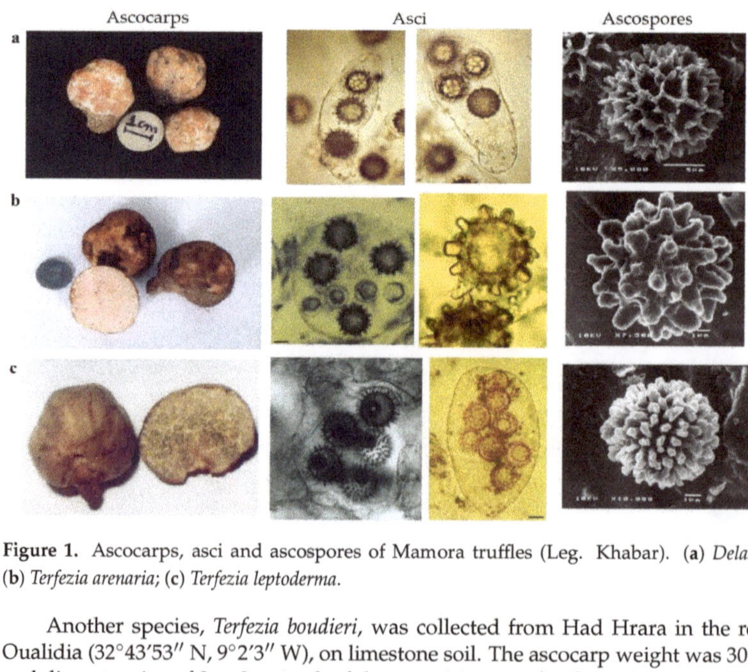

Figure 1. Ascocarps, asci and ascospores of Mamora truffles (Leg. Khabar). (**a**) *Delasria rosea*; (**b**) *Terfezia arenaria*; (**c**) *Terfezia leptoderma*.

Another species, *Terfezia boudieri*, was collected from Had Hrara in the region of Oualidia (32°43′53″ N, 9°2′3″ W), on limestone soil. The ascocarp weight was 30 to 100 g and diameter size of 3 to 8 cm, sub-globose, turbinate or fusiform. The sub-globose asci (66–80 × 56–60 µm) contained six globose ascospores 26 to 30 µm in diameter and yellow to brown in color at maturity (Figure 2).

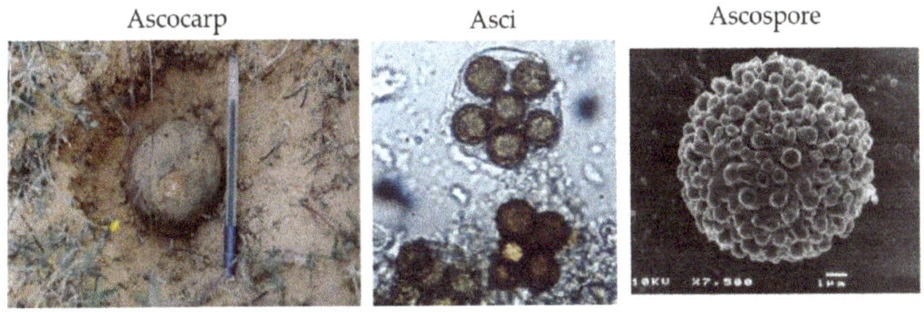

Figure 2. Ascocarps, asci and ascospores of *Terfezia boudieri* (Leg. Khabar).

In the oriental region of Morocco, other promising species were detected, such as *Terfezia claveryi, Picoa juniperi*, and *Tirmania* sp. *Terfezia claveryi*, which were collected close to *Helianthemum lippii* in the Arfoud region (31°26′20″ N, 4°14′37″ W) in the southeast of Morocco, called 'Red Terfess of Tafilalt'. The examined ascocarps had various shapes (cordate, ovoid, and rounded), measuring from 3 to 5 cm in diameter with a weight of 17 to 50 g. The asci were numerous and ovoid (64–68 × 72–84 µm). The ascospores were hyaline to grayish with a spherical shape (18–21 µm in diameter) (Figure 3a). *Picoa juniper* is a very rare species found in Ten Drara of Figuig province (33°03′ N, 2°00′ W). The ascocarps were very light and characterized by a modest size (3 to 6 cm). The asci were sessile and globose, 60 to 100 µm in diameter, containing eight smooth and slightly elliptical ascospores (24–28 × 28–30 µm) (Figure 3b). Otherwise, *Tirmania* sp., locally known as 'White Terfass of Tafilalt', was very abundant in the southeast, in arid and sub-Saharan climates. *Tirmania pinoyi* was collected in Bni Guil whilst *Tirmania nivea* was located in Bouarfa, both under *Helianthemum hirtum*.

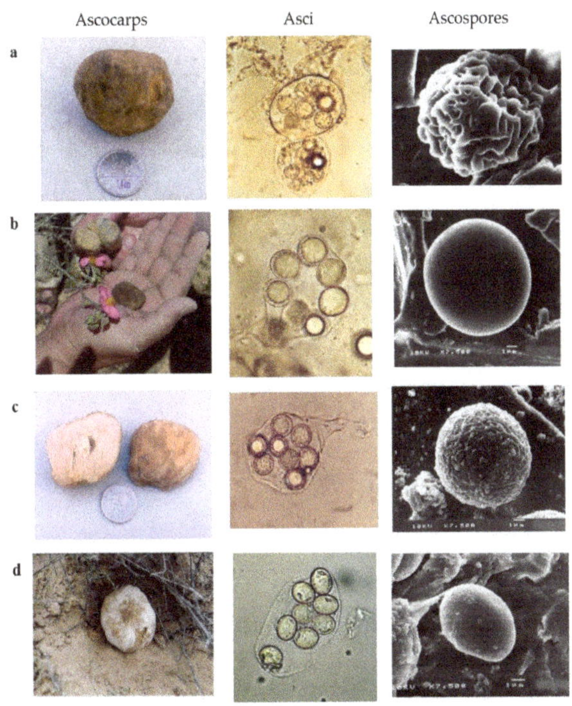

Figure 3. Ascocarps, asci and ascospores of Morocco Oriental truffles (Leg. Khabar). (**a**) *Terfezia claveryi*; (**b**) *Picoa juniperi*; (**c**) *Tirmania pinoyi*; (**d**) *Tirmania nivea*.

The ascocarps size of *Tirmania pinoyi* was from 3 to 8.5 cm in diameter. The asci were about 70–100 × 40–55 μm. The ascospores were globose and 16–24 μm in diameter (Figure 3c). The ascocarp size of *Tirmania nivea* was 3–12 cm in diameter and the asci were 60–80 × 30–50 μm. The ascospores were elliptical and 60–20 × 12–15 μm (Figure 3d).

Regarding tuber species in Morocco, five truffles were identified including *Tuber uncinatum/aestivum, Tuber brumal, Tuber excavatum, Tuber rufum* and *Tuber melanosporum*. This species was particularly found in mountainous regions, with pedoclimatic conditions similar to those of Europe. *Tuber melanosporum* (Figure 4) was found in the plateau of Debdou and in Immozer of Moyen Atlas. *Tuber uncinatum/aestivum, Tuber brumal, Tuber excavatum* and *Tuber rufum* (Figure 4) were collected in the middle Atlas under *Quercus ilex* and *Quercus faginea*. Finally, two other species were collected in semi-arid climates and in acidic sandy loam soil of Mamora forest, such as *Tuber asa* near *Helianthemum guttatum* and *Tuber oligospermum* under *Pinus pinaster* and *Pinus halepensis*.

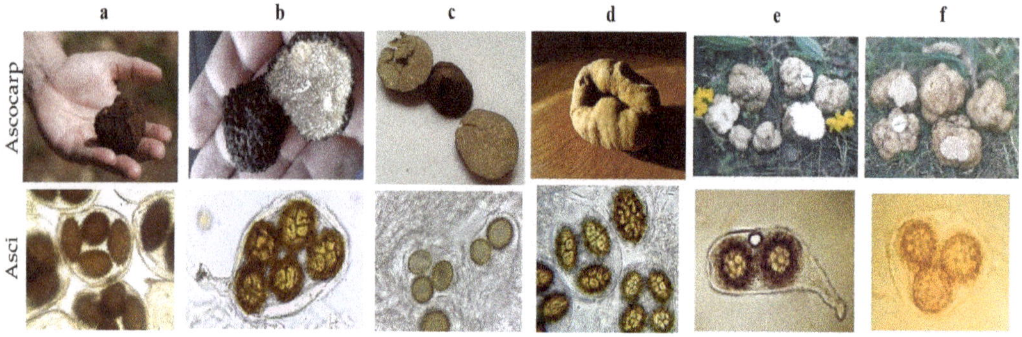

Figure 4. Ascocarps and asci of *Tuber* sp. (Leg. Khabar). (**a**) *Tuber melanosporum;* (**b**) *Tuber uncinatum/aestivum;* (**c**) *Tuber rufum;* (**d**) *Tuber excavatum;* (**e**) *Tuber asa;* (**f**) *Tuber oligospermum.*

4. Discussion

The species found in Morocco, such as *Terfezia arenaria, Terfezia leptoderma* and *Delastria rosea*, were also found in the south of Spain, the south of France and the south of Italy [7], and these species were in all countries of North Africa [4]. On the other hand, the species *Picoa juniperi, Terfezia claveryi, T. boudieri, Tirmania pinoyi,* and *T. nivea* collected in the Oriental regions of Morocco were also discovered in the deserts of Algeria [8] and Tunisia [9,10], whereas the black truffles collected in Moyen Atlas of Morocco were found in similar conditions of Europe truffles [7].

5. Conclusions

There are strong analogies between the species found/available in the Mediterranean countries. Still, the survey of Moroccan truffles is in progress, and further exploration will be performed in the southern and Saharan region of Morocco. We propose to develop methods of mycorrhization by several desert truffles to enhance the culture and rural development. We thought that desert truffles with their host plants would be an excellent alternative for recovering degraded soils.

Author Contributions: F.H. wrote the manuscript, L.K. conceived, supervised the work and carried out the experiment, C.M. and S.S. helped in performing the experiment. All authors have read and agreed to the published version of the manuscript.

Funding: This work was financed by Faculty of Sciences, University Mohammed V-Rabat Morocco.

Institutional Review Board Statement: Not applicable.

Informed Consent Statement: Not applicable.

Data Availability Statement: Not applicable.

Conflicts of Interest: The authors confirm no conflict of interest.

References

1. Khabar, L.; Najim, L.; Janex-Favre, M.C.; Parguey-Leduc, A. Contribution à l'étude de la flore mycologique du Maroc: Les truffes marocaines. *Bull. Soc. Mycol.* **2001**, *3*, 213–229.
2. Khabar, L. Etudes Pluridisciplinaires des Truffes du Maroc. Thèse D'état Es-Sciences, FSR-Rabat-Morocco, Rabat, Morocco, 2002; p. 137.
3. Khabar, L. Mediterranean Basin: North Africa. In *Desert Truffles*; Kagan-Zur, V., Roth-Bejerano, N., Sitrit, Y., Morte, A., Eds.; Springer: Berlin/Heidelberg, Germany, 2014; Volume 38, pp. 143–158.
4. Khabar, L. *Les Terfess et les Truffes au Maroc: Biodiversité et Valorisation*; Universitaires Européenes: Sarrebruck, Germany, 2016; p. 269. ISBN 978-3-639-54790-0.
5. Bermaki, F.Z.; Khabar, L.; Ezzanega, A. Bioecology of desert truffles in the province of Figuig in Eastern Morocco. *Rev. Mex. Mic.* **2017**, *46*, 29–36. [CrossRef]
6. Erdtman, G. The Acetolysis Method—A Revised Description. *Sven. Bot. Tidskr.* **1960**, *54*, 561–564.
7. Zambonelli, A.; Donnini, D.; Rana, G.L.; Fascetti, S.; Benucci, G.M.N.; Iotti, M.; Morte, A.; Khabar, L.; Bawadekji, A.; Piattoni, F.; et al. Hypogeous fungi in Mediterranean maquis, arid and semi-arid forests. *Plant Biosyst.* **2014**, *148*, 392–401. [CrossRef]
8. Bradai, L.; Bissati, S.; Chenchouni, H. Etude mycologique et bio-écologique de la Truffe blanche du désert (*Tirmania nivea* Desf. Trappe 1971) dans la région de Oued M'ya (Ouargla, Sahara algérien). *Rev. Des Bioressources* **2013**, *3*, 6–14. [CrossRef]
9. Sbissi, I.; Ghodhbane-Gtari, F.; Neffati, M.; Ouzari, H.I.; Boudabous, A.; Gtari, M. Diversity of the desert truffle Terfezia boudieri Chatin. in southern Tunisia. *Can. J. Microbiol.* **2011**, *57*, 599–605. [CrossRef] [PubMed]
10. Slama, A.; Fortas, Z.; Neffati, M.; Khabar, L. Etude taxinomique de quelques Ascomycota hypogés (Terfeziaceae) de la Tunisie méridionale. *Bull. Soc. Mycol. Fr.* **2006**, *122*, 187–195.

environmental sciences proceedings

MDPI

Proceeding Paper

Agronomic Practices and Performances of *Stevia rebaudiana bertoni* under Field Conditions: A Systematic Review †

El Hocine Hirich [1,2,*], Brahim Bouizgarne [1], Abdelmjid Zouahri [3] and Khalid Azim [2]

1 Laboratory of Plant Biotechnology, Team of Microbial Biotechnology, Faculty of Sciences of Agadir, Ibn Zohr University, Agadir 80000, Morocco; b.bouizgarne@uiz.ac.ma

2 Research Unit of Integrated Plant Production, Regional Centre of Agricultural Research, Avenue Ennasr, BP 415 Rabat Principale, Rabat 10090, Morocco; khalid.azim@inra.ma

3 Regional Centre of Agricultural Research, National Institute of Agronomic Research, Avenue Ennasr, BP 415 Rabat Principale, Rabat 10090, Morocco; abdelmjid.zouahri@inra.ma

* Correspondence: hirich.elhocine@gmail.com; Tel.: +212-697558737

† Presented at the 2nd International Laayoune Forum on Biosaline Agriculture, 14–16 June 2022; Available online: https://lafoba2.sciforum.net/.

Abstract: *Stevia (Stevia Rebauiana Bertoni)* is one of the most renowned medicinal plants for its low calorific value. Stevia's active components are steviol glycosides (SGs), which include Rebaudioside A, B, C, D, E, F, M, Stevioside, Steviolbioside, Dulcoside A, and Dulcoside C. These steviol glycosides are 150–300 times sweeter than sugar. The sweetening molecules stevioside and rebaudioside A are the most common. In this work, we performed a systematic review combined with a bibliometric analysis of stevia farming techniques in the field. The study is based on published literature data for the years 2000–2021. A sum of 54 articles was found, indicating that scientific study on stevia's agronomic techniques and productivity in the field is currently insufficient. Asia, Europe, and South America were the major research production sites in this domain, accounting for more than 90% of the research output. The number of articles dealing with density and planting that were examined was quite restricted. The principal themes covered in the scientific literature were the effects of "fertilization" and "irrigation", followed by plant growth-promoting rhizobacteria "PGPR" and fungi "PGPF", "salinity", and "harvest" on stevia yield and quality. The results of this research will allow us to highlight insufficient available research works and knowledge gaps and the agronomic treatments that had the greatest impact on productive response were fertilization, irrigation, and salinity.

Keywords: stevia; systematic review; bibliometric analysis; stevioside; rebaudioside-A; agronomic techniques

Citation: Hirich, E.H.; Bouizgarne, B.; Zouahri, A.; Azim, K. Agronomic Practices and Performances of *Stevia rebaudiana bertoni* under Field Conditions: A Systematic Review. *Environ. Sci. Proc.* **2022**, *16*, 44. https://doi.org/10.3390/environsciproc2022016044

Academic Editor: Abdelaziz Hirich

Published: 16 June 2022

1. Introduction

Stevia rebaudiana Bertoni is a sweet plant that belongs to the *Asteraceae* family and is a natural non-calorie bio-sweetener that can help people with diabetes and obesity. In the next years, global demand for high bio-sweeteners is likely to increase significantly [1]. Stevia leaves contain approximately ten steviol glycosides, the most significant of which are stevioside (3–10%), rebaudioside-A (13%), rebaudioside-B, rebaudioside-C, and rebaudioside-D [2].

Before the beginning of the 21st century, most stevia research was detailed in the so-called unpublished literature; numerous experimental trials were carried out in stevia's countries of origin, but not found in the worldwide database. Little research has been undertaken on stevia's performance and quality in field conditions when various agronomic management systems were applied, despite its global relevance, resilience to unfavorable environments, sweetening qualities, and pharmacological capabilities. To answer the question, "What are the knowledge gaps in agronomic management and the performance of stevia under field conditions?" a systematic review incorporating bibliometric analysis is required. In this study, a systematic review which is a high-level summary of fundamental research with the goal of discovering, selecting, synthesizing, and evaluating all high-quality data records was used [3]. The systematic review consisting of literature searching was combined with a bibliometric analysis of stevia farming techniques in field conditions as reflected in the worldwide database over the previous two decades (2000–2021).

2. Material and Methods

We used an approach based on a literature search, inclusion and exclusion criteria, bibliometric screening, and conceptual network analysis in this study, according to Sellami et al. [4], and other researchers [5–7].

2.1. Literature Research

Literature searching linked to the agronomic techniques of stevia (*Stevia rebaudiana bertoni*) production throughout the world was performed through a systematic review of two bibliographic databases: Web of Science and Scopus. The findings were published in English-language journals between 2000 and 2021. Academic database searches were carried out on 20 October 2021. The following search terms were used in bibliographic databases to search for "subject terms" coupled with Boolean operators: ((fertilization OR irrigation* OR harvest* OR salinity* OR PGPR/PGPF* OR planting) AND (yield OR quality OR stevioside* OR rebaudioside-A*) AND (stevia or (stevia and *rebaudiana*)). Any amount of characters can be represented using the wildcards.

2.2. Inclusion and Exclusion Criteria

To synthesize evidence from a variety of sources, we employed a highly robust and reasonable systematic review process. We limited the systematic review in this analysis by specifying boundaries that included: (I) studies conducted solely in the field, excluding glasshouse and pot studies; and (II) crop productivity studies, excluding forestry, fisheries, domestic animals, and other non-food farming. According to the systematic review evaluation criteria, search keywords were based on four PICO components: population, intervention, comparison, and outcome.

2.3. Bibliometric Screening

After deduplication, the data of the accepted articles were loaded into Endnote (online bibliographic management software; Clarivate Analytics, London, UK). All references were evaluated and were used to find and analyze all references based on the following methodology: The title, abstract, and complete text of each item were pre-selected three times each. At each level, files holding or possibly carrying critical information were identified and sent to the next level.

2.4. Bibliometric Analysis and Concept Network Analysis

The year of publishing, the journal, and the frequency with which authors utilized phrases and keywords were all evaluated in the systematic review meta-data. The box plots were created with MS-Excel software.

3. Results

Overall Yield across Factors of Variation

Results were obtained by the above-mentioned literature research methods, and variance was analyzed by MS-Excel software. Owing to the Köppen–Geiger climatic zone [8], the variance was extremely substantial, as shown in Figure 1a. The lowest yield response was 0.79 g/plant in the dry-humid subtropical climate (Cwa), whereas the maximum yield response was 142.8 g/plant in the warm-humid continental climate (Dfb). The highest yield was recorded in the warm-humid continental climate (Dfb), whereas the lowest was reported in the hot semi-arid climate (Bsh) with 142.8 g/plant and 0.79 g/plant respectively.

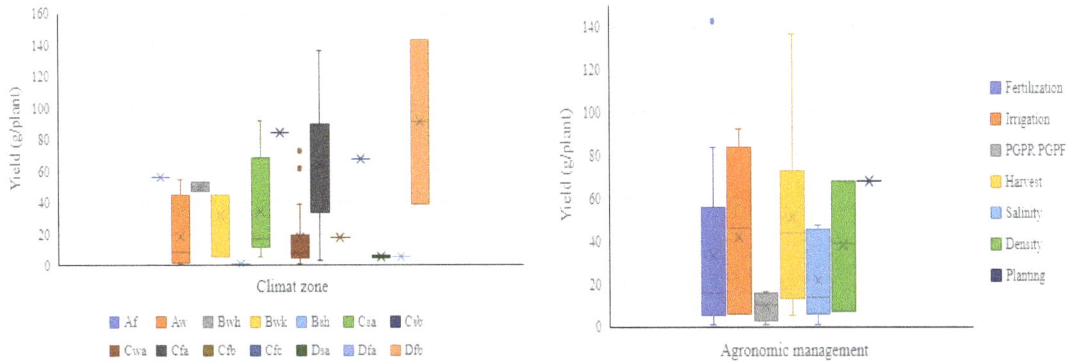

(**a**) (**b**)

Figure 1. Box plot showing yield (g/plant) trends for all articles (n = 54) across: (**a**) diverse climatic zones; (**b**) diverse agronomic practice groups. X: the average's representation. (-Köppen–Geiger climate zone are: Af: Equatorial climate Aw: tropical wet-dry climate; BSh: hot semi-arid climate; BWh: hot desert climate; BWk: cold desert climate; Cfa: humid subtropical; Cfb: maritime; Cfc: oceanic climate; Csa: interior Mediterranean; Csb: coastal Mediterranean; Cwa: dry-humid subtropical climate; Dsa: dry-hot continental climate; Dfa: hot-humid continental climate and Dfb: warm-humid continental climate. PGPR: plant growth promoting rhizobacteria, PGPF: plant growth promoting fungi-).

The yield variance between different agronomic management practices is shown in Figure 1b. The highest yield value above 80 g/plant was attained by the fertilization followed by the harvest with a yield value above 75 g/plant. The agronomic interventions that had the most influence on productive response were irrigation and harvesting time; PGPR/PGPF had less impact, with a yield value of 16.5 g/plant.

4. Discussion

The systematic review combined with the bibliometric analysis revealed that stevia can grow and ensure a high yield of leaves under different climatic zones in the world. Additionally, the reviewed studies have shown that climate zones like a warm-humid continental climate (Dfb) and humid subtropical climate (Cfa) produce the highest yield. However, a hot semi-arid climate (Bsh), dry-hot continental climate (Dsa), and hot-humid continental climate are much less productive. According to Ramesh et al. [9], stevia can be cultivated in a variety of climates, including semi-humid, subtropical, and temperate zones. Therefore, stevia yield has increased in temperate zones of Central and South Europe [10]. According to research conducted in Egypt, meteorological parameters such as temperature, as well as the length and intensity of the photoperiod, have a significant impact on stevia production, as seen by the considerable increase in yield during the summer compared to the winter [11].

The reviewed studies also have shown that agronomic methods like fertilization results in the highest yield. In addition to fertilization, irrigation is also a critical element that impacts stevia yield. Harvest and density are also important techniques to be considered. Biofertilizers by PGPR/PGPF are unfortunately less efficient.

The determination of the optimal type of fertilizer and/or irrigation regime is critical in the case of stevia introduction to new locations. Furthermore, in its native country (Paraguay), stevia yield under unirrigated conditions is 1500–2500 kg/ha, whereas the yield with irrigation is 4300 kg/ha [12]. According to Karimi et al. [13], the application rate of nitrogen (N) was raised from 30 to 90 kg/ha to boost stevia yield.

5. Conclusions

The investigation found an increased interest in research on fertilization, irrigation, and optimal harvest periods as long-term strategies to assure high yields even in disadvantaged locations like coastal and desert countries or those characterized by different biotic or abiotic stress. According to the geographical dispersion of the study, a stevia crop can adapt to a variety of climatic conditions. This systematic review may also be used by researchers to discover weaknesses and the best methodologies in a research procedure, improve cooperation, particularly among researchers from different countries, increase field studies, and maximize the application of research results. It would contribute significantly to the spread of stevia in many conditions all over the globe.

Author Contributions: A.Z. project coordinator; E.H.H. wrote the paper under the supervision of K.A. and B.B.; All authors have read and agreed to the published version of the manuscript.

Funding: The research was carried out as part of a project for the development and exploitation of stevia culture in organic mode at the Agadir regional center of the National Agricultural Research Institute and funded by the COSUMAR group.

Institutional Review Board Statement: Not applicable.

Informed Consent Statement: Not applicable.

Data Availability Statement: Not applicable.

Acknowledgments: We would like to thank COSUMAR for the financing of the project, and the Agadir regional center of the National Agricultural Research Institute for providing research facilities, material, and the supervision to undertake this study.

Conflicts of Interest: The authors declare no conflict of interest.

References

1. Takacs-Hajos, M.; Ruboczki, T.; Szabo, F.M.; Annamária, K.I.S.S. Effect of environmentally friendly nutrition supply on Stevia (*Stevia rebaudiana B.*) production. *Not. Bot. Horti Agrobot. Cluj Napoca* **2019**, *47*, 201–206. [CrossRef]
2. Bharathi, N. Stevia-The untapped sweetener market. *Plant Hort. Tech.* **2003**, *4*, 38–42.
3. Harris, J.D.; Quatman, C.E.; Manring, M.M.; Siston, R.A.; Flanigan, D.C. How to Write a Systematic Review. *Am. J. Sports Med.* **2013**, *42*, 2761–2768. [CrossRef] [PubMed]
4. Sellami, M.H.; Pulvento, C.; Lavini, A. Agronomic practices and performances of quinoa under field conditions: A systematic review. *Plants* **2020**, *10*, 72. [CrossRef] [PubMed]
5. Nasir, S.; Ahmed, J. *Stevia rebaudiana*: A Bibliometric Analysis from 1966–2019. *Adv. Life Sci.* **2021**, *8*, 195–201.
6. Youngblood, M.; Lahti, D. A bibliometric analysis of the interdisciplinary field of cultural evolution. *Palgrave Commun.* **2018**, *4*, 120. [CrossRef]
7. Zhang, X.; Estoque, R.C.; Xie, H.; Murayama, Y.; Ranagalage, M. Bibliometric analysis of highly cited articles on ecosystem services. *PLoS ONE* **2019**, *14*, e0210707. [CrossRef] [PubMed]
8. Köppen–Geiger Climate Zone. Available online: https://www.britannica.com/science/Koppen-climate-classification (accessed on 5 December 2021).
9. Ramesh, K.; Singh, V.; Megeji, N.W. Cultivation of stevia [*Stevia rebaudiana (Bert.) Bertoni*]: A comprehensive review. *Adv. Agron.* **2006**, *89*, 137–177.
10. Kumar, R.; Sharma, S.; Prasad, R. Yield, nutrient uptake, and quality of stevia as affected by organic sources of nutrient. *Commun. Soil Sci. Plant Anal.* **2013**, *44*, 3137–3149. [CrossRef]
11. Yadav, A.K.; Singh, S.; Dhyani, D.; Ahuja, P.S. A review on the improvement of stevia [*Stevia rebaudiana (Bertoni)*]. *Can. J. Plant Sci.* **2011**, *91*, 1–27. [CrossRef]
12. Karimi, M.; Moradi, K. The response of Stevia (*Stevia rebaudiana Bertoni*) to nitrogen supply under greenhouse condition. *J. Plant Nutr.* **2018**, *41*, 1695–1704. [CrossRef]
13. Mahajan, M.; Pal, P.K. Yield response, accumulation of bioactive ingredient and ion uptake of *Stevia rebaudiana* to different soil-moisture and nitrogen levels. *Agric. Water Manag.* **2022**, *264*, 107511. [CrossRef]

environmental sciences proceedings

MDPI

Proceeding Paper

How Does Compost Amendment Affect Stevia Yield and Soil Fertility? †

El Hocine Hirich [1,2,*], Brahim Bouizgarne [1], Abdelmjid Zouahri [3], Oumaima Ibn Halima [3] and Khalid Azim [2]

1 Laboratory of Plant Biotechnology, Team of Microbial Biotechnology, Faculty of Sciences, Ibn Zohr University, Agadir 80000, Morocco; b.bouizgarne@uiz.ac.ma

2 Research Unit of Integrated Plant Production, Regional Centre of Agricultural Research, Avenue Ennasr, BP 415 Rabat Principale, Rabat 10090, Morocco; khalid.azim@inra.ma

3 Regional Centre of Agricultural Research, National Institute of Agronomic Research, Avenue Ennasr, BP 415 Rabat Principale, Rabat 10090, Morocco; abdelmjid.zouahri@inra.ma (A.Z.); ibnhalima@gmail.com (O.I.H.)

* Correspondence: hirich.elhocine@gmail.com; Tel.: +212-697558737

† Presented at the 2nd International Laayoune Forum on Biosaline Agriculture, 14–16 June 2022; Available online: https://lafoba2.sciforum.net/.

Abstract: Stevia (*Stevia rebaudiana*) is a plant species belonging to the family *Asteraceae.* It contains natural intense sweeteners while presenting low carbohydrate content. Its insignificant effect on blood glucose makes its consumption possible for diabetic and hypoglycemic diets. Chemical fertilizers have a negative impact on the stevia leaves quality. Organic farming has become essential in producing medicinal plants such as Stevia to boost stevia growth with stevioside and rebaudioside-A content. The purpose of this study was to evaluate the effect of four compost doses, i.e., 0, 10, 20, and 30 t/ha on Stevia yield and soil fertility. This experiment was conducted in the Melk Zhar experimental domain, Belfaa of the National Institute of Agronomic Research (INRA), Agadir, Morocco during the period of February-June 2020. The obtained results revealed that increased compost doses significantly increased yield and soil fertility i.e., phosphorus; P, potassium; K and organic matter; OM contents. The highest yield (2.19 t/ha) was observed under treatment using 30 t/ha of compost followed by 20 t/ha (1.67 t/ha), and the lowest (1.50 t/ha) under control conditions. The highest OM content (1.02%) was found under treatment with 30 t/ha of compost while the lowest (0.85%) was under the control. The soil analysis also showed that the application of compost at 20 t/ha resulted in the highest P (28.68 ppm) and K (125.5 ppm). In the light of finding, it is concluded that the application of compost at 30 t/ha is the most effective recommended rate for improving Stevia yield and soil fertility.

Keywords: phosphorus; potassium; nutrients; organic matter; soil fertility; Stevia; yield

Citation: Hirich, E.H.; Bouizgarne, B.; Zouahri, A.; Ibn Halima, O.; Azim, K. How Does Compost Amendment Affect Stevia Yield and Soil Fertility? *Environ. Sci. Proc.* **2022**, *16*, 46. https://doi.org/10.3390/environsciproc2022016046

Published: 16 June 2022

1. Introduction

Stevia (*Stevia rebaudiana Bertoni*) is native of South American (Paraguay and Brazil), however, its cultivation as medicinal plant is spreading across the world. Worldwide, there are more than 100 species of Stevia are available. It produces sweet molecules called steviol glycosides (SG) [1]. The most important steviol glycosides in stevia leaf are stevioside (5–10% of dry leaf weight), which is approximately 300 times sweeter than sugar, and rebaudioside A (2–4%), which is better suitable for usage in food and beverage than Stevioside owing to its pleasant flavor [2].

In conventional agriculture systems, the application of agrochemicals has increased, which is resulting in increased residual content of those chemicals in therapeutic medicines. There is an increased danger of health risks and environmental issue because of the indiscriminate use of chemical fertilizers and pesticides. Compost and other organic fertilizers, when applied, can increase nutrient availability of the crops and assist to improve soil health. Organic fertilizers are good alternatives to inorganic fertilizers since they increase soil quality. One of the most recent breakthroughs is the adoption of organic fertilizers, which delays nitrification for an extended period and boosts soil fertility. The use of organic fertilizers in conjunction with integrated nutrition management has been shown to boost Stevia output [3]. Further, the use of organic fertilizers such as compost is beneficial in reducing environmental concerns caused by chemical fertilizers. Thus, the objective of this study was carried to determine the effect of various levels of compost on Stevia yield and soil fertility under arid climate of Souss Massa region in Morocco.

2. Material and Methods

2.1. Experimental Site and Design

This experiment was conducted in the Melk Zhar Experimental station, Belfaa of the National Institute of Agronomic Research (INRA), Agadir, Morocco (30°2′33″ N; 9°33′4″ W) The physico-chemical properties of the soil are presented in Table 1. The study focused on the growth of Stevia under four different levels of compost (0, 10, 20, and 30 t/ha) during the period of February–June 2020. These treatments were implemented in a randomized complete block design containing 4 plots with an area for each of (62 m × 34 m = 2108 m^2). The planting density is approximately 80,000 plant/ha.

Table 1. Physic-chemical characteristics of the soil of the experimental plots at the Melk Zhar experimental station.

Parameters	2020
Electric conductivity (dS/m)	0.10 ± 0.01
pH	8.24 ± 0.18
Clay (%)	8.80 ± 1.20
Fine silt (%)	4.73 ± 1.52
Coarse silt (%)	3.74 ± 1.75
Fine sand (%)	60.54 ± 1.65
Coarse sand (%)	22.45 ± 1.74
Organic matter (%)	0.86 ± 0.42
Potassium (ppm)	687.3 ± 765.25
Calcium (ppm)	23,548.5 ± 6258.25
Sodium (ppm)	1235.3 ± 525.68

2.2. Observations Recorded

2.2.1. Analytical Methods

Soil samples were collected from four plots at 25 cm deep using auger. The soil samples were spread and air dried for 48 h in the laboratory. Then, it was passed through a 2 mm mesh sieve to remove the coarse materials and retained only the fine soil to analyze available phosphorus (P), exchangeable potassium (K), and organic matter (OM). Table 2 shows the different methods used for the analysis of these elements.

Table 2. Methods used for soil fertility analyses.

Element	Method
Phosphorus	Olsen [4]
Potassium	Flame spectrophotometer [5]
Organic matter	Spectrometry [6]

2.2.2. Yield Parameters

To evaluate the effects of different doses of compost on yield (dry leaf yield), the plants from each plot were manually harvested 10 cm above the base of the stem at the end of the growth season (June 2020). Leaf samples were collected just before blooming to analyze steviol glycoside concentrations in the leaves, at this stage steviol content are at their highest [7]. Once cut, the fresh plants were dried immediately. The drying was carried out under natural conditions under the sun using a delta tunnel 9 for a period of 2 to 5 days depending on the climatic conditions.

3. Results

The analysis of variance statistical analysis showed a significant (<0.001) difference among the four compost treatments in the yield and P content, while no differences among the three doses and the control in OM and K levels were noted ($p < 0.000$). This indicates that compost affected yield and P content and did not affect K and OM contents. The results obtained (Figure 1a) showed that the highest yield (2.19 t/ha) was obtained after treatment with a compost dose of 30 t/ha followed by 20 t/ha (1.67 t/ha), and the lowest yield (1.50 t/ha) was observed under control conditions. Regarding soil fertility (Figure 1b), the highest OM content (1.02%) was found with 30 t/ha of compost, and the application of 20 t/ha resulted in the highest values of the remaining nutrients in the soil after harvest, equal to 28.68 and 125.5 ppm for P and K contents, respectively.

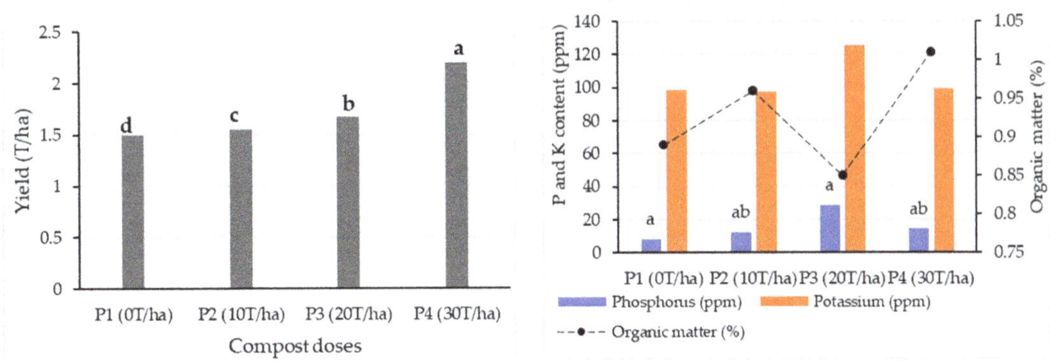

(a) (b)

Figure 1. Impact of compost amendment on Stevia yield (**a**) and soil fertility (**b**). The average values given by the same letters are not significantly different in ANOVA and the Tukey test ($p < 0.05$). Lowercase letters (a, b, c and d) indicate significant differences between treatments.

4. Discussion

Chemical fertilizers increased food production while posing a cost to the climate. Furthermore, Stevia produced through the application of chemical fertilizers has been linked to health issues according to some customers. As a result, research communities are searching for alternatives to agrochemicals [8]. Organic amendments such as composts, are an effective alternative to inorganic fertilizers and improve the production and development of Stevia in a sustainable way [9]. Compost enhanced the OM and nutritional content such as P and K of the soil, which had a good influence on Stevia yield, as had been reported in other crops [10]. Stevia has reduced nutritional requirements and can be easily adapted in poor soil quality [11]. On the other hand, a nutrient deficit can be harmful [12]. The higher yield of Stevia with compost might be due to better nutrient availability near the root zone. In fact, it is well recognized that increased productivity due to compost application helps to reduce the use of artificial fertilizers and hence boost the sustainability of stevia production [13].

5. Conclusions

The impact of compost on a crop is observed in the medium term and thereafter, concluding that recommendation of one of the three doses over the others is not supported for this first year of testing. The finding indicates clearly that Stevia responded positively to increased organic amendment dose and the optimal rate is 30 T/ha. As perspective, we recommend increasing further the amendment dose and minimizing the size of the experimental plot around 2000 m^2 and concentrate the factors studied while studying the economic impact of the tested solutions and estimating their cost.

Author Contributions: A.Z. project coordinator; E.H.H., O.I.H. and K.A. conceived and designed the experiments; E.H.H. performed the experiments and analyzed data under the supervision of K.A. and B.B.; E.H.H. wrote the paper under the supervision of K.A. and B.B.; All authors have read and agreed to the published version of the manuscript.

Funding: The research was carried out as part of a project for the development and exploitation of stevia culture in organic mode at the Agadir regional center of the National Agricultural Research Institute and funded by the COSUMAR group.

Institutional Review Board Statement: Not applicable.

Informed Consent Statement: Not applicable.

Data Availability Statement: Not applicable.

Acknowledgments: We would like to thank COSUMAR for the financing of the project, and the Agadir regional center of the National Agricultural Research Institute for providing research facilities, material and the supervision to undertake this study.

Conflicts of Interest: The authors declare no conflict of interest.

References

1. Díaz-Gutiérrez, C.; Hurtado, A.; Ortíz, A.; Poschenrieder, C.; Arroyave, C.; Peláez, C. Increase in steviol glycosides production from *Stevia rebaudiana Bertoni* under organo-mineral fertilization. *Ind. Crops Prod.* **2020**, *147*, 112220. [CrossRef]
2. Benhmimou, A.; Ibriz, M.; Al Faïz, C.; Gaboun, F.; Shaimi, N.; Amchra, F.Z.; Lage, M. Effects of water stress on growth, yield, quality and physiological responses of two stevia (*Stevia rebaudiana Bertoni*) varieties in Rabat region, Morocco. *Asian J. Agric. Biol.* **2018**, *6*, 21–34.
3. Coelho, L.; Osório, J.; Beltrão, J.; Reis, M. Organic compost effects on *Stevia rebaudiana* weed control and on soil properties in the Mediterranean region. *Rev. Bras. Cienc.* **2019**, *42*, 109–121.
4. Sims, J.T.; Edwards, A.C.; Schoumans, O.F.; Simard, R.R. Integrating soil phosphorus testing into environmentally based agricultural management practices. *J. Environ. Qual.* **2000**, *29*, 60–71. [CrossRef]
5. Kamentsky, L.A.; Melamed, M.R.; Derman, H. Spectrophotometer: New instrument for ultrarapid cell analysis. *Science* **1965**, *150*, 630–631. [CrossRef] [PubMed]
6. *ISO 14235:1998*; Qualité du sol—Dosage du Carbone Organique par Oxydation Sulfochromique. ISO: Geneva, Switzerland, 1998.
7. Guzman, R.D. Autoecological Role of Steviol Glycosides in *Stevia rebaudiana Bertoni*. Ph.D. Thesis, Central Queensland University, Rockhampton, Australia, 2011.
8. Bidabadi, S.S.; Afazel, M.; Poodeh, S.D. The effect of vermicompost leachate on morphological, physiological and biochemical indices of *Stevia rebaudiana Bertoni* in a soilless culture system. *Int. J. Recycl. Org. Waste Agric.* **2016**, *5*, 251–262. [CrossRef]
9. Sharma, S.; Walia, S.; Singh, B.; Kumar, R. Comprehensive review on agro technologies of low-calorie natural sweetener stevia (*Stevia rebaudiana Bertoni*): A boon to diabetic patients. *J. Sci. Food Agric.* **2016**, *96*, 1867–1879. [CrossRef] [PubMed]
10. Reis, R.V.; Chierrito, T.P.; Silva, T.F.; Albiero, A.L.; Souza, L.A.; Gonçalves, J.E.; Gonçalves, R.A. Morpho-anatomical study of *Stevia rebaudiana* roots grown in vitro and in vivo. *Rev. Bras. Farmacogn.* **2017**, *27*, 34–39. [CrossRef]
11. Ramesh, K.; Singh, V.; Megeji, N.W. Cultivation of stevia [*Stevia rebaudiana (Bert.) Bertoni*]: A comprehensive review. *Adv. Agron.* **2006**, *89*, 137–177.
12. Utumi, M.M.; Monnerat, P.H.; Pereira, P.R.G.; Fontes, P.C.R.; Godinho, V.D.P.C. Macronutrient deficiencies in Stevia: Visual symptoms and effects on growth, chemical composition, and stevioside production. *Pesqui. Agropecu. Bras.* **1999**, *34*, 1038–1043. [CrossRef]
13. Zaman, M.M.; Rahman, M.A.; Chowdhury, T.; Chowdhury, M.A.H. Effects of combined application of chemical fertilizer and vermicompost on soil fertility, leaf yield and stevioside content of stevia. *J. Bangladesh Agric. Univ.* **2018**, *16*, 73–81. [CrossRef]

environmental sciences proceedings

MDPI

Proceeding Paper

Characterization of Desert Truffles in the Great Moroccan Sahara: A Review †

Imane Zniber [1,*], Hassan Boukcim [2], Lahsen Khabar [3], Marc Ducousso [4], Fatima Henkrar [3], Ayoub El Mouttaqi [1] and Abdelaziz Hirich [1]

1 African Sustainable Agriculture Research Institute (ASARI), Mohammed VI Polytechnic University (UM6P), Laayoune 70000, Morocco; ayoub.elmouttaqi@um6p.ma (A.E.M.); Abdelaziz.hirich@um6p.ma (A.H.)

2 AgroBioSciences Research Division, Mohammed VI Polytechnic University, Lot 660–Hay Moulay Rachid, Ben-Guerir 43150, Morocco; hassan.boukcim@um6p.ma

3 Faculty of Sciences, Mohammed V University, Rabat 10000, Morocco; l.khabar@um5r.ac.ma (L.K.); f.henkrar@um5r.ac.ma (F.H.)

4 Laboratoire des Symbioses Tropicales et Méditerranéennes (LSTM, UMR113), Centre de Coopération Internationale en Recherche Agronomique pour le Développement (CIRAD), TA A-82/J, CEDEX 5, 34398 Montpellier, France; marc.ducousso@cirad.fr

* Correspondence: imane.zniber@um6p.ma

† Presented at the 2nd International Laayoune Forum on Biosaline Agriculture, 14–16 June 2022; Available online: https://lafoba2.sciforum.net/.

Abstract: Desert truffles are edible mushrooms compulsorily living in symbiosis with plants' roots. They are rich in essential nutrients and secondary metabolites, conferring therapeutic properties. In Morocco, around ten species can be found in semiarid and arid climate regions with low annual rainfalls and high average temperatures. They can be associated with the *Cistus* and *Pinus* species and some other species, but they are detected more often under the *Helianthemum* species. In this study, we present a short review of the distribution of Moroccan desert truffles in the Great Sahara, along with the progress achieved in their morphological and molecular identification and the evaluation of their nutritional content.

Keywords: desert truffles; distribution; morphological traits; molecular identification; nutritional value

Citation: Zniber, I.; Boukcim, H.; Khabar, L.; Ducousso, M.; Henkrar, F.; El Mouttaqi, A.; Hirich, A. Characterization of Desert Truffles in the Great Moroccan Sahara: A Review. *Environ. Sci. Proc.* **2022**, *16*, 55. https://doi.org/10.3390/environsciproc2022016055

Academic Editor: Redouane Choukr-Allah

Published: 16 June 2022

1. Introduction

Desert truffles or Terfess are mycorrhizal fungi classified into the Pezizales order (previously known as Tuberales), belonging to the Pezizaceae family in the phylum of Ascomycota, including the *Terfezia*, *Tirmania*, *Delastria*, *Picoa*, *Balsamia*, and *Melanogaster* genera. They produce edible hypogeous fruiting bodies called ascocarps that grow around the roots of some known desert plants as a result of the symbiotic association known as mycorrhiza [1]. In Arab countries, they are commonly called" Kamé", "Kholassi", and "Zoubaïdi" but also "truffles of sand" given that they are found in deserts' sandy soils; they are called "Criadilla de tierra" and "Criadilla vaquera" in Spain [2]. In Morocco, the discovered desert truffles are *Terfezia boudieri* Chatin, *T. leptoderma* (Tul. & C. Tul), *T. arenaria* (Moris) Trappe, and *T. claveryi* (Chatin) as well as *Tirmani nivea* (Desf.) Trappe and *Ti.pinoyi* (Maire) Malençon, particularly found in the southeast of the country [2]. In addition to Morocco, desert truffles can be found in all semi-arid and arid regions characterized by warm climates. They are distributed in North Africa, Middle Eastern countries, and the Mediterranean Basin [2]. The objective of this paper is to present a short review about the distribution of desert truffles in the Great Sahara and to explore the species discovered in Morocco along with their ecology, host plants, and the nutrient content of each species.

2. Ecology of Desert Truffles

2.1. In the Great Sahara of Morocco

According to [2], desert truffles are located in calcareous, aerated, sandy soil, characterized by a homogeneous texture and sometimes found in calcareous sandy loam soil but rarely in acidic soil. For host plants, most species that belong to the *Helianthemum* genus are able to grow in symbiosis with desert truffles, for example, *Helianthemum lipii* Pers., *H. ledifolium* (L.) Mill., *H. hirtum* Mill., *H. asperum* Lag. Ex Dunal, etc. The content of organic matter in the soils must be low, along with the fertility and exchange capacity. Regarding minerals, desert truffles require the presence of nitrogen, phosphate, potassium, calcium, and iron although it has been reported that a low content of phosphate and iron increases the mycorrhization potential. Similar environmental conditions have also been reported by [3] in Figuig regions. Concerning climatic conditions, the main limiting factor for desert truffles is rainfall, both the quantity and period of precipitation. For some species, precipitation must be in autumn, but generally, it is more needed during winter and spring. The bioclimatic stages where desert truffles may be found are diverse, from arid and semi-arid to sub-Saharan and Saharan. According to recent research [3], desert truffles are mostly found in the regions of Beni Guil and Abou Lakhal, characterized by moderate rainfall all year despite the desertic climate. The average temperatures are 17.9 and 18.6 °C, respectively. The annual rainfall in those regions is around 170 mm. These regions are also characterized by heavy rain in the evenings and early mornings, dewfalls, thunder, and lightning, which significantly contribute to truffles' growth. Most of these parameters can be found in the south regions of Morocco, including the Great Sahara.

2.2. In Other Countries

A study of desert truffle ecology in Saudi Arabia [4] also mentioned rainfall parameter as the important factor for desert truffle growth. A lower rate of precipitation may affect the physiology of the host plant and, consequently, decrease spore and further ascocarp production. Other factors are physiochemical properties of soil, meteorology, and ambient temperature. In basic soils, dominant host plants are *H. ledifolium* and *H. salicifolium* (L.) Mill. The climate in the desert truffle regions in Saudi Arabia is also hyper-arid to Saharan, with rainfall in October and December and a harvest period starting from the beginning of December.

3. Distribution of Desert Truffles

3.1. In the Great Sahara of Morocco

As already mentioned, some species of desert truffles discovered in Morocco are found in the south of the country. The first one is *T.claveryi* (*T. hafizi* Chat.), also called the "Red Terfess of Tafilalet". It is found in Aïn Beni Mather, Tendrara, Bouarfa, regions of Erfoud, and near Ksar es-Souk, Bou Bernous, Boudenib, and Figuig (Bni Guil and Abou Lakhal regions) [2,3]. It can be harvested after rainfall, from March to May, under the *H. lipii* as a host plant. Furthermore, *T.boudieri* is found in Had Hrara, the region of Oualidia, and a few kilometers to the east of Safi city in the Abda plains. It is also found in Aïn Beni Mather along with regions in Erfoud. Most importantly, it can be detected in desertic plains where it is harvested in March and April under the *Helianthemum* species *H. lipii* and *H. ledifolium*. Other species belonging to the *Tirmania* genus have been identified, namely *Ti. nivea* and *Ti. pinoyi*. They are distributed in the southeast, especially in regions of Aïn Beni Mather, Tendrara, Bouarfa, Erfoud, Figuig, and Rissani. The harvest period is from early December until March under the *H. hirtum* as a host plant.

3.2. In Other Countries

Desert truffles are distributed in Algeria, Tunisia, Libya, Egypt, Saudi Arabia, Iraq, Iran, Kuwait, Syria, and Palestine. They are also spread in the Mediterranean Basin, in the south of Italy, Spain, France, Portugal, and Greece [2]. *T.claveryi* is found in Algeria, Tunisia, Egypt [2], and Libia [3] as well as in the European countries France and Spain (especially in semi-desertic areas in the south) [2]. *T.boudieri* is also found in Algeria and Tunisia under the *Helianthemum lipii* and sometimes under the *Rhatherium suaveolens* Desf., particularly in Libya and Egypt.

4. Characterization of Desert Truffles

4.1. Morphological Traits

In Morocco, the collected ascocarps of *T.claveryi* (Figure 1a) had different shapes (cordiform, ovoid, and round) measuring 3 to 5 cm and fresh weights from 17 to 50 g. The color of the peridium was pinkish-white at the beginning and converted to brownish-black at maturity; there was a few millimeters of spotting to the gleba. The gleba was firm with a little spongy consistency, and its nodules were round, fertile, and pinkish then light brown, separated by pale veins. The asci were ovoid (64–68 × 72–84 μm) and contained hyaline, grey, spheric ascospores (18–21 μm) with an alveolus network. For *Tirmania* sp. ascocarps, the shapes and sizes changed (Figure 1b) from sub-globular and piriform to similar to that of a turbine, with lobes or irregular shapes along with a protruding basis that was easily detachable. The peridium was pale showing cracks and furrows, and the gleba was whitish-yellow, pulpy, and firm or a little spongy. Inside the ascocarp, fertile nodules were visible in labyrinth patterns, separated with sterile veins. The asci were round on the top and elongated at the basis, containing eight claviform, globular, or elliptic ascospores, with a double smooth membrane [2].

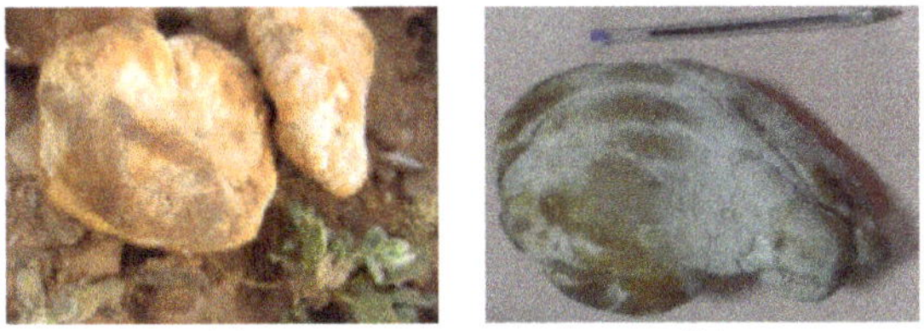

(a) (b)

Figure 1. Desert truffles species: (**a**) *Terfezia claveryi* [5] and (**b**) *Tirmania nivea* [3].

4.2. Molecular Traits and Genetic Diversity

Morphological observations of desert truffles are very informative in the identification of species although, sometimes, the morphology can vary. It depends on temperature, moisture, and the physicochemical properties of the soil but also on intraspecific variations. To face those challenges, the utilization of molecular approaches is useful in order to study intraspecific variability identification more closely [4]. To study the genetic diversity of Iraqi truffles, a team of researchers focused on a molecular study of the internal transcribed spacer (ITS) regions of ribosomal DNA by amplification with specific primers: ITS1-F and ITS4-R, followed by electrophoresis in BET-stained agarose gel, and, finally, observation with a special UV camera [6]. They proceeded with a purification of the PCR products, blasting on the NCBI GenBank database and comparing with other sequences of the same species. Then, the sequences obtained from the collected sample were deposited in the NCBI GenBank to obtain accession numbers. All accessions were studied for genetic diversity by generating a phylogenetic tree using MEGA 6.06. Indeed, all sequences from the collected samples were identified and belonged to two species, *T.claveryi* and *Ti.pinoyi*, as presumed. However, in the phylogenetic analysis, they were separated into two main groups, A and B, according to their rDNA ITS sequences of each species.

4.3. Chemical Composition

Although truffles are known to be rich in nutrients, the research in this field is scarce. Among the research, [7] conducted a study concerning the phytochemical composition and nutritional value of the *Ti.pinoyi* species in Morocco. After measuring different chemical compounds and nutrients from fresh material, dried material, and material macerated to powder, the following results were discovered: 81.5% of moisture and 64.7%, 26.2% and 3.0% of carbohydrates, crude proteins and lipids, respectively. For minerals, Cu measured about 65.3 mg/Kg, Zn about 38.1 mg/Kg, and Se about 35.3 mg/Kg; low quantities of Pb and Cd were detected. In Saudi Arabia [4], it was reported that desert truffles are rich in fibers and amino acids and have high quality proteins with a percentage between 16.3 and 18.5%. For other nutrients, lipid content is between 6.2 and 5.9%, and carbohydrates are between 67.2 and 65%. In addition, alkaline phosphatase (ALP) activity has been reported as a key indicator of metabolic activity of desert truffles as it is produced in drought conditions at a higher rate. Other compounds such as phenolic acids, carotenoids, flavonoids, antioxidants, and minerals can be found as well as different compounds conferring medical and therapeutic activities. Another study conducted on Iraqi *T.claveryi* [8] reported that the total proteins reached 17.6%, lipids arrived at 1.0%, twelve essential amino acids reached 1567.7 µg/g of dry weight, and carbohydrates arrived at 79.8%. The authors also obtained a moisture value of 82.0%, ash value of 1.5%, and a total phenolic compound value of about 6479.0 µg/g, which mainly included the rutin.

5. Conclusions

Desert truffles are a widely spread mushroom with a high nutritional, medical, and therapeutic value. In Morocco, their distribution is already known, but the characterization of their species and the evaluation of their chemical composition is barely studied. These aspects need to be explored further, given that the data available are not sufficient. Therefore, it is considered a promising area of study, especially in the regions of the Great Sahara, located in the south of the country, where optimal growth conditions of desert truffles can be found. As a perspective, we aim to continue the progress achieved in this field, focusing our work in south regions of Morocco. The study of growth conditions and the characterization of desert truffle species to achieve its domestication in the Great Sahara of Morocco represent a priority to preserve this desert diversity.

Author Contributions: Conceptualization, A.H. and H.B.; methodology, A.H., H.B. and M.D.; validation, A.H., H.B. and M.D.; investigation, I.Z., F.H., A.E.M. and L.K.; resources, A.H.; data curation, A.H. and I.Z.; writing—original draft preparation, I.Z.; writing—review and editing, A.H., L.K. and M.D.; supervision, A.H., H.B. and M.D.; project administration, A.H.; funding acquisition, A.H. All authors have read and agreed to the published version of the manuscript.

Funding: This research is funded by Mohammed VI Polytechnic University.

Institutional Review Board Statement: Not applicable.

Informed Consent Statement: Not applicable.

Data Availability Statement: Not applicable.

Conflicts of Interest: The authors declare no conflict of interest.

References

1. Zambonelli, A.; Donnini, D.; Rana, G.L.; Fascetti, S.; Benucci, G.M.N.; Iotti, M.; Morte, A.; Khabar, L.; Bawadekji, A.; Piattoni, F.; et al. Hypogeous fungi in Mediterranean maquis, arid and semi-arid forests. *Plant Biosyst.* **2014**, *148*, 392–401. [CrossRef]
2. Khabar, L. *Les Terfess et Truffes du Maroc: Biodiversité et Valorisation*; Éditions Universitaires Européennes: Sarrebruck, Germany, 2016.
3. Bermaki, F.Z.; Khabar, L.; Ezzanega, A. Bioecology of desert truffles in the province of Figuig in Eastern Morocco. *Rev. Mex. Micol.* **2017**, *46*, 29–36. [CrossRef]
4. Hashem, A.; Alqarawi, A.; Manzoor, A.S.; Wirth, S.; Egamverdieva, D.; Tabassum, B.; Abd_Allah, E.F. Desert Truffles in Saudi Arabia: Diversity, Ecology, and Conservation. In *Microbial Resource Conservation*; Sharma, S., Varma, A., Eds.; Springer: Cham, Switzerland, 2018; Volume 54, pp. 431–451. [CrossRef]
5. Akil, M.E.; Outcoumit, A.; Touhami, A.O.; Benkirane, R.; Douira, A. Study of eastern Morocco desert truffles. *Int. J. Curr. Res.* **2016**, *8*, 33922–33929.
6. Aish, A.A.; Abdulmalek, S.T.; Kareem, T.A.; Yasir, L.B.; Matny, O. Molecular identification and genetic diversity study of the Iraqi truffles. *J. Phytol.* **2020**, *12*, 121–126. [CrossRef]

7. Khlaif, D.K.; Kadum, H.; Abod, N.S. Nutritional and Chemical Compositions of the Desert Truffle (*Terfezia claveryi*) in Samawa City of Iraq. In Proceedings of the IOP Conference Series: Earth and Environmental Science, Jakarta, Indonesia, 25–26 September 2021; p. 923. [CrossRef]
8. Bouatia, M.; Touré, H.A.; Cheikh, A.; Eljaoudi, R.; Rahali, Y.; Oulad Bouyahya Idrissi, M.; Khabar, L.; Draoui, M. Analysis of nutrient and antinutrient content of the truffle (*Tirmania pinoyi*) from Morocco. *Int. Food Res. J.* **2018**, *25*, 174–178.

environmental sciences proceedings

MDPI

Proceeding Paper

Haloculture and Haloengineering in the Context of Water-Energy-Food Nexus †

Farhad Khorsandi

Department of Soil Science, Darab Branch, Islamic Azad University, Darab P.O. Box 7481783143-196, Iran; fkhorsandi@gmail.com

† Presented at the 2nd International Laayoune Forum on Biosaline Agriculture, 14–16 June 2022; Available online: https://lafoba2.sciforum.net/.

Abstract: The utilization of saline soil and water resources by Saline Agriculture systems is an economical and viable strategy for meeting the rising demand for global food. However, despite vast experiences in research and commercial accomplishments, Saline Agriculture have still not played a key role in the extension of sustainable agriculture in saline areas. This could be due to an insufficient devotion to the holistic needs (food, water and energy) of local societies in salt-affected regions. Consequently, Haloculture was developed as a new Saline Agriculture technology, which is defined as the sustainable production of biological and non-biological products in saline environments. The main objective of this paper is to describe and discuss Haloculture in the context of the Water-Energy-Food Nexus (WEFN). This technology integrates engineering tools and sciences referred to as Haloengineering with agricultural sciences to meet the holistic needs of human societies in salt-affected areas. Haloengineering attempts to produce water and energy by exploitation the inherent potentials of saline ecosystems. Haloculture tries to sustainably exploit the services and potentials of saline environments for the production of food, water and energy. It conforms to the contexts of WEFN, and promotes water and energy securities along with food security. Therefore, Haloculture can be a viable strategy for sustainable rural development in saline regions.

Keywords: Biosaline agriculture; ecosystem services; Saline agriculture; sustainable agriculture

Citation: Khorsandi, F. Haloculture and Haloengineering in the Context of Water-Energy-Food Nexus. *Environ. Sci. Proc.* **2022**, *16*, 63. https://doi.org/10.3390/environsciproc2022016063

Academic Editor: Abdelaziz Hirich

Published: 22 June 2022

1. Introduction

The alarming rise in the world population has imposed excessive pressure on soil, water and natural resources to provide the growing demands for food, water and energy. However, most of the productive soils are already being cultivated for agronomic purposes. In addition, the competition between agriculture and other economic sectors for water is escalating due to limitations on available global freshwater resources. Consequently, using salt-affected lands and saline waters for agricultural production through various Saline Agriculture systems (such as Haloculture) is essential for achieving rising human needs in the near future.

Saline Agriculture systems have adequate potentials for the development of sustainable agriculture in salt-affected ecosystems. However, despite rich research and commercial experiences, they have not yet had a significant impact on strengthening the food security in salt-affected regions, especially in less developed countries. This can be due to paying more attention to production aspects and to not paying enough attention to the extension and institutionalizing phases. The holistic needs of human societies and potentials of salt-affected ecosystems to meet those needs (water, energy and food), have not been addressed properly. Suitable measures for the enactment of saline production systems like Haloculture are necessary at regional and national levels. Haloculture is trying to follow the goals and objectives of the Water-Energy-Food Nexus (WEFN) concept. For this purpose, Haloengineering was proposed as a viable solution for the successful development and expansion of Haloculture in salt-affected regions [1]. Consequently, the objective of this paper is to discuss Haloculture and Haloengineering in the context of WEFN for the sustainable development of salt-affected areas.

2. Saline Agriculture Systems, Haloculture and Food Security

The common objective of Saline Agriculture systems is the utilization of saline resources for the improvement of food security. However, with the advancement of the technologies and experiences gained, some instructive differences between their concepts and views have evolved. Haloculture is a production system with emphasis on environmental stewardship and protection, as well as the sustainable and economic exploitation of saline soil and water resources. It started with the successful completion of an international Biosaline Agriculture project in several countries, including Iran [2]. After the end of the project, by considering the strengths and limitations of Biosaline Agriculture and the local and national capacities of the country, Haloculture was introduced [3].

Haloculture is the sustainable, economic production of agricultural (biologic) and industrial (non-biologic) products in saline environments [4]. The cutting edge between Haloculture and Biosaline agriculture is its emphasis on the holistic needs of human societies (i.e., water, energy and food). Thus, Haloculture considers the whole salt-affected ecosystem, rather than merely the saline soil and water resources, as a production resource.

Haloculture promotes integrated production systems (Figure 1). For example, saline aquaculture wastewater can be used to irrigate halophyte forage farms. In turn, the halophyte forages may be used as forages for animals such as cattle and sheep. The higher salinity drainage waters from halophyte farms can be utilized for higher halo-tolerant aquatics, such as microalga and Artemia. Therefore, both biologic (various halophytes and animals), and non-biologic or industrial, products (salts and minerals, drinking water and energy) are produced in the integrated system of Haloculture (Figure 1). Saline environments, which possess various natural capacities, are considered as a medium of production in Haloculture, which will be explained thoroughly later on. Haloculture thereby puts an emphasis on covering the holistic needs of rural communities in salt-affected environments (i.e., food, water and energy), which is more in agreement with the WEFN concept.

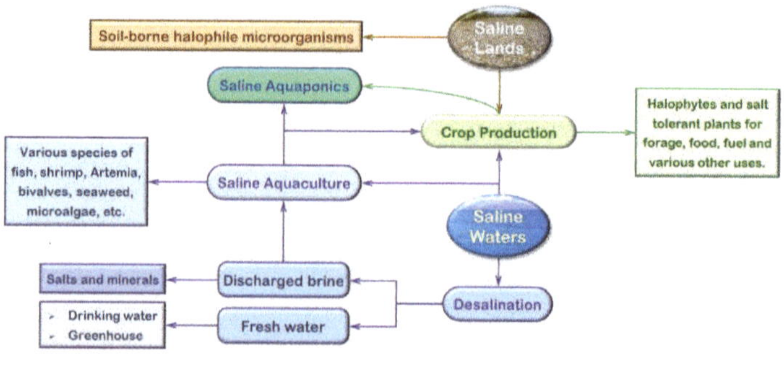

Figure 1. Haloculture concept for the sustainable and integrated use of saline resources [1].

Saline Agriculture systems (Seawater Agriculture, Biosaline Agriculture and Haloculture) have over six decades of national and international research and commercial experiences combined. It seems that research organizations have been fairly successful in raising the awareness of scientists and, to some extent, farmers and decision makers on the economic exploitation of saline soil and water resources. Private companies and organizations have validated the viability and profitability of saline agro-complexes. It seems that both research institutes and the private sector have not yet been successful in the widespread expansion of Saline Agriculture and enhancement of food security in arid, salt-affected regions [1]. Therefore, they have so far not played a significant role in the world food production, reduction of poverty and improvement of living conditions in less developed countries. It is evident that Saline Agriculture systems need to adopt new technical and/or social strategies to help more effectively in the actual implementation of sustainable development in salt-affected areas. This will in turn contribute much more effectively to global food security. Haloengineering is proposed as an effective contribution of Haloculture to the sustainable rural development of saline ecosystems, especially in less developed areas.

3. Haloengineering and Water-Energy-Food Nexus

Water, energy and food are essential needs for human health, poverty reduction and the sustainable development of societies. Consequently, the Water-Energy-Food Nexus (WEFN) has been presented as a framework for sustainable development [5]. The WEFN states that water security, energy security and food security are intricately linked to each other and that the actions taken in each one of them affect the other two [5,6]. According to WEFN, the future growth in world population will increase the need for food, water and energy simultaneously. Hence, systems that tackle these three basic human needs concurrently would promote their sustainable security more efficiently. Considering the definition and objectives of Haloculture, it is evident that Haloculture has tried to adhere to the views and objectives of WEFN by addressing the three basic human needs simultaneously.

Saline soil and water resources are usually abundant in salt-affected regions. However, a lack of vital infrastructures (such as road, electricity and drinking water), as well as harsh environmental and climatic conditions (such as excessive heat and cold, sand storms, windstorms and/or high humidity), are also apparent in such regions. These factors may discourage the large-scale acceptance and application of Haloculture by local farmers. Thus, along with the production of economic bio-products, the production of non-biologic commodities (i.e., energy and drinking water) is also emphasized in Haloculture.

Salt-affected ecosystems, depending on their geographic location, possess diverse capacities and features. These potentials may be managed for the production of water and energy. For example, coastal ecosystems have the potential to exploit their high humidity for freshwater harvesting. However, the appropriate technologies needed to achieve this goal are made possible by engineers, not agriculturalists. Consequently, cooperation between agriculturalists and engineering scientists is needed to fully utilize the natural capacities of saline environments. The applications of engineering concepts and technologies in Haloculture are called Haloengineering.

Haloengineering or *Haloculture engineering* is defined as the harmonic, interdisciplinary application of concepts, methods and tools from the engineering sciences for the sustainable development and enhancement of the living standards of mankind in salt-affected regions, by using locally available basic resources for the cost-effective and efficient production and management of energy and water [1]. Air, water, soil and biodiversity are the basic resources. The concepts and sustainability objectives of Haloengineering are illustrated in Figure 2. All saline ecosystems possess natural and inherent potentials and capabilities. These ecosystem services may include soil and water resources of different quality and quantity, genetic biodiversity, and climatic characteristics such as rainfall, sunshine hours, wind frequency and speed, and humidity. The tools and technologies developed by engineers may be used and/or developed specifically to exploit these inherent potentials, in order to produce the much needed drinking water and energy for human societies in the region (Figure 2). Thus, with the assistance of Haloengineering, Haloculture can contribute much more effectively to sustainable development programs in salt-affected ecosystems.

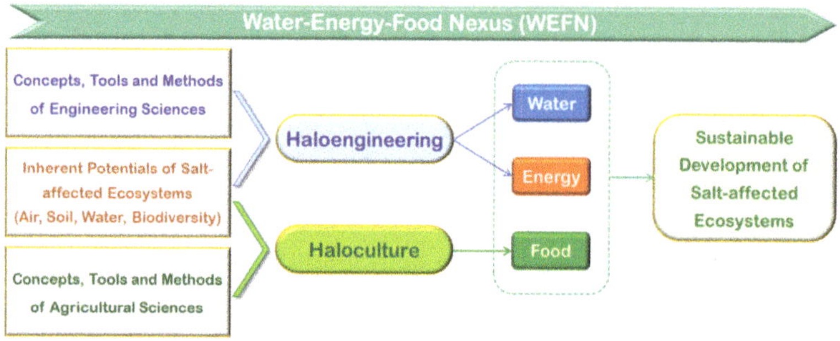

Figure 2. The principles of Haloengineering and Haloculture under the concept of WEFN.

4. Conclusions

Haloculture, and Haloengineering as its complementary component, were discussed under the framework of WEFN. Haloculture, by adopting and applying the idea of Haloengineering, ensures the concurrent production of water, energy and food; therefore, it may contribute much more proficiently to sustainable rural development in saline areas. Hence, a more comprehensive definition of Haloculture is given as the sustainable use of the potentials and capacities of salt-affected ecosystems for the production of biologic and non-biologic products.

Through Haloculture, significant portions of saline lands and waters are capable of having an economically viable food production. The production of all or some of the needed water and energy can be achieved through Haloengineering as the complementary component of Haloculture. These features of Haloculture are in accordance with the concepts of WEFN. Haloengineering will play a dynamic role in the expansion of Haloculture in saline regions.

Funding: This research received no external funding.

Institutional Review Board Statement: Not applicable.

Informed Consent Statement: Not applicable.

Data Availability Statement: Not applicable.

Conflicts of Interest: The authors declare no conflict of interest.

References

1. Khorsandi, F.; Siadati, S.M.H.; Rastegary, J. Haloengineering as a vital component of sustainable development in salt-affected ecosystems. *Environ. Dev.* **2020**, *35*, 100545. [CrossRef]
2. Pedraza, J.M. The use of nuclear techniques in Biosaline agriculture: The International Atomic Energy Agency (IAEA) experience. *Int. J. Energy Environ. Econ.* **2009**, *16*, 133–152.
3. Khorsandi, F.; Mirvakili, S.M.; Banakar, M.H. Haloculture: A strategy for sustainable utilization of saline land and water resources. In Proceedings of the Seventh International Conference on the Development of Drylands, Tehran, Iran, 14–17 September 2003; pp. 139–145.
4. Khorsandi, F. Haloculture: Strategy for sustainable utilization of saline land and water resources. *Iran. J. Earth Sci.* **2016**, *8*, 164–172.
5. Biggs, E.M.; Bruce, E.; Boruff, B.; Duncan, J.M.A.; Horsley, J.; Pauli, N.; McNeill, K.; Neef, A.; Van Ogtrop, F.; Curnow, J.; et al. Sustainable development and the water-energy-food nexus: A perspective on livelihoods. *Environ. Sci. Policy* **2015**, *54*, 389–397. [CrossRef]
6. Flammini, A.; Puri, M.; Pluschke, L.; Dubois, O. *Walking the Nexus Talk: Assessing the Water-Energy-Food Nexus in the Context of the Sustainable Energy for All Initiative*; Environment and Natural Resources Working Paper No. 58; FAO: Rome, Italy, 2014.

environmental sciences proceedings

MDPI

Proceeding Paper

Growth of Hybrids of *Sarotherodon melanotheron* (Rüppell, 1852) and *Oreochromis niloticus* (Linnaeus, 1758) in the Area of Lake Ahémé and Its Channels †

Christian Comlan Viaho [1,2,*], Hervé Hounsa [1], Delphine Adandedjan [1], Hyppolite Agadjihouede [3], Martin N. Gbedey [2] and Philippe A. Laleye [1]

1 Laboratory of Hydrobiology and Aquaculture, Faculty of Agricultural Sciences, University of Abomey-Calavi, Cotonou P.O. Box 526, Benin; hounsah@gmail.com (H.H.); adandedjandelph@gmail.com (D.A.); laleyephilippe@gmail.com (P.A.L.)

2 Agency for the Integrated Development of the Economic Zone of Lake Ahémé and Its Channels (ADELAC), Kpomassè P.O. Box 01, Benin; adelac.infos@gouv.bj

3 Aquaculture and Fisheries Management Research Unit, Animal and Fisheries Science Laboratory, Valley Aquaculture School, National University of Agriculture (UNA), Kétou P.O. Box 53, Benin; agadjihouede@gmail.com

* Correspondence: viahochristian@gmail.com; Tel.: +229-9611-2343

† Presented at the 2nd International Laayoune Forum on Biosaline Agriculture, 14–16 June 2022; Available online: https://lafoba2.sciforum.net/.

Abstract: The Ahémé Lake and its channels make up the saline area in Benin and do not facilitate the breeding of freshwater fish. The cross between the parents of *Oreochromis niloticus* and *Sarotherodon melanatheron* allowed us to obtain hybrids that grow in brackish water. It is a totally randomized device with four treatments and two repetitions. The larvae were brought back and fed to satiety for 6 weeks. The growth rates of the OS hybrids were 0.38 g/day, while those of the SO hybrids were 0.33 g/day. OS hybrids are an excellent candidate for fish farming in a brackish environment.

Keywords: growth performance; survival rate; aquaculture; brackish environment

Citation: Viaho, C.C.; Hounsa, H.; Adandedjan, D.; Agadjihouede, H.; Gbedey, M.N.; Laleye, P.A. Growth of Hybrids of *Sarotherodon melanotheron* (Rüppell, 1852) and *Oreochromis niloticus* (Linnaeus, 1758) in the Area of Lake Ahémé and Its Channels. *Environ. Sci. Proc.* **2022**, *16*, 43. https://doi.org/10.3390/environsciproc2022016043

Academic Editor: Abdelaziz Hirich

Published: 16 June 2022

1. Introduction

Fishing is a socio-economic activity that plays an important role in human nutrition [1]. Fish, the main source of protein for populations in southern Benin, is provided by water bodies [2]. However, ever-increasing demographic pressure, coupled with the erosion of banks and poor fishing practices have led to a degradation of the aquatic ecosystem and a drop in fish production [3]. Fish farming is, therefore, an essential alternative to meet the dietary needs of populations in fish protein. However, in the area of Lake Ahémé and its channels, the soils are salty for a good period of the year because of the connection of the environment with the sea [4]. However, the fast-growing fish species that are currently farmed in Benin are freshwater [5], which hinders the development of fish farming in the study area. Faced with this situation, the solution found is to cross the parents of a fast-growing freshwater species (*Oreochromis niloticus*) with the parents of a species supporting brackish water and medium growth (*Sarotherodon melanotheron*), whose goal is to obtain hybrids with rapid growth and supporting brackish water. However, what are the standards of this crossing? How should the hybrid larvae obtained be followed? This is the purpose of this article, which aims to improve fish production with hybrids resulting from the crossing of *S. melanotheron* and *O. niloticus* in the Lake Ahémé area and its channels. Specifically, it involves evaluating the optimal conditions for obtaining hybrids, to analyze the growth parameters of the hybrid larvae obtained and to identify the population of hybrids with the best zootechnical performance for a fish farm.

2. Material and Method

2.1. Study Area

The study area includes Lake Ahémé, the Ahô channel, the coastal lagoon of Grand-Popo, the coastal lagoon of Ouidah and the Tihimey channel. Lake Ahémé receives the waters of the Couffo River and flows into the Ahô channel and the Tihimey channel. The Ahô channel flows into the coastal lagoons of Grand-Popo and Ouidah [6]. The coastal lagoon of Grand-Popo receives the waters of the Mono River and flows into the sea through the mouth "Bocca Del Rio". The coastal lagoon of Ouidah, for its part, takes its source from the lagoon of Grand-Popo and the Ahô channel. The presence of the mouth makes the study area a very diversified lagoon estuary environment. Rainfall varies from 800 to 1200 mm per year, characterized by the alternation of two rainy seasons (April–July and October–November) and two dry seasons (December–March and August–September).

2.2. Biological Material

Table 1 summarizes the origins and number of broodstock used for the trial. The *S. melanotheron* come from Lake Ahémé, while those of *O. niloticus* are purchased from a fish farm located in the study area. They are divided by sex and quarantined for 15 days in small tanks measuring 1 m × 1 m × 0.8 m and fed with 3 mm granulated feed from the "le gouéssant" brand.

Table 1. Broodstock used and their origins.

Parents	Origins	Sex	Numbers	Average Weight (g)
Oreochromis niloticus	Fish pond at the "La main de Dieu" farm in Comê	Male	10	110
		Female	21	150
Sarotherodon melanotheron	Ahémé Lake	Male	50	53.2
		Female	22	45

2.3. Experimental Apparatus

Two basins 6 m long, 2 m wide with a depth of 1.5 m each were used. Four fish tanks of 1.5 m width with a depth of 1.5 m each were installed in each basin. These tanks were used to serve as protection for the various crossings. It is a totally randomized device with 4 treatments and 2 repetitions [7]. The fish were fed a diet composed of 9.00% fat, 32% crude protein, 2.94% crude ash, 0.88% calcium, 0.91% total phosphate and 0.15% sodium. The ration was 2.5%, served morning and evening. After obtaining the larvae, they were monitored with the same device at a rate of 120 larvae per tank. These larvae were then fed to satiety twice a day over a period of 6 weeks. The Kruskhal–Wallis test was used to verify the degree of significance of the variations. The test is a trial conducted in 2020.

3. Results

3.1. Obtaining Hybrid Larvae

The first larvae obtained are those of the control crosses; *O. niloticus* after 7 days and *S. melanotheron*, 15 days after crossing. As for the hybrid larvae, they are obtained 45 days after crossing.

3.2. Growth Parameters

Table 2 presents the evolution curves of the average growth of the larvae harvested after the various crosses carried out. We noticed:

- OO the fry from the control cross of the parents of *O. niloticus*.
- SS the fry from the control cross of the parents of *S. melanotheron*.
- OS the fry from the test cross "male *S. melanotheron* crossed by female *O. niloticus*".
- SO the fry from the test cross "male *O. niloticus* crossed by female *S. melanotheron*".

Table 2. Results from analysis of average fry growth.

Variables	Minimum	Maximum	Moy	StDev	Variance	CoefVar
OO	9.50	29.00	22.83	7.29	53.17	31.93
SS	10.00	20.25	16.49	3.63	13.18	22.01
OS	9.90	24.75	18.54	5.29	27.93	28.50
SO	9.85	22.50	17.58	4.45	19.81	25.32

OO = larvae from the control cross *O. niloticus*; SS = larvae from the control cross *S. melanotheron*; OS = hybrids from the OS test cross; SO = hybrids from the SO test cross; CoefVar = Coefficient of variation Avg. = Average; StDev = Standard Deviation.

This table shows an almost similar upward evolution of the growth curves. Nevertheless, we found that the fry growth curve of *O. niloticus* (OO) is well above the other three. It is followed by that of the hybrids from the test cross "male *S. melanotheron* crossed by female *O. niloticus*" (OS) then by that of the hybrids resulting from the test cross "male *O. niloticus* crossed by female *S. melanotheron*" (SO), and finally, that of the fry of *S. melanotheron* (SS). The results given by the statistical analysis also showed that the average growth of the OO fry (23.83 g) is higher than that of the other fry. That of OS fry (18.54 g) comes in second position followed by that of SO fry (17.58 g), which is ahead of SS fry (16.49 g).

4. Discussion

Regarding reproduction, crosses took a long time to produce larvae (45 days). This situation is due to *S. melanotheron*, which are found in a controlled environment. This time is greater than that observed by Chowdhury [8] in pure breeding ponds for the species. The survival rate of OS hybrid fry is higher than that of SO hybrid fry, which are 88.83% and 85%, respectively. These survival rates show us that the OS hybrid fry would have inherited a better survival performance from their parents, the females of *O. niloticus* and the males of *S. melanotheron*. In other hybrid growth monitoring studies, the survival rate of *O. niloticus* fry is relatively lower than that of *S. melanotheron* fry and SO and OS hybrids [9]. This is because the monitoring was carried out in the water of the lagoon in Côte d'Ivoire at a salinity higher than that of fresh water (0 to 5 g/L). In this case, we can say that the hybrids inherited the best adaptability from their *S. melanotheron* parents, which resist salinities ranging from 5 to 110 g/L. The calculated growth parameters show that during the 42 days of rearing, the fry of *O. niloticus* had a better average daily gain (0.46 g/d), the OS hybrids followed with an average daily gain of 0, 38 g/d, then the SO hybrids (0.33 g/d), and finally, the fry of *S. melanotheron*, with a daily weight gain of 0.29 g/d. Previous studies show strong growth of *O. niloticus* in ponds with an average daily gain (ADG) of 0.70 g/d [10]. The growth of the hybrids was, on the other hand, stronger than that of *S. melanotheron*. It was intermediate between those of *O. niloticus* and *S. melanotheron*, which confirms the previous work of Amon [11]. This last result is also in line with that of Toguyéni [12], on hybrids resulting from artificial insemination between *O. niloticus* and *S. melanotheron*.

5. Conclusions

The growth of *Sarotherodon melanotheron* (Rüppell, 1852) and *Oreochromis niloticus* (Linnaeus, 1758) hybrids in the Lake Ahémé area and its channels shows that these species manage to reproduce to provide hybrids that constitute a solution to the development of aquaculture in salty areas in Benin. In terms of survival in a lagoon environment, we can, therefore, conclude that the hybrids resulting from the crosses of *O. niloticus* and *S. melanotheron* will have a better survival performance compared to their *O. niloticus* parents. We can, therefore, affirm that the male hybrids result from the cross ♀*O. niloticus* × ♂S.

Author Contributions: Conceptualization, and methodology C.C.V.; software, and validation D.A.; formal analysis, H.A.; investigation, H.H.; data collected, H.H.; writing—original draft preparation, C.C.V.; visualization, M.N.G.; supervision, P.A.L.; funding acquisition, M.N.G. All authors have read and agreed to the published version of the manuscript.

Funding: This research received no external funding.

Institutional Review Board Statement: Not applicable.

Informed Consent Statement: Not applicable.

Data Availability Statement: The data can be obtained by writing to the corresponding author.

Conflicts of Interest: The authors declare no conflict of interest.

References

1. Viaho, C.C.; Adandedjan, D.; Montcho, S.A.; Gbedey, M.N.; Laleye, P.A. Inventory, description and analysis of fishing gear and techniques used in Lake Ahémé and its channels, facing the law on fishing in Benin (West-Africa). *World J. Adv. Res. Rev.* **2021**, *12*, 401–430. [CrossRef]
2. Amoussou, E. *Ancient Lakes and Lagoons in South-West Benin, Inventory of IUCN Benin (Study Report)*; IUCN: Gland, Switzerland, 2018; pp. 122–201.
3. Dessouassi, C.E.; Chikou, A.; Lederoun, D.; Adandedjan, D.; Gangbe, L.; Laleye, P. Diversity, biology and exploitation of brackish water crabs in West Africa: A review. *Int. J. Biol. Chem. Sci.* **2018**, *12*, 2355–2370. [CrossRef]
4. Viaho, C.C.; Ahouansou-Montcho, S.; Agblonon Houelome, T.M.; Adandedjan, D.; Agadjihouede, H.; Laleye, P.A. Caractérisation physico-chimique du lac Ahémé et ses chenaux au Sud-Ouest du Bénin. *Afr. Sci.* **2020**, *17*, 72–92.
5. Kpenavoun, S.; Gandonou, E.; Adegbidi, A.; Abokini, E. Mesure et déterminants de l'efficacité technique des pisciculteurs du Bénin. *Int. J. Biol. Chem. Sci.* **2018**, *11*, 2194–2208. [CrossRef]
6. Amoussou, O.T.; Youssao, I.A.K.; Toguyeni, A. Improving aquaculture production in the Kou Valley, Burkina Faso. *Agric. Innov. Sustain. Dev.* **2014**, *4*, 187–194.
7. Laporte, S.; Cucherat, M. Randomisation selon le schéma de Zelen fiche technique. *Lett. Pharmacol.* **2012**, *26*, 137–139.
8. Chowdhury, M.A.K.; Yi, Y.; Lin, C.K.; El-Haroun, E. Effect of salinity on carrying capacity of adult Nile tilapia *Oreochromis niloticus* L. in recirculating systems. *Aquac. Res.* **2006**, *37*, 1627–1635. [CrossRef]
9. Ipungu, L.; Ngoy, K.; Banze, K.; Lumfwa, K.; Kafund, M. L'étude de la croissance de Oreochromis niloticus par la fertilisation des étangs: Le cas de la ferme Naviundu Lubumbashi. *J. Appl. Biosci.* **2015**, *91*, 8503–8510. [CrossRef]
10. Lawson, E.O.; Anetekhai, M.A. Salinity tolerance and Preference of Hatchery Reared Nile Tilapia, *Oreochromis niloticus* (Linneaus 1758). *Asian J. Agric. Sci.* **2011**, *3*, 104–110.
11. Amon, Y.N.; Yao, K.; Ouattara, M.; Kouman, Y.C.; Atse, B.C. Morphologie des hybrides issus du croisement intergénérique entre *Oreochromis niloticus* (Linnaeus, 1758) et *Sarotherodon melanotheron* (Rüppel, 1852). *J. Appl. Biosci.* **2013**, *69*, 5475–5486. [CrossRef]
12. Toguyéni, A.; Fauconneau, B.; Mélard, C.; Fostier, A.; Lazard, J.; Baras, E.; Kühn, E.; Van Der Geyten, S.; Baroiller, J.-F. Sexual dimorphism in two pure Cichlid species, *Oreochromis niloticus niloticus* (Linnaeus, 1758) and *Sarotherodon melanotheron melanotheron* Rüppel 1852, and their intergeneric hybrids. *Afr. J. Aquat. Sci.* **2009**, *34*, 69–75. [CrossRef]

environmental sciences proceedings

MDPI

Proceeding Paper

Enhanced Salinity Tolerance of *Medicago sativa*, Roots AM Colonization and Soil Enzyme Activities by PGPR †

Nabil Tirry *, Mohamed Ferioun, Aziza Kouchou, Ghizlane Laghmari, Wifak Bahafid and Naïma El Ghachtouli

Laboratory of Microbial Biotechnology and Bioactive Molecules, Faculty of Sciences and Technology, Sidi Mohamed Ben Abdellah University, Fez 30000, Morocco; mohamed.ferioun@usmba.ac.ma (M.F.); kouchouaziza@gmail.com (A.K.); l.ghizlan@outlook.com (G.L.); wifak.bahafid@gmail.com (W.B.); naimaelghachtouli@yahoo.com (N.E.G.)

* Correspondence: nabiltirry@gmail.com

† Presented at the 2nd International Laayoune Forum on Biosaline Agriculture, 14–16 June 2022; Available online: https://lafoba2.sciforum.net/.

Abstract: Abiotic stresses such as salt are typical negative factors that have a considerable impact on agricultural output around the world. The goal of this study was to investigate the effect of halotolerant plant-growth-promoting rhizobacteria (PGPR) on plant growth and soil function under salinity stress. The consortium of four PGPR (*Pseudomonas putida*, *Alcaligenes* sp., *Klebsiella* sp., and *Pseudomonas cedrina*) was tested for its effect on growth, chlorophyll content, oxidative stress, and root arbuscular mycorrhizal (AM) colonization of *Medicago sativa* in pots experiment under salt stress. The bacteria's impact on soil enzyme activity was also investigated. Overall, in comparison to the non-inoculated control, inoculating *M. sativa* plants with the bacterial consortium allowed us to overcome the unfavorable effects of NaCl stress and enhanced plant growth, root AM colonization, and leaf chlorophyll content. It also reduced the levels of oxidative damage indicators such as malondialdehyde, hydrogen peroxide, and proline. Furthermore, the consortium had a beneficial effect on the activities of soil phosphatase, β-galactosidase, and arylamidase. The bacterial consortium has the potential to be employed as bio-inoculants for plants growing under salt stress.

Keywords: plant growth promoting rhizobacteria; salt stress; *Medicago sativa*; arbuscular mycorrhiza; soil enzyme activities

Citation: Tirry, N.; Ferioun, M.; Kouchou, A.; Laghmari, G.; Bahafid, W.; El Ghachtouli, N. Enhanced Salinity Tolerance of *Medicago sativa*, Roots AM Colonization and Soil Enzyme Activities by PGPR. *Environ. Sci. Proc.* **2022**, *16*, 14. https://doi.org/10.3390/environsciproc2022016014

Academic Editor: Abdelaziz Hirich

Published: 14 June 2022

1. Introduction

Salinity is a problem altering large land area and damages soils [1]. Salt stress reduces plant nutrient uptake, affect many morphological, physiological, and biochemical parameters in different parts of plants [2].

Plant growth promoting rhizobacteria (PGPR) could be an important technology to enhance plants tolerance against abiotic stresses including salinity [3]. Many studies show that various genera of PGPR can promote plants growth and enhance crops productivity at many plants species [4]. The cooperation between PGPRs and arbuscular mycorrhizal fungi (AMF) in soil and plant tissues can be beneficial for plants through optimizing nutrition and protection against biotic and abiotic stresses [5]. Positive interactions between PGPR and AMF in stimulating plant growth and salinity tolerance are well improved [6]. Otherwise, PGPRs can increase many enzymes activity of soil under saline conditions which promote plants growth [7,8].

Thus, in this study we aimed to analyze the effect of the interaction of a consortium of four potential PGPR with *M. sativa* to alleviate salt stress on plants growth and oxidative stress, as well as on plant roots mycorrhizal colonization and soil enzyme activities.

2. Material and Methods

The bacteria, *Pseudomonas putida*, *Alcaligenes* sp., *Klebsiella* sp., and *Pseudomonas cedrina*, used in this study were previously isolated from southern Morocco and characterized for their PGP traits and their resistance to salinity [9]. Pot experiments were designed in order to study the effect of the bacterial consortium on *M. sativa* (Alfalfa) under salt exposure. Experiments were conducted in plastic pots containing soil amended with 2% NaCl. AMF inoculum was represented by the indigenous population of soil AMF.

The plantlets were inoculated with 2 mL of a suspension of the consortium of bacterial culture (10^8 CFU mL^{-1}). For uninoculated plants, 2 mL of saline solution were added. Pots were placed in a greenhouse (about 16 h photoperiod, 26–30 °C day and 18–22 °C night) and watered daily. Plants were harvested after 30 days, subdivided in roots and shoots and washed with deionized water.

For plants morphological parameters, root and shoot dry weights were measured, after oven dying at 65 °C for 72 h. Chlorophyll content was quantified according to [10]. The proline content of the leaves was determined according to [11]. Hydrogen peroxide (H_2O_2) content of the leaves was determined according to [12]. The lipid peroxidation (malondialdehyde (MDA) content) in plant leaves tissues was measured according to [13].

Detection of AM colonization of roots was performed according to [14]. Percentage colonization of roots was recorded according to [15].

Determination of soil enzymes phosphatase (PHOS), β-galactosidase (GALA), arylamidase (AMID) and arylsulfatase (SULF) activities in the rhizospheric soils of *M. sativa* were performed, according to [16–19], respectively.

Data were submitted to ANOVA analysis for each data to determine significant differences between the means. A multiple range test at the 95% confidence level was performed using Tukey's method.

3. Results and Discussion

3.1. Effet of PGPR on Plant Tolerance and AM Colonization under NaCL Stress

Bacterial inoculation of *M. sativa* plants induced a positive impact on plant growth (Table 1). These results are in line with many studies improving a positive effect of PGPR inoculation in various plants cultures submitted to salt stress [4], and might be primarily linked to the PGP traits of the bacteria.

Table 1. Effect of bacterial inoculation on plant growth and mycorrhizal colonization (mycorrhizal frequency (F%), mycorrhizal intensity (M%) and richness of arbuscules (A%)) of *M. sativa* roots under NaCl stress. Values with different letters are significantly different (p = 0.05). Means that do not share a letter are significantly different.

Treatment	Plant Growth		Mycorrhizal Colonization		
	Shoot Dry Weight (mg)	Root Dry Weight (mg)	F%	M%	A%
Control	17.315 b	15.11 b	64.19 b	21.45 b	13.76 b
Consortium	26.611 a	18.050 a	74.55 a	37.06 a	21.58 a

A significant ($p < 0.05$) increase of roots mycorrhizal colonization was recorded in plants inoculated with the bacterial consortium (F%, M%, and A%) of salt stressed plants (Table 1). This result suggests that these bacteria might act as mycorrhizal helper bacteria (MHB). The ability of AMF to improve plants growth and support tolerance to biotic and abiotic stresses is well documented in literature [20]. The positive impacts of bacteria on enhancing plants tolerance against salinity could also be due to their effect on mycorrhization.

The results show a significant ($p < 0.05$) increase of total chlorophyll content and a significant reduction of the indicators of oxidative damage levels (malondialdehyde and hydrogen peroxide), and proline content under salt stress, compared to non-inoculated plants (Table 2). The reduction of MDA and H_2O_2 contents might be due to the activation of the antioxidant enzyme defensive pathway [21]. On the other hand, proline is linked with various functions (e.g., osmoregulation, scavenging free radicals, activation of detoxification pathways), and many plants accumulate this amino acid under stresses as a kind of adaptation against this condition [2]. The increase of proline content in inoculated plants might be due to high tolerance of inoculated *M. sativa* plant under salt stress.

Table 2. Effect of bacterial inoculation on chlorophyll, proline, MDA and H_2O_2 content of *M. sativa* leaves under NaCl stress. Values with different letters are significantly different (p = 0.05). Means that do not share a letter are significantly different.

Treatment	Chlorophyll (mg/g)	H_2O_2 (nmol/g)	MDA (μmol/g)	Proline (μmol/g)
Control	0.95 b	695.85 a	81.16 a	228.26 a
Consortium	1.32 a	499.60 b	61.55 b	109.21 b

3.2. Effet of PGPR on Soil Enzyme Activity

Enzymes activity of soil can be used as criteria to describe soil quality and fertility [22]. Our results showed that in NaCl treatment, the activities of phosphatase, galactosidase, and arylamidase were significantly ($p < 0.05$) increased in soils inoculated with bacteria (Table 3). This positive effect might be through direct enzymes production by the microbial biomass of PGPRs or result indirectly from their interactions with rhizosphere microorganisms, AMF in particular. Indeed, the ability of AMF to improve soil enzymatic activities is recently documented in the meta-analysis of [23].

Table 3. Effect of bacterial inoculation on soil enzymatic activities: galactosidase (GALA), phosphatase (PHOS), arylsulfatase (SULF) and arylamidase (AMID) of the rhizospheric soil of *M. sativa* plants under NaCl stress. Values with different letters are significantly different (p = 0.05). Means that do not share a letter are significantly different.

Treatment	Soil Enzyme Activities mU/g Dry Soil			
	GALA	PHOS	SULF	AMID
Control	1.04 b	15.19 b	74.92 a	36.99 b
Consortium	1.800 a	20.33 a	73.43 a	47.04 a

4. Conclusions

The use of PGPRs, tolerant under against salt stress, with effective PGP traits, able to cooperate positively with AM fungi, and to improve soil enzymes activities could optimize our ability to cultivate crops in saline environments.

Author Contributions: N.T.: conceptualization, methodology, formal analysis, writing—original draft, and visualization. M.F.: methodology, writing, and formal analysis. A.K.: methodology and formal analysis. G.L.: review and editing. W.B.: methodology. N.E.G.: conceptualization, supervision, review and editing, validation, resources, and funding acquisition. All authors have read and agreed to the published version of the manuscript.

Funding: This research did not receive any specific grant from funding agencies in the public, commercial, or not-for-profit sectors.

Institutional Review Board Statement: Not applicable.

Informed Consent Statement: Not applicable.

Conflicts of Interest: The authors declare that the research was conducted in the absence of any commercial or financial relationships that could be construed as a potential conflict of interest.

References

1. Ben Chikha, M.; Hessini, K.; Ourteni, R.N.; Ghorbel, A.; Zoghlami, N. Identification of barley landrace genotypes with contrasting salinity tolerance at vegetative growth stage. *Plant Biotechnol.* **2016**, *33*, 287–295. [CrossRef] [PubMed]
2. Sallam, A.; Alqudah, A.M.; Dawood, M.F.; Baenziger, P.S.; Börner, A. Drought stress tolerance in wheat and barley: Advances in physiology, breeding and genetics research. *Int. J. Mol. Sci.* **2019**, *20*, 3137. [CrossRef] [PubMed]

3. Sharma, S.; Kulkarni, J.; Jha, B. Halotolerant rhizobacteria promote growth and enhance salinity tolerance in peanut. *Front. Microbiol.* **2016**, *7*, 1600. [CrossRef] [PubMed]
4. Egamberdieva, D.; Wirth, S.; Bellingrath-Kimura, S.D.; Mishra, J.; Arora, N.K. Salt-Tolerant Plant Growth Promoting Rhizobacteria for Enhancing Crop Productivity of Saline Soils. *Front. Microbiol.* **2019**, *10*, 2791. [CrossRef]
5. Kapoor, R.; Evelin, H.; Devi, T.S.; Gupta, S. Mitigation of salinity stress in plants by arbuscular mycorrhizal symbiosis: Current understanding and new challenges. *Front. Plant Sci.* **2019**, *10*, 470.
6. Krishnamoorthy, R.; Shagol, C.; Kim, K.; Lee, S.; Yoon, S.; Sa, T. Isolation of salt tolerant arbuscular mycorrhizal fungi (AMF) from Saemangeum reclaimed soil and mass production in hairy roots. *Kor. S. Fert. Soc. Acad.* **2011**, *10*, 92–93.
7. Ren, H.; Lv, C.; Fernández-García, V.; Huang, B.; Yao, J.; Ding, W. Biochar and PGPR amendments influence soil enzyme activities and nutrient concentrations in a eucalyptus seedling plantation. *Biom. Conv. Bioref.* **2019**, *11*, 1865–1874. [CrossRef]
8. Pan, J.; Huang, C.; Peng, F.; Zhang, W.; Luo, J.; Ma, S.; Xue, X. Effect of arbuscular mycorrhizal fungi (AMF) and plant growth-promoting bacteria (PGPR) inoculations on ElaeagnusAngustifolia L. in saline soil. *Appl. Sci.* **2020**, *10*, 945.
9. Tirry, N.; Kouchou, A.; Laghmari, G.; Lemjereb, M.; Hnadi, H.; Amrani, K.; Bahafid, W.; El Ghachtouli, N. Improved salinity tolerance of Medicago sativa and soil enzyme activities by PGPR. *Biocatal. Agric. Biotechnol.* **2021**, *31*, 101914. [CrossRef]
10. Moran, R.; Porath, D. Chlorophyll determination in intact tissues using N, N-dimethylformamide. *Plant Physiol.* **1980**, *65*, 478–479. [CrossRef]
11. Bates, L.S.; Waldren, R.P.; Teare, L.D. Determination of proline for water-stress studies. *Plant Soil.* **1973**, *39*, 205–207. [CrossRef]
12. Jana, S.; Choudhuri, M.A. Glycolate metabolism of three submersed aquatic angiosperms: Effect of heavy metals. *Aquat. Bot.* **1981**, *11*, 67–77. [CrossRef]
13. Heath, R.L.; Packer, L. Photoperoxidation in isolated chloroplasts: I. Kinetics and stoichiometry of fatty acid peroxidation. *Arch. Biochem. Biophys.* **2016**, *125*, 189–198. [CrossRef]
14. Phillips, J.M.; Hayman, D.S. Improved procedures for clearing roots and staining parasitic and vesicular-arbuscularmycorrhizal fungi for rapid assessment of infection. *Trans. Br. Mycol. Soc.* **1970**, *55*, 158-IN18. [CrossRef]
15. Trouvelot, A.; Kough, J.L.; Gianinazzi-Pearson, V. Mesure du taux de mycorhization VA d'un système radiculaire. Recherche de méthode d'estimation ayant une signification fonctionnelle. In Proceedings of the 1st European Symposium on Mycorrhizae, Dijon, France, 1–5 July 1985; pp. 217–221.
16. Dick, R.P.; Breakwell, D.P.; Turco, R.F. Soil enzyme activities and biodiversity measurements as integrative microbiological indicators. *Methods Assess. Soil Qual.* **1997**, *49*, 247–271.
17. Sinsabaugh, R.L.; Reynolds, H.; Long, T.M. Rapid assay for amidohydrolase (urease) activity in environmental samples. *Soil Biol. Biochem.* **2000**, *32*, 2095–2097. [CrossRef]
18. Acosta-Martìnez, V.; Tabatabai, M.A. Inhibition of arylamidase activity in soils by toluene. *Soil Biol. Biochem.* **2002**, *34*, 229–237. [CrossRef]
19. Tabatabai, M.A.; Bremner, J.M. Arylsulfatase activity of soils. *Soil Sci. Soc. Am. J.* **1970**, *34*, 225–229. [CrossRef]
20. Gianinazzi, S.; Schüepp, H.; Barea, J.M.; Haselwandter, K. *Mycorrhizal Technology in Agriculture: From Genes to Bioproducts;* Birkhäuser: Basel, Switzerland, 2012.
21. Habib, S.H.; Kausar, H.; Saud, H.M. Plant growth-promoting rhizobacteria enhance salinity stress tolerance in okra through ROS-scavenging enzymes. *BioMed Res. Int.* **2016**, *2016*, 6284547. [CrossRef]
22. Sipahutar, M.K.; Piapukiew, J.; Vangnai, A.S. Efficiency of the formulated plant-growth promoting Pseudomonas fluorescens MC46 inoculant on triclocarban treatment in soil and its effect on Vigna radiata growth and soil enzyme activities. *J. Hazard. Mat.* **2018**, *344*, 883–892. [CrossRef]
23. Qin, M.; Zhang, Q.; Pan, J.; Jiang, S.; Liu, Y.; Bahadur, A.; Peng, Z.; Yang, Y.; Feng, H. Effect of arbuscular mycorrhizal fungi on soil enzyme activity is coupled with increased plant biomass. *Eur. J. Soil Sci.* **2020**, *71*, 84–92. [CrossRef]

environmental sciences proceedings

MDPI

Proceeding Paper

Evaluation of Young Date Palm Tolerance to Salinity Stress under Arbuscular Mycorrhizal Fungi and Compost Application †

Mohamed Ait-El-Mokhtar [1,2,*], Raja Ben-Laouane [2], Abderrahim Boutasknit [2], Mohamed Anli [2], Abdessamad Fakhech [2], Youssef Ait-Rahou [2], Toshiaki Mitsui [3], Said Wahbi [2], Marouane Baslam [3] and Abdelilah Meddich [2]

1 Laboratory of Biochemistry, Environment and Agri-Food URAC 36 (LBEA), Faculty of Science and Techniques—Mohammedia, Hassan II University of Casablanca, Casablanca 20000, Morocco
2 Laboratory of Agro-Food, Biotechnologies and Valorization of Plant Bioresources (AGROBIOVAL), Faculty of Science Semlalia, Cadi Ayyad University, Marrakesh 40000, Morocco; benlaouaneraja@gmail.com (R.B.-L.); abderrahim.boutasknit@gmail.com (A.B.); moh1992ali@gmail.com (M.A.); abdessamad.fakhech@edu.uca.ma (A.F.); youssefaitrahou41@gmail.com (Y.A.-R.); wahbi@ucam.ac.ma (S.W.); a.meddich@uca.ma (A.M.)
3 Laboratory of Biochemistry, Faculty of Agriculture, Niigata University, Niigata 1000, Japan; t.mitsui@agr.niigata-u.ac.jp (T.M.); mbaslam@gs.niigata-u.ac.jp (M.B.)
* Correspondence: mohamed.aitelmokhtar@gmail.com; Tel.: +212-671-492-144
† Presented at the 2nd International Laayoune Forum on Biosaline Agriculture, 14–16 June 2022; Available online: https://lafoba2.sciforum.net/.

Abstract: This study focused on the mitigation of the negative impact of salinity stress over time on the growth and development of *Phoenix dactylifera* plants using compost and/or arbuscular mycorrhizal fungi (AMF). The experiment had eight treatments in a randomized design. The treatments consisted of a control, AMF (native consortium) and compost (produced from green waste). The treatments were applied separately or in combination in the presence of 240 mM NaCl (saline condition) or 0 mM NaCl (non-saline condition) after 10 and 14 months of cultivation. Our results show that saline stress increased toxic ion (sodium and chlorine), proline, soluble sugars and stress marker (H_2O_2 and MDA) contents. At the same time, it lowered growth traits, mycorrhizal colonization, leaf water potential and nutrients (nitrogen (N), potassium (K), phosphorus (P), calcium (Ca)) and photosynthetic pigment concentrations. The application of compost and AMF individually or in combination alleviated salt-induced effects through mechanisms such as an increase in nutrient absorption (P, N, K and Ca), photosynthetic pigment content, relative water content, stomatal opening, leaf water potential, photosystem II efficiency, organic osmolyte content (proline and soluble sugars) and antioxidant enzyme activities (SOD, APX, CAT and POD) and reduction in lipid peroxidation and H_2O_2 content. Our results also show that the tolerance strategies of date palm to salinity were progressively improved over time in treated plants and especially in date palms grown in the presence of the two biofertilizers. Between 10 and 14 months of cultivation, growth parameters increased with a significant improvement in nutrient contents, a reduction in the concentrations of toxic ions and stress markers, and regulation of the antioxidant system (antioxidant enzyme activity and osmolyte content) in both leaves and roots. In conclusion, this investigation highlights the effectiveness of dual application of compost and native AMF in mitigating the deleterious effects of salinity on date palm with an improvement in the tolerance over time.

Keywords: date palm; arbuscular mycorrhizal fungi; compost; salinity; tolerance

Citation: Ait-El-Mokhtar, M.; Ben-Laouane, R.; Boutasknit, A.; Anli, M.; Fakhech, A.; Ait-Rahou, Y.; Mitsui, T.; Wahbi, S.; Baslam, M.; Meddich, A. Evaluation of Young Date Palm Tolerance to Salinity Stress under Arbuscular Mycorrhizal Fungi and Compost Application. *Environ. Sci. Proc.* **2022**, *16*, 15. https://doi.org/10.3390/environsciproc2022016015

Academic Editor: Abdelaziz Hirich

Published: 16 June 2022

1. Introduction

Soil salinization is one of the most critical environmental issues that damages agricultural land and leads to more than 20% decline in agricultural yield [1]. Increased soil salt concentrations induce osmotic, ionic and oxidative stress that generate deleterious effects on plants.

Arbuscular mycorrhizal fungi (AMF) figure among the most promising alternatives to enhance plant tolerance to salinity. AMF are telluric fungi that establish symbiotic associations with the roots of terrestrial plants. This symbiosis is known to mitigate the adverse effects of saline stress on plants, although it is itself affected by salinity [2]. Furthermore, the application of organic fertilizers to soils in arid and semi-arid areas is a key environmental strategy for soil restoration [3,4]. It has been reported that the use of compost was effective in the restoration of soils affected by salinity [5].

Heretofore, several scientific studies have investigated the responses of date palm to saline stress [6–8] and a few research studies have been conducted on the effect of AMF or compost application on mitigating the effects of saline stress in date palm. The present study was carried out to investigate the effects of the application of these two biofertilizers alone or combined on the growth, nutrient content, photosynthesis, water status and biochemical characteristics of date palm seedlings under saline conditions and the regulation of tolerance strategies between 10 and 14 months of cultivation.

2. Methods

2.1. Plant Material and Biological Treatments

Seeds of Boufeggous date palm cultivar were germinated and the seedlings were transplanted to black plastic bags containing sterile sandy substrate. AMF inoculum consists of the native Aoufous mycorrhizal consortium which was applied at a rate of 10 g of multiplication soil per bag. The organic amendment was obtained from the composting of quack grass waste and was applied to the corresponding bags at a rate of 5% (w/w).

2.2. Experimental Design and Growth Conditions

After germination, date palm transplants were regularly irrigated with distilled water for 5 months under a water regime of 75% of field capacity, then two different salinity concentrations were applied (0 and 240 mM NaCl) during the experiment period (14 months). The experiment design consisted of eight treatments crossing two levels of compost (−compost and +compost) and two levels of AMF inoculation (−AMF and +AMF) with two levels of salinity (0 and 240 mM NaCl). The bags were randomly arranged, and each treatment consisted of ten biological replicates for a total of 80 pots. The plants were grown for 14 months in the greenhouse.

2.3. Measured Parameters and Statistical Analyses

Plants of date palm at the stage of 10 and 14 months of cultivation were harvested. Growth, mycorrhization, physiological and water status parameters, mineral content, stress markers, organic osmolytes and antioxidant enzyme activity were assessed.

3. Results and Discussion

3.1. Growth, Nutritional and Physiological Parameters

In the presence of salinity, the largest increase in the growth parameters between the first and second harvest was recorded after the application of compost alone with a 76% increase compared to the values recorded at the first harvest (Table 1). Under salt stress, N levels showed the most important increase between 10 and 14 months of cultivation in the plants grown in the presence of compost + AMF, while Na and Cl levels significantly decreased in the treated plants (Table S1). The largest decline (17%) was recorded for Na uptake in plants treated with AMF alone. Considering the values of physiological traits in stressed plants between the two harvests, the chlorophyll b content recorded the greatest improvement (56%) in compost + AMF-treated plants (Table S2 and Figure S1). Our results show that the tolerance mechanism of date palm to salt stress was progressively improved over time in treated plants as reported by our previous studies [6,8,9] and particularly in compost + AMF-treated plants. Between 10 and 14 months of cultivation, growth parameters increased which could be explained by a significant improvement in nutrient contents (P, N and K^+) and a reduction in toxic ion concentrations (Na^+ and Cl^-) [5,10].

Table 1. Growth traits of date palm in the absence and presence of salinity with the application of compost and AMF after 10 and 14 months of cultivation.

NaCl Level	Treatments	10 Months of Cultivation				14 Months of Cultivation			
		Plant Height (cm)	Leaf Area (cm^2)	Shoot Dry Weight (g)	Root Dry Weight (g)	Plant Height (cm)	Leaf Area (cm^2)	Shoot Dry Weight (g)	Root Dry Weight (g)
0 mM	Control	30.06 ± 0.39 gh	18.95 ± 0.68 i	1.55 ± 0.06 gh	1.11 ± 0.10 h	34.52 ± 0.88 ded	24.19 ± 0.85 g	2.82 ± 0.11 f	1.92 ± 0.18 ef
	Compost	33.74 ± 0.43 ef	31.67 ± 0.27 d	3.59 ± 0.37 e	2.01 ± 0.10 e	38.76 ± 1.26 c	35.71 ± 1.05 c	6.65 ± 0.26 b	3.28 ± 0.36 b
	AMF	35.66 ± 0.54 d	32.40 ± 0.65 d	4.99 ± 0.89 d	3.36 ± 0.53 b	42.54 ± 1.18 a	38.00 ± 0.31 b	6.32 ± 0.27 bc	4.13 ± 0.30 a
	Compost + AMF	33.12 ± 0.54 f	34.85 ± 1.45 c	5.06 ± 0.28 d	2.45 ± 0.52 cd	40.70 ± 1.06 b	40.29 ± 0.65 a	8.83 ± 0.44 a	4.06 ± 0.44 a
240 mM	Control	27.32 ± 0.40 i	11.52 ± 1.20 l	1.09 ± 0.14 h	0.66 ± 0.09 i	29.62 ± 1.21 h	14.46 ± 0.39 k	1.85 ± 0.12 g	1.11 ± 0.06 h
	Compost	30.10 ± 0.64 h	17.01 ± 0.55 j	2.57 ± 0.50 f	1.29 ± 0.29 gh	33.24 ± 1.33 f	21.21 ± 0.51 h	4.51 ± 0.58 d	2.00 ± 0.31 e
	AMF	31.54 ± 0.34 h	20.62 ± 0.55 h	3.51 ± 0.47 e	1.68 ± 0.22 efg	34.76 ± 1.66 de	27.04 ± 0.50 f	4.63 ± 0.55 d	2.66 ± 0.33 c
	Compost + AMF	29.92 ± 0.80 h	23.31 ± 0.43 g	4.50 ± 0.27 d	1.53 ± 0.28 fg	33.96 ± 1.96 ef	27.97 ± 0.33 e	6.04 ± 0.74 c	2.09 ± 0.11 de

The values of each parameter labeled by different letters indicate significant differences assessed by Duncan's test after performing three-way ANOVA ($p < 0.05$).

3.2. *Stress Markers and Organic Osmolytes Content, and Activity of Antioxidant Enzymes*

Under salinity, the stress markers showed a significant reduction between 10 and 14 months of cultivation for most of the treated plants while the concentrations of proline and soluble sugars showed significant differences in most treatments (Figure 1 and Table S3). The root soluble sugar content showed the greatest evolution with an increase of 28% in plants treated with compost alone. The activity of the antioxidant enzymes mainly decreased between 10 and 14 months of cultivation, in the presence of saline stress, where the shoot SOD activity showed the greatest decline (25%) between the two harvests in compost-treated plants (Figures S2 and S3). The significant increase in antioxidant enzyme activity in treated plants in the presence of saline stress after 10 months of cultivation probably led to a significant decrease in stress markers [11,12]. This decrease was more significant after 14 months of cultivation. The low values obtained for the stress markers could have a negative feedback effect on the activity of antioxidant enzymes by reducing them after 14 months of cultivation and could also be the cause of the reduction in organic osmolyte content.

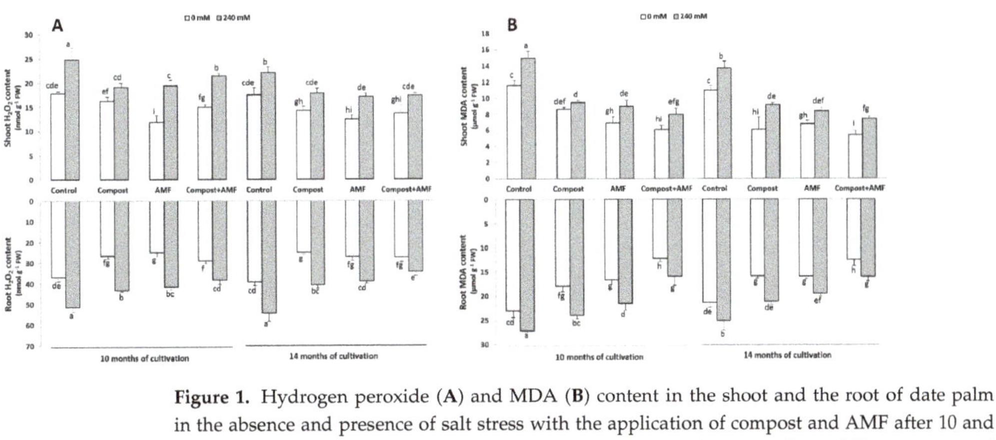

Figure 1. Hydrogen peroxide (**A**) and MDA (**B**) content in the shoot and the root of date palm in the absence and presence of salt stress with the application of compost and AMF after 10 and 14 months of cultivation. The bars labeled by different letters indicate significant differences assessed by Duncan's test after performing three-way ANOVA ($p < 0.05$).

4. Conclusions

The tolerance of date palm to salinity evolves over time with a decrease in organic osmolytes and antioxidant enzyme activity between 10 and 14 months of cultivation as a response to the decrease in H_2O_2 and MDA concentrations, which means an improvement in the plant's adaptation to salt stress. The present study strongly suggests the use of compost + AMF to mitigate the adverse impact of salt stress and to boost plant fitness in a saline environment in arid and semi-arid zones, particularly in Moroccan oasis ecosystems.

Supplementary Materials: The following supporting information can be downloaded at: https://www.mdpi.com/article/10.3390/environsciproc2022016015/s1, Figure S1. Photosynthetic pigment content (chlorophyll a (A), chlorophyll b (B), carotenoids (C) and total chlorophyll (D)) of mycorrhization of date palm plants grown in the absence (0 mM NaCl) and presence (240 mM NaCl) of salt stress with the application of compost and AMF alone or in combination after 10 and 14 months of cultivation, Figure S2. SOD (A) and CAT (B) antioxidant enzyme activity in the shoot and the root of mycorrhization of date palm plants grown in the absence (0 mM NaCl) and presence (240 mM NaCl) of salt stress with the application of compost and AMF alone or in combination after 10 and 14 months of cultivation, Figure S3 APX (A) and POD (B) antioxidant enzyme activity in the shoot and the root of mycorrhization of date palm plants grown in the absence (0 mM NaCl) and presence (240 mM NaCl) of salt stress with the application of compost and AMF alone or in combination after 10 and 14 months of cultivation, Table S1. Phosphorus (P), nitrogen (N), potassium (K), calcium (Ca), sodium (Na) and chlorine (Cl) uptake of date palm plants grown in the absence (0 mM NaCl) and presence (240 mM NaCl) of salt stress with the application of compost and AMF alone or in combination after 10 and 14 months of cultivation, Table S2. Physiological and water status of date palm plants grown in the absence (0 mM NaCl) and presence (240 mM NaCl) of salt stress with the application of compost and AMF alone or in combination after 10 and 14 months of cultivation, Table S3. Proline and soluble sugars content in shoot and root of date palm plants grown in the absence and presence of salt stress with the application of compost and AMF alone or in combination after 10 and 14 months of cultivation.

Author Contributions: Conceptualization, A.M., S.W. and M.B.; methodology, M.A.-E.-M., A.M., S.W. and M.B.; formal analysis, M.A.-E.-M., R.B.-L., M.A., A.F., Y.A.-R. and A.B.; writing—original draft preparation, M.A.-E.-M.; writing—review and editing, A.M., S.W., M.B. and T.M.; supervision and project administration, A.M., S.W. and M.B.; funding acquisition, M.B. and T.M. All authors have read and agreed to the published version of the manuscript.

Funding: This research received no external funding.

Conflicts of Interest: The authors declare no conflict of interest.

References

1. Muchate, N.S.; Nikalje, G.C.; Rajurkar, N.S.; Suprasanna, P.; Nikam, T.D. Plant salt stress: Adaptive responses, tolerance mechanism and bioengineering for salt tolerance. *Bot. Rev.* **2016**, *82*, 371–406. [CrossRef]
2. Ait-El-Mokhtar, M.; Fakhech, A.; Anli, M.; Ben-Laouane, R.; Boutasknit, A.; Wahbi, S.; Meddich, A. Infectivity of the palm groves arbuscular mycorrhizal fungi under arid and semi-arid climate and its edaphic determinants towards efficient ecological restoration. *Rhizosphere* **2020**, *15*, 100220. [CrossRef]
3. Anli, M.; Baslam, M.; Tahiri, A.; Raklami, A.; Symanczik, S.; Boutasknit, A.; Ait-El-Mokhtar, M.; Ben-Laouane, R.; Toubali, S.; Ait Rahou, Y.; et al. Biofertilizers as strategies to improve photosynthetic apparatus, growth, and drought stress tolerance in the date palm. *Front. Plant Sci.* **2020**, 1560. [CrossRef] [PubMed]
4. Boutasknit, A.; Anli, M.; Tahiri, A.; Raklami, A.; Ait-El-Mokhtar, M.; Ben-Laouane, R.; Ait Rahou, Y.; Boutaj, H.; Oufdou, K.; Wahbi, S.; et al. Potential effect of horse manure-green waste and olive pomace-green waste composts on physiology and yield of garlic (*Allium sativum* L.) and soil fertility. *Gesunde Pflanz.* **2020**, *72*, 285–295. [CrossRef]
5. Mbarki, S.; Skalicky, M.; Talbi, O.; Chakraborty, A.; Hnilicka, F.; Hejnak, V.; Zivcak, M.; Brestic, M.; Cerda, A.; Abdelly, C. Performance of *Medicago sativa* grown in clay soil favored by compost or farmyard manure to mitigate salt stress. *Agronomy* **2020**, *10*, 94. [CrossRef]
6. Ait-El-Mokhtar, M.; Ben Laouane, R.; Anli, M.; Boutasknit, A.; Wahbi, S.; Meddich, A. Use of mycorrhizal fungi in improving tolerance of the date palm (*Phoenix dactylifera* L.) seedlings to salt stress. *Sci. Hortic.* **2019**, *253*, 429–438. [CrossRef]
7. Al Kharusi, L.; Al Yahyai, R.; Yaish, M.W. Antioxidant response to salinity in salt-tolerant and salt-susceptible cultivars of date palm. *Agriculture* **2019**, *9*, 8. [CrossRef]

8. Ait-El-Mokhtar, M.; Baslam, M.; Ben-Laouane, R.; Anli, M.; Boutasknit, A.; Mitsui, T.; Wahbi, S.; Meddich, A. Alleviation of detrimental effects of salt stress on date palm (*Phoenix dactylifera* L.) by the application of arbuscular mycorrhizal fungi and/or compost. *Front. Sustain. Food Syst.* **2020**, *131*. [CrossRef]
9. Ait-El-Mokhtar, M.; Fakhech, A.; Ben-Laouane, R.; Anli, M.; Boutasknit, A.; Ait-Rahou, Y.; Wahbi, S.; Meddich, A. Compost as an eco-friendly alternative to mitigate salt-induced effects on growth, nutritional, physiological and biochemical responses of date palm. *Int. J. Recycl. Org. Waste Agric.* **2022**, *11*, 85–100.
10. Chaichi, M.R.; Keshavarz-Afshar, R.; Lu, B.; Rostamza, M. Growth and nutrient uptake of tomato in response to application of saline water, biological fertilizer, and surfactant. *J. Plant Nutr.* **2017**, *40*, 457–466. [CrossRef]
11. Ramzani, P.M.A.; Shan, L.; Anjum, S.; Ronggui, H.; Iqbal, M.; Virk, Z.A.; Kausar, S. Improved quinoa growth, physiological response, and seed nutritional quality in three soils having different stresses by the application of acidified biochar and compost. *Plant Physiol. Biochem.* **2017**, *116*, 127–138. [CrossRef]
12. Toubali, S.; Tahiri, A.; Anli, M.; Symanczik, S.; Boutasknit, A.; Ait-El-Mokhtar, M.; Ben-Laouane, R.; Oufdou, K.; Ait-Rahou, Y.; Ben-Ahmed, H.; et al. Physiological and biochemical behaviors of date palm vitroplants treated with microbial consortia and compost in response to salt stress. *Appl. Sci.* **2020**, *10*, 8665. [CrossRef]

environmental sciences proceedings

MDPI

Proceeding Paper

Effect of Phosphogypsum on Faba Bean Yield and Heavy Metals Content under Saline Conditions †

M Barka Outbakat [1], Mohamed El Gharous [1], Kamal El Omari [2] and Khalil El Mejahed [1,*]

1 Agricultural Innovation and Technology Transfer Center (AITTC), Mohammed VI Polytechnic University (UM6P), Benguerir 43150, Morocco; mbarka.outbakat@um6p.ma (M.B.O.); mohamed.elgharous@um6p.ma (M.E.G.)
2 Cherifian Phosphates Office (OCP), OCP sustainability platform, Casablanca 20200, Morocco; elomari@ocpgroup.ma
* Correspondence: khalil.elmejahed@um6p.ma
† Presented at the 2nd International Laayoune Forum on Biosaline Agriculture, 14–16 June 2022; Available online: https://lafoba2.sciforum.net/.

Abstract: Salinity is one of the most severe abiotic stresses which causes significant losses to agricultural production, especially in arid and semi-arid areas. In the present study, we conducted a pots experiment to evaluate Phosphogypsum (PG) and Gypsum (G) as amendments and their effect on faba bean shoot and grain yield under saline conditions (soil ECe = 11.17 mS/cm, water EC = 1.5 mS/cm and water SAR = 4.2 meq/L). In addition, we investigated the safety of their application based on heavy metals content in the harvested grain. Our findings demonstrate that the use of PG as amendment for saline soil reclamation improved faba bean grain and biomass yield without affecting grain quality regarding heavy metal content.

Keywords: salinity; phosphogypsum; yield parameters; heavy metals

check for updates

Citation: Outbakat, M.B.; El Gharous, M.; El Omari, K.; El Mejahed, K. Effect of Phosphogypsum on Faba Bean Yield and Heavy Metals Content under Saline Conditions. *Environ. Sci. Proc.* **2022**, *16*, 16. https://doi.org/10.3390/environsciproc2022016016

Academic Editor: Abdelaziz Hirich

Published: 16 June 2022

1. Introduction

Salinity is one of the main challenges facing agricultural production systems in arid and semi-arid regions. Currently, the total land area impacted by high salt levels is about 1 billion hectares, and the area of affected land is significantly increasing [1]. In Morocco, the total soil affected by salinity is about 1.148 Mha [2].

Salt-affected soils usually generate physical and chemical disorders in soil–plant water systems. The reclamation of salt-affected soils can be done using several amendments, such as phosphogypsum, a byproduct of the phosphate industry. It was reported that PG effectively mitigates soil salinity and enhances crop yields [3].

The phosphate industry in Morocco, a major phosphate-based fertilizer producer, generates around 25 Mt of PG annually [4]. However, studies on the valorization of PG as a soil amendment in salt-affected soils in Morocco are scarce. The objectives of this study were to investigate the effect of the PG on faba bean shoot and grain yield and to investigate the safety of its application based on heavy metals content in the harvested grain under saline conditions.

2. Material and Methods

2.1. Soil Sampling and Analysis

Soil from the Sidi Elmokhtar region of Morocco was identified from a soil database and then sampled, air-dried and ground to pass through a 2 mm mesh sieve. Soil pH was measured in 1:5 soil:water extract. ECe is the electrical conductivity of saturated paste which was prepared by hand mixing [5]. The available phosphorus was determined using the Olsen method [6]. Spectrophotometry (Cary 60 UV–Vis, Agilent Technologies, Santa Clara, CA, USA) was used to determine sulfate, ammonium, chlorine and nitrate contents. The exchangeable sodium, potassium, calcium and magnesium were determined by atomic absorption spectroscopy (200 Series AA, Agilent Technologies, Santa Clara, CA, USA). The soil is saline, Table 1 shows the chemical properties of the soil.

Table 1. Soil chemical properties.

Property/Element	pH	Ece (mS/cm)	P_2O_5 (mg/kg)	K_2O (mg/kg)	CaO (mg/kg)	Na_2O (mg/kg)	MgO (mg/kg)	SO_4 (mg/kg)	NO_3 (mg/kg)	NH_4 (mg/kg)
Value	8.1	11.17	67	308	7984	759	1067	3211	40.2	6.05

2.2. Irrigation Water

pH and Ec were measured directly by a pH meter (InoLab pH 7310) and a conductometer (Mettler Toledo. SevenCompact). The sulfate, ammonium, chlorine and nitrate contents were quantified by spectrophotometry (Agilent Technologies. Cary 60 UV–Vis). The phosphorus, sodium, potassium, calcium and magnesium were determined by ICP (Agilent Technologies. 5110 ICP-OES). The water had a moderate salinity (Table 2).

Table 2. Chemical properties of irrigation water.

Property/Element	pH	Ec (mS/cm)	SAR (meq/l)	K (mg/L)	Na (mg/L)	Ca (mg/L)	Mg (mg/)	NH_4 (mg/L)	Cl (mg/L)	SO_4 (mg/L)	NO_3 (mg/L)	P_2O_5 (mg/L)
Value	7.8	1.5	4.2	28.6	214.6	84.9	66.8	0.04	252.2	64.0	24.8	0.05

2.3. PG and G Analysis and Pot Preparation

The experiment was conducted in pots in a greenhouse at Mohammed VI Polytechnic University in Benguerir, Morocco. Each pot was filled with 10 kg of soil. The PG used is that of Jorf Lasfar Phosphate, situated near the city of El Jadida. The second amendment was natural gypsum for agricultural use. PG and G were incorporated with the top 9 cm. The treatments consisted of: Control, 15 t/h of G, 15, 30 and 45 t/ha of PG. pH and EC were measured in a ratio of 1:5 PG or G:water extract. The rest of the elements were quantified by ICP-OES. PG is more acidic than G, and it is richer in nutrients (Ca, S and P). The chemical compositions of PG and G are presented in Table 3.

Table 3. Chemical composition of PG and G.

Property/Element	pH	EC (mS/cm)	Ca (%)	S (%)	P (%)	K (ppm)	Mg (ppm)	Cd (ppm)	Zn (ppm)	Cu (ppm)	Fe (ppm)	Ni (ppm)	Pb (ppm)
Phosphogypsum	5.8	2.4	26.0	23.7	0.8	869.0	259.0	4.7	8.5	2.6	126.7	1.8	1.9
Gypsum	8.12	2.3	22.8	13.1	0.02	969.0	7587.0	<0.003	9.9	2.4	2606.4	4.2	1.4

2.4. Statistical Analysis

Data were subjected to statistical analyses using IMB SPSS 20 software. One-way ANOVA tests were performed to test the difference between treatments. When ANOVA was significant, Tukey's test was used to compare the means.

3. Results and Discussion

3.1. Yield Parameters

The amendments (G and PG) were associated with an increased number of grains and shoot yield. In pots, compared to the control, 30 t/ha and 45 t/ha of PG increased shoot dry matter by 52% and 54%, respectively (Table 4). Similar results were reported by [7] in maize trials where 20 t/ha and 40 t/ha of PG increased dry matter by 45% and 69%, respectively. Fresh biomass, fresh grain and dry weight were improved only with the highest rates of PG (30 and 45 t/ha) (Figure 1 and Table 4). The application of 30 and 45 t/ha increased grain dry weight by 52% and 62%, respectively, when compared to the control. In previous works, 10t/ha of G was enough to significantly increase faba bean grain yield [8]. PG application increased the yield of spring chickpeas by 50% and lentils by 27% [9]. The application of 30 and 45 t/ha also significantly increased thousand grain weight. However, neither PG nor G had a significant effect on the harvest index (Table 4).

Table 4. Yield parameters of faba bean (mean ± standard deviation). Means followed by the same letter within the same column are not significantly different.

Treatments	Number of Grains (m^2)	Plant Fresh Weight (g/m^2)	Plant Dry Weight (g/m^2)	Thousand Grain Weight (g)	Harvest Index (%)
Control	106 ± 7 c	464 ± 34 b	198 ± 14 c	1073 ± 71 b	37% ± 1% a
15 t/h of Gypsum	118 ± 7 bc	504 ± 23 b	227 ± 7 b	1070 ± 50 b	36% ± 2% a
15 t/h of Phosphogypsum	118 ± 7 bc	506 ± 13 b	225 ± 8 b	1152 ± 68 ab	38% ± 2% a
30 t/h of Phosphogypsum	140 ± 13 ab	628 ± 24 a	300 ± 11 a	1236 ± 38 a	37% ± 2% a
45 t/h of Phosphogypsum	153 ± 18 a	634 ± 14 a	305 ± 12 a	1215 ± 62 a	38% ± 1% a

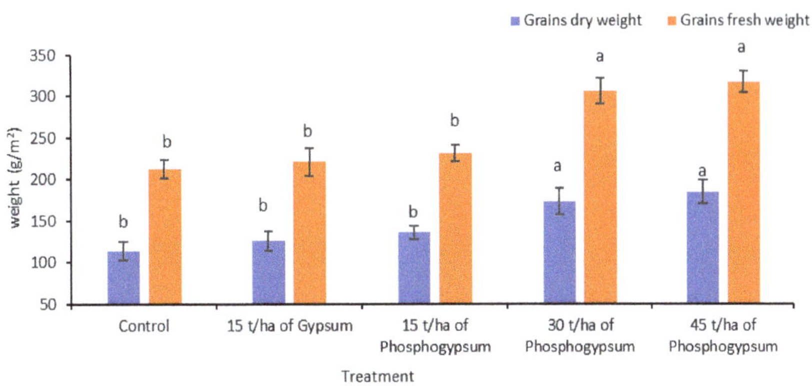

Figure 1. Dry and fresh grain weight of faba bean (mean ± standard deviation). Same letters in a column series indicate no significant differences among treatments.

Yield increases observed with PG compared to G may have been due to PG acidity and its calcium, sulfur and phosphor contents. In addition, PG was reported to dissolve faster and produce an acidic reaction in the rhizosphere, thus positively influencing nutrient availability [10].

3.2. *Heavy Metals Content*

This study showed that the application of the PG did not affect the grain quality. The grain heavy metals contents were below the recommended levels (Table 5) and were not affected by the application of PG and G. This finding is in agreement with a previous study where PG did not have an accumulative impact on plant heavy metal content [11]. This result can be explained by the alkaline soil pH as reported by [12] who indicated that transfer of heavy metals from soil to plants was negatively correlated with soil pH.

Table 5. Grain heavy metals content (mean ± standard deviation). Means followed by the same letter within the same column are not significantly different.

Treatments	Cu	Fe	Zn	Ni	Pb	Cd
	ppm					
Control	12.7 ± 1.5 a	62.7 ± 16.3 a	72.7 ± 12 a	3.4 ± 0.3 a	<0.01	<0.003
15 t/h of Gypsum	11.2 ± 1.7 a	85.2 ± 19.9 a	67.2 ± 7.8 a	3.6 ± 0.9 a	<0.01	<0.003
15 t/h of Phosphogypsum	11.3 ± 2.3 a	74.2 ± 28.8 a	69.2 ± 8.5 a	3.0 ± 0.3 a	<0.01	<0.003
30 t/h of Phosphogypsum	12.0 ± 1.5 a	78.7 ± 26.0 a	82.2 ± 21.7 a	3.0 ± 0.3 a	<0.01	<0.003
45 t/h of Phosphogypsum	11.2 ± 1.7 a	79.7 ± 20.8 a	70.2 ± 6.6 a	3.0 ± 0.2 a	<0.01	<0.003
Recommended Limits [13]	73.3	425.5	99.4	67.9	0.3	0.2

4. Conclusions

Results from this study have shown that the effectiveness of the treatments, on shoot and grain yield, was in the order of: 45 t/ha of PG > 30 t/ha of PG >15 t/ha of PG >15 t/ha of G > control. The application of 30 and 45 t/ha of PG increased grain dry weight by 52% and 62%, respectively, when compared to the control. Grain heavy metal contents were below the recommended limits and similar across treatments. Our findings suggest that the application of PG increased faba bean yield and can be considered as a safe amendment for reclaiming salt-affected soils.

Author Contributions: Conceptualization, K.E.M., M.E.G. and K.E.O.; methodology, K.E.M. and M.B.O.; software, M.B.O.; validation, K.E.M. and M.E.G.; formal analysis, M.B.O.; investigation, M.B.O.; resources, K.E.M., M.E.G. and K.E.O.; data curation, K.E.M.; writing—original draft preparation, M.B.O.; writing—review and editing, K.E.M.; visualization, M.B.O.; supervision, K.E.M. and M.E.G.; project administration, K.E.M., M.E.G. and K.E.O.; funding acquisition, K.E.O. All authors have read and agreed to the published version of the manuscript.

Funding: This research was financially supported by OCP sustainability platform under the specific agreement "AS 49–P1".

Institutional Review Board Statement: Not applicable.

Informed Consent Statement: Not applicable.

Data Availability Statement: Not applicable.

Acknowledgments: The authors would like to gratefully acknowledge the AITTC Soil, Water and Plant Analysis Laboratory, the AITTC experimental farm for providing the facilities and OCP sustainability platform for financial support to carry out the present investigation.

Conflicts of Interest: The authors declare no conflict of interest.

References

1. Ivushkin, K.; Bartholomeus, H.; Bregt, A.K.; Pulatov, A.; Kempen, B.; De Sousa, L. Global mapping of soil salinity change. *Remote Sens. Environ.* **2019**, *231*, 111260. [CrossRef]
2. FAO. *Salt-Affected Soils and Their Management Soils Bulletin 39*. 1988. Available online: http://www.fao.org/3/x5871e/x5871e00.htm (accessed on 28 October 2021).
3. Saadaoui, E.; Ghazel, E.N.; Ben Romdhane, C.; Massoudi, N. Phosphogypsum: Potential uses and problems—A review. *Int. J. Environ. Stud.* **2017**, *74*, 558–567. [CrossRef]
4. Harrou, A.; Gharibi, E.K.; Taha, Y.; Fagel, N.; El Ouahabi, M. Phosphogypsum and black steel slag as additives for ecological bentonite-based materials: Microstructure and characterization. *Minerals* **2020**, *10*, 1067. [CrossRef]
5. Richards, L.A. Diagnosis and improvement of saline and alkaline soils. *Soil Sci. Soc. Am. J.* **1954**, *18*, 348. [CrossRef]
6. Olsen, S.; Cole, C.; Watandble, F.; Dean, L. Estimation of available phosphorus in soils by extraction with sodium bicarbonate. *J. Chem. Inf. Model.* **1954**, *53*, 1689–1699.
7. El Mejahed, K.; Zeroual, Y. Valorization of phosphogypsum as fertilizer—Results of agronomic field trials: Morocco. In *Phosphogypsum Leadership innovation Partnership*; Hiltonno, J., Ed.; IFA: Paris, France, 2020; pp. 54–59.

8. Smith, A.P.; Chen, D.; Chalk, P.M. N_2 fixation by faba bean (*Vicia faba* L.) in a gypsum-amended sodic soil. *Biol. Fertil. Soils* **2009**, *45*, 329–333. [CrossRef]
9. El Mejahed, K.; Zeroual, Y. Valorisation of phosphogypsum as amendment for the reclamation of saline/sodic soils: Agronomic field trials—Morocco. In *Phosphogypsum Leadership Innovation Partnership*; Hiltonno, J., Ed.; IFA: Paris, France, 2020; pp. 60–69.
10. Gharaibeh, M.A.; Rusan, M.J.; Eltaif, N.I.; Shunnar, O.F. Reclamation of highly calcareous saline-sodic soil using low quality water and phosphogypsum. *Appl. Water Sci.* **2014**, *4*, 223–230. [CrossRef]
11. Esawy, M.; Nasser, A. Heavy metal immobilization in contaminated soils using phosphogypsum and rice straw compost. *L. Degrad. Dev.* **2015**, *26*, 819–824.
12. Ben Chabchoubi, I.; Bouguerra, S.; Ksibi, M.; Hentati, O. Health risk assessment of heavy metals exposure via consumption of crops grown in phosphogypsum-contaminated soils. *Environ. Geochem. Health* **2021**, *43*, 1953–1981. [CrossRef] [PubMed]
13. FAO; WHO. Toxicological Evaluation of Certain Food Additives and Contaminants. 1989. Available online: https://apps.who.int/iris/handle/10665/37491 (accessed on 10 November 2021).

environmental sciences proceedings

MDPI

Proceeding Paper

Desert Actinobacterial Strains Increase Salt Stress Resilience in Crops †

Celia Alvarez-Sastre and Lorena Carro *

Lab 230, Plaza Doctores de la Reina, Departamento de Microbiología y Genética, Universidad de Salamanca, 37007 Salamanca, Spain
* Correspondence: lcg@usal.es
† Presented at the 2nd International Laayoune Forum on Biosaline Agriculture, 14–16 June 2022; Available online: https://lafoba2.sciforum.net/.

Abstract: The adaptation of crops to saline stress conditions generated by changes in rainfall and the current production systems is essential for maintaining many of them and minimizing possible damage or reductions in their productivity. The use of microorganisms to improve the conditions of plants from extreme environments, increasing their resilience, appears to be a possible alternative. In this work, we isolated strains from samples obtained in extreme environments, such as the Atacama Desert and Sahara Desert, and evaluated their capacity to promote the growth of plants directly and under stress conditions. We studied their ability to grow under salinity, and we selected some of these strains for their capacity to improve plant resilience.

Keywords: desert; soil; plant–bacteria interaction; climate change; microbiology; actinobacteria; *Micromonospora*; taxonomy; biodiversity; resilience

Citation: Alvarez-Sastre, C.; Carro, L. Desert Actinobacterial Strains Increase Salt Stress Resilience in Crops. *Environ. Sci. Proc.* **2022**, *16*, 17. https://doi.org/10.3390/environsciproc2022016017

Academic Editor: Abdelaziz Hirich

Published: 16 June 2022

1. Introduction

One major problem facing the world is climate change. The decrease in rainfall in some areas and some of the agricultural practices developed for years are generating changes in the land of many countries, increasing the areas affected by high salinity [1].

Soils, in general, and desert soils, in particular, harbor a wide bacterial diversity with an immense potential for agriculture and biotechnology [2]. Soil microbes play essential roles in maintaining soil fertility through recycling nutrients, improving soil structure, degrading pollutants, and supporting healthy plant growth. Recent metagenomic and subsequent culturing studies on one of the most extreme biomes on Earth, the Atacama Desert, have revealed an abundant diversity of microbiological life harboring a remarkable potential to improve plant growth [3,4]. In general, microorganisms from deserts have evolved to develop a wide variety of strategies to survive under limiting conditions, including high salt tolerance [5]. On the other hand, plants' evolution and adaptation capacity are slower, and it is possible that they rely on their microbiome to survive [6]. Therefore, the application of microorganisms with different origins and capacities to determine their effects on plants, both in generalized improvement of development [7] and in the increased tolerance to limiting conditions [8], are nowadays of strong interest.

During this work, several soil and plant tissues' isolates from desertic areas were obtained, determining their capacity to grow under saline environments and their effect on the growth of *Medicago sativa* plants in the presence and absence of salt stress.

2. Materials and Methods

2.1. Strains of the Work

Strains from Atacama Desert were obtained from the collection of the team obtained in a previous work [9]. Other strains were recovered from nodules and roots of legumes recovered at the Sahara Desert. Those tissues were surface sterilized, as previously described [10], and strains were isolated on YMA media and subcultured on GYM media (M65 DSMZ medium).

2.2. Evaluation of In Vitro Salt Tolerance

Isolates were tested for their tolerance to grow over several NaCl (*w*/*v*) percentages (1, 3, 5, 7, 9, 12, 15, 18, and 21%) using GYM as basal media. For inoculation, strains were grown for 7 days at 28 °C and inocula were prepared in saline solutions with concentrations of around 1×10^8 UFC/mL. Ten microliters were inoculated in triplicate for each strain and kept at 28 °C for 1 month; we evaluated the capacity for growth weekly.

2.3. Plant Growth under Greenhouse Conditions

Medicago sativa seeds were germinated axenically, being sequentially immersed in 70% (*v*/*v*) ethanol for 30 s and 2.5% $HgCl_2$ (*w*/*v*) for 2 min, followed by several rinses with sterile distilled water. The seeds were then transferred to tap-water agar plates in the dark. After germination, the seedlings were placed on pots containing Rigaud and Puppo nutrient agar [11]. Three conditions of saline stress were tested (0, 1, and 3% NaCl (*w*/*v*)). The seedlings were kept in a plant growth chamber with fluorescent lighting with a photoperiod of 16 h light/8 h dark, a constant temperature of 21 to 22 °C, and 70% relative humidity. Plants were inoculated with the appropriate bacterial suspensions (1×10^8 CFU/mL), prepared as described before, and sterile distilled water as control. Plants were grown for 40 days with weekly evaluations of their development (stabilization, number of leaves, size) as well as a final evaluation, including wet and dry weights.

2.4. DNA Extraction and Identification

Extraction of genomic DNA for 16S ribosomal RNA (rRNA) gene sequence identification was carried out, as previously described [4]. A multiple sequence alignment of 16S rRNA genes was performed using DNASTAR with default options, and the strains were compared with type strains deposited in public databases using EzBioCloud. A phylogenetic tree was built using MEGA10. The 16S rRNA gene alignment was used to compute the genetic distances between all pairs of isolates.

3. Results

3.1. Isolation and Identification of Bacterial Strains

Fifteen strains presenting an actinobacterial morphology were isolated from nodules of autochthonous legumes growth in the Sahara Desert, while 17 strains were isolated from roots of the same plants. Nine strains previously recovered from Atacama Desert soils were also evaluated and analyzed. From them, a phylogenetic analysis of the 16S rRNA gene sequences showed that all the isolates from the Atacama Desert belonged to the *Micromonospora* genus, while, for the Sahara Desert, the strains belonged to several genera including *Micromonospora, Dermacoccus, Micrococcus,* and *Microbacterium* (Table 1).

Table 1. Classification of the isolates obtained in the study, grouped by genera depending on the origin of the samples.

	Atacama Desert Soil	Sahara Desert Nodules	Sahara Desert Roots
Dermacoccus sp.	-	3	-
Micromonospora sp.	9	9	-
Micrococcus sp.	-	1	13
Microbacterium sp.	-	1	2

3.2. Tolerance to Saline Stress

All the strains were able to tolerate a minimum of 1% NaCl (*w*/*v*). In general, *Micromonospora* strains showed a lower tolerance to salt, with a maximum of 5%, while some strains of the genus *Micrococcus* reached 18%. Based on the results obtained here, six strains were selected for plant inoculation: three strains belonging to *Micromonospora* (S30, A3, and A9) that tolerated until 3, 5, and 5%, respectively, and three strains belonging to the *Micrococcus* genus (S27, S14, and SR36) that tolerated to 12, 15, and 18%, respectively.

3.3. Plant Growth Promotion Evaluation

Based on the data obtained, we were able to verify that some of the strains evaluated exerted a promoting effect on plant growth regardless of the presence of salt, improving and accelerating the development of plants (Figure 1), a result obtained mainly for the strains from the Atacama Desert (A). In addition, an improvement in the development of plants was observed in the presence of a moderate concentration of salinity of 1% in several of the strains analyzed, highlighting, in this case, the results obtained for the Sahara strain S14. On the other hand, it was observed that the negative effects of the high salt concentration could not be mitigated by the presence of the inoculated bacteria or a very slight mitigation was obtained.

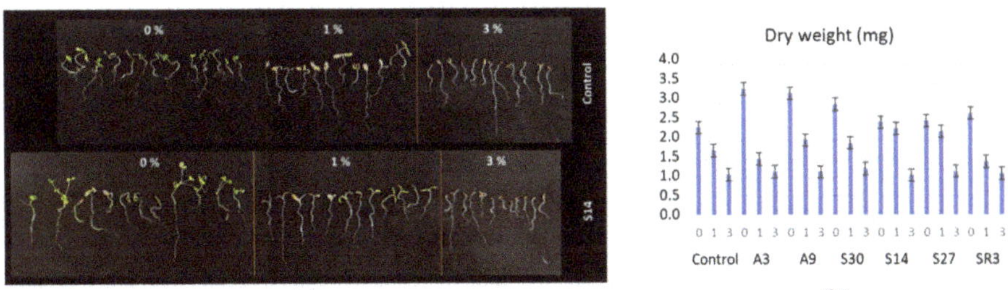

(a) (b)

Figure 1. Plant growth promotion effects under salt stress of strains isolated from desert samples. (**a**) Comparison of the development of plants at 0, 1, and 3% of NaCl (w/v) in control vs. S14-inoculated plants; (**b**) differences observed in dry weight for all the samples tested under several saline concentrations.

4. Discussion

Our results showed the capacity of desertic strains to survive high concentrations of salinity in the media, with percentages as high as 18% for a root endophytic actinobacteria. In addition, we confirmed the strong relationship of actinobacteria with plants in desertic environments and their implication in increasing plant tolerance to salinity to similar levels as was described for the other type of bacteria: *Halomonas* and *Bacillus* [12]. *Micromonospora*, *Micrococcus*, and *Microbacterium* were previously described as endophytes [13]; however, this is the first time *Dermacoccus* strains were obtained from internal plant tissues. The study of microorganisms from extreme environments, such as deserts, in which the knowledge of their diversity is still very limited [3], is of special relevance, especially the determination of those that have a direct relationship with the development of plants under limiting conditions. The use of microorganisms to improve the development of plants and their adaptation to new stressful situations appears as a possibility to fight climate change consequences.

5. Conclusions

Our results show how several actinobacterial strains isolated from deserts present a high capacity to tolerate salinity as well as improving the plant's tolerance to this stress. Based on the data presented here, we can say that, even if the *Micromonospora* isolates do not have the highest tolerant to salinity, they present a high capacity to improve plant growth and are also able to increase plant resilience to saline stress. The capacity of *Micromonospora* and *Micrococcus* as plant growth promotors has been exposed in several works, showing their capacity to produce IAA, chitinases, siderophores, or phosphatases as well as their ability to inhibit the growth of several pathogens; however, this is the first time that their capacity to increase plant tolerance to saline stress is described.

Author Contributions: Conceptualization, L.C.; methodology, C.A.-S. and L.C.; software, L.C.; validation, C.A.-S. and L.C.; formal analysis, L.C.; investigation, L.C.; resources, L.C.; data curation, L.C.; writing—original draft preparation, L.C.; writing—review and editing, L.C.; visualization, L.C.; supervision, L.C.; project administration, L.C.; funding acquisition, L.C. All authors have read and agreed to the published version of the manuscript.

Funding: This research was funded by Universidad de Salamanca, grant number PIC2-2020-04, and the Agencia Estatal de Investigación, grant number RYC2019-028311I.

Institutional Review Board Statement: Not applicable.

Informed Consent Statement: Not applicable.

Data Availability Statement: Not applicable.

Conflicts of Interest: The authors declare no conflict of interest.

References

1. Aragüés, R.; Medina, E.T.; Zribi, W.; Clavería, I.; Álvaro-Fuentes, J.; Faci, J. Soil salinization as a threat to the sustainability of deficit irrigation under present and expected climate change scenarios. *Irrig. Sci.* **2015**, *33*, 67–79. [CrossRef]
2. Molina-Menor, E.; Gimeno-Valero, H.; Pascual, J.; Peretó, J.; Porcar, M. High Culturable Bacterial Diversity From a European Desert: The Tabernas Desert. *Front. Microbiol.* **2021**, *11*, 583120. [CrossRef] [PubMed]
3. Idris, H.; Goodfellow, M.; Sanderson, R.; Asenjo, J.A.; Bull, A.T. Actinobacterial Rare Biospheres and Dark Matter Revealed in Habitats of the Chilean Atacama Desert. *Sci. Rep.* **2017**, *7*, 8373. [CrossRef] [PubMed]
4. Carro, L.; Castro, J.F.; Razmilic, V.; Nouioui, I.; Pan, C.; Igual, J.M.; Jaspars, M.; Goodfellow, M.; Bull, A.T.; Asenjo, J.A.; et al. Uncovering the potential of novel micromonosporae isolated from an extreme hyper-arid Atacama Desert soil. *Sci. Rep.* **2019**, *9*, 4678. [CrossRef] [PubMed]
5. Kothari, V.V.; Kothari, R.K.; Kothari, C.R.; Bhatt, V.D.; Nathani, N.M.; Koringa, P.G.; Joshi, C.G.; Vyas, B.R.M. Genome Sequence of Salt-Tolerant Bacillus safensis Strain VK, Isolated from Saline Desert Area of Gujarat, India. *Genome Announc.* **2013**, *1*, e00671-13. [CrossRef] [PubMed]
6. Carro, L.; Nouioui, I. Taxonomy and systematics of plant probiotic bacteria in the genomic era. *AIMS Microbiol.* **2017**, *3*, 383–412. [CrossRef] [PubMed]
7. Ferreira, C.M.H.; Soares, H.M.V.M.; Soares, E.V. Promising bacterial genera for agricultural practices: An insight on plant growth-promoting properties and microbial safety aspects. *Sci. Total Environ.* **2019**, *682*, 779–799. [CrossRef] [PubMed]
8. del Carmen Orozco-Mosqueda, M.; Glick, B.R.; Santoyo, G. ACC deaminase in plant growth-promoting bacteria (PGPB): An efficient mechanism to counter salt stress in crops. *Microbiol. Res.* **2020**, *235*, 126439. [CrossRef] [PubMed]
9. Carro, L.; Razmilic, V.; Nouioui, I.; Richardson, L.; Pan, C.; Golinska, P.; Asenjo, J.A.; Bull, A.T.; Klenk, H.-P.; Goodfellow, M. Hunting for cultivable Micromonospora strains in soils of the Atacama Desert. *Antonie Van Leeuwenhoek* **2018**, *111*, 1375–1387. [CrossRef] [PubMed]
10. Benito, P.; Carro, L.; Bacigalupe, R.; Trujillo, M.E. From roots to leaves: The colonization capacity of an endophytic filamentous actinobacteria. *Phytobiomes J.* **2022**, *6*, 35–44. [CrossRef]
11. Rigaud, J.; Puppo, A. Indole-3-acetic Acid Catabolism by Soybean Bacteroids. *J. Gen. Microbiol.* **1975**, *88*, 223–228. [CrossRef]
12. Kearl, J.; McNary, C.; Lowman, J.S.; Mei, C.; Aanderud, Z.T.; Smith, S.T.; West, J.; Colton, E.; Hamson, M.; Nielsen, B.L. Salt-tolerant halophyte rhizosphere bacteria stimulate growth of alfalfa in salty soil. *Front. Microbiol.* **2019**, *10*, 1849. [CrossRef] [PubMed]
13. Parte, A.C.; Sardà Carbasse, J.; Meier-Kolthoff, J.P.; Reimer, L.C.; Göker, M. List of Prokaryotic names with Standing in Nomenclature (LPSN) moves to the DSMZ. *Int. J. Syst. Evol. Microbiol.* **2020**, *70*, 5607–5612. [CrossRef] [PubMed]

environmental sciences proceedings

MDPI

Proceeding Paper

Study of the Sodicity of Phosphate By-Products and Sludge Mixture for Large-Scale Application in Mine Site Reclamation †

Yao Kohou Donatien Guéablé [1], Obinna Dominic Uke [2], Youssef Bezrhoud [3], Haitam Moulay [1], Lhoussaine Moughli [4], Mohamed Hafidi [5,6], Mohamed El Gharouss [1] and Khalil El Mejahed [1,*]

1 Agricultural Innovation and Technology Transfer Center (AITTC), Mohammed VI Polytechnic University (UM6P), Ben Guerir 43150, Morocco; donatien.gueable@um6p.ma (Y.K.D.G.); haitam.moulay@um6p.ma (H.M.); mohamed.elgharous@um6p.ma (M.E.G.)
2 School of Agriculture, Fertilizers and Environmental Sciences, Mohammed VI Polytechnic University (UM6P), Ben Guerir 43150, Morocco; dominic.uke@um6p.ma
3 Digital for Research (D4R), Mohammed VI Polytechnic University (UM6P), Ben Guerir 43150, Morocco; youssef.bezrhoud@um6p.ma
4 Institut Agronomique et Vétérinaire Hassan II, Rabat 16000, Morocco; l.moughli@iav.ac.ma
5 Laboratory of Microbial Biotechnologies, Agrosciences and Environment (BioMAgE), Cadi Ayyad University, Marrakesh 40000, Morocco; hafidi@uca.ac.ma
6 Agrobioscience Program, Mohammed VI Polytechnic University (UM6P), Ben Guerir 43150, Morocco
* Correspondence: khalil.elmejahed@um6p.ma

† Presented at the 2nd International Laayoune Forum on Biosaline Agriculture, 14–16 June 2022; Available online: https://lafoba2.sciforum.net/.

Abstract: Morocco has a very long mining tradition, and is threatened by ground salinization. The objective of this study was to evaluate the salinity level in the mixture of phosphate mining by-products and sludge prior to its use to reclaim a mine site or for soil remediation. The experiment was conducted with Italian ryegrass in 4 months under greenhouse. The design was a randomized complete block with 10 treatments and 4 replications. The results revealed that treatments containing phosphogypsum helped to reduce the effect of sodicity on soil. Thus, phosphogypsum associated with sludges can be used as an amendment to reclaim mine soil affected by sodicity.

Keywords: salinity; substrates; phosphate by-products; sludge

Citation: Guéablé, Y.K.D.; Uke, O.D.; Bezrhoud, Y.; Moulay, H.; Moughli, L.; Hafidi, M.; El Gharouss, M.; El Mejahed, K. Study of the Sodicity of Phosphate By-Products and Sludge Mixture for Large-Scale Application in Mine Site Reclamation. *Environ. Sci. Proc.* **2022**, *16*, 18. https://doi.org/10.3390/environsciproc2022016018

Academic Editor: Abdelaziz Hirich

Published: 16 June 2022

1. Introduction

Morocco, a Mediterranean country located in arid and semi-arid climate zones with a very long mining tradition, is threatened by ground salinization. More than 5% of Moroccan areas are already affected by salinity to various degrees [1]. Thus, on a global scale, after erosion in Morocco, salinity is ranked as the second threat to crop production [2]. Faced with the degradation and scarcity of land and the difficulties of management, it is becoming urgent to implement cost-effective, efficient, and less expensive techniques through the valorization of by-products of the phosphate industry and sewage sludge by revegetation of mining sites. This new approach has been performed through the recent study [3]. These studies showed that phosphate and sewage sludge by-products could be used as substrates for mine site reclamation, but no study has evaluated the effect of phosphogypsum on the mixture before large-scale application.

The objective of the study was to evaluate the sodicity level in the mixture of phosphate mining by-products and sludge prior to its use to reclaim a mine site or for soil remediation.

2. Materials and Methods

2.1. Substrates

The substrates used for this study constituted the by-products from phosphate mining and wastewater treatment plants. The by-products from phosphate mining were topsoil (TS) from the overburden, gathered and put aside prior to mining; phosphogypsum (PG); ground phosphate waste rocks (PWR); and phosphate sludge (PS). The by-product from wastewater treatment plants is sewage sludge (SS). The different substrates from phosphate mining were from the complexes El Jadida and Youssoufia, whereas the substrate from wastewater treatment plants was from Ben Guérir. Control (TS) used corresponds to topsoil. Samples (PG, PWR and TS) were sieved through 2 mm mesh sieve before the experiment.

2.2. Pot Preparation and Experimental Layout

The experiment was conducted with Italian ryegrass (*Lolium multiflorum*) in 4 months under greenhouse conditions at the Agriculture Innovation Transfer Technology Center-Ben Guérir (AITTC) of the University Mohammed 6 polytechnic at Ben Guérir, Morocco. The pots used for the experiment had a volume of 12 L. The experimental layout was a randomized complete block design with 5 substrates defined in different proportions, 10 treatments and 4 replications. Fertilizers were applied in treatments not receiving sludge as described by Guéablé et al., (2021) (Table 1). Water from the well of the experimental farm of AITTC was used for irrigation, and the seedlings were watered daily.

Table 1. Treatments and substrate composition and additions.

Treatments	PG	PWR	TS	SS	PS	Texture	Bulk Density (g/cm^3)	EC (dS/cm^3)	SAR (Cmol(+)/kg)
T1	-	100%	-	-	-	Sandy silt	1.20 ± 0.08	1.31 ± 0.03	0.23 ± 0.03
T2	-		100%	-	-	Clay	1.38 ± 0.80	3.19 ± 0.04	0.33 ± 0.02
T3	65%	30%	-	5%			1.02 ± 0.22	3.36 ± 0.33	0.26 ± 0.03
T4	65%		-	5%	30%	Sandy silt	1.02 ± 0.06	3.36 ± 0.17	0.30 ± 0.02
T5	65%		35%	-	-		1.05 ± 0.08	3.12 ± 0.75	0.24 ± 0.04
T6	65%	35%	-	-	-	Sandy silt clay	1.01 ± 0.08	3.13 ± 0.47	0.20 ± 0.18
T7	65% *	35% *	-	-	-	Sandy silt	1.02 ± 0.82	3.13 ± 1.43	0.24 ± 0.02
T8	-	95%	-	5%	-		1.22 ± 0.82	1.63 ± 0.09	0.24 ± 0.14
T9	-	65%	35%	-	-	Sandy silt clay	1.26 ± 0.28	1.37 ± 0.09	0.21 ± 0.04
T10	-	-	-	5%	95%		1.22 ± 0.04	1.18 ± 0.55	0.45 ± 0.42

* Addition of phosphorus in DAP (Diammonium phosphate) form at a rate of 1.08 g/pot (50 ppm), and potassium in the form of potassium sulphate at a rate of 8.3 g/pot (400 ppm). PG: phosphogypsum; PWR: phosphate waste rock; TS: topsoil; SS: sewage sludge; PS: phosphate sludge.

2.3. Data Collection

The experiment was conducted under greenhouse at AITTC for 4 months using Ital-ian ryegrass. Samples of the different substrate components and water irrigation were analyzed at the soil, plant, and water laboratory of AITTC, and the data analyzed were obtained after the experiment. Thus, pH measurements, the electrical conductivity (EC) and the cation exchange capacity (CEC) were analyzed as described by [3].

2.4. Sodium Adsorption Ratio (SAR)

One of the criteria currently recognized in the scientific literature as indices of soil and substrate sodicity is the Sodium Adsorption Ratio (SAR) with a reported threshold of 13 cmol(+)/kg. For irrigation water, SAR represent the hazard of soil sodicity following the use of the water for irrigation. [4]. It is defined by Equation (1).

$$\text{SAR} = \text{millimoles of Na}^{+}/[\text{millimoles of } 0.5(\text{Ca}^{2+} + \text{Mg}^{2+})/2]^{-1/2} \quad (1)$$

where: SAR = Sodium adsorption ratio, (Cmol(+)/kg).

Na^{+}, Ca^{2+}, Mg^{2+} = Exchangeable cations (Cmol(+)/kg).

2.5. *Characteristics of the Water Irrigation*

Water irrigation used in this study was from the experimental farm of AITTC (Table 2). The irrigation rate used for each pot was 500 mL. According to [3], this water had moderately saline water. Thus, this water could have negative effects on many crops, but it could be utilized with careful management practices [4]. Furthermore, the mean value SAR value of 3.63 cmol(+)/kg indicated a low risk of sodalisation with a sodicity class S1. A similar result was obtained by [5], where SAR was 8.69 cmol(+)/kg.

Table 2. Characteristics of the water irrigation used.

Parameters	Values
EC (dS/m)	2.58
pH	7.35
SAR (Cmol(+)/kg)	3.63

3. Results

The mean pH values of the different treatments varied between 7.77 and 8.20; therefore, the pH was alkaline (Table 3). However, the lowest pH values were for treatments (T3, T4, T5, T6 and T7) containing PG. Furthermore, the mean EC values of all treatments ranged 1.30 to 3.50 dS/m. Thus, these treatments were lower than 4 dS/m. In addition, the SAR values of different treatments ranged from 0.21 to 1.51 cmol(+)/kg, and were less than 13 cmol(+)/kg. The SAR values containing PG (T3, T4, T5, T6 and T7 treatments) were lowest, ranging from 0.21 to 0.29 cmol(+)/kg. However, for the treatments (T1, T2, T8, T9 and T10) without PG, the SAR values were greater, ranging from 0.95 to 1.51 cmol(+)/kg.

Table 3. Selected chemical properties of different treatments.

Treatments	pH	EC (dS/m)	SAR (Cmol(+)/kg)
T1	8.2 ± 0.16	1.41 ± 0.44	1.17 ± 0.15
T2	7.77 ± 0.08	3.05 ± 0.82	1.51 ± 0.19
T3	7.46 ± 0.06	3.5 ± 0.58	0.29 ± 0.09
T4	7.46 ± 0.02	3.32 ± 0.69	0.28 ± 0.05
T5	7.06 ± 0.02	3.31 ± 0.55	0.23 ± 0.04
T6	7.37 ± 0.02	3.21 ± 0.60	0.21 ± 0.04
T7	7.47 ± 0.12	3.11 ± 0.42	0.23 ± 0.02
T8	7.88 ± 0.02	2.38 ± 0.41	1.39 ± 0.20
T9	7.94 ± 0.03	1.38 ± 0.09	0.95 ± 0.48
T10	7.96 ± 0.02	1.30 ± 0.47	1.06 ± 0.14

4. Discussion

The mean EC values of all treatments lower than 4 dS/m. In addition, the SAR values of different treatments less than 13 cmol(+)/kg. According to [4,6], these treatments are classified as sodics because their EC are lower than 4 dS/m and their SAR are less than 13 cmol(+)/kg. However, the treatments (T3, T4, T5, T6 and T7) containing PG had a very low sodicity compared with the treatments (T1, T2, T8, T9 and T10) without PG. Thus, PG associated with PS and/or SS helped to reduce the effect of sodicity in soil. PS and SS can be considered as amendments, and according to several authors, PG associated with amendments was more effective than that with simple gypsum [6,7]. This combination can decrease the sodicity of the soils or substrates. Indeed, PG has a very variable composition which depends to a large extent on the composition of the natural phosphates from which it is obtained. However, it is mainly a source of calcium (CaO) and sulfur (S). PG, thanks to the addition of calcium, allows to fight against the harmful effects of sodium on the soil structure and infiltration capacity. The calcium (Ca^{2+}) brought to the soil is then fixed on the soil colloids, and each time a calcium ion is fixed, there is a sodium ion (Na^{+}) which is progressively evacuated from the soil into solution. This allows for the removal of sodium during drainage. The presence of significant amounts of calcium will allow the evacuation of sodium ions to continue over time and restructure the soil [6]. However, in the absence of drainage, sodium sulfate could cause a drop in pH. Therefore, water management is essential in the reclamation of soils in order to maintain the soil pH in neutrality.

5. Conclusions

The results showed that the soils were sodics. However, the treatments containing PG helped to reduce the effect of sodicity on soil. Thus, phosphogypsum associated with sludges can be used as an amendment to reclaim mine soil affected by sodicity.

Author Contributions: Conceptualization, Y.K.D.G., O.D.U., Y.B., H.M., L.M. and M.H.; methodology, L.M. and K.E.M.; software, Y.K.D.G.; validation, L.M., M.H., M.E.G. and K.E.M.; formal analysis, Y.K.D.G. and O.D.U.; investigation, Y.K.D.G., Y.B. and H.M.; data curation, Y.K.D.G. and K.E.M.; writing—original draft preparation, Y.K.D.G. and O.D.U.; writing—review and editing, Y.K.D.G., O.D.U., L.M. and M.H.; visualization, L.M.; M.H.; M.E.G. and K.E.M.; supervision, L.M.; M.H.; M.E.G. and K.E.M.; project administration, K.E.M.; funding acquisition, M.E.G. and K.E.M. All authors have read and agreed to the published version of the manuscript.

Funding: This research was financially supported by the Gantour OCP phosphate production site of Ben Guerir, Morocco under the "specific agreement OCP-UM6P no. RE02".

Institutional Review Board Statement: Not applicable.

Informed Consent Statement: Not applicable.

Data Availability Statement: Data can be provided upon request from the corresponding author.

Acknowledgments: Authors would like to greatly acknowledge the OCP Gantour production site of Ben Guerir for the financial support under the Specific Agreement RE02. The AITTC Soil, Water and Plant Analysis Laboratory and the AITTC experimental farm for providing the facilities.

Conflicts of Interest: The authors declare no conflict of interest.

References

1. Barbouchi, M.; Abdelfattah, R.; Chokmani, K.; Ben Aissa, N.; Lhissou, R.; El Harti, A. Soil salinity characterization using polarimetric inSAR coherence: Case studies in Tunisia and Morocco. *IEEE J. Sel. Top. Appl. Earth Obs. Remote Sens.* **2015**, *8*, 3823–3832. [CrossRef]
2. El Hasini, S.; Iben. Halima, O.; El. Azzouzi, M.; Douaik, A.; Azim, K.; Zouahri, A. Organic and inorganic remediation of soils affected by salinity in the Sebkha of Sed El Mesjoune—Marrakech (Morocco). *Soil Tillage Res.* **2019**, *193*, 153–160. [CrossRef]
3. Guéablé, Y.K.D.; Bezrhoud, Y.; Moulay, H.; Moughli, L.; Hafidi, M.; El Gharouss, M.; El Mejahed, K. New approach for mining site reclamation using alternative substrate based on phosphate industry by-product and sludge mixture. *Sustainability* **2021**, *13*, 10751. [CrossRef]
4. Zaman, M.; Shahid, S.A.; Heng, L. *Guideline for Salinity Assessment, Mitigation and Adaptation Using Nuclear and Related Techniques*; Springer: Berlin/Heidelberg, Germany, 2018; pp. 14–20.
5. Bradaï, A.; Douaoui, A.; Marlet, S. *Qualité des Eaux Souterraines Utilisées en Irrigation et Risques de Dégradation des Sols Dans la Plaine*, 4th ed.; Cirad: Mostaganem, Algérie, 2008; pp. 26–28.
6. Diop, T.; Ndiaye, R.; Sow, S.A.; Ba, D.D. Analyse des effets du phophogypse et du fumier sur la salinité de la cuvette de Ndiol dans le Delta du fleuve Sénégal Résumé. *Afrique Sci.* **2019**, *15*, 71–80.
7. Prapagar, K.; Indraratne, S.P.; Premanandharajah, P. Effect of soil amendments on reclamation of saline-sodic soil. *Trop. Agric. Res.* **2012**, *23*, 168–176. [CrossRef]

environmental sciences proceedings

MDPI

Proceeding Paper

Effects of Different Rates of Liquid Sewage Sludge Amendment on Nutrient Content of the Soil in Rabat, Morocco †

Abdelmajid Achkir [1,*], Abdelhakim Aouragh [1], Mohammed El Mahi [1], El Mostapha Lotfi [1], Najoua Labjar [1], Taruvinga Badza [2] and Tawfik El Moussaoui [3]

1 Laboratory of Spectroscopy, Molecular Modeling, Materials, Nanomaterials, Water and Environment, CERN2D, ENSAM, Mohammed V University in Rabat, Rabat 10100, Morocco; aouragh.abdelhakim@gmail.com (A.A.); m.elmahi@um5r.ac.ma (M.E.M.); lotfi58@yahoo.fr (E.M.L.); najoualabjar@hotmail.com (N.L.)

2 School of Agricultural, Earth and Environmental Sciences, University of KwaZulu-Natal, Private Bag X01, Scottsville, Pietermaritzburg 3209, South Africa; tarubadza@yahoo.com

3 Resources Valorization, Environment and Sustainable Development Research Team (RVESD), Department of Mines, Mines School of Rabat, Ave Hadj Ahmed Cherkaoui–BP 753, Agdal, Rabat 53000, Morocco; elmoussaoui.ing@gmail.com

* Correspondence: abdelmajid_achkir@um5.ac.ma or achkirabdelmajidensam@gmail.com; Tel.: +212-660-363-317

† Presented at the 2nd International Laayoune Forum on Biosaline Agriculture, 14–16 June 2022; Available online: https://lafoba2.sciforum.net/.

Abstract: The objective of this study was to evaluate different rates of the liquid sewage sludge fertilizer developed by a treatment plant in Skhirat, Rabat, Morocco on improving soil fertility indicators. The results revealed that the application of liquid sludge on the soil increased soil pH from 7.3 to 7.7, electrical conductivity from 227.1 to 416.21 $\mu S.cm^{-1}$, and other nutrients such as soil organic matter from 9.0 to 21.4%, Kjeldahl nitrogen (TKN) from 0.1 to 0.7%, and total organic carbon (TOC) from 2.4 to 16.5%; the phosphorus ranged 79.1 to 127.8 $mg.kg^{-1}$ in a dose-dependent manner in amended soil compared to untreated controls. However, the results also showed increase in heavy metal content in the following order: Zn > Cu > Pb > Ni > Cd, (Zn = 136.69 $mg.kg^{-1}$, Cu = 69.05 $mg.kg^{-1}$, Pb = 17.91 $mg.kg^{-1}$, Ni = 4.73 $mg.kg^{-1}$, Cd = 0.03 $mg.kg^{-1}$); nevertheless, we noticed that their concentrations were lower than the critical values established by the European Union for the agronomic use of the soil.

Keywords: soil; sludge; electrical conductivity; agriculture

Citation: Achkir, A.; Aouragh, A.; El Mahi, M.; Lotfi, E.M.; Labjar, N.; Badza, T.; El Moussaoui, T. Effects of Different Rates of Liquid Sewage Sludge Amendment on Nutrient Content of the Soil in Rabat, Morocco. *Environ. Sci. Proc.* **2022**, *16*, 19. https://doi.org/10.3390/environsciproc2022016019

Academic Editor: Abdelaziz Hirich

Published: 13 June 2022

1. Introduction

Management and disposal of sewage sludge from municipal wastewater treatment plants is a growing concern, especially with the increase in the world's population [1]. Sewage sludge is the waste product of wastewater treatment plants, produced by the primary (physical and/or chemical), secondary (biological) and tertiary (complementary to secondary) treatment of wastewater [2]. The amount of sludge generated and its composition depend on the properties of the influent and the wastewater treatment process used [3], type of land where sewage sludge is used for soil amendment, and lead improvement in soil structure, fertility, and soil porosity. It also provides an effective means of recycling organic matter and other plant nutrients, including nitrogen and phosphorus [4]. The organic carbon (OC) content in soil amended with sludge can be higher than that of soils amended by inorganic fertilizers [5], while the electrical conductivity (EC) of the soil (with or without lime) increases due to the presence of higher levels of salts [6]. Despite several benefits, sludge application could also increase the accumulation of toxic metals in receiving soils [7]. Sludge may contains trace metals from domestic, commercial industrial, and surface water runoff [8]. In areas of high urbanization and industrialization, sludge can contain high levels of potentially toxic metals, for example, Hg, Ni, Pb, Cd, Cr, Cu, and Zn [9]. Trace metal content in sludge is one of the limiting factors for its application on a bigger scale as a soil amendment. The soil threshold level (content) of metals such as Hg, Ni, Pb, Cd, Cr, Cu, and Zn have been established by European legislation (European Directive 86/278/EEC) to regulates the application, and maintaining these elements at lower levels than the threshold when using sewage sludge for agricultural purposes [10].

The aim of this study was to investigate the liquid sludge sourced from the wastewater treatment plant of the activated sludge lagoon type, located in the city of Skhirat, Rabat, Morocco for its suitability to apply as an amendment for agriculture. It is believed that soil application of liquid sludge can improve soil properties and enhance crop growth and yield.

2. Materials and Methods

Experimental Design and Physico-Chemical Investigation

This experiment was designed to determine the effect of different proportions of liquid sludge (generated by the WWTP Skhirat Morocco) mixed with the soil. Different mixtures of sludge with clay soil were prepared in plastic pots and experimentally monitored for two years. The physicochemical parameters of the sludge and the soil were analyzed and compared with a control soil. The experiment was carried out in plastic pots with a capacity of 5 L. Five treatments were evaluated in a randomized complete block design. The treatments comprised: (1) 20% liquid sludge and 80% soil, (2) 40% liquid sludge and 60% soil, (3) 60% liquid sludge and 40% soil, (4) 80% liquid sludge and 20% soil, (5) 100% soil (control); the texture of soil was determined according to [11]. pH (1:2 w/v soil water) and electrical conductivity (EC) (1:4 w/v soil water) were performed using a multiparameter Type HACHLANG/Model: sensi ON + MM374. The chloride (Cl^-) was determined by titration with AgNO3 [12]. Available phosphorus (P) was measured by NaHCO3-ascorbic acid method [13]. Total Khjeldal nitrogen (TKN) was determined using the Kjeldahl method (Bremner 1996) [14] and total organic carbon (TOC) was determined using titration [15]. Heavy metals such as Cu, Ni, Zn, Pb, and Cd were extracted with acids and determined using a microwave plasma atomic emission spectrometry: MP-AES 4210, at the National Laboratory of Studies and Surveillance of Pollution Rabat Morocco.

3. Results and Discussion

The physicochemical properties and heavy metals of the soil after the experiments in pots are presented in Table 1. Soil pH is one of the most important and widely measured chemical properties of soil that directly influence plant growth. It is a major factor affecting the availability of elements for plant uptake. In the present work, the pH of the liquid sludge applied amended to the soil was within the threshold level for plant growth and ranged from 7.32 to 7.75. Similar results were obtained by other researchers and they also noticed an increase in soil pH after sludge application [16]. With regard to salinity, the measured electrical conductivity (EC) (Table 1) showed that application of the liquid sludge, did not lead to soil salinization, and EC values remained generally in the accepted range of $\geq$4000 μS cm^{-1} according to Cantrell and Linderman, 2001 [17]. The concentrations of macronutrients K, Na+, and P were increased with sludge treatments (20% liquid sludge +80% soil) (Table 1). With the application of sludge, TOC increased from 0.64% (soil control) to 25.34% in 80% liquid and sludge and 20% soil. TOC is directly related to soil fertility and agricultural productivity potential [18]. Most of the sludge treatments decreased C/N ratio, indicating favorable mineralization of sludge. The trace metal content in sludge is one of the limiting factors for its use as a soil amendment. The threshold limits for Cd, Cr, Cu, Hg, Ni, Pb, and Zn have been set by European legislation (European Directive 86/278/EEC). The liquid sludge added to the soil resulted in an increase in heavy metals content in the receiving soil. However, this was mainly a dose-dependent increase (highest increase in the highest proportion of sludge). For the heavy metal content from the 80% sludge treatment, the average values recorded in the treatments was in the following order

Zn > Cu > Pb > Ni > Cd (Table 1). Their concentration in all treatments with sludge was not higher than the critical values defined by the European Union for agronomy.

Table 1. Physio-chemical properties of experimental soils and liquid sludge from the treatment plant in Skhirat, Rabat, Morocco.

Parameter	LS	LS 1/S1	LS2/S2	LS3/S3	LS4/S4	CS	Limits of Heavy Metal Concentrations in Soil
Clay %	-	-	-	-	-	56.12	
Sand %	-	-	-	-	-	12.03	
Silts %	-	-	-	-	-	31.85	
pH	7.86	7.32	7.58	7.69	7.75	7.24	
EC (µS cm^{-1})	3091.08	227.1	249.84	294.15	416.21	213.52	
Moisture (%)	95.14	16.24	38.64	48.98	69.78	14.98	
OM (%)	83.78	9.03	16.78	19.10	21.43	0.71	
TOC (%)	41.71	2.45	4.29	9.12	25.34	0.64	
TKN (%)	1.23	0.19	0.31	0.49	0.73	0.082	
C/N ratio	33.91	12.89	13.83	18.61	22.6	7.8	
P Olsen (mg kg^{-1})	189.47	79,10	86.78	113.29	127.89	26.03	
K+ (mg kg^{-1})	56.47	21,85	29.69	35.90	38.07	4.92	
Na+ (mg kg^{-1})	46.09	13.56	19.37	24.83	27.17	2.4	
Cl- (mg kg^{-1})	2149.16	143.71	194.26	234.51	678.47	8.13	
Zn (mg kg^{-1})	289.84	19.04	26.75	37.11	136.69	7.71	150–300
Pb (mg kg^{-1})	59.03	9.58	11.14	14.98	17.91	4.11	750–1200
Ni (mg kg^{-1})	8.45	2.8	3.1	3.89	4.73	0.4	300–400
Cd (mg kg^{-1})	<0.003	0.001	0.001	0.002	0.03	0.001	1–3
Cu (mg kg^{-1})	116.64	25.49	32.18	43.98	69.05	9.68	1000–1750

LS; liquid sludge, LS1/S1: 20% liquid sludge +80% soil, LS2/S2:40% liquid sludge +60% soil, LS3/S3: 60% liquid sludge+40% soil, LS4/S4: 80% liquid sludge +20 soil, CS: control soil.

4. Conclusions

The potential use of liquid sludge as a soil amendment was evaluated by mixing in different proportions with soil. The highest concentrations of organic matter and macro- and micronutrients (NPK) in the treatments with sludge application indicate its potential to replace mineral fertilizers. The nutrient and heavy metal parameters (trace metals) determined are below the European limits, which indicates it is possible and suitable to apply as a soil amendment. However, a future study on thorough microbial study is suggested to prevent its possible negative effect on human and environmental health.

Author Contributions: A.A. (Abdelmajid Achkir): Conducted the experiments, performed analysis, and wrote the original manuscript. A.A. (Abdelhakim Aouragh): Participated in data analysis and reviewing. M.E.M.: Supervision, editing, and reviewing. E.M.L.: co-supervising, editing, and reviewing. N.L.: Reviewing. T.B.: Reviewing. T.E.M. Participated in data analysis, reviewing, and editing. All authors have read and agreed to the published version of the manuscript.

Funding: This research received no external funding.

Institutional Review Board Statement: Not applicable.

Informed Consent Statement: Not applicable.

Data Availability Statement: Data sharing not applicable.

Acknowledgments: This work was realized at the National Laboratory for Pollution Studies and Monitoring—Ministry of Energy, Mines and Environment—Department of Environment, Rabat—Morocco. All the team of the laboratory are warmly thanked for their efforts and support during the period of training. A. Achkir is grateful to the director of WWTP Skhirat, Rabat for the support.

Conflicts of Interest: The authors declare no conflict of interest.

References

1. Kelessidis, A.; Stasinakis, A.S. Comparative Study of the Methods Used for Treatment and Final Disposal of Sewage Sludge in European Countries. *Waste Manag.* **2012**, *32*, 1186–1195. [CrossRef] [PubMed]
2. Fytili, D.; Zabaniotou, A.Ã. Utilization of Sewage Sludge in EU Application of Old and New Methods—A Review. *Renew. Sustain. Energy Rev.* **2008**, *12*, 116–140. [CrossRef]
3. Von Sperling, M.; Gonçalves, R.F. *Sludge Characteristics and Production*; Sludge Treatment and Disposal-IWA Publishing: London, UK, 2007.
4. Heimersson, S.; Svanström, M.; Cederberg, C.; Peters, G. Improved Life Cycle Modelling of Benefits from Sewage Sludge Anaerobic Digestion and Land Application. *Resour. Conserv. Recycl.* **2017**, *122*, 126–134. [CrossRef]
5. Nyamangara, J.; Mzezewa, J. Effect of Long-Term Application of Sewage Sludge to a Grazed Grass Pasture on Organic Carbon and Nutrients of a Clay Soil in Zimbabwe. *Nutr. Cycl. Agroecosystems* **2001**, *59*, 13–18. [CrossRef]
6. Mtshali, J.S.; Tiruneh, A.T.; Fadiran, A.O. Characterization of Sewage Sludge Generated from Wastewater Treatment Plants in Swaziland in Relation to Agricultural Uses. *Resour. Environ.* **2014**, *4*, 190–199.
7. Topcuoglu, B. Chemical Speciation, Bioavailability and Environmental Pollution Risks of Heavy Metals in the Greenhouse Soil Amended with Sewage Sludge and Municipal Solid Waste Compost. *Int. J. Adv. Agric. Environ. Eng.* **2015**, *2*, 72–78.
8. Fuerhacker, M.; Haile, T.M. Treatment and Reuse of Sludge. In *Waste Water Treatment and Reuse in the Mediterranean Region*; Springer: Berlin/Heidelberg, Germany, 2010; pp. 63–92.
9. Aljerf, L. High-Efficiency Extraction of Bromocresol Purple Dye and Heavy Metals as Chromium from Industrial Effluent by Adsorption onto a Modified Surface of Zeolite: Kinetics and Equilibrium Study. *J. Environ. Manag.* **2018**, *225*, 120–132. [CrossRef] [PubMed]
10. Fuentes, A.; Lloréns, M.; Sáez, J.; Aguilar, M.I.; Ortuño, J.F.; Meseguer, V.F. Phytotoxicity and Heavy Metals Speciation of Stabilised Sewage Sludges. *J. Hazard. Mater.* **2004**, *108*, 161–169. [CrossRef] [PubMed]
11. Bouyoucos, G.J. Hydrometer Method Improved for Making Particle Size Analyses of Soils 1. *Agron. J.* **1962**, *54*, 464–465. [CrossRef]
12. Karaivazoglou, N.A.; Papakosta, D.K.; Divanidis, S. Effect of Chloride in Irrigation Water and Form of Nitrogen Fertilizer on Virginia (Flue-Cured) Tobacco. F. *Crop. Res.* **2005**, *92*, 61–74. [CrossRef]
13. Watanabe, F.S.; Olsen, S.R. Test of an Ascorbic Acid Method for Determining Phosphorus in Water and NaHCO3 Extracts from Soil. *Soil Sci. Soc. Am. J.* **1965**, *29*, 677–678. [CrossRef]
14. Bremner, J.M. Nitrogen-total. Methods soil Anal. Part 3 Chem. *Methods* **1996**, *5*, 1085–1121.
15. Walkley, A.; Black, I.A. An Examination of the Degtjareff Method for Determining Soil Organic Matter, and a Proposed Modification of the Chromic Acid Titration Method. *Soil Sci.* **1934**, *37*, 29–38. [CrossRef]
16. Zhang, S.; Wang, S.; Shan, X.; Mu, H. Influences of Lignin from Paper Mill Sludge on Soil Properties and Metal Accumulation in Wheat. *Biol. Fertil. Soils* **2004**, *40*, 237–242. [CrossRef]
17. Cantrell, I.C.; Linderman, R.G. Preinoculation of Lettuce and Onion with VA Mycorrhizal Fungi Reduces Deleterious Effects of Soil Salinity. *Plant Soil* **2001**, *233*, 269–281. [CrossRef]
18. Veeresh, H.; Tripathy, S.; Chaudhuri, D.; Ghosh, B.; Hart, B.; Powell, M. Changes in Physical and Chemical Properties of Three Soil Types in India as a Result of Amendment with Fly Ash and Sewage Sludge. *Environ. Geol.* **2003**, *43*, 513–520. [CrossRef]

environmental sciences proceedings

MDPI

Proceeding Paper

The Rapid Identification of Solid Materials Using the ACP Method †

Otmane Sadek *, Samira Touhtouh and Abdelowahed Hajjaji

Laboratoire des Sciences de l'Ingénieur Pour l'Energie (LabSIPE), Ecole Nationale des Sciences Appliquées d'El Jadida, Université Chouaib Doukkali, BP 1166, EL Jadida 24002, Morocco; samira.touhtouh@gmail.com (S.T.); hajjaji.a@ucd.ac.ma (A.H.)

* Correspondence: otmansadek@gmail.com

† Presented at the 2nd International Laayoune Forum on Biosaline Agriculture, 14–16 June 2022; Available online: https://lafoba2.sciforum.net/.

Abstract: In this work, we will present a new protocol to identify solid matter quickly, and at low cost, by physico-chemical analysis methods. The fast and less expensive methods for identification that we investigate, using principal component analysis (ACP), are hydrogen potential (pH), electrical conductivity, density and solubility.

Keywords: rapid identification; solid materials; the coefficient of variation; principal component analysis (ACP)

check for updates

Citation: Sadek, O.; Touhtouh, S.; Hajjaji, A. The Rapid Identification of Solid Materials Using the ACP Method. *Environ. Sci. Proc.* **2022**, *16*, 22. https://doi.org/10.3390/environsciproc2022016022

Academic Editor: Abdelaziz Hirich

Published: 16 June 2022

1. Introduction

Principal component analysis (ACP) is one of the multivariate data analysis methods, and is very useful when there are large amounts of quantitative data to process and interpret. ACP is a tool, whereby it is possible to make correlations between different variables by their projections in reduced spaces made up of axes designated as principal components [1]. The methods for identifying solid compounds include Infrared Spectroscopy, Scanning Electron Microscopy and X-ray Diffraction. However, these methods are very expensive and difficult to handle. In this article we study and validate different rapid techniques using ACP, that facilitate identification of solid compounds at low cost [2–4]. The identification methods which will be studied are solubility, hydrogen potential (pH), electrical conductivity and density.

2. Materials and Methods

The study was carried out on four phosphate fertilizer materials purchased from Sigma-Aldrich to test and validate the method for each material We prepared four series of measurements under repeatability conditions. These four series corresponded, respectively, to hydrogen potential (pH), electrical conductivity, density and solubility [5]. In order to determine the solubility of the material, we prepared a saturated solution of the sample. Then, the prepared solution was filtered under vacuum. The solid part of the extraction was recovered and put in an oven to dry until solid. Finally, the solubility was the ratio of the soluble mass to the volume [6]. To measure the pH and the electrical conductivity of the samples, we measured the solubility pH and the conductivity of the filtra (saturated solution), respectively, by making use of a pH-meter (PREMIUM) and a conductivity meter [7,8]. The density was measured by filling a cylinder (test tube) with a determined volume, having a determined mass, and calculating the density as mass over volume [9].

3. Results and Discussion

At the end of validating the identification, the solubility, the pH, the electrical conductivity and the volumetric mass, we calculated the coefficient of variation (CV (%)) and the lower and upper repeatability limits for each series of measurements, with a probability of 95% [10]. The results show that for the four phosphate fertilizers and for the four methods, all the values did not exceed the lower and upper values of repeatability. Furthermore, the relative error in all the cases was lower than 5%. Then, the repeatability of the four methods was checked and validated. After checking the validation of the four methods, we applied the ACP method using the results of measurements for the different materials, with the aim of immersing two relevant pieces of information. For each material, ten measurements were included in the analysis and were introduced without any specification of the type considered. The samples were numbered from 1 to 40 and the matrix, thus formed. contained 160 pieces of data to be processed.

Figure 1 presents the eigenvalues obtained in the data processing. The first two main axes mainly carry the relevant information representing about 98% of the cumulative variability [11]. In this case, the ACP application allowed us to project the 160 pieces of data on a two-dimensional space.

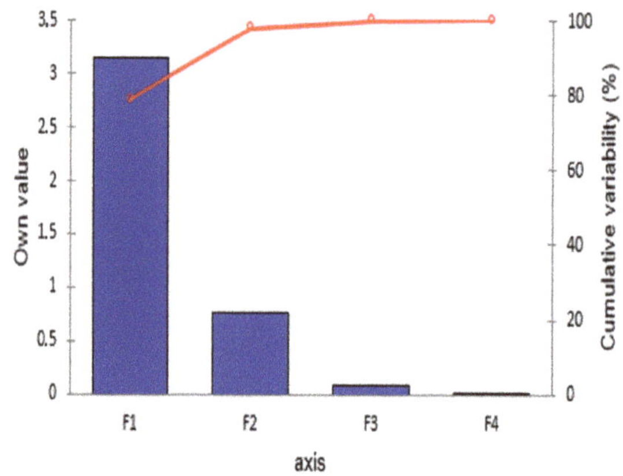

Figure 1. Representation of the eigenvalues and the cumulative variability.

Figure 2 shows that the four variables are well represented in the space of the two components F1 and F2. From this figure the representation of the samples shows four distinct groups; each linked to the type of material considered. The possibility of discriminating between the types of materials must be exploited in the attribution of any type of sample analyzed.

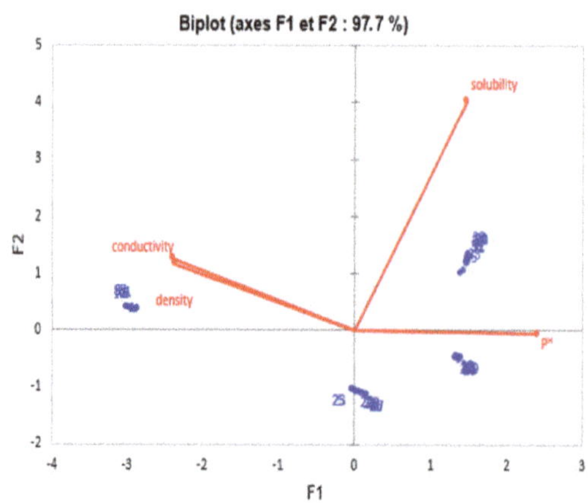

Figure 2. Representation of variables and tests in the plane of the main component.

4. Conclusions

In this work, the following physico-chemical parameters were measured for four types of phosphate fertilizer materials, with the aim of being able to set up a rapid identification method for these materials: solubility, pH, conductivity and density. Validity tests applied to the results of these measurements validated the four methods. Data from repeatability tests were processed using the Principal Component Analysis method. We noticed that the tests of each type of material formed a group. Therefore, thanks to the analysis of the principal components (PCA) we were able to discriminate any unknown analyzed sample and were, therefore, able to recognize it and easily attribute it to the type of matching material, quickly and unambiguously.

Author Contributions: O.S.: Conceptualization, Methodology, Investigation, Experimental investigation, Formal analysis, Data curation and Writing—original draft. S.T.: Methodology, Supervision, Validation, Formal analysis and Writing—review & editing. A.H.: Supervision and Validation. All authors have read and agreed to the published version of the manuscript.

Funding: This research received no external funding.

Conflicts of Interest: The authors declare that they have no known competing financial interests or personal relationships that could have appeared to influence the work reported in this paper.

References

1. Lotfi, D.H.; Izbaim, H.; Bita, H.; Mesbah, H.; Banouni, H. Assessment of ultrasonic data of signals backscattered by mortar using the principal component analysis. *Data Brief* **2021**, *34*, 106741. [CrossRef] [PubMed]
2. Ramalingam, V.V.; Ramasamy, P. Influence of friction stir processing parameters on the wear resistance of aluminium alloy AA5083. *Mater. Today Proc.* **2018**, *5 Pt 2*, 7437–7446. [CrossRef]
3. Dubey, R.S.; Rajesh, Y.B.R.D.; More, M.A. Synthesis and Characterization of SiO2 Nanoparticles via Sol-gel Method for Industrial Applications. *Mater. Today Proc.* **2015**, *2*, 3575–3579. [CrossRef]
4. Sokolov, A.N.; Roberts, M.E.; Bao, Z. Fabrication of low-cost electronic biosensors. *Mater. Today* **2009**, *12*, 12–20. [CrossRef]
5. Myz, S.A.; Mikhailovskaya, A.V.; Mikhailenko, M.A.; Bulina, N.V.; Kuznetsova, S.A.; Shakhtshneider, T.P. New crystalline betulin-based materials: Improving betulin solubility via cocrystal formation. *Mater. Today Proc.* **2019**, *12*, 82–85. [CrossRef]
6. Lakkavarapu, S.; Bondili, J.; Brahman, P. Development of proof of concept for prostate cancer detection: An electrochemical immunosensor based on fullerene-C60 and copper nanoparticles composite film as diagnostic tool. *Mater. Today Chem.* **2020**, *16*, 100257. [CrossRef]
7. Ismail, I.; Bakar, N.F.A.; Ling, T.H.; Ideris, N.; Zain, Z.H.M.; Radacsi, N. Morphology and Conductivity Evaluation of Electrospun Polyacrylic Acid (PAA) Microfiber. *Mater. Today Proc.* **2019**, *17*, 574–583. [CrossRef]
8. Barton, J.; Guillemet, C. Masse volumique et dilatabilité. In *Masse Volumique et Dilatabilité*; EDP Sciences: Les Ulis, France, 2021; pp. 39–44. [CrossRef]
9. Olivier, A.; Pierrefeu, G.; Scotti, M.; Blanquart, B. Incertitudes des mesures de débit réalisées à l'ADCP. *J. Hydraul.* **2008**, *30*, 1–7. [CrossRef]
10. Kumar, A.; Singh, K.; Singh, K.; Thakur, K.; Singh, H.; Singh, J. Product performance optimization in an automotive manufacturing plant using shainin DOE tools: A case study. *Mater. Today Proc.* **2021**, *50 Pt 5*, 1679–1682. [CrossRef]
11. Archana, T.; Vijayakumar, K.; Arivanandhan, M.; Jayavel, R. TiO_2 nanostructures with controlled morphology for improved electrical properties of photoanodes and quantum dot sensitized solar cell characteristics. *Surf. Interfaces* **2019**, *17*, 100350. [CrossRef]

environmental sciences proceedings

MDPI

Proceeding Paper

A New Organic Amendment Based on Insect Frass for Zucchini (*Cucurbita pepo* L.) Cultivation †

Jamaa Zim [1,*], Amlal Aitikkou [2], Moulay Hicham EL Omari [2], Soumia EL Malahi [3,4], Khalid Azim [5], Abdelaziz Hirich [6], Abdelaziz Nilahyane [6] and Ali Oumouloud [2]

1 Department Plant Protection, Agronomic and Veterinary Institute Hassan II, Agadir 80000, Morocco
2 Department of horticulture, Agronomic and Veterinary Institute Hassan II, Agadir 80000, Morocco; aitikkouamlal98@gmail.com (A.A.); h.elomari@iav.ac.ma (M.H.E.O.); ali.oumouloud@gmail.com (A.O.)
3 Faculty of Science, Ibn Zohr University, Agadir 80000, Morocco; s.elmalahi@uca.ma
4 Higher School of Technology, University of Cadi Aayad, El Kelâa des Sraghna 43000, Morocco
5 Integrated Crop Production Research Unit, Regional Center of Agricultural Research of Agadir, National Institute of Agricultural Research, Avenue Ennasr, BP 415 Rabat Principale, Rabat 10090, Morocco; azim.khalid@yahoo.fr
6 African Sustainable Agriculture Research Institute (ASARI), Mohammed VI Polytechnic University (UM6P), Laayoune 70000, Morocco; abdelaziz.hirich@um6p.ma (A.H.); abdelaziz.nilahyane@um6p.ma (A.N.)
* Correspondence: j.zim@iav.ac.ma

† Presented at the 2nd International Laayoune Forum on Biosaline Agriculture, 14–16 June 2022; Available online: https://lafoba2.sciforum.net/.

Abstract: Insect frass is considered as a sustainable resource for plant nutrition. This new amendment is characterized by a high salinity (EC) which constitutes an issue for its agronomic utilization. The objective of this work is to evaluate the effect of four organic amendments with different EC: frass of *Hermetia illucens* (15.19 mS/cm), frass of *Tenebrio molitor* (6.47 mS/cm), vermicompost (2.07 mS/cm), and cattle manure (1.33 mS/cm), on zucchini crop. The results showed a great improvement of the agronomic parameters of zucchini using frass of *T. molitor* insects. This study shows that insect frass could be potential alternatives of conventional amendments and may have positive impact on sustaining agricultural production.

Keywords: cattle manure; *Cucurbita pepo* L.; insect frass; plant nutrition; salinity; vermicompost

check for updates

Citation: Zim, J.; Aitikkou, A.; EL Omari, M.H.; EL Malahi, S.; Azim, K.; Hirich, A.; Nilahyane, A.; Oumouloud, A. A New Organic Amendment Based on Insect Frass for Zucchini (*Cucurbita pepo* L.) Cultivation. *Environ. Sci. Proc.* **2022**, *16*, 28. https://doi.org/10.3390/environsciproc2022016028

Academic Editor: Redouane Choukr-Allah

Published: 16 June 2022

1. Introduction

Global population growth requires the development of new strategies for sustainable food production [1]. Mass rearing of insects could be a sustainable and ecological alternative for animal protein production [2]. However, the insect farming activity generates large amounts of excreta called insect frass [3]. The use of these frass as biofertilizers could represent an opportunity to decrease the use of chemical fertilizers and thus contribute to the development of resilient and sustainable agriculture. Studies have highlighted the use of insect frass as organic fertilizers, such as those of *Tenebrio molitor* [4] and the fly *Hermetia illucens* [5]. These studies have shown that insects frass are rich in plant-available nutrients and microorganisms and promote plant growth and natural defense. The salinity of these new amendments poses a real problem in their use especially at high doses. In this context, the present work consists in determining the effects of four organic amendments, namely: vermicompost, cattle manure and the frass of *Hermetia illucens* and *Tenebrio Molitor*, on growth and production of zucchini cultivated under greenhouse.

2. Materials & Method

2.1. Location of the Trial and Plant Material

This study was conducted between June and September 2021 at the experimental farm of the Horticultural Complex of Agadir. The trial took place in a greenhouse type "multi

Chapel" with an area of 48 m^2. A hybrid F1 variety "Leen" of zucchini was used as plant material of the study.

2.2. Preparation of the Culture Substrates

The four applied organic amendments are cattle manure, vermicompost and two biofertilizers based on the frass of *Tenebrio molitor* and *Hermetia illucens*. Sand was used as a control substrate. Four mixtures were prepared based on sand to which, one of the organic amendments was added. The mixtures were made at a ratio of 1 (organic amendment): 20 (sand) in terms of volume [6]. A chemical analysis was made to investigate the effect of the organic amendments on the physico-chemical characteristics of the mixtures, Table 1 below summarizes the different chemical characteristics of the tested amendments.

Table 1. Chemical characteristics of the tested amendments.

Chemical Characteristics	Composition				Unit
	Hermetia illucins Frass	*Tenebrio molitor* Frass	Vermicompost	Cattle Manure	
OM	8.02	7.65	7.87		%
C/N	4.02	3.84	20.20	20.00	-
pH	7.05	5.97	8.60	8.13	-
EC	15.19	6.47	2.07	1.33	(mS/cm)
N	1.18	1.16	0.23	0.62	%
K	0.38	0.35	0.38	0.61	%
Ca	3.13	2.44	46.28	261.30	Meq/100 g
Mg	6.30	66.36	22.86	84.70	Meq/100 g
Na	60.23	11.92	7.52	-	Meq/100 g
Zn	17.51	10.34	2.41	1.49	mg/Kg
Mn	1.87	1.87	1.87	2.68	mg/Kg
Fe	6.20	15.77	3.48	64.00	mg/Kg
Cu	0.92	1.05	0.16	0.20	mg/Kg

2.3. Experimental Design

The trial was conducted in a randomized complete block design with five replications. Each block was composed of 20 zucchini plants, planted in 8-L containers. The experimental unit corresponds to 4 plants with a spacing of 50 cm between the pots and 80 cm between the blocks.

2.4. Statistical Analysis

The analysis of variance (ANOVA) was performed with the Minitab 16 (Minitab Inc. State College, PA, USA). Tukey's test was adopted to compare the means between different treatments, at 5% significance level.

3. Results

Growth Attributes

Statistical analysis of data related to leaf area, leaf dry weight, number of fruits, and average fruit weight per plant, fresh root weight and plant height revealed that the differences among treatments were very highly significant. Comparison of the means by Tukey's test distinguished different groups. For leaf area *Tenebrio molitor* frass (FTM) followed by Cattle manure (Cm) showed good vegetative growth. FTM and Vermicompost (V) showed the best results for dry leaf biomass with means of 42.26 and 40.60 g/plant, respectively, followed by Cm and *Hermetia illucins* frass (FBSF) which represented 36.79 and 36.48 g/plant respectively and finally, the control (T) with 30.94 g/plant. The number of fruits per plant was maximum for the FTM treatment, followed by V, Cm, FBSF and T with average values of 2.95, 2.65, 2.5, 2.45 and 2 fruits/plant respectively. For the fruits weight, two homogeneous groups were observed, the FTM and V treatments were represented by the first group and the Cm, FBSF and T treatments were included in the second group. The FTM amendment resulted in the highest fresh weight of the zucchini plant roots, which was around 34.45 g. The V, Cm and FTM amendments resulted in average values of 33.95; 30.49; and 27.47 and 24.05 g/plant respectively. The highest value of final plant height of zucchini was recorded in the plants amended by FTM followed by V, Cm, FBSF and finally by T with values of 113.4, 111, 106.6, 102.6 and 98.4 cm respectively. Table 2 summarizes the effect of the different organic amendments on different agronomic parameters.

Table 2. Effect of different organic amendments on growth attributes. The means followed by the same letter are not statistically different at $p < 5\%$ according to the Tukey test.

	Number of Fruits per Plant	Plant Height (cm)	Dry Weight of Leaves (g)	Leaf Area (cm^2)	Fresh Root Weight (g)	Average Fruit Weight (g)
Control (T)	2.45 (b)	102.7 (e)	30.94 (c)	16,956.75 (d)	24.06 (d)	292.55 (b)
Cattle manure (Cm)	2.95 (a)	106.6 (c)	36.79 (b)	18,839.46 (b)	27.48 (c)	299.47 (b)
Vermicompost (V)	2.65 (ab)	111.0 (b)	40.61 (a)	19,541.45 (a)	33.95 (a)	320.42 (a)
Hermetia illucins frass (FBSF)	2.50 (b)	106.1 (d)	36.48 (b)	18,049.94 (c)	30.50 (b)	298.42 (b)
Tenebrio molitor frass (FTM)	2.00 (c)	114.4 (a)	42.26 (a)	19,905.52 (a)	34.45 (a)	323.42 (a)

4. Discussion

The obtained results show that the substrates amended by FTM recorded the highest values in terms of plant height, leaf area and fresh and dry weight of the vegetative part. This can be explained by the availability of mineral elements easily assimilated by the plant in the substrates amended by the insect's frass, as well as by the heterogeneity of the quantities of fertilizing elements in the soil. This could also be explained by the rate of absorption of mineral salts by the roots and by the quantities of nitrogen and phosphorus consumed, which stimulate the synthesis of proteins and consequently the accumulation of dry matter as observed under the insect's frass and vermicompost treatments. While for the control treatment (100% sand), the low values of the growth parameters recorded could be explained by the use of plant assimilates in fruit enlargement rather than in vegetative growth. Indeed, similar results to the effect of FTM were found by [5] who revealed that FTM may have some success in improving lettuce yield. Fresh weight, length of aerial part, basal stem width, and chlorophyll content of leaves all indicated significant increases with Molitor's tapeworm taper application compared to the negative control. Also, Przemieniecki et al. [7] found that soil fertilization with meal from *Tenebrio larvae* increased fresh and dry weight of biomass compared to mineral nitrogen fertilizer. It is noted that excessive Na^+ uptake can cause evident leaf damage, reduction in leaf number, leaf area recession and reduction in fresh and dry weight [8], which may explain the reduction in leaf number and fresh and dry weight of the black soldier treatment, since it has a higher Na^+ content. Graifenberg et al. [9] reported that in response to salinity there is a reduction in dry weight mainly in the aerial part of the zucchini plant and that the ratio of the dry weight of the aerial part to the dry weight of the root part decreases under salt stress conditions, which means that the dry weight of the roots is less affected by salinity. The positive results obtained by vermicompost amendment in terms of number of fruits per plant and average fruit weight of zucchini are confirmed by [10] who showed that vermicompost had significant effects on fruit weight.

5. Conclusions

Hermetia illucens frass is very saline due to the high content of mineral salts. This negatively influenced the growth and development of zucchini plants with a low yield compared to plants fertilized with *Tenebrio molitor* frass which recorded a high yield. The two amendments that provided the best vegetative growth of the zucchini plants were the *Tenebrio molitor* frass and vermicompost. These amendments showed adequate performance in terms of veg-etative growth parameters and fruit size. The insect frass-based organic amendments (e.g., *Tenebrio molitor* frass) improved the yield of zucchini plants by 62% compared to control, suggesting that these products could be potential alternatives to chemical fertilizers.

Author Contributions: Conceptualization, J.Z. and A.A.; methodology, M.H.E.O.; software, S.E.M.; validation, A.N. and A.H.; formal analysis, A.A.; resources, J.Z.; data curation, A.A.; writing—original draft preparation, J.Z., A.A. and S.E.M.; writing—review and editing, A.O., K.A. and A.N.; supervision, J.Z. and M.H.E.O.; project administration, A.O. All authors have read and agreed to the published version of the manuscript.

Funding: This research received no external funding.

Conflicts of Interest: The authors declare no conflict of interest.

References

1. Crist, E.; Mora, C.; Engelman, R. The interaction of human population, food production, and biodiversity protection. *Science* **2017**, *356*, 260–264. [CrossRef] [PubMed]
2. Chia, S.Y.; Tanga, C.M.; van Loon, J.J.; Dicke, M. Insects for sustainable animal feed: Inclusive business models involving smallholder farmers. *Curr. Opin. Environ. Sustain.* **2019**, *41*, 23–30. [CrossRef]
3. Chavez, M.; Uchanski, M. Insect left-over substrate as plant fertilizer. *J. Insects Food Feed* **2021**, *7*, 683–694. [CrossRef]
4. Poveda, J.; Jiménez-Gómez, A.; Saati-Santamaría, Z.; Usategui-Martín, R.; Rivas, R.; García-Fraile, P. Mealworm frass as a potential biofertilizer and abiotic stress tolerance-inductor in plants. *Appl. Soil Ecol.* **2019**, *142*, 110–122. [CrossRef]
5. Schmitt, E.; de Vries, W. Potential benefits of using *Hermetia illucens frass* as a soil amendment on food production and for environmental impact reduction. *Curr. Opin. Green Sustain. Chem.* **2020**, *25*, 100335. [CrossRef]
6. Temple, W.D.; Radley, R.; Baker-French, J.; Richardson, F. *Use of Enterra Natural Fertilizer (Black Soldier Fly Larvae Digestate) as a Soil Amendment*; Enterra Feed Corporation: Langley City, BC, Canada, 2013.
7. Przemieniecki, S.W.; Kosewska, A.; Purwin, C.; Zapałowska, A.; Mastalerz, J.; Kotlarz, K.; Kolaczek, K. Biometric, chemical, and microbiological evaluation of common wheat (*Triticum aestivum* L.) seedlings fertilized with mealworm (*Tenebrio molitor* L.) larvae meal. *Appl. Soil Ecol.* **2021**, *167*, 104037. [CrossRef]
8. West, D.; Taylor, J. Response of six grape cultivars to the combined effects of high salinity and rootzone waterlogging. *J. Am. Soc.* **1984**, *109*, 844–851.
9. Graifenberg, A.; Botrini, L.; Giustiniani, L.; Lipucci Di Paola, M. Yield, growth and element content of zucchini squash grown under saline-sodic conditions. *J. Hortic. Sci.* **1996**, *71*, 305–311. [CrossRef]
10. Razzaghifard, S.; Gholipouri, A.; Tobeh, A.; Meshkini, S. Effect of mycorrhiza, vermicompost and nanofertilizer on quantitative and qualitative characteristics of *Cucurbita pepo* L. *Eur. J. Hortic. Sci.* **2017**, *82*, 105–114. [CrossRef]

environmental sciences proceedings

MDPI

Proceeding Paper

Effect of Humic Acid on Soil Properties and Productivity of Maize Irrigated with Saline Water †

Hassan Dergam * and Omar Abdulrazzak

The Arab Center for the Studies of Arid Zones and Dry Lands (ACSAD), Damascus 96311, Syrian Arab Republic; omar354691@gmail.com

* Correspondence: hassandergam@hotmail.com

† Presented at the 2nd International Laayoune Forum on Biosaline Agriculture, 14–16 June 2022; Available online: https://lafoba2.sciforum.net/.

Abstract: The aim of this search was to understand the effect of humic acid on soil properties and the growth and salinity tolerance of maize irrigated with saline water. The experiment was conducted in the Sabboura Research Station, Syria in 2019–2020. As treatments, humic acid was added to the soil at concentrations of 0, 1, and 2 g/L·m^2, and the crop was irrigated using saline water with 0.5, 2, 4, and 6 ds/m salinity, measured by electrical conductivity (EC). The results showed that treatment with 2 ds/m irrigation water salinity and 2 g/L·m^2 of humic acid achieved the highest fresh biomass production and plant height compared to the control and other treatments, while 4 ds/m irrigation water salinity and 0 g/L·m^2 of humic acid gave the highest productivity measured by dry mass. In addition, the pH value increased up to 9.15 for the treatment with 6 ds/m irrigation water salinity and 2 g/L·m^2 of humic acid, while the EC values increased with an increase in salinity in the irrigation water, and the highest value (2.59 ds/m) occurred in the treatment with 6 ds/m irrigation water salinity and 2 g/L·m^2 of humic acid. The use of highly saline irrigation water (6 ds/m) led to an increase in EC and caused a noticeable decrease in plant height and fresh biomass accumulation in all treatments, indicating that maize cannot tolerate more than 6 ds/m irrigation water salinity.

Keywords: salinity; humic acid; maize; saline water; Sabboura

Citation: Dergam, H.; Abdulrazzak, O. Effect of Humic Acid on Soil Properties and Productivity of Maize Irrigated with Saline Water. *Environ. Sci. Proc.* **2022**, *16*, 32. https://doi.org/10.3390/environsciproc2022016032

Academic Editor: Abdelaziz Hirich

Published: 16 June 2022

1. Introduction

The use of irrigation water with salinity below a threshold value is the key to meeting agricultural needs with economic and environmental sustainability [1]. Especially in the context of climate change, leading to increasing water demand for several purposes, freshwater needs to be used judiciously to preserve the soil, water, and climate [2–5]. At the same time, soil health is a key issue to consider for the long-term sustainability of crop production, especially in arid and saline drylands. The addition of organic matter or its extracts to the soil has a very important effect on the various physical, chemical, and biological properties of the soil, as it is an important source of many nutrients and an essential source of energy for soil organisms [5–9]. Some organic extracts (chelates) such as humic acid have a significant role in improving plant growth and increasing the efficiency of roots in absorbing water and dissolved nutrients, increasing the plant's ability to tolerate salt stress, and improving various physical, chemical, and biological properties of the soil. Thus, this experiment was conducted to study the effect of humic acid with saline irrigation water application on soil properties, to determine their effect on the growth and productivity of maize, and to understand the maize plant's tolerance for salinity.

2. Materials and Methods

The experiment was carried out in pots at ACSAD Research Center in Al-Saboura, Syria using a factorial randomized complete block design. Two factors, i.e., three levels of irrigation water salinities, A_0 = 0.5 ds/m, A_1 = 2 ds/m, A_2 = 4 ds/m, and A_3 = 6 ds/m, and three rates of humic acid, B_0 = 0 g/L·m^2, B_1 = 1 g/L·m^2, and B_2 = 2 g/L·m^2 were applied as treatments. The data were analyzed as a factorial experiment and the significant differences were estimated using the F-test. The treatments' mean separation was determined using the LSD test at the 5% level of significance for both main and interaction (A*B) effects.

3. Results and Discussion

Table 1 shows the measured results for the fresh weight, dry weight, and plant height. We found the superiority of all treatments with humic acid over the control treatment (No. 1) in producing fresh weight, whereas treatment No. 6 resulted in the highest production of vegetative mass.

Table 1. Fresh and dry weight and plant height as affected by the application at different rates.

Treatment No.	Irrigation Water Salinity (ds/m)	Humic Acid Rates (g/L·m^2)	Fresh Weight (g)	Increase Over Control %	Dry Weight (g)	Increase Over Control %	Plant Height (cm)	Increase Over Control %
1	0.5	0	184	100	43.18	100	42.25	100
2	0.5	1	305.66	166.	77.29	178.99	70.5	166.86
3	0.5	2	269.33	146.37	74.43	172.37	63.16	149.49
4	2	0	308.	167.88	98.36	227.79	65.75	155.62
5	2	1	314.33	170.8	86.84	201.11	67.5	159.76
6	2	2	335.6	182.42	98.24	227.51	74	175.14
7	4	0	296.6	161.2	111.77	258.84	62	146.74
8	4	1	251	136.41	68.74	159.19	58.75	139.05
9	4	2	232.	126.3	57.7	133.76	49	115.97
10	6	0	240	130.43	49.89	115.53	60.75	143.78
11	6	1	275.	149.45	105.88	245.20	43.25	102.36
12	6	2	201	109.23	41.40	95.87	42.55	100.7
LSD (5%) (A × B)			12.3		4.5		3.6	

This was followed by the two treatments No. 5 and No. 4. Regarding the dry weights of plants, we note that the highest dry weight was recorded for plants with treatment No. 7, which amounted to 111.77 g per pot, compared to the control whose average dry weight did not exceed 43.18 g. Likewise, with regard to the average plant height, we find that treatment No. 6 resulted in the highest average height of plants (74 cm), with an increase over the height of the control plants of 75.14%. Humic acid has an important role in improving the physical, chemical, and biological properties of the soil as a store for many necessary nutrients for plant growth and has a role in maintaining the buffering capacity and improving the strength of plant growth [2,5,10,11].

In general, the differences were very significant between most of the treatments and the control, while we note that the use of water with high salinity (6 ds/m) led to a significant decrease in both the wet weight of the product and the height of the plant. The reason for this decrease is the negative impact of salts on plant growth, as well as the shortage of water absorbed as a result of osmotic stress [8–12].

Table 2 shows a decrease in the value of electrical conductivity with the use of non-saline water (0.5 ds/m) and an increase for the two treatments, i.e., 4 and 6 ds/m. The maximum value reached was in treatment No. 11, with a score of 2.597, compared to the control with a score of 0.255. Here, we note that in the treatment irrigated by water (0.5 ds/m), the value of the electrical conductivity decreased with an increase in the amount of humic acid used, with a value of 0.199 for treatment No. 3, compared to a value of 0.255 for treatment No. 1. The decreasing EC values are suspected to occur because humic acid exchanges H^+ ions with soil Na^+, so that the Na^+ content decreases and the H^+ levels increase, causing the soil EC to decrease [2,5,10]. From the table, we also note fluctuations in the values of organic matter, which decreased significantly in the humic acid treatments, regardless of the saline concentration of the irrigation water. [5–8]. We can also see that in the treatment irrigated by water (6 ds/m), the amount of organic matter remained close to the control in treatment No. 1, which is evidence of the effect of a high salt concentration on the activity of microorganisms, slowing down the decomposition of the organic matter. Regarding the changes in the concentration of elements N, P, and K, they were limited and to some extent correlated with the rate of organic decomposition and plant production.

Table 2. Soil chemical properties as affected by the application of different rates of humic acid and irrigation water salinity.

Treatment No	Irrigation Water Salinity (ds/m)	Humic Acid Rates (g/L·m^2)	pH (kcl) (1/5)	EC (ds/m) (1/5)	C-org (%)	OM (%)	N-No3 (mg/kg)	N (%) Total	P (mg/kg)	K (mg/kg)
1	0.5	0	8.50	0.255	0.508	0.876	11.50	0.057	7.85	314
2	0.5	1	8.59	0.214	0.536	0.924	5.13	0.082	7.60	312
3	0.5	2	8.37	0.199	0.438	0.755	5.70	0.053	6.03	308
4	2	0	8.80	0.537	0.464	0.800	3.26	0.055	7.73	297
5	2	1	8.76	0.675	0.431	0.743	5.03	0.063	5.90	288
6	2	2	8.83	0.858	0.417	0.719	5.13	0.068	5.90	306
7	4	0	8.78	1.597	0.483	0.833	5.96	0.059	5.56	285
8	4	1	8.88	1.861	0.457	0.788	7.10	0.083	6.40	311
9	4	2	8.95	1.258	0.411	0.709	14.7	0.069	5.90	322
10	6	0	9.04	1.989	0.588	1.014	14.8	0.070	7.80	355
11	6	1	8.94	2.597	0.582	1.003	13.4	0.074	7.80	386
12	6	2	9.15	1.758	0.598	1.031	11.35	0.075	6.40	387
LSD (5%) (A × B)			0.23	0.23	n.s.	0.2	1.6	n.s.	0.3	6.5

4. Conclusions

In conclusion, the results indicate that humic acid applications can improve growth, fresh and dry biomass accumulation, and plant height. A slight decrease in the EC value with an increase in the amount of humic acid was recorded. The amount of soil organic matter remained close to the control in treatments irrigated with highly saline water. This experiment shows the importance of using organic fertilizers in improving soil properties and increasing the tolerance of maize to salt stress when saline water is used for irrigation.

Author Contributions: Conceptualization, O.A. and H.D.; methodology, O.A. and H.D.; formal analysis, O.A. and H.D.; writing—original draft, O.A. and H.D.; writing and editing, H.D. and O.A.; funding acquisition, O.A. All authors have read and agreed to the published version of the manuscript.

Funding: This research received no external funding.

Institutional Review Board Statement: Not applicable.

Informed Consent Statement: Not applicable.

Conflicts of Interest: The authors declare no conflict of interest.

References

1. Devkota, K.P.; Devkota, M.; Rezaei, M.; Oosterbaan, R. Managing salinity for sustainable agricultural production in salt-affected soils of irrigated drylands. *Agric. Syst.* **2022**, *198*, 103390. [CrossRef]
2. Hamdani, F.M. Interaction between Salinity of Irrigation Water and Phosphate Fertilizer and its Relation with Some Chemical Soil Properties and Wheat Plant Extract. Ph.D. Thesis, University of Baghdad, Baghdad, Iraq, 2000.
3. Shukri, H.M. Effect of Alternating and Mixed Saline Water on Wheat Growth and Salt Accumulation in Soil. Ph.D. Thesis, University of Baghdad, Baghdad, Iraq, 2002.
4. Atti, A.S.; Naser, K.M.; Albandawy, B.B. Water Use Efficiency, Fruit Yield and Quality of Squash Cucurbita pepo L. with Different Irrigation Quantities and Humic Acid Fertilization in a Semi-arid Agricultural Area. *Anbar J. Agric. Sci.* **2017**, *15*, 362–370.
5. Tan, K.H. *Humic Matter in Soil and the Environment: Principles and Controversies*; CRC Press Taylor & Francis Group: Boca Raton, FL, USA, 2014.
6. Al-Khafaji, M. Effect of Concentration and spraying date of Humic acid on Growth and yield of Zea mays L. *Kufa J. Agric. Sci.* **2015**, *7*, 156–170.
7. Abd El-Monem, E.A.A.; Saleh, M.M.S.; Mostafa, E.A.M. Minimizing the quantity of mineral nitrogen fertilizers on grapevine by using humic acid, organic and biofertilizers. *Res. J. Agric. Biol. Sci* **2008**, *4*, 46–50.
8. Jasim, A.H.; Alryahii, I.; Abed, H.M.; Badry, A.N. Estimation Effect of some treatments on alleviating of environmental stress on growth and yield of squash (cucurbittapepo L.). *Mesop. Environ. J.* **2015**, *1*, 67–74.
9. Ragab, A.A.M.; Hellal, F.A.; Abd El-Hady, M. Water salinity impacts on some soil properties and nutrients uptake by wheat plants in sandy and calcareous soil. *Aust. J. Basic Appl. Sci.* **2008**, *2*, 225–233.
10. Al-Alawi, H.H.M. Effect of irrigation water source and nitrogen on growth of millet panicum and some soil properties. Master's Thesis, University of Baghdad, Baghdad, Iraq, 2003.
11. Mindari, W.; Sasongko, P.E.; Kusuma, Z.; Syekhfani, S.; Aini, N. Efficiency of various sources and doses of humic acid on physical and chemical properties of saline soil and growth and yield of rice. *J. Degrad. Min. Land Manag.* **2014**, *2*, 259–268.
12. Barker, A.V.; Pilbeam, J. *Hand Book of Plant Nutrition*; CRC Press Taylor & Francis Group: Boca Raton, FL, USA, 2008; pp. 123–164.

environmental sciences proceedings

MDPI

Proceeding Paper

Reclamation of Sodic Soils and Improvement of Corn Seed Germination Using Spent Grains, Cheese Whey, Gypsum, and Compost †

Emad Aboukila [1,*] and Abdelaziz Nilahyane [2]

1 Department of Natural Resources and Agricultural Engineering, College of Agriculture, Damanhour University, Damanhour 22511, Egypt
2 African Sustainable Agriculture Research Institute, Mohammed VI Polytechnic University, Laayoune 43150, Morocco; abdelaziz.nilahyane@um6p.ma
* Correspondence: emad@agr.dmu.edu.eg
† Presented at the 2nd International Laayoune Forum on Biosaline Agriculture, 14–16 June 2022; Available online: https://lafoba2.sciforum.net/.

Abstract: Incubation and germination experiments were carried out to evaluate spent grain, cheese whey, gypsum, and compost for reclamation of sodic soils and enhancing corn (*Zea mays* L.) germination. Results indicated that all organic amendments effectively reduced exchangeable sodium percent (ESP), sodium adsorption ratio (SAR), and soil pH, while they enhanced soil organic matter, macronutrients, and corn germination percentages compared with gypsum and control. The positive impacts of all amendments followed the arrangement: spent grain > cheese whey > compost > gypsum > control. Moreover, one-month incubation was enough time for amendments to mitigate soil sodicity before crop plantation.

Keywords: sodic soils; industrial by-products; cheese whey; organic amendments; *Zea mays* L.

Citation: Aboukila, E.; Nilahyane, A. Reclamation of Sodic Soils and Improvement of Corn Seed Germination Using Spent Grains, Cheese Whey, Gypsum, and Compost. *Environ. Sci. Proc.* **2022**, *16*, 36. https://doi.org/10.3390/environsciproc2022016036

Academic Editor: Abdelaziz Hirich

Published: 16 June 2022

1. Introduction

Globally, more than 60% of salt-affected soils are classified as sodic soils [1]. Sodic soils reclamation is generally more expensive compared with saline–sodic or saline soils [2]. However, sodic soil could be remediated by incorporation of organic matter or chemical amendments. Because of the lack of organic farm waste in many regions around the world, industrial organic by-products, such as brewer's spent grain (SG) and cheese whey, could be used as alternatives to conventional organic fertilizers. SG is a byproduct of beer production, constituting almost 85% of the total byproducts produced matters [3]. Cheese whey is a liquid byproduct of cheese and cottage cheese manufacture. For each kg of cheese produced, 9 kg of cheese whey is generated [4].

Spent grain and cheese whey disposal are often environmental issues. The properties of SG and cheese whey make them ideal candidates for recycle in agriculture. Developing this practice may mitigate the economic and environmental issues of degraded soil and excess wastes. The specific objectives of this research were to assess the capacity of SG, cheese whey, compost, and gypsum to reduce soil sodicity, improve soil sequestration of macronutrients, and enhance maize growth on sodic soils.

2. Materials and Methods

Sodic soils used for the experiment (*Typic Torrifluvents*) with clay loam texture were collected from the region of Abees, Alexandria, Egypt. Four types of amendments were applied in this study and included spent grain, cheese whey, compost, and gypsum. The characterizations of the SG, cheese whey and compost were calculated using methods outlined in ref. [5]. The amendments characterizations are shown in Table 1.

Table 1. Characterization of organic amendments used in this study (oven dry weight basis).

Analyte	Compost	Spent Grain	Cheese Whey
pH (1:5 w:w)	6.95	4.16	3.82
EC (dS m^{-1}, 1:5 w:w)	5.91	1.45	5.12
Organic Matter (%)	42.5	75	3.15
Total N (%)	2.26	6.12	2.63
Total P (%)	1.13	1.86	0.14
Total K (%)	0.59	2.74	0.33
Organic carbon (%)	24.65	43.5	1.83
C:N ratio	10.91	7.1	0.69
Moisture (%)	23	75	96
Fe (mg kg^{-1})	930	1130	nd [1]
Zn (mg kg^{-1})	215	368	nd
Mn (mg kg^{-1})	98	210	nd
Cu (mg kg^{-1})	67	98	nd
B (mg kg^{-1})	23	39	nd

[1] Not determined.

2.1. Incubation Experiment

Seven treatments were applied in this research and included 2 levels of cheese whey (W1 and W2), 2 levels of compost (C1 and C2), gypsum (G), spent grain (SG), and control (Ctrl). The application rates of cheese whey were 1% and 2% dry matter content. The base level of spent grain (SG) and compost (C1) was the quantity of spent grain or compost required to increase soil organic matter (SOM) by 1%. Compost was also added twice the base level (C2). Gypsum was added at equivalent rate of gypsum requirement to reduce ESP to 10%. Amendments were incorporated with 5 kg of sodic soil and packed in polyethylene pots. All pot treatments were incubated under natural field conditions, without plants, for a month. After incubation, sub-samples were analyzed for selected chemical properties (Table 2).

Table 2. Change in EC_e (ds m^{-1}), soil pH, SAR, ESP (%), OM (%), soil available N, P, K (mg kg^{-1}), and biomass (g pot^{-1}) after soil incubation with various treatments. Same letters within columns indicate no significant differences ($p < 0.05$, Tukey's test).

	EC_e	pH	SAR	ESP	OM	N	P	K	Biomass
Initial	3.91	8.75	26.9	31.0	1.66	78.5	4.45	352	-
Ctrl	3.55a	8.60a	20.1a	23.7a	1.57d	75.8c	4.62e	341c	2.25d
W1	3.28ab	7.86de	9.2b	11.0b	1.90cd	115.7c	5.33de	426bc	5.09b
W2	3.23b	7.71e	8.2b	9.5b	2.32abc	125.0bc	5.96cd	554ab	4.81bc
C1	2.48c	8.07c	11.0b	12.3b	1.98bcd	183.2d	6.64c	415bc	2.92d
C2	2.41c	7.92cd	10.3b	11.9b	2.52ad	288.3a	8.80b	484bc	3.32cd
G	2.59c	8.32b	11.0b	12.4b	1.47d	87.7c	4.85de	369c	2.28d
SG	2.70bc	7.48f	9.1b	10.9b	2.80a	325.0a	10.85a	669a	7.03a

2.2. Germination Experiment

After incubation, the pots were sown with 4 corn seeds (*Zea mays* L.). The germination experiment continued for 15 days. At the end of the germination experiment, the germination percentage was determined, and then dry weight was recorded.

3. Results and Discussion

3.1. Incubation Experiment

All treatments produced significant decreases in soil EC_e compared with the respective initial values (Table 2). This reduction of salt concentration is attributed to the high leaching of solute in the treated soil. These reductions in soil EC_e can be illustrated by the mobilization of Ca via increasing the dissolution of soil calcite, which exchanges Na at cation exchange complex and forms Na_2SO_4, $MgSO_4$, and other high solubility salts. Moreover, these reactions promote water infiltration, soil flocculation, and stability [6]. As a result, the majority of these soluble salts leached with the drainage water. Post incubation soil analysis showed that soil pH was significantly decreased in all treatments compared with control (Table 2). Soil pH reduced to less than 8.5, the critical value of pH for salt-affected soil. Spent grain and cheese whey are acidic. Thus, they might be preferably used than compost as organic amendments for decreasing soil pH and reclamation of sodic soils. It is likely that incubation of organic treatments probably enhanced the partial pressure of CO_2 because of increases of the microbial activity. This possibly caused by the formation of organic and inorganic acids, which lead to decreasing pH in organic treated soils [7]. Furthermore, solubilization of minerals such as Ca and Mg [8] because of microbial activity assists the decrease of pH in sodic soils by exchanging with Na from the cation exchange complex [9].

The SAR levels reduced by 25.3%, 65.7%, 69.4%, 59.2%, 61.5%, 59.0%, and 66.1% for Ctrl, W1, W2, C1, C2, G, and SG, respectively. The releasing of Ca and Mg by organic treatments increased their contents in soil solution and facilitated the exchange with exchangeable Na on the cation exchange complex and release Na to soil solution. Thus, it helped Na loss by leaching. As a result, soil SAR reduced. Furthermore, extra Ca release from gypsum promoted the exchangeable Na release to soil solution, and consequently enhanced additional decreases of soil SAR [10]. All treatments, except for control, were effective in decreasing the soil ESP to less than 15% (Table 2). The significant decrease in the exchangeable Na levels led to the reduction in the soil ESP. Organic amendments and gypsum were statistically equally successful in decreasing the soil ESP. This reduction could be attributed to the releasing of Ca from gypsum, organic amendments, and the solubilization of soil calcite by organic amendments that probably promote the Na–Ca exchange rate among the cation exchange complex and soil solution.

All organic treatments produced significant increases in SOM. The greatest increases were recorded with the SG and C2 treatments compared with control (Table 2). It is noted that SOM in gypsum treated soil decreased by 6.6%. The size of the increase in SOM tends to correlate with the quantities of organic matter applied by each amendment. Comparable results were observed by ref. [11] who reported a significant level of SOM in calcareous soil treated with SG or compost compared with the untreated ones. Similarly, ref. [12] stated that increasing application rate of mozzarella cheese whey led to increasing SOM in both clay and calcareous soil. Soil-available N, P, and K concentration were increased using all amendments compared with the Ctrl. Therefore, all of amendments were efficient at contributing soil available N, P, and K to sodic soil (Table 2). The SG significantly produced larger levels of N, P, and K compared with other organic amendments. Mineralization of organic materials present in the added organic treatments leads to increases in the soil available N, P, and K. Furthermore, organic amendments could be added before sowing to have enough time for the mineralization processes of organic compounds, thus increasing the availability of plant nutrients. Comparable conclusions were reported by ref. [12].

3.2. Germination Experiment

No significant differences in the final germination percent (FGP) were found between treatments in sodic soil. However, there were significant differences in biomass yield among the treatments (Table 2). Spent grain (SG) was the most effective one in enhancing seed germination in sodic soil. The biomass yield of SG significantly increased to 312% in comparison with untreated soil (Ctrl). This was due to the superior effect of the spent grain (SG) in ameliorating soil sodicity by decreasing soil pH, exchangeable Na, soluble Na, ESP, and SAR, and increasing N, P, and K. Gypsum treated soil was a less effective amendment and produced poor seed germination as control. This may be due to the fact that the gypsum was a less effective treatment in decreasing soil pH. Moreover, it does not support sodic soil with any remarkable amount of OM, N, P, or K, compared with organic treatments.

4. Conclusions

All organic amendments examined in the present study were efficient at remediating sodic soil properties and improving corn seed germination. The commonly used amendment gypsum was less effective than organic amendments in ameliorating sodicity and improved corn seed germination in sodic soils. Hence, use of such industrial organic wastes as spent grain and cheese whey in sodic soil reclamation provides an environmentally friendly and economic practice of disposal and consequently enhancing soil quality and crop yield. The addition of spent grain or cheese whey to sodic soil was confirmed to be a very efficient method of decreasing pH, ESP, and SAR and increasing organic matter, nitrogen, phosphorous, potassium, and corn yield. Increasing the application rates of cheese whey and compost did not significantly enhance biomass yield or reduce sodicity compared to lower application rates. The economics of sodic soil reclamation require a low-cost method for successful implementation. Spent grain and cheese whey avoid the cost of composting. Thus, they are more economical amendments than compost.

Author Contributions: Conceptualization, E.A.; methodology, E.A.; validation, E.A. and A.N.; writing—review and editing, E.A. and A.N. All authors have read and agreed to the published version of the manuscript.

Funding: This research received no external funding.

Institutional Review Board Statement: Not applicable.

Acknowledgments: The authors thank college of agriculture, Damanhour University, Egypt for funding this research.

Conflicts of Interest: The authors declare no conflict of interest.

References

1. Choudhary, O.P. Use of Amendments in Ameliorating Soil and Water Sodicity. In *Bioremediation of Salt Affected Soils: An Indian Perspective*; Arora, S., Singh, A., Singh, Y., Eds.; Springer: Cham, Swizerland, 2017; pp. 195–210.
2. Seelig, B.D.; Richardson, J.L.; Knighton, R.E. Comparison of statistical and standard techniques to classify and delineate sodic soils. *Soil Sci. Soc. Am. J.* **1991**, *55*, 1042–1048. [CrossRef]
3. Mussatto, S.I.; Dragone, G.; Roberto, I.C. Brewers' spent grain: Generation, characteristics and potential applications. *J. Cereal Sci.* **2006**, *43*, 1–14. [CrossRef]
4. Robbins, C.W.; Lehrsch, G.A. Cheese whey as a soil conditioner. In *Handbook of Soil Conditioners: Substances That Enhance the Physical Properties of Soil*; Wallace, A., Terry, R.E., Eds.; Marcel Dekker Inc.: New York, NY, USA, 1998; pp. 167–186.
5. Kehres, B. *Methods Book for the Analysis of Compost*; Federal Compost Quality Assurance Organization (FCQAO); Bundesgutegemeinschaft Kompost e.V. (BGK): Koln-Gremberghoven, Germany, 2003.
6. Clark, R.B.; Ritchey, K.D.; Baligar, V.C. Benefits and constraints for use of FGD products on agricultural land. *Fuel* **2001**, *80*, 821–828. [CrossRef]
7. Wong, V.N.L.; Dalal, R.C.; Greene, R.S.B. Carbon dynamics of sodic and saline soils following gypsum and organic material additions: A laboratory incubation. *Appl. Soil Ecol.* **2009**, *41*, 29–40. [CrossRef]
8. Qadir, M.; Oster, J.; Schubert, S.; Noble, A.; Sahrawat, K. Phytoremediation of sodic and saline–sodic soils. *Adv. Agron.* **2007**, *96*, 197–247.
9. Chaganti, V.N.; Crohn, D.M. Evaluating the relative contribution of physiochemical and biological factors in ameliorating a saline-sodic soil amended with composts and biochar and leached with reclaimed water. *Geoderma* **2015**, *259–260*, 45–55. [CrossRef]
10. Chaganti, V.N.; Crohn, D.M.; Simunek, J. Leaching and reclamation of a biochar and compost amended saline–sodic soil with moderate SAR reclaimed water. *Agric. Water Manag.* **2015**, *158*, 255–265. [CrossRef]
11. Aboukila, E.F.; Nassar, I.N.; Rashad, M.; Hafez, M.; Norton, J.B. Reclamation of calcareous soil and improvement of squash growth using brewers' spent grain and compost. *J. Saudi Soc. Agric. Sci.* **2018**, *17*, 390–397. [CrossRef]
12. Aboukila, E.F.; Abdelraouf, E.A.; Gomma, I. Effects of cheese whey on some chemical and physical properties of calcareous and clay soils. *Int. J. Plant Soil Sci.* **2018**, *21*, 1–12. [CrossRef]

environmental sciences proceedings

Proceeding Paper

Autochthonous Biostimulants as a Promising Biological Tool to Promote Lettuce Growth and Development under Salinity Conditions †

Redouane Ouhaddou [1,2], Raja Ben-Laouane [1,2], Aiman Slimani [1,2,3], Abderrahim Boutasknit [1,2], Mohamed Anli [1,2], Khalid Oufdou [3], Marouane Baslam [4] and Abdelilah Meddich [1,2,*]

1 Center of Agrobiotechnology and Bioengineering, Research Unit Labelled National Center for Scientific Research and Technology (Centre AgroBiotech-URL-CNRST-05), Cadi Ayyad University, Marrakesh 40000, Morocco; ouhadou.redouan@gmail.com (R.O.); benlaouaneraja@gmail.com (R.B.-L.); slimani.aiman@gmail.com (A.S.); abderrahim.boutasknit@gmail.com (A.B.); moh1992anli@gmail.com (M.A.)
2 Laboratory of Agro-Food, Biotechnologies and Valorization of Plant Bioresources (AGROBIOVAL), Department of Biology, Faculty of Science Semlalia, Cadi Ayyad University (UCA), Marrakesh 40000, Morocco
3 Laboratory of Microbial Biotechnologies, Agrosciences, and Environment (BioMAgE), Labeled Research Unit-National Center for Scientific Research and Technology N°4, Faculty of Science Semlalia, Cadi Ayyad University (UCA), Marrakesh 40000, Morocco; oufdou@uca.ac.ma
4 Laboratory of Biochemistry, Faculty of Agriculture, Niigata University, Niigata 950-2181, Japan; mbaslam@gs.niigata-u.ac.jp
* Correspondence: a.meddich@uca.ma

† Presented at the 2nd International Laayoune Forum on Biosaline Agriculture, 14–16 June 2022; Available online: https://lafoba2.sciforum.net/.

Abstract: An adapted sustainable management program was used to evaluate lettuce tolerance to salt stress using autochthonous biostimulants (arbuscular mycorrhizal fungi (AMF), plant growth-promoting rhizobacteria (PGPR), and compost). Salinity harmed plant growth, root colonization, and physiology. However, biostimulants application, especially AMF and PGPR treatments, significantly improved lettuce growth and salinity tolerance (120% and 50%, respectively, for biomass; 60% and 20%, respectively, for stomatal conductance; and 1.5% and 1.3%, respectively, for chlorophyll fluorescence) compared to non-inoculated and compost-free controls under stressed conditions.

Keywords: salt stress; endogenous mycorrhiza; PGPR; compost; stress tolerance

Citation: Ouhaddou, R.; Ben-Laouane, R.; Slimani, A.; Boutasknit, A.; Anli, M.; Oufdou, K.; Baslam, M.; Meddich, A. Autochthonous Biostimulants as a Promising Biological Tool to Promote Lettuce Growth and Development under Salinity Conditions. *Environ. Sci. Proc.* **2022**, *16*, 41. https://doi.org/10.3390/environsciproc2022016041

Academic Editor: Abdelaziz Hirich

Published: 16 June 2022

1. Introduction

Salt-affected soils are constantly increasing worldwide, especially in arid and semi-arid areas. Salinity-related impacts include low agricultural productivity, low economic returns, and soil erosion; these threaten global food productivity and challenge the sustainability of cultivating crops under saline conditions. New directions towards natural biological resources (e.g., AMF, PGPR, and organic amendments) are promising and environmentally friendly strategies to improve the growth and development of plants under abiotic stress. Our objectives are the following: (1) mitigate the harmful effects of salinity by using biostimulants with indigenous plant-associated microbiomes and compost; and (2) understand the mechanisms underlying plant-microbe-compost interactions that confer salt stress tolerance.

2. Materials and Methods

2.1. Biological Materials and Cultivation Conditions

Four bacterial strains (Z1, Z2, Z4, and ER21) were used in this experiment, which were isolated from arid and semi-arid regions of southeast Marrakesh, Morocco. The bacterial suspension inoculation was carried out near the lettuce (*lactuca sativa* L.) roots. Similarly, the native fungal complex of AMF was isolated from the same region. This AMF consortium is based on a mixture of native species [1]. Inoculation of lettuce plants was performed by adding a quantity of the inoculum next to the root system of lettuce seedlings. The compost used underwent composting processes for 3 months, of which no other inorganic products were added. At the three leaves stage, lettuce seedlings were transplanted into plastic bags filled with sterilized soil. Plants were grown in a controlled greenhouse at 25.5 C (16/8 h light/dark) with supplemented light and average relative humidity of 68.5%.

2.2. Treatments and Study Design

The experimental design consisted of 12 treatments (10 replicates each): control, seedlings treated with AMF (M), PGPR (R), or compost (C) grown under 0, 50, and 100 mM NaCl. The salt stress was applied 15 days after transplantation. In order to avoid osmotic shock to the plant, NaCl doses were applied progressively.

2.3. Mycorrhization Parameters

After harvesting, root samples were washed with distilled water and treated according to the Phillips and Hayman method [2]. Mycorrhizal structures' rate of root infection was assessed by microscopic observation (ZEISS, Model Axioskop 40) according to the technique described by Trouvelot et al. [3]. The mycorrhization frequency (MF) and intensity (MI) were calculated using the following equations:

$$\text{Mycorhization Frequency MF } (\%) = \left(\frac{\text{Infected root segments}}{\text{Total root segments}}\right) \times 100$$

$$\text{Mycorhization Intensity MI } (\%) = \frac{(95\text{n}5 + 70\text{n}4 + 30\text{n}3 + \text{n}1)}{\text{Total root segments}}$$

2.4. Growth Assessment

The growth performance of lettuce plants was assessed by measuring the total dry matter (TDM; obtained after drying the samples at 80 °C until the weight remained constant).

2.5. Photosynthetic Efficiency and Gas Exchanges

Chlorophyll fluorescence was measured by a fluorometer (OPTISCIENCE, OS30p). Dark adaptation was made on the upper side of the leaf by obscuring for 30 min. This parameter was measured by transmission at 650 nm on a leaf area of 12.5 mm^2. The fluorescence signal was recorded at an acquisition speed of 10 ms for a second. Stomatal conductance (g_s) was measured using a porometer system (Leaf Porometer LP1989, Decagon Device, Inc., Washington, DC, USA).

3. Results

3.1. Mycorrhizal Symbiosis

Our results showed that no mycorrhizal structure was observed in the roots of non-treatment controls nor PGPR- and compost-treated plants. AMF infection frequency decreased significantly under 100 mM NaCl (Figure 1A). Mycorrhizal lettuce roots showed the lowest root colonization intensity under salt stress conditions (Figure 1B).

3.2. Growth Assessment, Photosynthetic Efficiency and Gas Exchanges

The application of biostimulants improved TDW compared to untreated control, independently of salt levels (Figure 2A). Total biomass was significantly affected by salt stress. The 100-millimolar NaCl treatment reduced this parameter substantially. R and M treatments improved the biomass under moderate salt stress (50 mM NaCl), while only M enhanced this trait under 100 mM NaCl compared to the control plants.

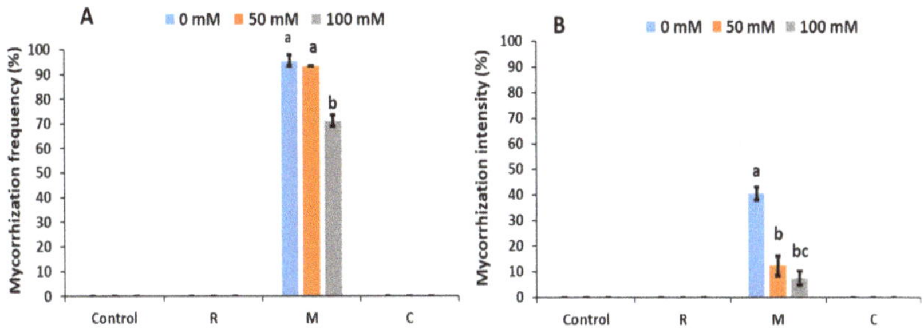

Figure 1. Influence of salinity levels (0, 50, and 100 mM NaCl) on (**A**) mycorrhization frequency and (**B**) intensity in lettuce control plants (non-amended, non-inoculated), and plants inoculated with plant growth promoting rhizobacteria (R), or arbuscular mycorrhizal fungi (M), or amended with composts (C). Data are mean ± SE of 10 biological replicates. Means followed by the same letters are not significantly different at $p < 0.05$ (Tukey's HSD).

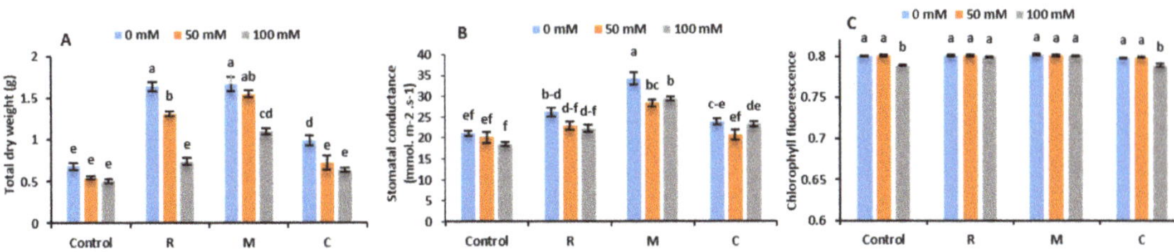

Figure 2. (**A**) total dry weight, (**B**) stomatal conductance and (**C**) chlorophyll fluorescence of lettuce plants grown without (0 mM NaCl) or with (50 and 100 mM NaCl) salt stress, and subjected to different biostimulant treatments. Control: untreated plants, R: inoculated with PGPR, M: inoculated with arbuscular mycorrhizal fungi, C: amended with compost. Data are mean ± SE of 10 biological replicates. Means followed by the same letters are not significantly different at $p < 0.05$ (Tukey's HSD).

The overall results revealed that salinity caused a significant decline in stomatal conductance (g_s) in all treatments and chlorophyll fluorescence (F_v/F_m), both in controls and compost-treated plants under 100 mM NaCl. Application of biostimulants, especially M and R, significantly improved g_s, independently of the salinity (Figure 2B). Under 100 mM NaCl, plants inoculated with R or M yielded an improvement in F_v/F_m compared to untreated plants (Figure 2C).

4. Discussion

The ability of plants to tolerate salinity stress is usually evaluated in terms of biomass produced, and several studies have highlighted that biofertilizers impart salinity tolerance in host plants by virtue of higher biomass compared to untreated plants. Our results showed that M and R were more effective, under both normal and salt stress conditions, despite the reduction of AMF root colonization (Figure 1). This reduction might be due to the adverse effects of salinity on both the host plant and fungal growth, and/or on establishing arbuscular mycorrhiza. Our results align with the study of Santander et al. [4], which showed that mycorrhizal infection and intensity decrease when host plants are exposed to salt stress.

Native AMF or PGPR strains were very effective in helping lettuce attenuate salt stress' detrimental effects on growth and photosynthetic machinery. In the absence of biofertilizers, lettuce biomass production was negatively affected by increasing NaCl stress levels (Figure 2). This reduction in untreated plants is a result of both the osmotic phase during which growth inhibition is caused by the difficulty for the plant to absorb water, and an ionic phase due to the toxic effect of the salt within the plant, observed with the higher

Na^+ and Cl^-. Interestingly, we found that AMF, or PGPR, or organic amendments significantly promoted lettuce biomass under both control and saline conditions as compared to the untreated controls. This could be justified by the stimulation of phytohormones that accelerate cell division processes [5], improving nutrient acquisition (especially P) and ionic homeostasis, plant growth, photosynthesis machinery, osmoregulation, water status, alteration in root architecture, and/or protection against ROS-induced oxidative stress [6]. Salt-treated plants' lowered leaf area, coupled with a decrease in stomatal and mesophyll conductance, limits CO_2 availability and assimilation, which results ultimately in decreased CO_2 supply to RuBisCO, and/or accumulation of excess energy; this in turn leads to increased accumulation of electrons in the thylakoid membranes. Our data showed that AMF and PGPR bolster these mechanisms and alleviate the negative effects of salinity on plant photosynthetic capacity by increasing the g_s and net photosynthetic rate, resulting in more quantum yield (F_v/F_m) compared to untreated control plants.

5. Conclusions

Inoculation with selected AMF and PGPR, or amendment with organic compost, can improve the performance of lettuce under salt stress conditions. Data imply that AMF, PGPR, or compost deploy an array of biochemical and physiological mechanisms that act in a concerted manner to provide more salinity tolerance to the host plant. Altogether, the use of biostimulants is considered to be an efficient approach for bio-amelioration of salinity stress.

Author Contributions: A.M. and M.B. designed and supervised the research. R.O., R.B.-L. and A.S. performed the experiments and carried out the analyses. R.O., R.B.-L., M.A. and A.B. performed the data analysis, interpretation and contributed to analytic tools. A.M., M.B. and K.O. contributed to the conception and design of the work. R.O., R.B.-L. and M.A. wrote the manuscript. A.M., M.A. and A.B. revised and finalized the manuscript. All authors have read and agreed to the published version of the manuscript.

Funding: The present study was supported by the FOSC project (Sus-Agri-CC) from the European Union's Horizon 2020 research and innovation program under grant agreement N°862555 and Grant-in-Aid for Early-Career Scientists to MB (JSPS KAKENHI Grant Number 20K15425).

Institutional Review Board Statement: Not applicable.

Informed Consent Statement: Not applicable.

Data Availability Statement: Not applicable.

Conflicts of Interest: The authors declare no conflict of interest.

References

1. Meddich, A.; Jaiti, F.; Bourzik, W.; El Asli, A.; Hafidi, M. Use of mycorrhizal fungi as a strategy for improving the drought tolerance in date palm (Phoenix dactylifera). *Sci. Hortic.* **2015**, *192*, 468–474. [CrossRef]
2. Phillips, J.M.; Hayman, D.S. Improved procedures for clearing roots and staining parasitic and vesicular-arbuscular mycorrhizal fungi for rapid assessment of infection. *Trans. Br. Mycol. Soc.* **1970**, *55*, 158–161. [CrossRef]
3. Trouvelot, A.; Kough, J.L.; Gianinazzi-Pearson, V. Mesure du taux de mycorhization VA d'un système radiculaire. Recherche de méthodes d'estimation ayant une signification fonctionnelle. *Mycorhizes Physiol. Génétique* **1986**, 217–220. [CrossRef]
4. Santander, C.; Aroca, R.; Cartes, P.; Vidal, G.; Cornejo, P. Aquaporins and cation transporters are differentially regulated by two arbuscular mycorrhizal fungi strains in lettuce cultivars growing under salinity conditions. *Plant Physiol. Biochem.* **2021**, *158*, 396–409. [CrossRef]
5. Khalloufi, M.; Martínez-Andújar, C.; Lachaâl, M.; Karray-Bouraoui, N.; Pérez-Alfocea, F.; Albacete, A. The interaction between foliar GA3 application and arbuscular mycorrhizal fungi inoculation improves growth in salinized tomato (*Solanum lycopersicum* L.) plants by modifying the hormonal balance. *J. Plant Physiol.* **2017**, *214*, 134–144. [CrossRef] [PubMed]
6. Ben-Laouane, R.; Baslam, M.; Ait-El-mokhtar, M.; Anli, M.; Boutasknit, A.; Ait-Rahou, Y.; Toubali, S.; Mitsui, T.; Oufdou, K.; Wahbi, S.; et al. Potential of native arbuscular mycorrhizal fungi, rhizobia, and/or green compost as alfalfa (Medicago sativa) enhancers under salinity. *Microorganisms* **2020**, *8*, 1695. [CrossRef] [PubMed]

environmental sciences proceedings

MDPI

Proceeding Paper

Salt-Affected Soil Management Utilizing an Innovative Coated Sand Material—"Breathable Sand" †

Yuming Su [1,*], Shengyi Qin [2] and Chandra Dake [3]

1 State Key Laboratory of Silica Sand Utilization—US Office, College Park, ML 20742, USA
2 Beijing Rechsand Science & Technology Group Co., Ltd., Beijing 100083, China; qinsy@rechsand.com
3 Dake Rechsand LLC, Dubai P.O. Box 82071, United Arab Emirates; Chandra@dake.ae
* Correspondence: yumingsu@yahoo.com; Tel.: +1-(484)535-0664
† Presented at the 2nd International Laayoune Forum on Biosaline Agriculture, 14–16 June 2022; Available online: https://lafoba2.sciforum.net/.

Abstract: An innovative air-permeable watertight material (hereinafter "Breathable Sand") was invented and has been used in various agricultural applications. Proprietary coating and surface modification processes are applied on desert sand particles, and thus the air permeability can be retained, while the material is watertight due to its water repellency properties. Breathable Sand has been used as a liner in salt-affected fields in Inner Mongolia and Tianjin in China. After years of its initial application, no salt intrusion in the root zone was observed. In addition, studies in Zhejiang Province showed that about a 30% water saving can be achieved with 3 cm of Breathable Sand applied below the roots in rice fields, compared to conventional rice fields. Roots were stronger and the average leaf count, grain weight, grain per ear, matured grain count, and grain maturity rates were also higher.

Keywords: breathable sand; water conservation; saline/sodic soil; coated sand

check for updates

Citation: Su, Y.; Qin, S.; Dake, C. Salt-Affected Soil Management Utilizing an Innovative Coated Sand Material—"Breathable Sand". *Environ. Sci. Proc.* **2022**, *16*, 49. https://doi.org/10.3390/environsciproc2022016049

Academic Editor: Abdelaziz Hirich

Published: 16 June 2022

1. Introduction

Worldwide, over 1100 million hectares of soils are affected by salinity and sodicity, including saline, sodic, and saline-sodic affected lands [1]. In China, according to the results of the Second National Soil Survey, there are more than 33 million hectares of salt-affected land, of which about 13 million hectares have the potential for agricultural development [2].

Modern coated sand has been studied and used in various fields, including some agricultural applications. Some studies showed that specially coated sand with surface modifications, when applied under plant root zones, can significantly reduce water consumption in arid land [3]. However, applications are sporadic.

In this study, aeolian sand was treated and a new engineered sand material was invented and developed. State-of-the-art treatment processes were developed, which allowed sand from the desert to meet various engineering needs. Thanks to the coating, surface roughening, and surface modification processes, a new air-permeable watertight sand material (hereinafter "Breathable Sand") was invented by Rechsand Science & Technology Group (hereinafter "Rechsand") [4]. It has been used in stormwater harvesting, building construction, agricultural soil modification, water treatment, and reverse desertification projects in China, including salt-affected soil management sites in Dengkou City, Inner Mongolia, and Tianjin City.

2. Materials

Aeolian sand (typical particle size D_{50} = 65–149 microns) was used as the core material, and a series of proprietary pre-treatments and surface modification processes were used to form at least two nanoscale layers on individual sand particles (Figure 1).

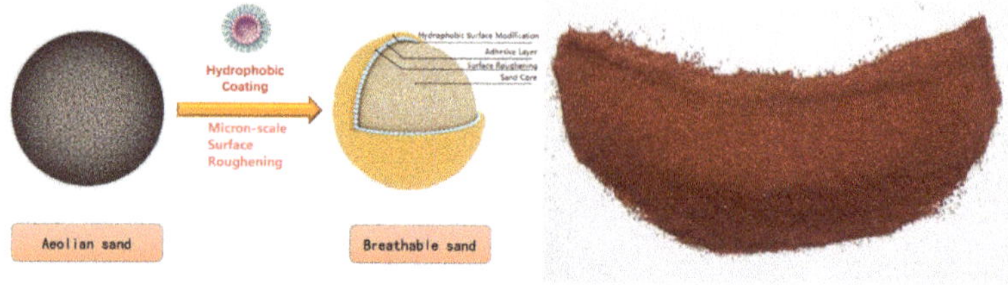

(**a**) (**b**)

Figure 1. Breathable Sand: (**a**) conceptual illustration of the coating, surface roughening, and surface modification processes; (**b**) Breathable Sand product.

3. Methodology

3.1. Use Breathable Sand as a Liner—Agricultural Applications

Breathable Sand has been used in agricultural and afforestation applications under different climate and soil conditions in China, such as Inner Mongolia, Ningxia, Gansu, Xinjiang, Beijing, Zhejiang, etc. A typical sequence of agricultural land preparation is shown in Figure 2.

Figure 2. A typical sequence of agricultural land preparation: excavation, Breathable Sand layer preparation, backfilling, and planting.

In this study, water usage and plant growth were monitored to evaluate the effects of utilizing Breathable Sand as a liner. Hybrid late rice in three adjacent lots (one control/without a liner, one with a geotextile liner, and one with approximately 3 cm of Breathable Sand liner) in a demonstration farm in Zhejiang Province was monitored in the summer of 2019. Irrigation water use, root and leaf development, and rice quality and yield from the three lots were monitored and compared.

3.2. Case Studies of Salt-Affected Soil Management

Salt-affected soil applications with Breathable Sand are similar to regular agricultural land applications, but ex situ cultivatable soil is typically needed during backfills. Salt-affected soil management projects have been conducted in Dengkou City, Inner Mongolia, and Tianjin City in China since 2017, which are introduced in this paper.

4. Results

4.1. Breathable Sand Properties

The finished Breathable Sand retains its air permeability feature due to its unchanged porous nature from untreated sand, while can withstand certain water pressure without becoming wet due to its water repellency properties. Tests were conducted and the results are summarized in Table 1. These features also allow Breathable Sand to be applied in salt-affected lands as a barrier to prevent salts in the deeper sodic soils from reaching the plant root zone.

Table 1. Water pressure resistance and air permeability of Breathable Sand.

	Results	Results in Other Units	Remarks
Static water pressure resistance	0.1 Mpa [1]	10.2 m of H_2O or 1ATM	No leakage observed
Air permeability	1934 mL/20 s	96.8 mL/s or 1.23 cm/s	At 0.05 Mpa

[1] Per Chinese standard DL/T 5150-2001. Specimens in a column (φ100 mm × 60 mm) were used.

4.2. Breathable Sand in Agricultural Applications

The Breathable Sand liner has a positive impact on the plant's root and leaf development during the growing season, indicating that the air-permeable feature may potentially enhance nutrient uptake and plant growth. The quality and quantity of rice yield with Breathable Sand also exceeded those without a liner. The average root weight, leaf count, grain weight, grain per ear, matured grain count, and grain maturity rates were 1.0%, 5.5%, 0.6%, 1.3%, 4.8%, and 3.3% higher than those without a liner, respectively (See Table 2). Moreover, monitored water use showed that 29% water saving was achieved with Breathable Sand, compared to that without a liner, which is close to that achieved with a geomembrane liner.

Table 2. Rice leaf and root development, yield, and water use comparison.

	Root Weight (g)	Average Leaf Count per Plant	Grain Weight (g/1000 Grains)	Grain per Ear	Matured Grain Count	Grain Maturity Rates (%)	Water Use July to September (mm)
Conventional/No Liner	58.2	16.3	23.12	530.6	369.0	69.8	536
w/Breathable Sand	58.8	17.2	23.26	537.4	386.8	72.1	382
Change	1.0%	5.5%	0.6%	1.3%	4.8%	3.3%	−29%

4.3. Breathable Sand Applications—Salt-Affected Soil Management Results

Breathable Sand has been used as a liner in a salt-affected field in Dengkou, Inner Mongolia since 2017. Both desert sand and clean ex situ sandy soil from adjacent land were used in the backfill. Hybrid rice, as well as local grass, was used in the demonstration. The before-and-after photos are shown in Figure 3.

Figure 3. Salt-affected land, rice field site before and after Breathable Sand treatment (Inner Mongolia, China).

After four years of the initial application, there are no visual signs of plant depression or salt intrusion observed. Similar results were also achieved in Tianjin City, as shown in Figure 4.

Figure 4. Salt-affected land, during and after the Breathable Sand treatment (Tianjin, China).

5. Conclusions

A coated sand-based material with surface modifications, called Breathable Sand, was invented. The material is watertight yet air-permeable. These features allow it to be used as a liner in many agricultural projects, including salt-affected land management in Inner Mongolia and Tianjin in China, which are introduced in this study. Breathable Sand installed under the plant root zone can significantly reduce water infiltration, and avoid salt intrusion from the subsoil. Field studies also showed that 29% of water-saving, stronger root and leaf development, and higher crop quantity and quality were achieved from rice fields with 3 cm of Breathable Sand liner, compared to those from conventional fields.

Author Contributions: Conceptualization, S.Q.; methodology, Y.S. and S.Q.; writing—original draft preparation, Y.S. and C.D.; writing—review and editing, C.D. All authors have read and agreed to the published version of the manuscript.

Funding: This research received no external funding.

Institutional Review Board Statement: Not applicable.

Informed Consent Statement: Not applicable.

Data Availability Statement: No new data were created or analyzed in this study. Data sharing is not applicable to this article.

Conflicts of Interest: The authors declare no conflict of interest.

References

1. Food and Agriculture Organization of the United Nations. *Harmonized World Soil Database (Version 1.0)*; FAO: Rome, Italy; IIASA: Laxenburg, Austria, 2008.
2. Wang, Y.; Kong, X. The Protective Development of Saline-Alkali Land is Urgent. *The Economic Daily*, 13 April 2021. Available online: https://www.chinanews.com/gn/2021/04-14/9454569.shtml (accessed on 24 January 2021). (In Chinese)
3. Salem, M.; Al-Zayadneh, W.; Cheruth, A. Water Conservation and Management with Hydrophobic Encapsulation of Sand. *Water Res. Manag.* **2010**, *24*, 2237–2246. [CrossRef]
4. Rechsand's Breathable Sand in Agricultural Applications. Available online: https://www.rechsand.com/index.php/Pr_index_gci_40.html (accessed on 24 January 2022). (In Chinese)

environmental sciences proceedings

MDPI

Proceeding Paper

How Does Organic Amendment and NPK Fertilization Improve Forage Yield of Cereals under Salinity and Arid Conditions?: Case of Moroccan Sahara †

Ayoub El Mouttaqi *, Ihssane Mnaouer, Abdelaziz Nilahyane, Mohamed Belcaid, Mohamed Ibourki, Karima Lazaar, Lamine Diatta, Krishna Prasad Devkota, Lamfeddal Kouisni and Abdelaziz Hirich

Biosaline Agriculture Research Program, African Sustainable Agriculture Research Institute (ASARI), Mohammed VI Polytechnic University (UM6P), Laayoune 70000, Morocco; ihssane.mnaouer@um6p.ma (I.M.); abdelaziz.nilahyane@um6p.ma (A.N.); med.belcaid80@gmail.com (M.B.); mohamed.ibourki@um6p.ma (M.I.); karima.lazaar@um6p.ma (K.L.); diattalamine29@gmail.com (L.D.); krishna.devkota@um6p.ma (K.P.D.); lamfeddal.kouisni@um6p.ma (L.K.); abdelaziz.hirich@um6p.ma (A.H.)

* Correspondence: ayoub.elmouttaqi@um6p.ma; Tel.: +212-675673127

† Presented at the 2nd International Laayoune Forum on Biosaline Agriculture, 14–16 June 2022; Available online: https://lafoba2.sciforum.net/.

Abstract: The experiment on effect of organic amendment and NPK fertilizer of forage yield of cereals using saline water irrigation (12.4 dS/m) was conducted in Es-Smara province in the South of Morocco during 2020–2021. We found that organic amendments have a great positive impact on increasing forage yield of selected cereals grown under salinity conditions. Compared to the control, the application of amendments improved dry biomass yield by 55, 101, 126%, through the application of compost, sheep manure, and NPK, respectively. The highest increment in dry matter productivity was through the combined application of organic amendment and NPK fertilization: clearly showed that a judicious combined application could improve forage supply in the salt-affected irrigated drylands.

Keywords: salinity; aridity; amendments; cereals; forage

Citation: El Mouttaqi, A.; Mnaouer, I.; Nilahyane, A.; Belcaid, M.; Ibourki, M.; Lazaar, K.; Diatta, L.; Devkota, K.P.; Kouisni, L.; Hirich, A. How Does Organic Amendment and NPK Fertilization Improve Forage Yield of Cereals under Salinity and Arid Conditions?: Case of Moroccan Sahara. *Environ. Sci. Proc.* **2022**, *16*, 51. https://doi.org/10.3390/environsciproc2022016051

Academic Editor: Redouane Choukr-Allah

Published: 16 June 2022

1. Introduction

The Southern region of Morocco is mostly the Desert which covers more than 317,000 km^2 area. Water scarcity, drought, high temperatures, poor soil health are significantly affecting the region and causing land degradation and desertification [1]. Moreover, with a groundwater salinity exceeding 2 g/L [2] combined with poor irrigation water quality leading to an increase of salts accumulation in the surface horizons, and therefore, a decrease in agricultural production. In the Sahara region of Morocco forage production is highly insufficient and most of the supply meets through transport (>1500 km) from the North at a very high price. Low production, high cost of transport, and poor supply are putting significant pressure to increase the supply through improving local production.

In the Sahara region, cereals are an essential source of energy for animal diets and have been used as grain, hay, silage, or for grazing. In the region, opportunity to increase forage production from cereal crops such as wheat, barley, triticale exists and several agronomic practices have been recommended to improve the production of those crops in salt-affected, saline, and low fertile lands. Among them, application of organic amendments could be a key factor in long-term for improving forage supply alleviating the problem of soil salinization through improving physical, chemical, and biological properties of soils [3]. Thus, the objective of this study was to investigate the effect of different organic amendments and mineral fertilizers to find out the best combination for improving forage supply with improving soil quality (reducing soil salinity and improving fertility) in the Southern region of Morocco.

2. Materials and Methods

2.1. Site Characteristics

On-farm experiment was conducted in Es-smara province, Morocco (Longitude = 12°07′22″ W; Latitude = 26°32′43″ N; Altitude = 313 m) during 2020–2021. The soil of the experimental site was a sandy loam with an electrical conductivity (ECe) of >10 dS/m. Soil analysis showed that the contents of major elements are 0.03% for total nitrogen and 28, and 408 ppm for P_2O_5 and K_2O, respectively.

2.2. Experimental Design, Crop Management and Measurement

The experiment was implemented from November 2020 to July 2021. Six major cereal species viz., Barley (*Hordeum vulgare* L.), triticale (*x Triticosecale* Wittmack), oat (*Avena sativa* L.), forage corn (*Zea mays* L.), pearl millet (*Pennisetum glaucum* L.), and sorghum (*Sorghum bicolor* L.) were evaluated under three organic and mineral amendments. The evaluated organic and mineral amendments were sheep manure (30 $t{\cdot}ha^{-1}$), compost (5 $t{\cdot}ha^{-1}$), and 10-30-10 NPK fertilizer (100 $kg{\cdot}ha^{-1}$). During the autumn season, the experiment was conducted using a latin square design with four replications for the three cereals (barley, triticale, and oat). During the spring season, the experiment was conducted with a randomized complete block design with four replications for forage corn, pearl millet, and sorghum. The crops were sown in November 2020 (autumn planting) and March 2021 (spring planting). The trials were irrigated using a drip irrigation system with 50 cm row-to-row and 20 cm dripper spacing. The irrigation water (groundwater) used for irrigation was highly saline (EC = 12.4 dS/m). The recorded amount of irrigation (based on spacing of the dripper and the discharge rate), autumn crops received an amount of 840 mm and spring crops 390 mm irrigation water. Biomass of barley, triticale, and oat was harvested twice, i.e., at flowering and maturity stages, while forage corn, pearl millet, and sorghum were harvested once at maturity (in July 2021). Several agro-morphological parameters were monitored during the crop growing period and at harvest. The measured parameters include plant height, number of tillers, number of leaves, root length, plant fresh weight, and fresh biomass. For the determination of the dry biomass, harvested plants were oven-dried at 60 °C for 72 h. Fresh leaves sampling was carried out to assess chlorophyll and proline content.

2.3. Statistical Analysis

Statistical analysis was performed using R 4.0.5 software. A one-way analysis of variance (ANOVA) was used to assess the effects of amendments on different crop species. The level of significance was set to $p \leq 0.05$. For each crop, when the *p*-value in ANOVA was significant (<0.05), mean differences between treatments were identified using Tukey's pairwise comparisons test ($p \leq 0.05$).

3. Results

Table 1 shows the effect of organic amendments on different parameters on different crop species during autumn and spring season. Organic amendment had a significant effect on plant height and root length for all crops studied except for sorghum. Plant height was heights under manure combined with NPK fertilizer application for autumn cereals, while effect of different organic amendments was not significant for the height of pearl millet and forage corn. Root length was significantly affected due to the application of organic amendments, where it was longest under manure combined with NPK. In addition, a significant effect of amendments on irrigation water productivity was documented for all evaluated cereal species.

Table 1. Effect of organic amendments and NPK fertilizers on plant height, root length, and irrigation water productivity (IWP) of different crops planted during autumn and spring seasons. Results presented are means $\pm$ standard deviation. Means sharing same letters do not differ significantly at $p = 0.05$.

		Autumn			Spring					
Traits	**Treatment**	**Barley**	**Oat**	**Triticale**	**Barley**	**Oat**	**Triticale**	**Maize**	**Pearl Millet**	**Sorghum**
Plant height (cm)	Control	49.4 $\pm$ 11.5 c	24.5 $\pm$ 0.5 b	62.8 $\pm$ 14 b	39.25 $\pm$ 8.17 b	20.5 $\pm$ 0.9 c	54.5 $\pm$ 10.64 a	88.3 $\pm$ 4.9 b	48.5 $\pm$ 5.5 b	81 $\pm$ 0.8 a
	Manure	70.5 $\pm$ 12.9 a	54 $\pm$ 9 a	83.9 $\pm$ 9.3 a	45.25 $\pm$ 3.9 ab	51 $\pm$ 4 ab	59.5 $\pm$ 5.02 a	107.7 $\pm$ 6.2 a	114 $\pm$ 9 a	89 $\pm$ 5.7 a
	NPK	63.6 $\pm$ 11 b	52.3 $\pm$ 5.7 a	79.3 $\pm$ 13.8 a	49.25 $\pm$ 6.76 a	50.33 $\pm$ 6.7 ab	58.5 $\pm$ 9.71 a	99.5 $\pm$ 1.5 ab	111 $\pm$ 10 a	95 $\pm$ 5.7 a
	Manure $\times$ NPK	71.2 $\pm$ 8.3 a	57 $\pm$ 2.2 a	86.3 $\pm$ 14.1 a	50 $\pm$ 3.54 a	51 $\pm$ 2.6 a	60 $\pm$ 4.64 a	107 $\pm$ 5.7 a	106 $\pm$ 11 a	92.3 $\pm$ 12.7 a
	Compost	-	-	-	41.5 $\pm$ 4.09 ab	38.33 $\pm$ 2.3 bc	53.75 $\pm$ 7.98 a	105.7 $\pm$ 5.3 a	102 $\pm$ 6.5 a	84.3 $\pm$ 9.7 a
	Compost $\times$ NPK	-	-	-	45.25 $\pm$ 1.92 ab	49.33 $\pm$ 1.7 ab	65.5 $\pm$ 9.21 a	104.7 $\pm$ 3.9 ab	122.7 $\pm$ 3.4 a	85 $\pm$ 7.9 a
Root length (cm)	Control	9.9 $\pm$ 2.5 ab	4 $\pm$ 0 c	8.2 $\pm$ 1.8 b	7.25 $\pm$ 3.27 a	4.2 $\pm$ 0.18 a	5.75 $\pm$ 1.64 a	31.3 $\pm$ 1.7 ab	12.0 $\pm$ 1.0 b	23.7 $\pm$ 1.3 a
	Manure	11 $\pm$ 3 a	6.5 $\pm$ 0.5 b	10.2 $\pm$ 2.7 ab	6.5 $\pm$ 1.5 a	5.3 $\pm$ 0.4 a	5 $\pm$ 1.87 a	26.3 $\pm$ 1.9 ab	25.0 $\pm$ 3.0 a	25.3 $\pm$ 4.1 a
	NPK	8.3 $\pm$ 1.9 c	7.3 $\pm$ 0.5 ab	10.8 $\pm$ 4 a	6 $\pm$ 1.22 a	6.81 $\pm$ 0.53 a	6.25 $\pm$ 1.79 a	19.5 $\pm$ 4.5 b	27.0 $\pm$ 3.0 a	29 $\pm$ 8 a
	Manure $\times$ NPK	9.7 $\pm$ 2.7 bc	8.7 $\pm$ 0.5 a	10.4 $\pm$ 2.1 ab	7.25 $\pm$ 2.49 a	7.5 $\pm$ 0.71 a	6.75 $\pm$ 2.86 a	34 $\pm$ 3.7 a	23.0 $\pm$ 1.0 a	24 $\pm$ 12.1 a
	Compost	-	-	-	7 $\pm$ 2.12 a	7 $\pm$ 2.45 a	7 $\pm$ 1.58 a	35.3 $\pm$ 2.1 a	22.0 $\pm$ 1.6 a	20 $\pm$ 2.9 a
	Compost $\times$ NPK	-	-	-	6.25 $\pm$ 3.11 a	7.33 $\pm$ 1.25 a	5.25 $\pm$ 1.79 a	25.7 $\pm$ 4.5 ab	26.0 $\pm$ 2.5 a	19.7 $\pm$ 2.4 a
IWP (Kg/m3)	Control	0.9 $\pm$ 0.33 c	0.56 $\pm$ 0.1 ab	0.42 $\pm$ 0.16 b	0.67 $\pm$ 0.18 b	0.24 $\pm$ 0.08 b	0.35 $\pm$ 0.24 b	0.7 $\pm$ 0.17 d	0.99 $\pm$ 0.65 b	0.81 $\pm$ 0.73 b
	Manure	1.55 $\pm$ 0.34 b	0.51 $\pm$ 0.15 b	0.95 $\pm$ 0.45 a	1.11 $\pm$ 0.03 ab	0.5 $\pm$ 0.09 b	0.62 $\pm$ 0.14 b	1.62 $\pm$ 0.46 cd	2.22 $\pm$ 0.9 ab	2.71 $\pm$ 0.1 a
	NPK	1.63 $\pm$ 0.33 ab	1.3 $\pm$ 0.29 a	1.33 $\pm$ 0.43 a	1.19 $\pm$ 0.47 ab	0.56 $\pm$ 0.14 b	0.87 $\pm$ 0.39 b	2.96 $\pm$ 0.01 ab	4.41 $\pm$ 0.46 a	1.57 $\pm$ 0.44 ab
	Manure $\times$ NPK	1.92 $\pm$ 0.33 a	0.93 $\pm$ 0.3 ab	1.35 $\pm$ 0.35 a	1.43 $\pm$ 0.27 ab	0.94 $\pm$ 0.22 a	1.23 $\pm$ 0.28 a	2.17 $\pm$ 0.17 bc	3.48 $\pm$ 0.5 ab	2.73 $\pm$ 0.76 a
	Compost	-	-	-	1.12 $\pm$ 0.4 ab	0.24 $\pm$ 0.07 b	0.2 $\pm$ 0.02 b	1.78 $\pm$ 0.09 cd	2.31 $\pm$ 0.4 ab	1.8 $\pm$ 0.33 ab
	Compost $\times$ NPK	-	-	-	1.6 $\pm$ 0.15 a	0.58 $\pm$ 0.1 ab	1.13 $\pm$ 0.2 ab	3.27 $\pm$ 0.39 a	3.63 $\pm$ 0.1 ab	2 $\pm$ 0.19 ab

4. Discussion

Results indicate that organic amendments had a great positive impact on growth and productivity of cereals grown under salinity conditions (Figure 1). Barley and pearl millet performed better compared to other cereals. The combined application of sheep manure and NPK fertilizer doubled the dry biomass yield of barley to reach 16 T/ha, while the use of deep fertilization on pearl millet increased the production 4 times (17 t/ha) compared to the control.

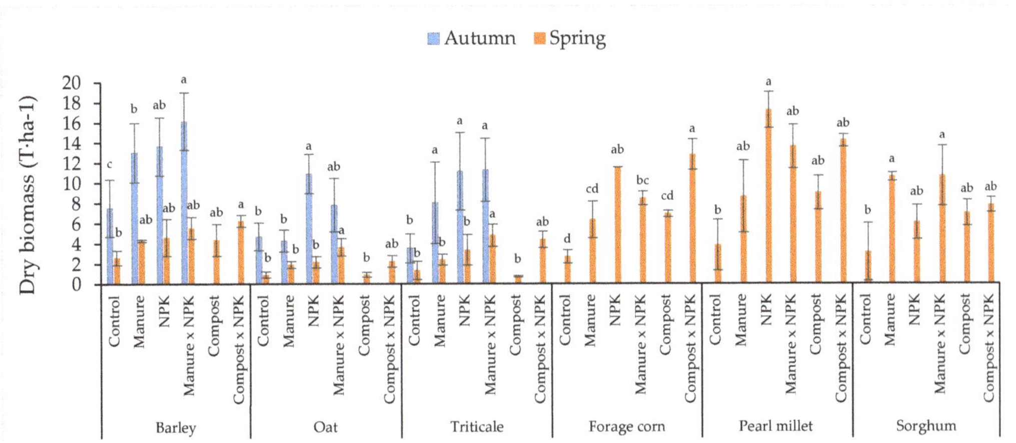

Figure 1. Total dry matter yield of different cereals as affected by different organic and inorganic amendments. Means sharing same letters do not differ significantly at $p = 0.05$. Vertical lines indicate the standard deviation.

Our results are consistent with the results of Saudi et al. [4] who found improved height and dry biomass through the application of organic amendment on barley in salt-affected drylands. Combined application might be preferred over sole application of organic amendment. Inorganic N fertilizer, enhances the forage production of pearl millet by 43% [5]. The high negative charge in animal manure increases the organic carbon content which is associated with a greater cation exchange capacity and, therefore, decrease the loss of nutrients. This is important to enhance organic carbon content, available phosphorus, extractable potassium, and make them available for plants [3].

5. Conclusions

Results of the experiments carried out in Es-smara province indicate that the use of sustainable cropping practices, such as the use of farmyard manure combined with mineral fertilizer, could be a key-element to ensure forage availability in salt-affected lands. There is an immense scope of improving forage productivity and forage supply in such region (Sahara) through improving soil quality especially by the application of organic amendments.

Author Contributions: Conceptualization, A.H.; methodology, A.H., A.E.M. and I.M.; software, A.E.M.; validation, A.H., A.N. and K.P.D.; formal analysis, A.E.M.; investigation, A.E.M., M.B., M.I., K.L. and L.D.; resources, A.H.; data curation, A.H.; writing—original draft preparation, A.H., A.E.M. and I.M.; writing—review and editing, A.H., A.N., L.K. and K.P.D.; visualization, A.E.M.; supervision, A.H.; project administration, A.H.; funding acquisition, A.H. All authors have read and agreed to the published version of the manuscript.

Funding: This research was funded by OCP Phosboucraa Foundation within the project titled "Scaling up the cultivation of alternative crops in salt-affected farms in southern provinces of Morocco".

Institutional Review Board Statement: Not applicable.

Informed Consent Statement: Not applicable.

Data Availability Statement: Not applicable.

Acknowledgments: The authors gratefully acknowledge the OCP-Phosboucraa Foundation, for providing funds to conduct the project activities. We also appreciate the efforts of the cooperative Ramth El Wastani to provide us with farmlands to carry out the crop introduction trials.

Conflicts of Interest: The authors declare no conflict of interest.

References

1. Daoud, S.; Lyagoubi, A.; Harrouni, M. Moroccan Agriculture Facing Climate Change: Adaptation and Local Distribution of the Value Added. In *Science, Policy and Politics of Modern Agricultural System*; Behnassi, M., Shahid, S., Mintz-Habib, N., Eds.; Springer: Dordrecht, The Netherlands, 2014; pp. 83–95. [CrossRef]
2. Hssaisoune, M.; Bouchaou, L.; Sifeddine, A.; Bouimetarhan, I.; Chehbouni, A. Moroccan Groundwater Resources and Evolution with Global Climate Changes. *Geosciences* **2020**, *10*, 81. [CrossRef]
3. Diacono, M.; Montemurro, F. Long-term effects of organic amendments on soil fertility. A review. In *Agronomy for Sustainable Development*; Springer: Dordrecht, The Netherlands, 2010; Volume 30, pp. 401–422. [CrossRef]
4. Saudi, A.R.; Mahrous, Y.M.A.; Sabry, A.H.; Mamdouh, A.E. Effect of some organic amendments on barley plants under saline condition. *J. Plant Nutr.* **2020**, *43*, 1840–1851. [CrossRef]
5. Hassan, K.; Abd El-Maaboud, M.; Draz, M.; El Shaer, H. Performance of sorghum and pearl millet forage crops productivity by using different agricultural managements under salinity conditions. *J. Plant Prod.* **2016**, *7*, 311–316. [CrossRef]

environmental sciences proceedings

MDPI

Proceeding Paper

Reclamation of a Saline-Sodic Soil with Organic Amendments and Leaching †

Demis Andrade Foronda [1,2]

1 Soil-Water-Plant Exchanges, Gembloux Agro-Bio Tech, University of Liège, 5030 Gembloux, Belgium; dn.andrade@doct.uliege.be
2 Facultad de Ciencias Agrícolas y Pecuarias, Universidad Mayor de San Simón, Cochabamba 4926, Bolivia
† Presented at the 2nd International Laayoune Forum on Biosaline Agriculture, 14–16 June 2022; Available online: https://lafoba2.sciforum.net/.

Abstract: Excessive amounts of Na^+ and soluble salts are characteristics of saline-sodic soils. Loss of soil structure and osmotic stress in plants are negative effects of salinity-sodicity. This study evaluated the effect of cattle manure, biochar and tropical peat at 1 and 2% (w/w) with leaching, on the exchangeable sodium percentage (ESP), electrical conductivity (EC_e) and pH of a saline-sodic soil from the High Valley of Cochabamba (Bolivia). The soil was placed in simulated soil columns and two lixiviations were applied. The initial values of soil were as follows: ESP of 66.6%, EC_e of 20.5 dS m^{-1}, and pH of 8.55. Results after leaching differed significantly (p = 0.05) among the interactions. Cattle manure at 2% was the most effective in reducing soil ESP to 27.6%, followed by the rest of the treatments. The three amendments at any level were efficient in lowering EC_e below 4 dS m^{-1}. Peat at 2% decreased the soil pH to 7.76. The superiority of cattle manure can be explained by the improvement of soil aggregation and leaching efficiency, through its OM and Ca^{2+} + Mg^{2+} contribution. Overall, cattle manure was superior in reclaiming the soil salinity-sodicity, and only the EC_e threshold value from the US Salinity Lab classification was reached by any amendment, indicating that cattle manure, biochar or tropical peat with leaching, can be used to reclaim some saline-sodic soils.

Keywords: saline-sodic soil; soil remediation; manure; biochar; peat

Citation: Andrade Foronda, D. Reclamation of a Saline-Sodic Soil with Organic Amendments and Leaching. *Environ. Sci. Proc.* **2022**, *16*, 56. https://doi.org/10.3390/environsciproc2022016056

Academic Editor: Abdelaziz Hirich

Published: 16 June 2022

1. Introduction

As a category of salt-affected soils, saline-sodic soils are characterized by an excessive amount of soluble salts, and sodium (Na^+) in the soil solution and cation exchange complex. Loss of soil structure and osmotic stress in plants are some of the negative effects of salinity-sodicity. Soil salinity can be measured through electrical conductivity (EC), and sodicity by the exchangeable sodium percentage (ESP) or the sodium adsorption ratio (SAR). Saline-sodic soils can be classified using the threshold values from the US Salinity Lab (USSL) classification [1], as follows: ESP > 15%, EC_e > 4 dS m^{-1} and pH < 8.5. Saline-sodic soils can be reclaimed by leaching with non-saline water and adding chemical/organic amendments.

The addition of organic amendments in sodic soils binds the small soil particles together into large water-stable aggregates, increases porosity and thus improves the soil physical properties [2]. Using organic amendments instead of inorganic amendments can reduce input cost savings as a sustainable and efficient management method to reclaim salt-affected soils [3], besides the beneficial impacts on nutritional and biological soil properties.

A soil-column experiment was carried out to evaluate the reclamation effect of cattle manure, biochar and tropical peat at two rates with leaching, on the ESP, EC_e and pH of a saline-sodic soil.

2. Materials and Methods

The soil (Table 1) was collected from the High Valley of Cochabamba (Bolivia) at a depth of 25 cm. It should be noted that the soil pH is slightly higher than the threshold value of the USSL classification. The organic amendments (Table 2) used to reclaim the soil were: cattle manure (CM) collected locally, tropical peat (PE) as tree fern fiber from the tropical area, and biochar (BI) branded by Greenpoch SA (Belgium).

Table 1. Chemical and physical parameters of the saline-sodic soil, before reclamation.

Property	Value	Property	Value	Property	Value
TOC (%)	0.3	EC_e (dS m^{-1})	20.5	K^+ ($mmol_c$ L^{-1})	1.5
Clay (%)	18.2	ESP (%)	66.6	HCO_3^- ($mmol_c$ L^{-1})	40.3
Silt (%)	52.1	Na^+ ($mmol_c$ L^{-1})	339.2	CO_3^{2-} ($mmol_c$ L^{-1})	20.0
CEC (cmol kg^{-1})	5.0	Ca^{2+} ($mmol_c$ L^{-1})	0.5	Cl^- ($mmol_c$ L^{-1})	185.0
pH	8.55	Mg^{2+} ($mmol_c$ L^{-1})	0.7	SO_4^{2-} ($mmol_c$ L^{-1})	71.1

TOC: total organic carbon, CEC: cation exchange capacity, EC_e: electrical conductivity (paste extract).

Table 2. Some chemical properties and TOC of the organic amendments.

Property	Cattle Manure	Biochar	Tropical Peat
Na^+ (mmol kg^{-1})	1.4	0.1	0.0
Ca^{2+} (mmol kg^{-1})	46.7	5.1	15.5
Mg^{2+} (mmol kg^{-1})	77.4	4.0	30.9
EC (d Sm^{-1})	3.7	0.3	0.7
pH	8.5	9.7	3.6
TOC (%)	23.7	33.0	22.5

Following the protocol of [4], simulated soil columns were assembled with PVC tubes (Ø of 15 cm), and each was filled with 6.7 kg of soil sieved at 4 mm, and then the upper layer was mixed with the respective amendment. The dose of amendments was calculated on a dry weight basis to reach 1 and 2% of organic matter (OM). To simulate the water from the rain, distilled water was used for the leaching process. The volume of water was calculated as a pore volume (PV) using the formula provided by Kahlon et al. [5] and Ahmad et al. [4]. After an initial soil saturation with 3/4 PV, two lixiviations were applied, each with one PV for two to four weeks. Response parameters were soil ESP, EC_e, and pH. The ESP was calculated using Equation 3 in Qadir et al. [6]. The design was completely randomized and the treatments were: CM-1%, CM-2%, BI-1%, BI-2%, PE-1%, PE-2% and control (only leaching). The results were evaluated using LSM-Tukey adjustment.

3. Results and Discussion

The results after leaching showed that soil ESP, EC_e and pH, differed significantly ($p < 0.05$) among the interactions. CM-2% was the best treatment for reducing the initial soil ESP by 39%, followed by CM-1% (by 31.5%), and lastly the rest of the treatments with a similar effect (Figure 1a). CM-1% and CM-2% were as effective as BI-2% and PE-2% for lowering EC_e by over 16 dS m^{-1} concerning the initial soil, while BI-1% and PE-1% showed a lower efficiency but higher than that of the control (Figure 1b). PE-2% decreased the initial pH to 7.76, followed by CM-1%, CM-2% and PE-1% in equal magnitude; in contrast, BI maintained a pH around the initial value (Figure 1c). Although organic amendments were effective in reclaiming this saline-sodic soil, the ESP and pH threshold values from the USSL classification were not reached. It should be pointed out that the percolation time of PE and BI was double that of CM.

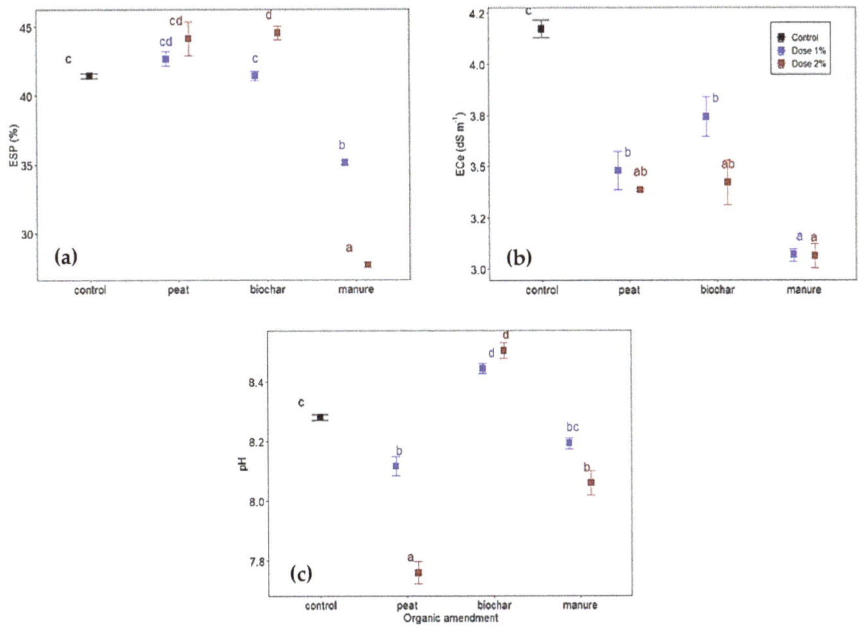

Figure 1. Soil ESP (**a**), EC_e (**b**) and pH (**c**), for the interactions between organic amendments and doses. Means sharing a letter are not significantly different according to pairwise comparisons of LSM with Tukey adjustment ($p < 0.05$). The bars indicate the standard error.

The superiority of CM in decreasing ESP and EC_e can be partly attributed to its initial amounts of TOC, Ca^{2+} and Mg^{2+}, contributing to the improvement of soil structure and infiltration, thus displacing Na^+ from the soil. The lower effectiveness of PE in reducing ESP was likely due to its swelling capacity which interacted with soil dispersion leading to a slowdown of the leaching process. In this regard, [7] reported that reclaimed soil with bentonite showed a lower decrease in salinity and sodicity levels and a higher percolation time due to the swelling capacity. The BI also showed a weak effect on sodicity potentially due to its insufficient ability to influence soil structure, and since, as [8] indicated, the mode of action of BI is physiochemical while composts provide a comprehensive reclamation when biological and physiochemical factors act together. In contrast to BI, the PE significantly reduced the soil pH due to its very low pH, causing an acidic counteracting effect, as [3] found that composts significantly improved soil CEC and pH values but the BI did not.

Water by itself was less effective in decreasing Na^+, but lowered EC_e to 4.2 dS m^{-1}, coinciding with [9], which found that EC decreased significantly even for the unamended soil possibly caused by solute leaching; moreover, [10] stated that flushing water reduced salinity with and without the application of manure.

Overall, the results suggest that CM, BI and PE enhanced the reclamation effect of leaching in remediating soil salinity and/or sodicity, through the positive impact of their OM on soil structure and infiltration, thus improving Na^+ displacement, agreeing with the following findings: organic amendments significantly lowered the level of soil EC_e, ESP and SAR compared to the control soils, improved soil structure, aggregate stability and saturated hydraulic conductivity, even more in compost treated soils [3]. The physical properties of the salinized soil, such as structural stability, infiltration rate, water-holding capacity and washing capacity were considerably improved by OM from the solid waste application [11]. Water hyacinth and rice straw compost singly or combined showed a pronounced decrease in EC, pH, SAR, and ESP compared with control [12].

4. Conclusions

Cattle manure at 2% was the best treatment for decreasing soil ESP to 27.6%, and any treatment was more effective than control in lowering EC_e below 4 dS m^{-1}. Peat at 2% showed a higher reduction in the soil pH (to 7.76). The superiority of cattle manure in reducing ESP and EC_e may be due to the improvement of the soil structure and infiltration through its OM and divalent cations contribution, whereas peat and biochar were less effective possibly due to the swelling capacity and insufficient rate, respectively, which in addition to the soil dispersion led to a slowdown of leaching. Overall, cattle manure with leaching was more efficient in ameliorating soil salinity-sodicity, and any amendment was effective in lowering salts. However, the ESP and pH threshold values from the USSL classification were not reached. This study suggests that some saline-sodic soils can be reclaimed by adding cattle manure, biochar or tropical peat, with leaching.

Funding: This research was funded by the 'Académie de recherche et d'enseignement supérieur-ARES' AI-11420 (Belgium), and provided by the 'Universidad Mayor de San Simón-UMSS' (Bolivia).

Institutional Review Board Statement: Not applicable.

Informed Consent Statement: Not applicable.

Acknowledgments: The Soil-Water Lab (UMSS) and Jose Luis Mamani.

Conflicts of Interest: The author declares no conflict of interest.

References

1. Richards, L.; Allison, L.; Bernstein, C.; Bower, J.; Brown, M.; Fireman, J.; Richards, W. *Diagnosis and Improvement of Saline Alkali Soils*; United States Salinity Laboratory Staff—Department of Agriculture, Agricultural Research Service: Washington, DC, USA, 1954; 169p.
2. Srivastava, P.K.; Gupta, M.; Shikha; Singh, N.; Tewari, S.K. Amelioration of Sodic Soil for Wheat Cultivation Using Bioaugmented Organic Soil Amendment. *Land Degrad. Dev.* **2016**, *27*, 1245–1254. [CrossRef]
3. Chaganti, V.; Crohn, D.; Simunek, J. Leaching and Reclamation of a Biochar and Compost Amended Saline-Sodic Soil with Moderate SAR Reclaimed Water. *Agric. Water Manag.* **2015**, *158*, 255–265. [CrossRef]
4. Ahmad, S.; Ghafoor, A.; Akhtar, M.E.; Khan, M.Z. Implication of Gypsum Rates to Optimize Hydraulic Conductivity for Variable-Texture Saline-Sodic Soils Reclamation. *Land Degrad. Dev.* **2016**, *27*, 550–560. [CrossRef]
5. Kahlon, U.Z.; Murtaza, G.; Murtaza, B.; Hussain, A. Differential response of soil texture for leaching of salts receiving different pore volumes of water in saline-sodic soil column. *Pak. J. Agric. Sci.* **2013**, *50*, 191–198.
6. Qadir, M.; Oster, J.D.; Schubert, S.; Noble, A.D.; Sahrawat, K.L. Phytoremediation of sodic and saline-sodic soils. In *Advances in Agronomy*; Elsevier: Amsterdam, The Netherlands, 2007. [CrossRef]
7. Shaygan, M.; Reading, L.P.; Baumgartl, T. Effect of Physical Amendments on Salt Leaching Characteristics for Reclamation. *Geoderma* **2017**, *292*, 96–110. [CrossRef]
8. Chaganti, V.N.; Crohn, D.M. Evaluating the Relative Contribution of Physiochemical and Biological Factors in Ameliorating a Saline–Sodic Soil Amended with Composts and Biochar and Leached with Reclaimed Water. *Geoderma* **2015**, *259–260*, 45–55. [CrossRef]
9. Mahmoodabadi, M.; Yazdanpanah, N.; Sinobas, L.R.; Pazira, E.; Neshat, A. Reclamation of Calcareous Saline Sodic Soil with Different Amendments (I): Redistribution of Soluble Cations within the Soil Profile. *Agric. Water Manag.* **2013**, *120*, 30–38. [CrossRef]
10. Manzano Banda, J.I.; Rivera Ortiz, P.; Briones Encinia, F.; Zamora Tovar, C. Rehabilitación de suelos salino-sódicos: Estudio de caso en el distrito de riego 086, Jiménez, Tamaulipas, México. *Terra Latinoam.* **2014**, *32*, 211–219.
11. Lax, A.; Díaz, E.; Castillo, V.; Albaladejo, J. Reclamation of Physical and Chemical Properties of a Salinized Soil by Organic Amendment. *Arid Soil Res. Rehabil.* **1994**, *8*, 9–17. [CrossRef]
12. Abdel-Fattah, M.K. Role of Gypsum and Compost in Reclaiming Saline-Sodic Soils. *IOSR J. Agric. Vet. Sci.* **2012**, *1*, 30–38. [CrossRef]

environmental sciences proceedings

MDPI

Proceeding Paper

Eco-Efficient Approach for Wastewater Treatment and Agricultural Valorization: Fertigation Effect on Soil and Plant †

Tawfik El Moussaoui [1,2,*], Mohamed Oussama Belloulid [2], Laila Mandi [2,3], Said Wahbi [4], Salvatore Masi [5] and Naaila Ouazzani [2,3]

1 Resources Valorization, Environment and Sustainable Development Research Team (RVESD), Department of Mines, Mines School of Rabat, Ave Hadj Ahmed Cherkaoui, Agdal, P.O. Box 753, Rabat 10100, Morocco
2 Laboratory of Water, Biodiversity and Climate Change, Faculty of Sciences Semlalia, Cadi Ayyad University, Av. Prince My Abdellah, P.O. Box 2390, Marrakech 40000, Morocco; belloulid1@gmail.com (M.O.B.); mandi@uca.ac.ma (L.M.); ouazzani@uca.ac.ma (N.O.)
3 National Center for Studies and Research on Water and Energy (CNEREE), BP/511, Cadi Ayyad University, Av. Abdelkrim Elkhattabi, P.O. Box 511, Marrakech 40000, Morocco
4 Laboratoire de Biotechnologie et Physiologie Végétale, Faculty of Sciences Semlalia, Cadi Ayyad University, Av. Prince My Abdellah, P.O. Box 2390, Marrakech 40000, Morocco; wahbi@ucam.ac.ma
5 Sanitary-Environmental Engineering Laboratory, School of Engineering, University of Basilicata, viale dell'Ateneo Lucano 10, 85100 Potenza, Italy; salvatore.masi@unibas.it
* Correspondence: elmoussaoui@enim.ac.ma; Tel.: +212-661-836-360

† Presented at the 2nd International Laayoune Forum on Biosaline Agriculture, 14–16 June 2022; Available online: https://lafoba2.sciforum.net/.

Abstract: In the present research, an alternative eco-efficient biotechnology system, both selective and environmentally viable, is proposed and compared to a conventional scheme. The results showed an optimal process function and a significant difference ($p < 0.001$) in wastewater measured quality parameters compared to the conventional scheme. Indeed, the treated wastewater from the eco-efficient process revealed significant concentrations of organic matter and nutrients. Accordingly, there was an improvement in soil and plant quality parameters as a result of irrigation with raw urban wastewater and treated wastewater from the eco-efficient system compared to ground water.

Keywords: urban wastewater; activated sludge; fertigation; organic matter; sustainable irrigation

Citation: El Moussaoui, T.; Belloulid, M.O.; Mandi, L.; Wahbi, S.; Masi, S.; Ouazzani, N. Eco-Efficient Approach for Wastewater Treatment and Agricultural Valorization: Fertigation Effect on Soil and Plant. *Environ. Sci. Proc.* **2022**, *16*, 62. https://doi.org/10.3390/environsciproc2022016062

Academic Editor: Abdelaziz Hirich

Published: 21 June 2022

1. Introduction

In arid and semi-arid regions, the demand for clean water is continuously increasing. Therefore, wastewater is considered a valuable and attractive source of irrigation water and a fertilizing material for these regions [1,2]. Although wastewater reuse in irrigation is a common practice in these regions, there is a widespread and growing concern about its opposing impact on the environment and health [3]. Thus, inappropriate handling and management of wastewater reuse for irrigation can create serious environmental and health risks [4].

The application of advanced technologies in wastewater treatment, including several physical, chemical and biological methods as well as various combinations of them, lead to efficient process performances [5]. Therefore, regarding the high concentration of organic matter and nutrients content in raw wastewater, this wastewater constitutes a significant source of beneficial compounds for agriculture [6,7]. However, advanced abetment of organic matter and nutrients is not frequently suitable from wastewater, particularly if the objective is reuse in agriculture. Thus, the removal of sanitary risks simultaneously and the conservation of specific nutrients such as nitrogen and phosphorus during wastewater treatment could be interesting and beneficial for both soil and plant. However, most existing wastewater treatment approaches are not selective in terms of organic matter and nutrient abatement, and it is crucial to explore alternative selective approaches and environmentally viable options.

In this optic, we investigated an innovative and sustainable alternative methodology for wastewater treatment and valorization in agriculture. To this end, physicochemical and bacteriological parameters of raw and treated urban wastewater from both systems were investigated. Thereby, we evaluated the fertigation effect of irrigation with raw and treated wastewaters on soil and plants.

2. Materials and Methods

2.1. Experimental Setup and Procedure

Two experimental systems were assessed during the operation of this experiment (Figure 1). (a) A conventional activated sludge system composed of an anoxic basin, aeration tank and a settling tank. (b) An eco-efficient system in which total flow rate was split into two portions, the first portion collected from the anoxic basin (Split 1) and the second collected from the aerobic basin (Split 2). After that, effluent was sent to a secondary settling tank, filtered through a send filter (SF), and finally disinfected in a disinfection tank using PAA.

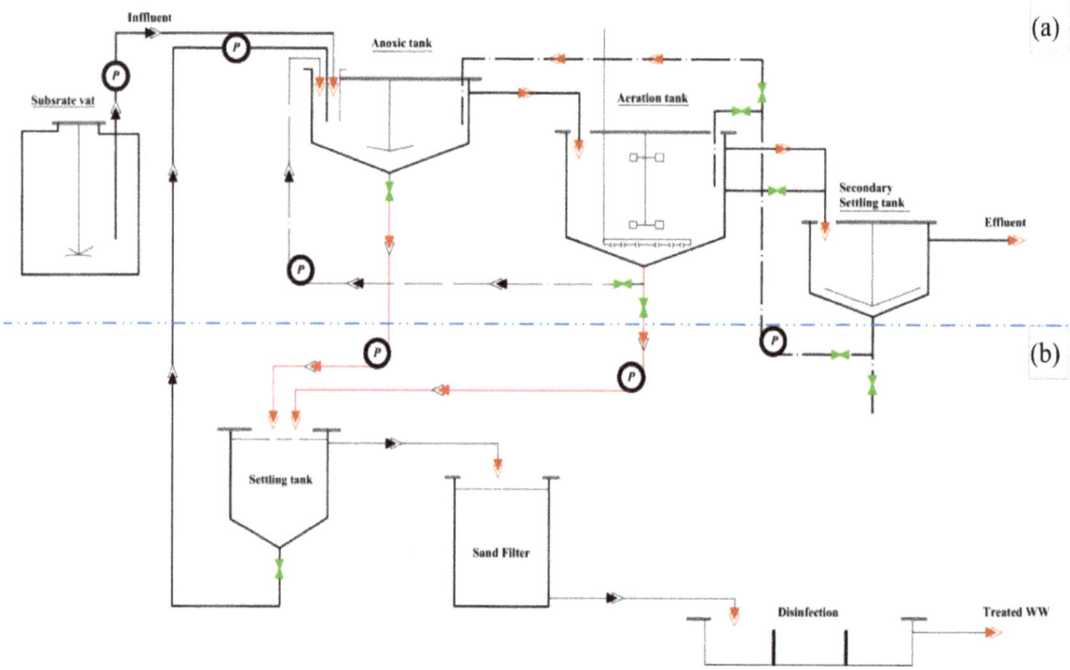

Figure 1. Schematic diagram of the conventional and eco-efficient schemes. (**a**) Conventional activated sludge system; (**b**) eco-efficient activated sludge system.

2.2. Raw and Treated Urban Wastewaters Characteristics

Physicochemical characteristics of raw and treated wastewaters as well as groundwater were determined according to standard methods [8–10]. The monitored parameters were the electrical conductivity (EC), total suspended solids (TSS), chemical oxygen demand (COD), five-day biological oxygen demand (BOD5), nitrogen (TKN and NH4), phosphorus (TP and PO_4^{3-}), chloride (Cl^-), sodium (Na^+) and calcium (Ca^{2+}).

2.3. Soil and Plant Sampling and Analytical Methods

Soil and plant quality parameters were performed according to standards methods [11–13]. The monitored parameters were pH, EC, TKN Phosphorus, Total organic carbon (TOC), Macro-elements and microelements, and were analyzed by X-ray fluorescence spectrometry and flame photometry.

2.4. Statistical Analysis

One-way ANOVA was used to evaluate the effects of water irrigation on the assessed parameters in soil and plants. All differences were considered significant at 5% ($p < 0.05$). Principal component analysis (PCA) was performed to obtain data using R program.

3. Results

Figure 2 shows the Organic matter (COD, BOD_5) and nitrogen (NTK and NH_4^+) in two systems during the experimental period.

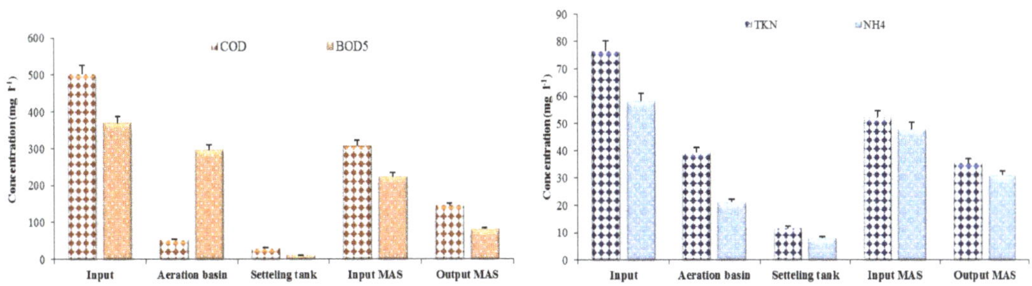

Figure 2. Organic matter and nutrient load evaluation from both systems: (**a**) BOD5 and COD evolution in conventional and eco-efficient systems; (**b**) TN and NH_4^+ evolution in both systems during this experiment.

4. Discussions

The obtained results from this research study suggest that the eco-efficient process allows for partial abatement of organic matter (BOD_5 and COD) and nutrients (nitrogen and phosphorous). This propose that the eco-efficient process is the optimal choice regarding agriculture reuse. Indeed, this process provides hygienic effluent with high organic matter and nutrient content. If the goal is to discharge into receiving environmental systems, the convention process is the best choice by ensuring significant abetment of organic load. Hence, COD and BOD_5 abatement levels were around 55% and 66% on average, respectively, and thus lower than the conventional system, at 94% and 97%, respectively (Figure 2a). Concerning nutrients, such as nitrogen, the eco-efficient scheme leads to a partial removal of nitrogen (41% of TKN and 40% NH4). By contrast, the conventional system provides high nitrogen-removal efficiency of about 85% and 86% for TKN and NH4, respectively (Figure 2b). In addition, the output effluent concentrations from conventional process were in accordance with the Moroccan applied discharges standards [14]. Assessment of the fertigation effect of irrigation water quality on soil and plants from both process shows a significant correlation between irrigation water and quality parameters of plants and soil (Figure 3). Indeed, soil and plants irrigated with eco-efficient-treated wastewaters showed an improvement in their proprieties compared to control soil and ground water. There was a significant impact ($p < 0.05$) on soil-fertility-indicating parameters OM, TKN, TP, and micro- and macro-nutrients. Compared to control soil, Na and EC concentrations in soil irrigated with raw and different treated wastewaters were significant. Consequently, amplified salinity in soils could decrease the potential for soil dispersion due to enhanced Na concentration [15]. This result was sustained with PCA analysis, as presented in Figure 3.

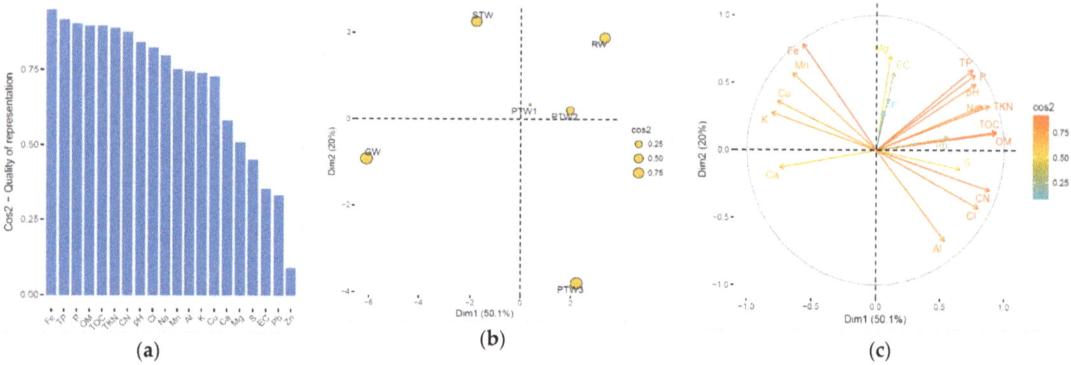

Figure 3. Scatter plot based on principal component analysis (PCA): (**a**) Cos^2 of variables' contribution; (**b**) PCA of treated wastewater from both systems; (**c**) PCA analysis using all using all the measured soil parameters.

5. Conclusions

The results reported in this experiment suggest that the developed innovative schemes provide a partial elimination of organic matter and nutrients. Indeed, the results showed a great positive impact of treated urban wastewaters from the eco-efficient activated sludge process on plant productivity and soil proprieties.

Author Contributions: T.E.M.: Conceptualization, methodology and writing—original draft preparation; M.O.B.: investigation and formal analysis; L.M., S.W. and S.M.: visualization; N.O.: supervision and validation. All authors have read and agreed to the published version of the manuscript.

Funding: This research received no external funding.

Institutional Review Board Statement: Not applicable.

Informed Consent Statement: Not applicable.

Data Availability Statement: Data sharing not applicable.

Acknowledgments: The authors would like to knowledge RADEEMA and CNEREE for their technical support.

Conflicts of Interest: The authors declare no conflict of interest.

References

1. Al-Rashed, M.F.; Sherif, M.M. Water resources in the GCC countries: An overview. *Water Resour. Manag.* **2000**, *14*, 59–73. [CrossRef]
2. Mohammad, M.J.; Mazahreh, N. Changes in soil fertility parameters in response to irrigation of forage crops with secondary treated wastewater. *Commun. Soil Sci. Plant Anal.* **2003**, *34*, 1281–1294. [CrossRef]
3. Jiries, A.; Hussain, H.; Lintelmann, J. Determination of polycyclic aromatic hydrocarbons in wastewater: Sediments, sludge and plants in Karak Province, Jordan. *Water Air Soil Pollut.* **2000**, *121*, 217–228. [CrossRef]
4. Haruvy, N.; Offer, R.; Hadas, A.; Ravina, I. Wastewater irrigation economic concerns regarding beneficiary and hazardous effects of nutrients. *Water Resour. Manag.* **1999**, *13*, 303–314. [CrossRef]
5. El Moussaoui, T.; Wahbi, S.; Mandi, L.; Masi, S.; Ouazzani, N. Reuse study of sustainable wastewater in agroforestry domain of Marrakesh city. *J. Saudi Soc. Agric. Sci.* **2019**, *18*, 288–293. [CrossRef]
6. Belloulid, M.O.; El Moussaoui, T.; Latrach, L.; Saykouk, J.; El Azzouzi, F. National Program of Sanitation and Wastewater Treatment in Morocco: Objectives, Achievements and Challenges. *Environ. Water Sci. Public Health Territ. Intell. J.* **2018**, *2*, 67–76. [CrossRef]

7. WHO. *Guidelines for the Safe Use of Wastewater, Excreta and Greywater*, 3rd ed.; World Health Organization: Geneva, Switzerland, 2006; ISBN 9789241546867.
8. Oller, I.; Malato, S.; Sánchez-Pérez, J.A. Combination of Advanced Oxidation Processes and biological treatments for wastewater decontamination-A review. *Sci. Total Environ.* **2011**, *409*, 4141–4166. [CrossRef] [PubMed]
9. Metcalf, W. *Metcalf and Eddy Wastewater Engineering: Treatment and Reuse*, 4th ed.; McGraw-Hill: New York, NY, USA, 2003.
10. French Standardization Association. *Qualité de l'eau. recueil des normes françaises, evironnement [Water Quality. Collection of French Standards, Environment]*; Tomes 1, 2, 3 and 4; AFNOR: Paris-La Defense, France, 1997; 1372p. (In French)
11. Rodier, J. *L'analyse de l'eau*, 9th ed.; DUNOD: Paris, France, 2009; 1579p.
12. APHA. *Standard Methods for the Examination of Water and Wastewater*; American Public Health Association: Washington, DC, USA, 2005.
13. Bremner, J.M. Nitrogen-Total. In *Methods of Soil Analysis, Part 3, Chemical Methods*; SSSA Book Set: 5; Sparks, D.L., Ed.; American Society of Agronomy Inc.: Madison, WI, USA, 1996; pp. 1085–1121.
14. Moroccan Discharge Standards. [arrêté conjoint 2013] Ministre de l'intérieur, du Ministre de l'énergie, des Mines, de l'eau et de l'environnement, du Ministre de l'industrie, du Commerce et des Nouvelles Technologies et du Ministre de l'artisanat N° 2942-13 du 1er Hija 1434 (7 Octobre 2013) Fixant les Valeurs Limites Générales de Rejet dans les Eaux Superficielles ou Souterraines. Bulletin Officiel du Royaume du Maroc N° 6202-3. 2013; pp. 2456–2458. Available online: http://www.environnement.gov.ma/fr/78-cat1/1012-valeurs-limites-des-rejets (accessed on 30 May 2022).
15. Heidarpour, M.; Mostafazadeh-Fard, B.; Abedi Koupai, J.; Malekian, R. The effects of treated wastewater on soil chemical properties using subsurface and surface irrigation methods. *Agric. Water Manag.* **2007**, *90*, 87–94. [CrossRef]

environmental sciences proceedings

MDPI

Proceeding Paper

Screening of Plant-Growth-Promoting Bacterial Isolates from Rhizosphere Soil of *Prosopis cineraria* from U.A.E. †

Neeru Sood [1,*], Shreya Padmesh Prajapat [1], Naushin Sadiq Shaikh [1], Trupti Gokhale [1] and Sumitha Thushar [2]

1 Department of Biotechnology, Birla Institute of Technology and Science, Pilani, Dubai Campus, P.O. Box Dubai 345055, United Arab Emirates; f20170009@alumni.bits-pilani.ac.in (S.P.P.); shaikh@gmail.com (N.S.S.); trupti@dubai.bits-pilani.ac.in (T.G.)

2 International Center of Biosaline Agriculture, P.O. Box Dubai 14660, United Arab Emirates; s.thushar@biosaline.org.ae

* Correspondence: sood@dubai.bits-pilani.ac.in

† Presented at the 2nd International Laayoune Forum on Biosaline Agriculture, 14–16 June 2022. Available online: https://lafoba2.sciforum.net/.

Abstract: Desert regions occupy 33% of the Earth's land surface. Despite the harsh climatic conditions, different types of plants and microorganisms have adapted to survive and proliferate in these arid extreme conditions and developed associations. *Prosopis cineraria* is able to flourish in spite of the harsh climatic conditions of high salinity in the desert soils of the United Arab Emirates (U.A.E.), the present study therefore aimed at isolating, screening and understanding the diversity of microflora associated with the rhizosphere region of *Prosopis cineraria* grown at 15 ds/m salinity. Eleven morphologically distinct purified isolates from rhizosphere soil were screened for IAA production, phosphate solubilization, antibacterial activity and nitrogen-fixing activity. Four of the isolates exhibited nitrogen-fixing activity, two antibacterial activity and five phosphate-solubilizing activity. The plant-growth-promoting bacterial (PGPB) isolates can be used as appropriate bioinoculants which can then in turn be used for cultivation of other plants in the region.

Keywords: PGPB; salinity; rhizosphere; *Prosopis cineraria*; *IAA*

Citation: Sood, N.; Prajapat, S.P.; Shaikh, N.S.; Gokhale, T.; Thushar, S. Screening of Plant-Growth -Promoting Bacterial Isolates from Rhizosphere Soil of *Prosopis cineraria* from U.A.E. *Environ. Sci. Proc.* **2022**, *16*, 69. https://doi.org/10.3390/environsciproc2022016069

Academic Editor: Abdelaziz Hirich

Published: 1 July 2022

1. Introduction

Desert regions represent one of the harshest ecosystems and are characterized by arid climate conditions with extreme temperatures, low rainfall and high solar radiation. At the same time, the desert soils are also sandy in nature, with poor water retention capacities, low fertility and high salinity [1,2]. Due to such adverse climatic conditions and poor soil characteristics, desert regions do not support agriculture. Several studies have reported the role of bacteria that are able to survive, proliferate and play a role in promoting plant growth by enhancing the availability of nutrients by fixing atmospheric nitrogen, solubilizing inorganic phosphate and modulating the plant hormones. Plant-growth-promoting bacteria (PGPB) also indirectly prevent the attack of fungal pathogens by synthesizing antibiotics and antifungal compounds [3]. Rhizosphere haloarchaea have helped the desert plants in conferring tolerance to high temperatures, salt stress, pH and water stress [4].

Prosopis cineraria is native to the desert ecosystem of the U.A.E., able to stand tall in the harsh climatic conditions and in the loose sandy saline soil of this region. Not much research has been carried out on the plant–microbe association of *Prosopis*. Discovering their soil microbiota under salinity stress and the PGPB involved would be of great importance, which could then be further used for various applications. Identification of such PGPB could pave a new way and help improve cultivation of various crops that is limited because of the current environmental abiotic conditions. Keeping this in mind, the following study aimed to isolate the PGPB from the rhizosphere area of salinity stress of *P. cineraria* to explore their possible applications in desert soil for sustainable agriculture production.

2. Materials and Method

2.1. Rhizospheric Soil Sampling

Rhizospheric soil sample of *Prosopis cineraria* tree located at 25°05′59″ N and 55°23′42″ E was collected from the International Center for Biosaline Agriculture, Dubai, U.A.E., during the month of February. Around 20 g rhizosphere soil was collected near the root system at about 30 cm depth. The soil sample was stored in sterile, zip-lock bags and immediately brought back to the laboratory. The electrical conductivity of the soil sample was determined using the saturated paste method [5].

2.2. Sample Processing for Isolation of Bacterial Communities

Soil sample was suspended in standard phosphate-buffered saline and vortexed. The soil suspension was serially diluted six times (10^{-1}, 10^{-2}, 10^{-3}, 10^{-4}, 10^{-5} and 10^{-6}), and 0.1 mL of each dilution was spread-plated on nutrient agar plates and incubated at 28 °C for 24 to 48 h. For screening of Azotobacter, two rounds of soil enrichments were carried out in Ashby's mannitol nitrogen-free media supplemented with 2% NaCl for 7 days at room temperature. The enriched culture was then spread-plated on Ashby agar plates and incubated for a week at 25 °C.

2.3. Morphological and Biochemical Characterization of Bacterial Isolates

Morphologically distinct colonies obtained on the spread plates were counted and assigned numbers. The colony characteristics of each isolated colony were identified in accordance with Bergey's manual. All the isolated colonies were studied for Gram staining [6].

2.4. Screening of Phosphate Solubilizers

Pikovskaya's Agar was used for the screening of phosphate solubilizers. All eleven purified isolates were spot-inoculated on the Pikovskaya's agar plates and incubated at 30 °C overnight. Plates were observed for bacterial growth and development of a zone of clearance [7].

2.5. Indole Acetic Acid Production Test

The Indole Acetic Acid test was performed by culturing the isolates on nutrient media supplemented with 2% NaCl and 100 mg/L tryptophan and incubating at 28 °C for 3 days on a shaking incubator. The bacterial cultures were centrifuged and tested for production of IAA using Salkowski's reagent. The tube was observed for the development of pink color in the dark, which would indicate the sample was positive for the production of IAA [8].

2.6. Screening for Antibiotic Production

The bacterial isolates were spot-inoculated on LB agar plates and incubated at room temperature for 24 h. Soft agar seeded with test pathogens, namely *E. coli* and *S. aureus*, was overlayered on the grown culture, and further incubated at room temperature for 24 h. A zone of clearance around the isolate indicated the production of the antibiotic.

3. Results

Eleven morphologically distinct colonies were isolated on nutrient agar (PR(HS)-5, 6, 7, 8, 10, 11 and 12), and four isolates (AZ-1, 2, 3 and 4) were able to grow on the Ashby mannitol agar plates (Figure 1), exhibiting nitrogen-fixing activity.

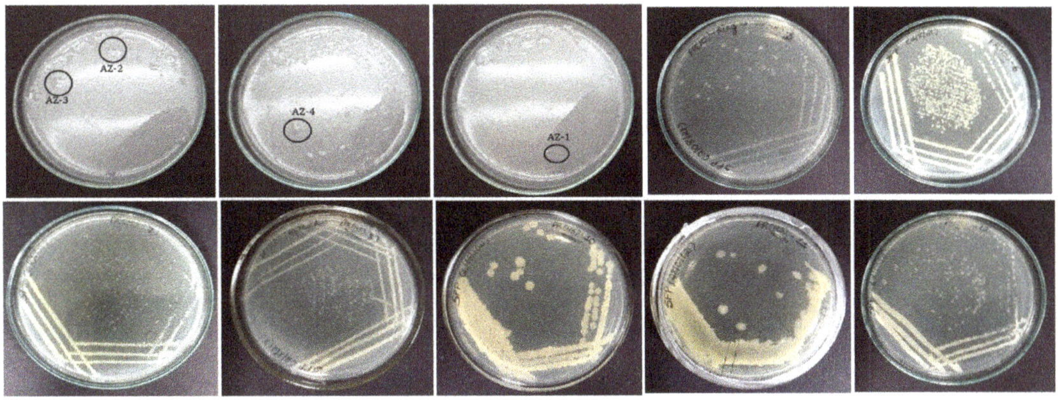

Figure 1. Pure colony of bacteria isolated on Ashby agar plates (AZ-1, 2, 3 and 4), and nutrient agar plates (PR-HS-5, 6, 7, 8, 10, 11 and 12).

The colony morphology studied is represented in Table 1. The Gram staining indicates that the isolates (PR(HS)-5, 6, 7, 8, 10, 11 and 12) were Gram-positive while the four N2-fixer isolates were found to be Gram-negative (Figure 2a). Two isolates exhibited antibacterial activity (isolate PR(HS)5 against *E. coli* and isolate PR(HS)-6 exhibited activity against *S. aureusI*. The isolates PR(HS)-3, 4, 6, 8 and 9 were able to form a clear halo zone on Pikovskaya's agar indicating they were positive for phosphate-solubilizing activity (Figure 2b). The qualitative IAA test results carried out on the isolates cultured in salt medium, however, showed they failed to produce a pink color upon addition of Salkowski's reagent.

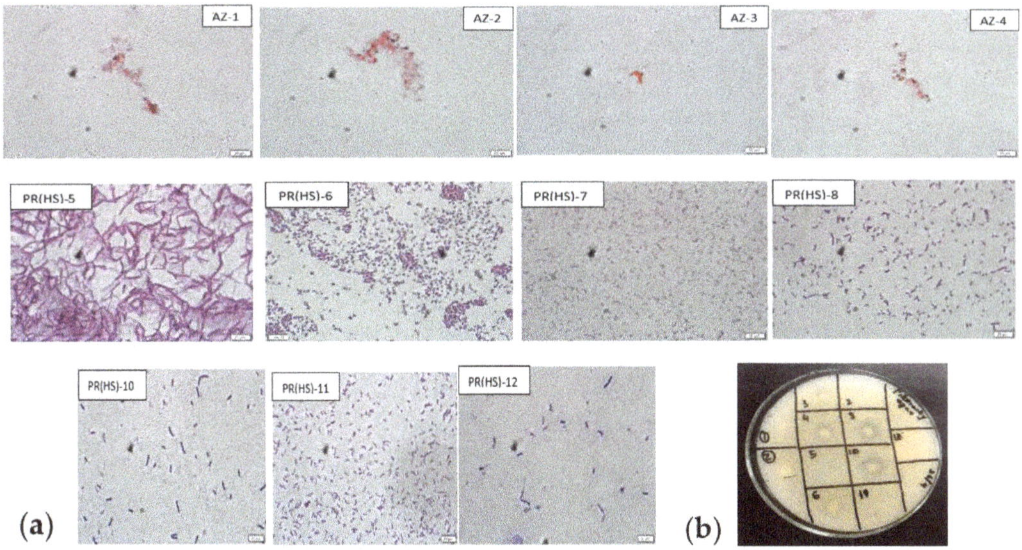

Figure 2. (**a**) Microscopic images of Gram-stained bacterial isolates (1st row (left to right)—Ashby isolates, AZ-1, 2, 3, 4 with negative Gram character. Second and third row—PGPR isolates, PR(HS)-5, 6, 7, 7, 10, 11 and 12 with positive Gram character). (**b**) Phosphate solubilization test for isolates PR(HS)-1, 2, 3, 4, 5, 6, 7, 8, 10,11 and 12. Clear halo zones around the isolates PR(HS)-3, 4, 6, 8 and 10.

Table 1. Morphological studies and Gram staining of bacterial isolates.

Colony No.	Form	Size	Margin	Surface	Opacity	Color	Gram Character and Shape
AZ-1, 2, 3, 4	Circular	Medium to large	Entire	Mucoid	Transparent		Gram negative—Cocci
PR(HS)-5	Circular with ring	Small	Entire	Smooth, shiny	Opaque	Cream-white	Gram positive—Bacillus
PR(HS)-6	Circular	Medium	Entire	Smooth, shiny	Opaque	Cream-white	Gram positive—Bacillus
PR(HS)-7	Circular	Tiny	Entire	Smooth, shiny	Translucent	Pale yellow	Gram positive—Bacillus
PR(HS)-8	Circular	Small	Entire	Smooth, shiny	Translucent	Cream	Gram positive—Bacillus
PR(HS)-10	Circular	Large	Undulate	Mucoid, wrinkled	Opaque	White surface	Gram positive—Bacillus
PR(HS)-11	Irregular swarming	Medium–Large	Undulate	Mucoid, glistening	Opaque	Off-white with green tinge	Gram positive—Bacillus
PR(HS)-12	Circular	Medium	Entire	Smooth, viscous	Opaque	Cream-whitish	Gram positive—Bacillus

4. Discussion

Soil salinity has been a major concern in arid and semi-arid regions, which in turn limits the cultivation and agricultural productivity. Soil also has a large diversity of microorganisms, ranging from bacteria to archaea to fungi, which are known to have a significant impact on the plant growth and proliferation under a variety of abiotic stresses [9]. In the present study, the preliminary experimental results from the colony morphology and Gram staining show that the purified isolates are morphologically distinct, Gram-positive bacilli and Gram-negative azotobacter spp. Some of the isolates have been shown to solubilize and exhibit their potential as phosphate solubilizers in the salinity-stressed desert soils. *Bacillus megaterium* has been reported to ameliorate saline conditions in tomatoes with the help of phosphate solubilization and bears the potential for use as a biofertilizer in arid and salinity-affected areas [10]. *Azotobacter* spp. also act as an excellent candidate of PGPR-related activities via acceleration of IAA production, nitrogen fixation and heavy metal metabolization and thereby finds use in reclamation of soil [11]. Co-inoculation of *Azotobacter chroococcum* and *Alcaligenes faecalis* in canola plants augmented the production of antioxidant enzymes, carotenoids, soluble sugars and protein contents under salinity stress conditions [12]. Though the isolates (AZ-1, 2, 3, 4) were able to grow on Ashby's mannitol agar, the production was not detected, probably due to low production. Hence, these need to be further tested as bioinoculants for their potential to support plant growth in saline soils. These studies therefore highlight the role of rhizosphere bacteria as a potent biofertilizer, especially under salinity stress conditions.

5. Conclusions

In the present study, screened bacterial isolates from rhizosphere soil of *Prosopis cinreria* were found to be both Gram-positive and Gram-negative PGPB with nitrogen-fixing (AZ-1,2,3 and 4) and phosphate-solubilizing activity. This association might be helping the plants to have better nutrient availability and salt tolerance. Several studies have reported the role of Bacillus spp. and other PGBP as plant-growth-promoting bacteria and their use as bioinoculants. This is a preliminary study, and further studies may lead to these isolates becoming promising biofertilizers, biostimulants and growth promoters for other plants/crops in the region, contributing to sustainable agriculture in the U.A.E.

Author Contributions: Conceptualization, N.S. and S.T.; methodology, N.S., S.T. and T.G.; validation, N.S., T.G. and S.T.; formal analysis, N.S., S.P.P. and T.G.; investigation, S.P.P. and N.S.S.; resources, N.S. and S.T.; data curation, N.S.; writing—original draft preparation, S.P.P. and N.S.; writing—review and editing, S.P.P., N.S. and T.G.; supervision, N.S. and T.G.; project administration N.S. All authors have read and agreed to the published version of the manuscript.

Funding: The authors thank BITS Pilani Dubai Campus and ICBA for partially funding the work.

Institutional Review Board Statement: Not applicable.

Informed Consent Statement: Not applicable.

Data Availability Statement: Not applicable.

Acknowledgments: The authors are grateful to the BITS Pilani Dubai campus and ICBA, Dubai, for the facilities and providing the support to conduct this research.

Conflicts of Interest: The authors declare no conflict of interest.

References

1. Eida, A.A.; Ziegler, M.; Lafi, F.F.; Michell, C.T.; Voolstra, C.R.; Hirt, H.; Saad, M.M. Desert Plant Bacteria Reveal Host Influence and Beneficial Plant Growth Properties. *PLoS ONE* **2018**, *13*, e0208223. [CrossRef] [PubMed]
2. Alsharif, W.; Saad, M.M.; Hirt, H. Desert Microbes for Boosting Sustainable Agriculture in Extreme Environments. *Front. Microbiol.* **2020**, *11*, 1666. [CrossRef] [PubMed]
3. Soussi, A.; Ferjani, R.; Marasco, R.; Guesmi, A.; Cherif, H.; Rolli, E.; Mapelli, F.; Ouzari, H.I.; Daffonchio, D.; Cherif, A. Plant-Associated Microbiomes in Arid Lands: Diversity, Ecology and Biotechnological Potential. *Plant Soil* **2016**, *405*, 357–370. [CrossRef]

4. Yadav, A.N.; Gulati, S.; Sharma, D.; Singh, R.N.; Rajawat, M.V.S.; Kumar, R.; Dey, R.; Pal, K.K.; Kaushik, R.; Saxena, A.K. Seasonal Variations in Culturable Archaea and their plant growth promoting attributes to predict their role in establishment of vegetation in Rann of Kutch. *Biologia* **2019**, *74*, 1031–1043. [CrossRef]
5. McKenzie, R.C.; Chomistek, W.; Clark, N.F. Conversion of Electromagnetic Inductance Readings to Saturated Paste Extract Values in Soils for Different Temperature, Texture, and Moisture Conditions. *Can. J. Soil Sci.* **1989**, *69*, 25–32. [CrossRef]
6. Coico, R. Gram Staining. *Curr. Protoc. Microbiol.* **2006**, A.3C.1–A.3C.2. [CrossRef] [PubMed]
7. Zerihun, T.; Birhanu, G.; Genene, T.; Adey, F.; Solomon, C. Isolation and Biochemical Characterization of Plant Growth Promoting (PGP) Bacteria Colonizing the Rhizosphere of Tef Crop during the Seedling Stage. *J. Plant Sci. Phytopathol.* **2019**, *3*, 013–027. [CrossRef]
8. Rahman, A.; Sitepu, I.R.; Tang, S.Y.; Hashidoko, Y. Salkowski's Reagent Test as a Primary Screening Index for Functionalities of Rhizobacteria Isolated from Wild Dipterocarp Saplings Growing Naturally on Medium-Strongly Acidic Tropical Peat Soil. *Biosci. Biotechnol. Biochem.* **2010**, *74*, 2202–2208. [CrossRef] [PubMed]
9. Egamberdieva, D.; Wirth, S.; Bellingrath-Kimura, S.D.; Mishra, J.; Arora, N.K. Salt-Tolerant Plant Growth Promoting Rhizobacteria for Enhancing Crop Productivity of Saline Soils. *Front. Microbiol.* **2019**, *10*, 2791. [CrossRef] [PubMed]
10. Chookietwattana, K.; Maneewan, K. Screening of Efficient Halotolerant Phosphate Solubilizing Bacterium and Its Effect on Promoting Plant Growth under Saline Conditions. *World Appl. Sci. J.* **2012**, *16*, 1110–1117.
11. Sumbul, A.; Ansari, R.A.; Rizvi, R.; Mahmood, I. Azotobacter: A Potential Bio-Fertilizer for Soil and Plant Health Management. *Saudi J. Biol. Sci.* **2020**, *27*, 3634–3640. [CrossRef]
12. Latef, A.A.H.A.; Omer, A.M.; Badawy, A.A.; Osman, M.S.; Ragaey, M.M. Strategy of Salt Tolerance and Interactive Impact of Azotobacter Chroococcum and/or Alcaligenes Faecalis Inoculation on Canola (*Brassica napus* L.) Plants Grown in Saline Soil. *Plants* **2021**, *10*, 110. [CrossRef]

environmental sciences proceedings

MDPI

Proceeding Paper

Groundwater Contamination Due to Landfill Leachate—A Case Study of Tadla Plain †

Yousra El Mouine [1,2,3,*], Amal El Hamdi [1,2,3], Moad Morarech [2], Vincent Valles [1,4], Hasna Yachou [3] and Houria Dakak [3]

1 Laboratory Mediterranean Environment and Agro-Hydrosystem Modelling (UMR 1114 EMMAH), Hydrogeology Department, Avignon University, 84000 Avignon, France; amalgoaea@gmail.com (A.E.H.); vincent.valles@univ-avignon.fr (V.V.)
2 Laboratory in Applied and Marine Geosciences, Geotechnics and Georisk (LR3G), Faculty of Science Tetouan, Abdelmalek Essaâdi University, Tetouan 93002, Morocco; morarech2000@gmail.com
3 National Institute of Agricultural Research (INRA), CRRAR, URECRN, Rabat 10000, Morocco; hasnayachou@gmail.com (H.Y.); dakak_h@yahoo.fr (H.D.)
4 Faculty of Sciences and Technology (FSTBM), Sultan Moulay Sliman University, BP 523, Béni Mellal 23000, Morocco
* Correspondence: yousra.elmouine@gmail.com

† Presented at the 2nd International Laayoune Forum on Biosaline Agriculture, 14–16 June 2022; Available online: https://lafoba2.sciforum.net/.

Abstract: In many parts of the world, the impact of open landfills on soils, the biosphere, and groundwater has become a major concern. This study was carried out on an uncontrolled landfill in the Tadla plain, Morocco's main agricultural region. The study of physicochemical parameters of water resources sample suggests that the evaluation of water quality parameters as well as water quality management practices should be carried out periodically to protect water resources and, on the other hand, to confirm groundwater pollution due to a pollution plume. Physicochemical analyses of leachate were thus carried out in order to determine the leachate nature of the landfill

Keywords: physicochemical parameters; landfill; water

Citation: El Mouine, Y.; El Hamdi, A.; Morarech, M.; Valles, V.; Yachou, H.; Dakak, H. Groundwater Contamination Due to Landfill Leachate—A Case Study of Tadla Plain. *Environ. Sci. Proc.* **2022**, *16*, 53. https://doi.org/10.3390/environsciproc2022016053

Academic Editor: Abdelaziz Hirich

Published: 16 June 2022

1. Introduction

Open dumps have been identified as one of the main threats to groundwater sources [1]. The open waste disposal system is the most common disposal method in Morocco [2] and specifically in the Tadla region. The Tadla region represents a pilot region in agriculture, but there are several problems that threaten the sustainability of agriculture in this region, such as salinity [3] and an open dumpsite. It is a kind of system where waste is disposed of by backfilling depressions on land that may include valleys and excavations, regardless of the composition of the waste [4]. Waste disposed of in an open landfill can be subject to precipitation infiltration. As water seeps into the waste, it absorbs a variety of organic and inorganic compounds that drain from the waste and accumulate at the bottom of the landfill [5]. The resulting contaminated leachate can migrate to soil and contaminate groundwater if not managed properly [6]. Such contamination of groundwater resources can pose significant health risks, including waterborne diseases, such as typhoid, cholera, and infectious dysentery, to local groundwater users.

Monitoring the physicochemical quality of the surrounding waters is therefore essential in order to determine the impact of these discharges on the waters, and thus good management of water resources.

2. Materiels and Methods

2.1. Study Area

Souk Sebt Ouled Nemma is a small town that is becoming very populated, located in the plain of Tadla in Morocco, the main region of agricultural production, with a dense population and fast. These changes have often been made at the expense of environmental balances. Among these balances is the abundance of wild landfills [7]. The city of Souk Sebt Ouled Nemma, like most other towns in the region, relies on agriculture.

On a regional scale, the Tadla plain is characterized by an arid to semi-arid climate; the amount of precipitation varies between 300 and 750 mm depending on the year [7]. The summer is very hot because of the hot winds of the southwest–east, known as 'chergui'. Water from the mountains is fed from the Atlas to the south. In addition to this natural supply, there is a powerful irrigation system from the Bin El Ouidane Dam, located about 100 km upstream of the region.

The Tadla plain, a syncline gutter, is filled with mio-plio-quaternary deposits consisting of alternating conglomerates, sandstone, and marls and ending in lacustrine deposits [8].

The plain is drained by the river Oum Er Rbia, which crosses it in its center from east to west (Figure 1). The study site is on the left bank, and the water table, at a depth of 5 to 6 m under an indurated limestone crust, flows regionally to the north–northwest. Regionally, it fluctuates due to recharge during irrigation periods. On the surface, the soils have a fine texture and moderate permeability. Land use is intense, and the population in the scattered colonies also uses well water for washing, drinking, and cooking. There are many unauthorized landfills, and poor soil and groundwater quality can affect agricultural production in the region and irrigation water quality [9]. Water quality monitoring and study are important.

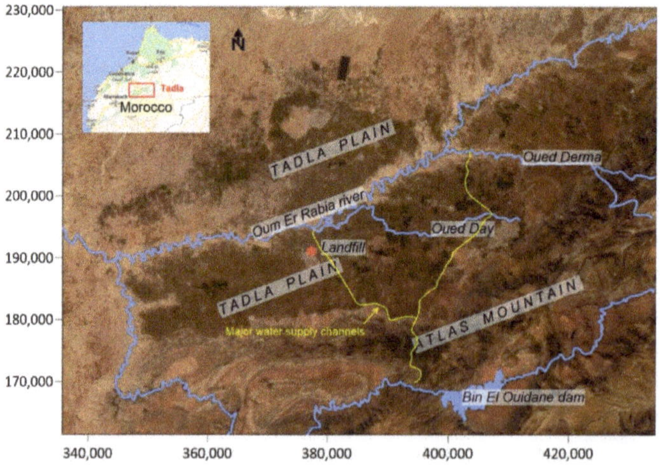

Figure 1. Geographical framework and location of the Souk Sebt landfill in the Tadla plain. UTM coordinates in meters using Lambert conical conformal projection (Morocco zone 1).

2.2. Sampling Methodology

During the first campaign (June 2019 and September 2019), a sampling network was selected and consisted of 6 wells and boreholes symbolized by W and L (Leacheat Sample), respectively, and spread across the entire study area, around the wild landfill, and along the direction of groundwater flow. These well samples were selected according to their availability, proximity, and their distribution upstream and downstream of the landfill. This choice allowed for a general, credible characterization of pollutants and an identification of the sectors where water quality is the most problematic, as well as high representativeness of the results in addition to landfill leachate samples.

For the second sampling companion, we collected samples in February 2020 to cover the dry and rainy seasons of the leachate symbolized by L′ and surrounding surface water (irrigation canal and collector drain water), symbolized by SW. In situ parameters (T, Eh, pH) were measured during the first measurement.

Groundwater samples were also collected from surrounding boreholes. Leachate samples were taken at the bottom of the landfill. Plastic containers of 100 cl and 50 cl were used to collect groundwater and leachate samples, respectively. Before the samples were taken, the containers were washed with acid water to sterilize them, and then they were rinsed thoroughly with distilled water. The samples collected were sealed and properly labeled. The samples were stored at 4 °C and then taken to the laboratory for analysis. All samples were analyzed for physicochemical parameters and heavy metals.

3. Results and Discussion

3.1. Leachate Characterisation

Table 1 shows the results of physicochemical analyses (mg/L) of the leachate samples.

Table 1. Physicochemical characterization of leacheate in Souk Sebt landfill (in summer period, July 2019, and winter period, February 2021).

Parametrs	pH	EC (ms/cm)	$HCO3^-$	F^-	Cl^-	$PO4^-$	$SO4^-$	Na^+	$NH4^+$	K^+	Mg^{++}	Ca++	TOC	SiO^2	COD	NO^{2-}, NO^{3-}′
July 2019	7.5	42.5	-	-	2178	78.42	-	1083	677.9	1181	64.13	113.3	-	-	31,929	0.02
February 2021	7	33.3	7110	395	3268	102	0.02	1741	696	1838	315	593	441	348	-	0.02

The leachate of Souk Sebt has a brownish-black color, and the strong smell of the landfill leachate may present the first pollution indicator, accompanied by strong mineralization by chemical elements (Table 1). The characterization of the leachate has shown the presence of mature leachates with high mineral and organic loads. Organic load is translated by the high values of COD and TOC. On the other hand, the mineral pollution is shown by the high values of $NH4^+$ (mg/L), $NO3^-$ (mg/L), and $PO4^-$ (mg/L).

3.2. Groundwater and Surface Water

The relatively high value of EC and dissolved solids in the samples indicated the presence of inorganic material in both seasons. Higher concentrations of chlorides were observed in wells close to the dumping site. The highest value was recorded in well located downstream hydrogeologically, which is 106 m away from the site. Pollution sources, such as domestic effluents and fertilizers, as well as natural sources, such as rainfall, can lead to high Cl (Figure 2).

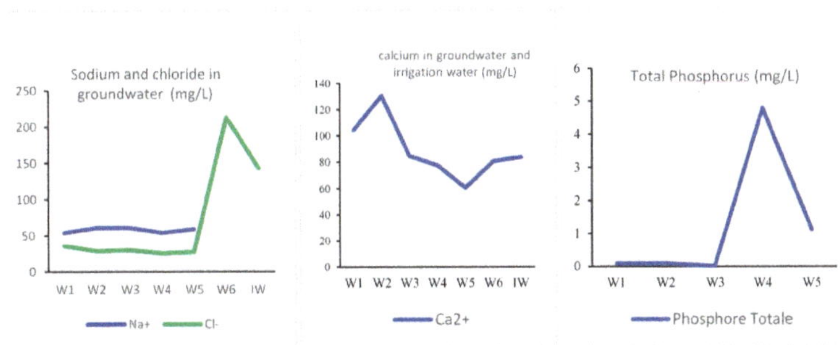

Figure 2. Sodium, Chloride, calcium and total phosphorus of groundwater samples in Souk-Sebt.

Nitrogen is mainly present in water as nitrate (NO_3^-), but under reducing conditions, it can be present as ammonium ($NH4^+$). Phosphorus is also present in water as a phosphate ion ($PO4^-$). Both nitrate and phosphate are an environmental concern as they are potential sources of nutrient enrichment in rivers, lakes, and wetlands (Figure 2). The presence of nitrates (0.0–3.17 mg L^{-1}), ammonia (0–0.6 mg L^{-1}), and phosphates (0.65–9.9 mg L^{-1}) in groundwater are related to contamination by anthropogenic activities or infiltration of residual water [10,11]. These results can be confirmed by the delineation study of the pollution plume at the landfill level at Souk Sebt, which shows the geometry of the pollution plume in a north–north–west direction [12].

Author Contributions: The authors have contributed in several ways to the success of this work. project administration, Y.E.M.; conceptualization, formal analysis, funding acquisition, resources; Y.E.M., M.M. and A.E.H.; methodology, V.V., M.M.; software, Y.E.M., A.E.H. and H.D.; investigation, V.V.; data curation, Y.E.M., A.E.H. and M.M.; writing—original draft preparation and editing review, Y.E.M., V.V., M.M., H.Y. and H.D.; visualization and validation; M.M., H.D. and V.V.; supervision, M.M., H.D. and V.V. All authors have read and agreed to the published version of the manuscript.

Funding: This research received no external funding.

Institutional Review Board Statement: Not applicable.

Informed Consent Statement: Not applicable.

Data Availability Statement: The data presented in this study are available on request from the corresponding author.

Acknowledgments: The authors thank Avignon and Abdelmalek Essaadi University, as well as the Dean of the National Institute of Agriculture Researchers. The authors also acknowledge the help and support of the Dean and staff of The Scientific Research Deanship at Hydrogeology Laboratory of Avignon for their critical and constructive review of the manuscript.

Conflicts of Interest: The authors declare no conflict of interest.

References

1. Jhamnani, B.; Singh, S.K. Groundwater Contamination due to Bhalaswa Landfill Site in New Delhi. *Int. J. Environ. Sci. Eng.* **2009**, *1*, 121–125.
2. Longe, E.O.; Balogun, M.R. Groundwater Quality Assessment near a Municipal Landfill, Lagos, Nigeria. *Res. J. Appl. Sci. Eng. Technol.* **2010**, *2*, 39–44.
3. El Hamdi, A.; Morarech, M.; El Mouine, Y.; Rachid, A.; El Ghmari, A.; Yameogo, S.; Chalikakis, K.; Yachou, H.; Kacimi, I.; Zouahri, A.; et al. Sources of spatial variability of soil salinity: The case of Beni Amir irrigated command areas in the Tadla Plain, Morocco. *Arid. Land Res. Manag.* **2022**, 1–20. [CrossRef]
4. Han, Z.; Ma, H.; Shi, G.; He, L.; Wei, L.; Shi, Q. A review of groundwater contamination near municipal solid waste landfill sites in China. *Sci. Total Environ.* **2016**, *569–570*, 1255–1264. [CrossRef] [PubMed]
5. Dvorski, S.E.-M.; Gonsior, M.; Hertkorn, N.; Uhl, J.; Müller, H.; Griebler, C.; Schmitt-Kopplin, P. Geochemistry of Dissolved Organic Matter in a Spatially Highly Resolved Groundwater Petroleum Hydrocarbon Plume Cross-Section. *Environ. Sci. Technol.* **2016**, *50*, 5536–5546. [CrossRef] [PubMed]
6. Gu, X.; Xiao, Y.; Yin, S.; Shao, J.; Pan, X.; Niu, Y.; Huang, J. Groundwater level response to hydrogeological factors in a semi- arid basin of Beijing, China. *J. Water Supply Res. Technol.* **2017**, *66*, 266–278. [CrossRef]
7. Arora, T.; Linde, N.; Revil, A.; Castermant, J. Non-intrusive characterization of the redox potential of landfill leachate plumes from self-potential data. *J. Contam. Hydrol.* **2007**, *92*, 274–292. [CrossRef] [PubMed]
8. Gallas, J.D.F.; Taioli, F.; Malagutti Filho, W. Induced polarization, resistivity, and self-potential: A case history of contamination evaluation due to landfill leakage. *Environ. Earth Sci.* **2011**, *63*, 251–261. [CrossRef]
9. Martínez-Pagán, P.; Jardani, A.; Revil, A.; Haas, A. Self-potential monitoring of a salt plume. *Geophysics* **2010**, *75*, WA17–WA25. [CrossRef]
10. El Mouine, Y.; El Hamdi, A.; Morarech, M.; Kacimi, I.; Touzani, M.; Mohsine, I.; Tiouiouine, A.; Ouardi, J.; Zouahri, A.; Yachou, H.; et al. Landfill Pollution Plume Survey in the Moroccan Tadla Using Spontaneous Potential. *Water* **2021**, *13*, 910. [CrossRef]
11. Naudet, V.; Revil, A. A sandbox experiment to investigate bacteria-mediated redox processes on self-potential signals. *Geophys. Res. Lett.* **2005**, *32*. [CrossRef]
12. Naudet, V.; Revil, A.; Bottero, J.-Y.; Bégassat, P. Relationship between self-potential (SP) signals and redox conditions in contaminated groundwater. *Geophys. Res. Lett.* **2003**, *30*. [CrossRef]

environmental sciences proceedings

MDPI

Proceeding Paper

Soil Salinity Assessment and Characterization in Abandoned Farmlands of Metouia Oasis, South Tunisia †

Khaled Ibrahimi [1,*], Nabil Khader [2] and Lotfi Adouni [3]

1 Higher Agronomic Institute of Chott-Mariem, The University of Sousse, Sahloul 4, Sousse 4011, Tunisia
2 Sol-Sud Engineering Consultant Office, El Hamma 6020, Tunisia; pedosolsud@hotmail.com
3 Regional Office for Agricultural Development of Gabes, Gabes 6000, Tunisia; lotfi.adouni130@gmail.com
* Correspondence: ibrahimi.isacm@gmail.com

† Presented at the 2nd International Laayoune Forum on Biosaline Agriculture, 14–16 June 2022; Available online: https://lafoba2.sciforum.net/.

Abstract: In Tunisia, the phenomenon of abandonment of agricultural fields due to soil salinization is becoming common in oasis systems. However, little is known about the salinity level and the geochemical composition of abandoned oasis soils. A total of 156 soil samples were collected from abandoned plots in the Metouia Oasis, south-east of Tunisia, and characterized for electrical conductivity and cation and anion composition. In addition, a spatial analysis using spatial data and field verification was carried out. The results can be used to establish a management plan to address the problem of farmland abandonment associated with soil salinization and ensure the viability and sustainability of oasis systems.

Keywords: soil salinity; farmland abandonment; spatial analysis; anions; cations; oasis system

check for updates

Citation: Ibrahimi, K.; Khader, N.; Adouni, L. Soil Salinity Assessment and Characterization in Abandoned Farmlands of Metouia Oasis, South Tunisia. *Environ. Sci. Proc.* **2022**, *16*, 3. https://doi.org/10.3390/environsciproc2022016003

Academic Editor: Abdelaziz Hirich

Published: 16 June 2022

1. Introduction

In recent years, agricultural land abandonment has become a serious problem in many countries worldwide. Land abandonment due to soil salinization has been reported in China [1], Saudi Arabia [2], and most European countries [3]. This phenomenon is leading to the loss of agricultural soils and threatening food security and local livelihoods. In Tunisia, the phenomenon of abandonment of agricultural fields is becoming a common occurrence in oasis systems where it is usually associated with soil salinization. However, little is known about the salinity level and the geochemical composition of abandoned oasis soils. Such information would enable a better understanding of salinization sources, which is crucial for amelioration and for the development of sustainable management of salt-affected soils. Therefore, the objectives of our study were: (1) to assess the salinity of abandoned soils in a coastal oasis of south Tunisia, (2) to characterize the salt profiles of the abandoned soils.

2. Materials and Methods

The Metouia Oasis is part of the coastal oases located in the south-east of Tunisia. It covers an area of about 270 ha divided into 2132 plots distributed among 1450 farmers. The average sized are 0.1862 ha/owner and 0.1266 ha/plot. The oasis of Metouia is characterized by an arid climate where precipitation is rare (164.86 mm/year on average over the 10-year period) and irregular. The maximum annual average temperature is 31.5 °C, while the minimum annual average temperature is 11.23 °C. The average monthly evaporation varies from 62.65 mm in January to 210.45 mm in August, which shows a water deficit over the whole year of about 1000 mm, given an average rainfall of 164.86 mm.

In order to characterize soil salinity in the abandoned farmlands, soil samples were collected from 26 locations from six depths. A total of 156 soil samples were collected from abandoned plots. The electrical conductivity (EC) of the soil extracts was determined using a conductivity meter. The cations (Ca^{2+}, Mg^{2+}, Na^+, K^+) and anions (Cl^-, $SO_4{}^{2-}$, $NO_3{}^-$, Br^-) were determined by the simultaneous determination of anions and cations using ion chromatography [4]. In order to characterize the chemical properties of the saturated paste extracts of the abandoned soils, a Piper diagram was used for presenting their anionic and cationic compositions. Additionally, spatial analysis using spatial data and field investigation was carried out to identify and characterize abandoned farmlands in the Metouia Oasis. In this study, high resolution Google Earth satellite images of the year 2020 covering the study area were utilized to identify the abandoned farmlands through onscreen digitizing, using ArcGIS software. The outcomes of this image-based identification were then verified and corrected through several field visits. The plot boundaries were set based on the plot map of the Metouia Oasis.

3. Results

3.1. Salinity Status of Abandoned Soils

The soil surface salinity of the abandoned plots showed a very high salinity value, exceeding 60 dS m^{-1} in all sampled soils. The mean EC values of abandoned soils varied between a maximum of 76 dS m^{-1} at the soil surface (0–20 cm) to a minimum of 27 dS m^{-1} in the bottom layer (100–120 cm). In most of the locations, the EC values showed a decreasing trend from the surface to the bottom of the soil profile (Figure 1). Additionally, the correlation analysis showed high and positive correlation coefficients of soil salinity between all soil layers ($p < 0.05$).

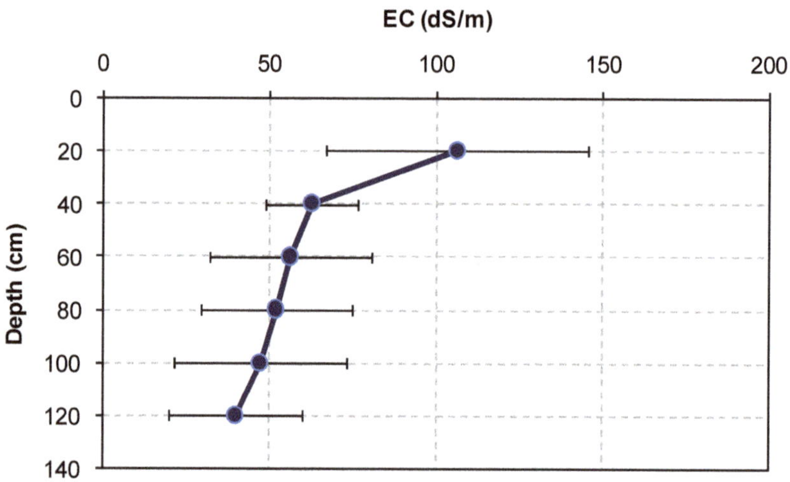

Figure 1. Soil salinity profile at abandoned farmlands in Metouia Oasis.

3.2. Chemical Properties of Abandoned Soils

At the surface layer, cation concentrations were in the order $Na^+ > Mg^{2+} > Ca^{2+} > K^+$ while anions were in the order $Cl^- > SO_4{}^{2-} > NO_3{}^- > Br^-$. At the deepest sampled layer (100–120 cm) cation concentrations were in the decreasing order of $Na^+ > Ca^{2+} > Mg^{2+} > K^+$ and anions had similar patterns as the surface layer. From a cationic point of view, saturation extracts have dominant sodium content at the surface and bottom layers. As for the anionic composition, in most of the studied profiles, the saturation extracts were chlorine dominant (Figure 2).

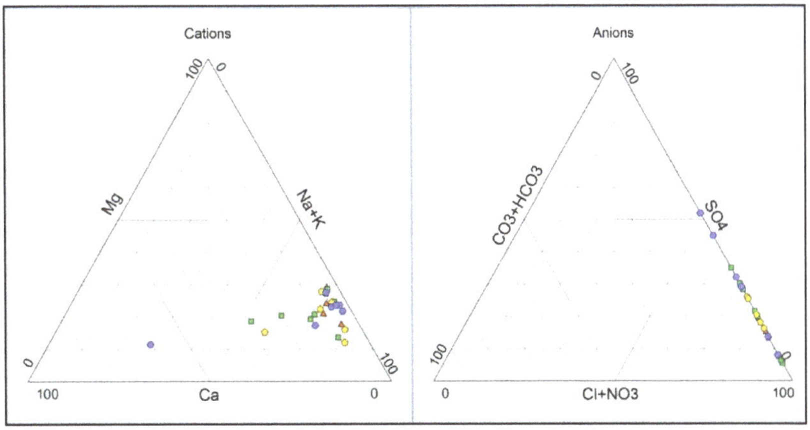

Figure 2. Piper diagram presenting anionic and cationic compositions of the saturated paste extracts of abandoned soils.

3.3. *Spatial Distribution of Abandoned Farmlands*

The spatial distribution of abandoned farmlands in Metouia Oasis is shown in Figure 3. We found that there are more abandoned farmlands throughout the oasis area. In total, there are 312 abandoned farmlands occupying 66 ha, which represent 24% of the total oasis area.

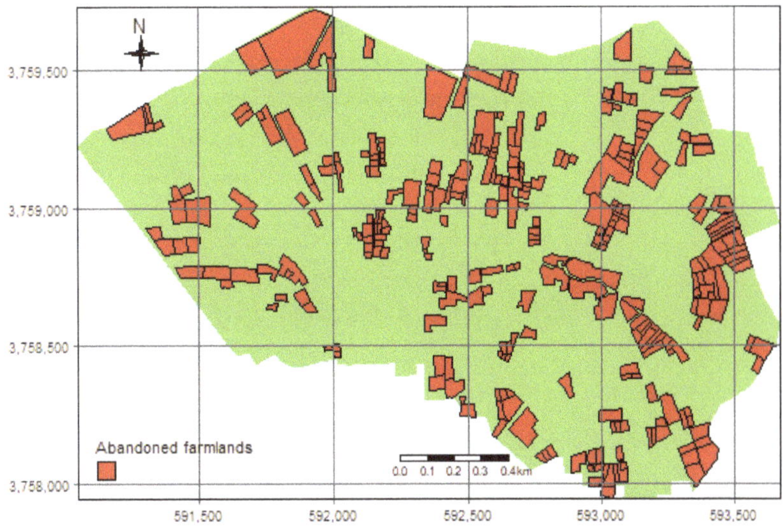

Figure 3. Spatial distribution of abandoned farmlands in Metouia Oasis.

4. Discussion

The EC data showed that soils of abandoned plots in the Metouia Oasis were characterized as highly saline. Our results are consistent with those of Costa et al. [5], who reported extremely high values of salinity in abandoned soils of the oasis of Masafi in the United Arab Emirates. Gopalakrishnan and Kumar [6] also reported extremely saline soils in permanently abandoned paddy lands in the semi-arid region of the Jaffna Peninsula, Sri Lanka. In our study area, it was observed that soil salinity gradually decreased with increasing depth. This pattern of salt profiles in the abandoned plots can be attributed to the effect of evaporation, leading to continuous and significant salt accumulation in the soil surface layer [7]. In our study site, the aridity of the climate and the absence of leaching of salts have favored the existence of this upward salinity gradient of the abandoned soils. Previous studies carried out by Ibrahimi et al. [8] in the Metouia Oasis showed that groundwater depth and salinity are the major factors affecting soil salinization. Considering the soil profile, the adjacent layers were better correlated in terms of EC levels than distant ones, suggesting the more pronounced mutual effect of soil salinity between soil layers which are close to each other. Similarly, Wang et al. [9] reported higher correlation coefficients of soil salinity between two adjacent soil layers than between non-adjacent ones. Moreover, in our study area, geochemical analysis showed that abandoned soils appear enriched in Na and Cl, suggesting that the main sources of soil salinity in the abandoned plots of Metouia Oasis were sodium and chloride.

5. Conclusions

Abandoned farmlands are spread throughout Metouia Oasis. Soils at these plots were characterized by very high salinity rates. The salt profiles in these soils showed salt accumulation at the surface layer with an upward gradient of salinity. Saturation extracts of abandoned soils showed dominant sodium and chloride ions at surface and bottom layers of the soil profile. The results of this study can be used to establish a management plan to address the problem of farmland abandonment and associated soil salinization in order to ensure the viability and the sustainability of oasis systems.

Author Contributions: Methodology, K.I. and N.K.; software, N.K. and L.A.; formal analysis, K.I., N.K. and L.A.; data curation, K.I.; writing—original draft preparation, K.I.; writing—review and editing, K.I. All authors have read and agreed to the published version of the manuscript.

Funding: This research received no external funding.

Institutional Review Board Statement: Not applicable.

Informed Consent Statement: Not applicable.

Data Availability Statement: Not applicable.

Acknowledgments: Authors are thankful to the Soil Analysis Laboratory staff (CRDA Gabes) for their technical support.

Conflicts of Interest: The authors declare no conflict of interest.

References

1. Li, Y.; Shan, Y.; Chen, Y. Analysis of Farmland Abandonment and Government Supervision Traps in China. *Int. J. Environ. Res. Public Health* **2021**, *18*, 1815. [CrossRef] [PubMed]
2. Al-Rowaily, S.L.; Al-Dosari, D.H.; Assaeed, A.M.; Abd-ElGawad, A.M.; El-Sheikh, M.A.; El-Bana, M.I.; Al-Taisan, W.A. Native Perennial Plants Colonizing Abandoned Arable Fields in a Desert Area: Population Structure and Community Assembly. *Agriculture* **2020**, *10*, 550. [CrossRef]
3. Kozak, M.; Pudełko, R. Impact Assessment of the Long-Term Fallowed Land on Agricultural Soils and the Possibility of Their Return to Agriculture. *Agriculture* **2021**, *11*, 148. [CrossRef]
4. Nesterenko, P.N. Simultaneous separation and detection of anions and cations in ion chromatography. *Trends Anal. Chem.* **2001**, *20*, 311–319. [CrossRef]
5. Costa, S.; Purdue, L.; Dufour, A.; Charbonnier, J. An oasis soil reference collection for the identification and study of ancient cultivated soils in arid environments (Oasis of Masafi, United Arab Emirates). *Geoarchaeology* **2021**, *36*, 404–428. [CrossRef]
6. Gopalakrishnan, T.; Kumar, L. Linking Long-Term Changes in Soil Salinity to Paddy Land Abandonment in Jaffna Peninsula, Sri Lanka. *Agriculture* **2021**, *11*, 211. [CrossRef]
7. Liu, S.; Huang, Q.; Ren, D.; Xu, X.; Xiong, Y.; Huang, G. Soil evaporation and its impact on salt accumulation in different landscapes under freeze–thaw conditions in an arid seasonal frozen region. *Vadose Zone J.* **2021**, *20*, e20098. [CrossRef]
8. Ibrahimi, K.; Miyazaki, T.; Nishimura, T.; Imoto, H. Contribution of shallow groundwater rapid fluctuation to soil salinization under arid and semiarid climate. *Arab. J. Jeosci.* **2014**, *7*, 3901–3911. [CrossRef]
9. Wang, Z.; Zhao, G.; Gao, M.; Chang, C. Spatial variability of soil salinity in coastal saline soil at different scales in the Yellow River Delta, China. *Environ. Monit. Assess.* **2017**, *189*, 80. [CrossRef] [PubMed]

environmental sciences proceedings

MDPI

Proceeding Paper

Spatial Variability of Soil Salinity: The Case of Beni Amir in the Tadla Plain of Morocco †

Amal El Hamdi [1,2,3,*], Yousra El Mouine [1,2,3], Moad Morarech [2], Vincent Valles [1,4], Hasna Yachou [3] and Houria Dakak [3]

1 Laboratory Mediterranean Environment and Agro-Hydrosystem Modelling (UMR 1114 EMMAH), Hydrogeology Department, Avignon University, 84000 Avignon, France; yousra.elmouine@gmail.com (Y.E.M.); vincent.valles@univ-avignon.fr (V.V.)
2 Laboratory of Applied and Marine Geosciences, Geotechnics and Geohazards (LR3G), Faculty of Sciences (FS), University Abdelmalek Essaadi, Tetouan 93000, Morocco; morarech2000@gmail.com
3 National Institute of Agricultural Research (INRA), CRRAR, URECRN, Rabat 10000, Morocco; hasnayachou@gmail.com (H.Y.); dakak_h@yahoo.fr (H.D.)
4 Faculté des Sciences et Techniques (FSTBM), Sultan Moulay Slimane University, BP 523, Beni Mellal 23000, Morocco
* Correspondence: amalgoaea@gmail.com
† Presented at the 2nd International Laayoune Forum on Biosaline Agriculture, 14–16 June 2022; Available online: https://lafoba2.sciforum.net/.

Abstract: The control of soil salinity and the achievement of sustainable agricultural development goals require knowledge of the soil salinity spatial distribution and its temporal evolution. The objective of this study was to understand the spatial variations of soil salinity in the geomorphologic zone in Beni Amir, Tadla region of Morocco. Electromagnetic induction was used to measure the apparent electrical conductivity (ECa) of the soil. The results showed that the distribution of ECa varies across the distance from the drain, and suggests the movement of salts from upstream to downstream. The vertical-horizontal electromagnetic readings show that the saline profile is descending. This study shows that the spatial variability of salinity in a geomorphological site is related to the position in the landscape

Keywords: salinization; geomorphology; irrigation; electromagnetic induction

Citation: El Hamdi, A.; El Mouine, Y.; Morarech, M.; Valles, V.; Yachou, H.; Dakak, H. Spatial Variability of Soil Salinity: The Case of Beni Amir in the Tadla Plain of Morocco. *Environ. Sci. Proc.* **2022**, *16*, 9. https://doi.org/10.3390/environsciproc202201609

Academic Editor: Abdelaziz Hirich

Published: 16 June 2022

1. Introduction

The irrigated perimeter of Tadla, located in the center of Morocco, is among the oldest and most important agricultural production regions in the country. The region includes a 300,000-ha agricultural area of which 124,600 ha is irrigated [1,2]. The study region Beni Amir is an irrigated sub-perimeter of Tadla. It irrigates with the waters from Oued Oum Er Rbia, which contains 1 g/L of salt in the groundwater. The practice of irrigation using poor-quality (high-salinity) water coupled with non-maintenance of drainage canals leads to the deterioration of soil and groundwater salinization [3,4]. These unsustainable practices impact agricultural productivity and hence lead to land abandonment and desertification.

Several studies have been conducted to minimize soil salinization in the irrigated perimeter of Tadla. Some have improved knowledge of the nature of degradation mechanisms [3]. Previously, a few soil salinity mapping works have been conducted. Some of them used many measurements, but on a small area; others tried to draw a salinity map of Tadla [5,6], but with a number of weak measures. The aim of this study was to understand the mechanisms for spatial variations of soil salinity in natural areas as affected by geomorphology via the use of a geophysical method.

2. Materials and Methods

2.1. Study Site

The study region Beni Amir lies under the irrigated perimeter of the Tadla plain (Figure 1). It irrigates using the saline waters from Oum Er Rbia River, which contains salinity of 2.25 dS/m. The overall salinity of the soil in the study region varies between 0.35 and 13 dS/m [4].

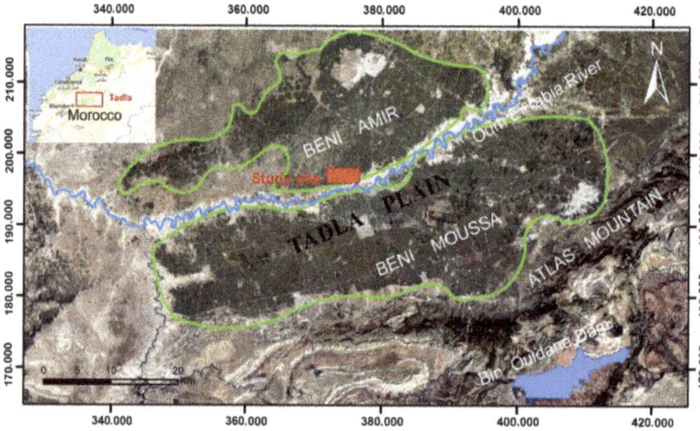

Figure 1. Study area under the irrigated perimeter of Tadla, Morocco.

The study region has regular topography with an altitude of 400 m. Its latitude ranges from 32.06 to 32.56 and its longitude ranges from −7.09 to −6.24. The climate of the study site is Mediterranean type arid to semi-arid with a continental character marked by a cold and humid winter and a hot and dry summer and annual precipitation of 350 mm, irregularly distributed. The summer season is characterized by the absence of precipitation with the exception of rare brutal showers. Temperatures vary greatly seasonally. Winter minimum temperature ranges from 0 °C to 5 °C, while the summer maximums range from 38 °C to 42 °C. The average annual evaporation is 1796 mm and varies between 1500 mm and 2000 mm.

2.2. Data Collection

The Geonic EM-38RT Instrument was used to map and diagnose highly conductive soil such as salt soils [7]. Its easy and fast implementation allows for the collection of a large amount of data. The field measurement was carried out over two periods. In the first measurement period of July 2019, a total of 320 measurements of electromagnetic conductivity were carried out in half a day and detected by the GPS survey. The second measurement period was in the month of July 2021; during this period, a total of 365 measurements of electromagnetic conductivity were detected by the GPS survey. For both campaigns, a calibration data set corresponded to five measurement points selected from the measurement points mentioned above so as to spatially represent the entire study area. For each point, the GPS position, the deep conductivity VD (0–0.80 m), and surface conductivity HD (0–0.40 m) were measured. Upstream, the measurements were carried out along paths that intersect the valleys with a distance of about 7 m between the measurements, intensifying the measurements closer to the downstream.

3. Results and Discussion

3.1. Calibration of Electromagnetic Response

In order to be able to express the electromagnetic measurements (EC_a) in terms of the soil salinity of the root zone, sampling was carried out for calibration of the electromagnetic conductivity measurements using data from five points scanning the entire range of salinity.

A linear relation was established between EC_{aH} obtained in the field and EC_e measured on the saturated extracts, with a correlation coefficient (r^2) of 0.94.

3.2. *Apparent Electromagnetic Conductivity*

The EC_{aV} and EC_{aH} salinity maps of the two components (Figures 2 and 3) show a certain salinity gradient oriented from north-west to south-east where (1) non-saline soils (less than mS/m) occupy most of the Talweg head area (northwest part); (2) the slightly saline soils (between 2 mS/m and 4 mS/m) occupy a large part mainly in cultivated plots (between the Talweg head and Oued Oum Er Rbia = terrace); (3) saline soils (above 4 mS/m) occupy a large part along the Oued Oum Er Rbia. The distribution of salinity is a function of the distance from the drains, suggesting the existence of movement of salts from upstream to downstream.

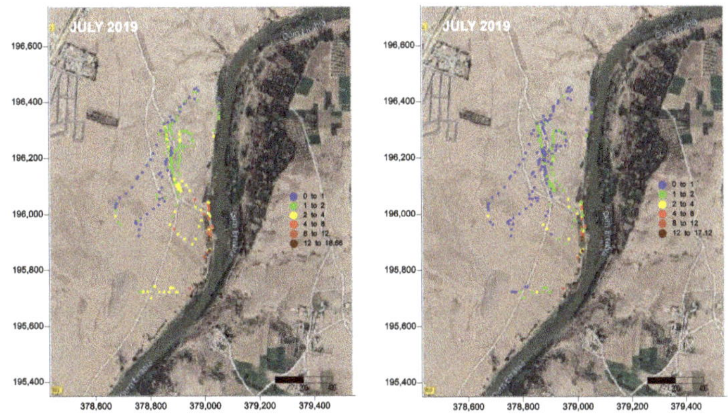

(a) (b)

Figure 2. Distribution of EC_{aV} (**a**) and EC_{aH} (**b**) in July 2019.

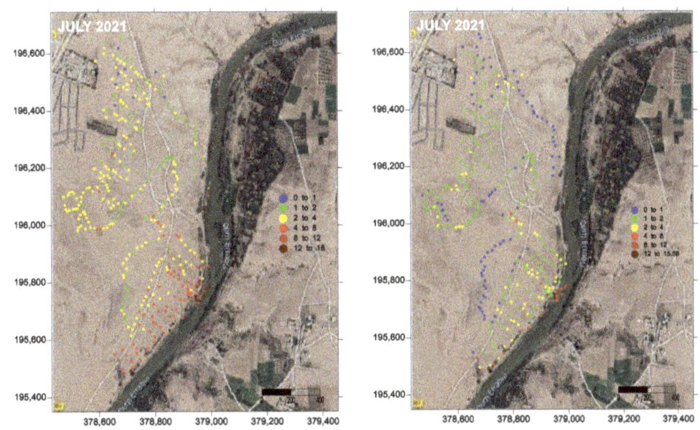

(a) (b)

Figure 3. Distribution of EC_{aV} (**a**) and EC_{aH} (**b**) in July 2021.

Salt distribution showed a systematic increase in ECa from the watershed to the valley, with salinity increasing downstream of geological structures. Salt distribution has been controlled by the geomorphologic properties of the basin, which control hydrological and hydrogeological processes, and by the recharge discharge mechanisms that influence salt mobilization and accumulation.

3.3. A Negative Hydrogeological Balance and Increasing Groundwater Salinity

In our study area, the depth of the water table has been increasing over the last 24 years. This trend is in agreement with the observations of farmers who mention episodes wells drying up in recent years. This result highlights the effect of water withdrawal from wells on the local hydrogeological balance as observed in many irrigated plains in arid regions [2–8].

4. Conclusions

Groundwater dynamics and management are essential determinants of salinity. In the absence of a functional drainage system or in the case of excessive irrigation using poor-quality water, the downstream areas are subjected to congestion and the process of salt concentration, which is enhanced due to evaporation. Conversely, when drainage is efficient, large quantities of salts can then be exported.

Author Contributions: Project administration: A.E.H.; conceptualization, formal analysis, funding acquisition, and resources: A.E.H., Y.E.M., V.V. and M.M.; methodology: V.V.; software: A.E.H., Y.E.M. and H.D.; investigation: M.M.; data curation: A.E.H., Y.E.M. and M.M.; writing—original draft preparation and editing review: A.E.H., Y.E.M., M.M., V.V., H.Y. and H.D.; visualization and validation: V.V. and M.M.; supervision: V.V., M.M., H.D. and H.Y. All authors have read and agreed to the published version of the manuscript.

Funding: This research received no external funding.

Institutional Review Board Statement: Not applicable.

Informed Consent Statement: Not applicable.

Data Availability Statement: The data presented in this study are available on request from the corresponding author.

Acknowledgments: The authors wish to thank Avignon and Abdelmalek Essaadi Universites, and the Dean of the National Institute of Agricultural Research. The authors are also grateful to the staff of the Hydrogeology Laboratory of Avignon for their help and support. Critical and constructive reviews of the manuscript by the reviewers were greatly treasured.

Conflicts of Interest: The authors declare no conflict of interest.

References

1. Jebbour, S. Combined Use of Surface Water and Groundwater in the Perimeter of the Tadla: Quantitative and Qualitative Aspect. Third Cycle Dissertation in Rural Engineering, Agronomic and Veterinary Institute Hassan II, Rabat, Morocco, 1995; p. 177.
2. Bouchaou, L.; Choukr-Allah, R.; Hirich, A.; Ennasr, M.S.; Malki, M.; Abahous, H.; Bouaakaz, B.; Nghira, A. Climate Change and Water Valuation in Souss-Massa Region: Management and Adaptive Measures. *Eur. Water* **2017**, *60*, 203–209.
3. El Hamdi, A.; El Mouine, Y. Sources of spatial variability of soil salinity: The case of Beni Amir irrigated command areas in the Tadla Plain, Morocco. *Arid. Land Res. Manag.* **2022**, *6*, 140–147. [CrossRef]
4. Devkota, K.P.; Devkota, M.; Rezaei, M.; Oosterbaan, R. Managing salinity for sustainable agricultural production in salt-affected soils of irrigated drylands. *Agric. Syst.* **2022**, *198*, 103390. [CrossRef]
5. Dakak, H.; Huang, J.; Zouahri, A.; Douaik, A.; Triantafilis, J. Mapping Soil Salinity in 3-Dimensions Using an EM38 and EM4Soil Inversion Modelling at the Reconnaissance Scale in Central Morocco. *Soil Use Manag.* **2017**, *33*, 553–567. [CrossRef]
6. Amezketa, E. Soil salinity assessment using directed soil sampling from a geophysical survey with electromagnetic technology: A case study. *Span. J. Agric. Res.* **2007**, *5*, 91–101. [CrossRef]
7. Soudi, B.; Badraoui, M. The Storage of Organic Carbon in Dryland Soils of Africa Constraints and Opportunities. In *Soil and Fertilizers Managing the Environmental Footprint*; Lal, R., Ed.; CRC Press: Boca Raton, FL, USA, 2020; p. 372.
8. Qadir, M.; Sharma, B.R.; Bruggeman, A.; Choukr-Allah, R.; Karajeh, F. Non-Conventional Water Resources and Opportunities for Water Augmentation to Achieve Food Security in Water Scarce Countries. *Agric. Water Manag.* **2007**, *87*, 2–22. [CrossRef]

environmental sciences proceedings

MDPI

Proceeding Paper

Spatial Distribution of Desert Plant Species According to Soil Salinity †

Hafsa Debbagh-Nour [1,*], Mohamed Belcaid [1], Mohamed Ibourki [1], Hassan Boukcim [2], Marc Ducousso [3], Youssef El Kharrassi [1], Ayoub El Mouttaqi [1] and Abdelaziz Hirich [1]

1 African Sustainable Agriculture Research Institute (ASARI), Mohammed VI Polytechnic University (UM6P), Laayoune 70000, Morocco; med.belcaid80@gmail.com (M.B.); mohamed.ibourki@um6p.ma (M.I.); youssef.elkharrassi@um6p.ma (Y.E.K.); ayoub.elmouttaqi@um6p.ma (A.E.M.); abdelaziz.hirich@um6p.ma (A.H.)

2 AgroBioSciences Research Division, Mohammed VI Polytechnic University, Lot 660–Hay Moulay Rachid, Ben-Guerir 43150, Morocco; hassan.boukcim@um6p.ma

3 Laboratoire des Symbioses Tropicales et méditerranéennes (LSTM, UMR113), French Agricultural Research Centre for International Development (CIRAD), TA A-82/J, CEDEX 5, 34398 Montpellier, France; marc.ducousso@cirad.fr

* Correspondence: hafsa.debbagh-nour@um6p.ma

† Presented at the 2nd International Laayoune Forum on Biosaline Agriculture, 14–16 June 2022; Available online: https://lafoba2.sciforum.net/.

Abstract: The arid climate of the Saharan regions is accentuated even more by the consequences of climate change and environmental threats, thus affecting normal plant development. The salinization of soils is one of the major stresses in the Moroccan Sahara Desert that strongly penalizes the production. In this sense, it would be interesting to explore the native flora of the target regions, and using it remains the best option to provide sustainable agriculture. The present study is part of a collaborative project with the aim of exploring desert plant species. Different parameters were taken for each species, including soil salinity and density. Two zones showed the highest levels of salinity, Daoura and Akhfennir, both in the Tarfaya province with 19.38 and 24.33 mS/cm, respectively. Although, several species were highly tolerant to salinity and were present at moderate to high densities. Among them, one can cite *Aeluropus littoralis* (Gouan) Parl., *Halocnemum strobilaceum* (Pall.) M. Bieb. and *Suaeda ifniensis* Caball. ex Maire.

Keywords: spatial distribution; soil salinity; desert plant; Sahara Desert; Morocco

Citation: Debbagh-Nour, H.; Belcaid, M.; Ibourki, M.; Boukcim, H.; Ducousso, M.; El Kharrassi, Y.; El Mouttaqi, A.; Hirich, A. Spatial Distribution of Desert Plant Species According to Soil Salinity. *Environ. Sci. Proc.* **2022**, *16*, 10. https://doi.org/10.3390/environsciproc2022016010

Academic Editor: Redouane Choukr-Allah

Published: 16 June 2022

1. Introduction

With each passing year, climate change is becoming more intense, especially in desert ecosystems, which strongly impacts major crops. Native desert plant species seem an interesting alternative, as they may play an important role in sustainable development thanks to their environmental and economical values. In addition, they will allow for the exploitation of the existing biodiversity. The target Sahara Desert covers an area of about 270 km^2. The lasted botanical expeditions held in this area in 2021 revealed that it shelters around 360 species of which 328 are entirely identified. Those species belong to 29 orders, 58 families, and 207 genera. The most common families include *Asteraceae*, *Poaceae*, *Leguminosae*, *Amaranthaceae*, and *Brassicaceae* and their prevalence is mainly due to the adaptability to desert conditions. Native species have the ability to survive in areas with harsh conditions: irregular rainfall, high temperatures, saline soils, and scarce water resources. Unfortunately, the large majority of these species are underused or not even known by the inhabitants of the regions.

Salinity is among the major threats in desert areas. Based on soil salinity classes defined by the Food and Agriculture Organization (FAO) [1], 14% of the soils in our study area are very strongly saline. They are mainly concentrated in the coastal zones and the Sabkhas.

With the aim of preserving the rich floral diversity, but also valorizing many species, identification and characterization are deemed to be advantageous. A collaborative project was launched to elaborate an atlas book containing relevant information related to the characteristics, virtues, and indigenous knowledge of native plants. Botanical expeditions were carried out in 2021 and targeted 120 sites in the Sahara Desert.

2. Material and Methods

2.1. Ethnobotanical Field Trips

The ethnobotanical field trips covered 120 sites in the two regions of Laayoune Sakia El Hamra and Dakhla Oued Dahab (Figure 1) which span an area of 270 km^2. They are bordered on the north by the Guelmim Oued Noun region and on the south by Mauritania. The itinerary was set up in advance on the basis of remote sensing and with the help of guides who knew the area very well. Whenever a remarkable site was reached, samples of the species and the soil were taken in addition to other crucial parameters for identification: plant density, geographic coordinates, etc.

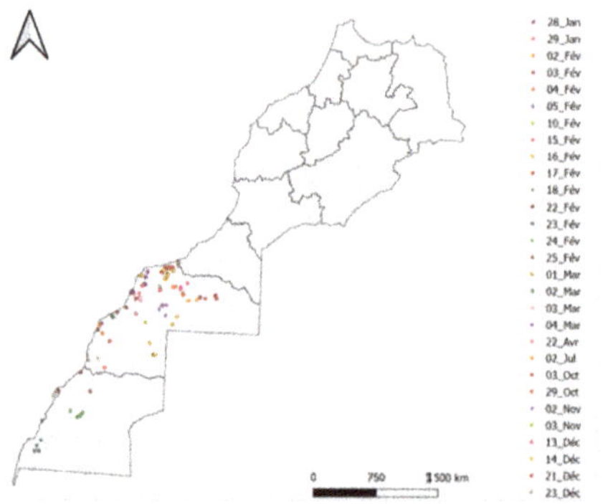

Figure 1. Geographic distribution of the visited sites on the map of Morocco.

2.2. Climate

Based on Köppen climate classification [2], the climate in the study area is considered arid. It is characterized by cold winters and dry and very hot summers. On the coastal strip, the temperature tends to be moderate due to the proximity to the Atlantic Ocean. As we move inside the land to the east, the climate becomes drier. As for daylight, it is almost permanent. In terms of rainfall, it is irregular and scarce. The average annual precipitation is around 60 mm [3,4].

2.3. Measured Parameters

2.3.1. Soil Salinity Estimation

Soil samples were taken along the route. Salinity was determined by the electrical conductivity (EC) according to the international standard ISO 11265 [5]. This method consists of using a soil:water ratio of 1:5. The sample was mixed to allow the dissolution of electrolytes, then sieved. The EC was measured using a multiparameter meter (Mettler Toledo, Columbus, OH, USA). The results were displayed in millisiemens per centimeter (mS/cm).

2.3.2. Density

The density of vegetation was represented by a 1–5 scale: 1 being very low density and 5 being very high.

3. Results & Discussion

Soil salinity varied greatly between locations. Figure 2 shows the distribution of the soils by salinity classes according to FAO [1]. Over one-half of the Sahara Desert soils are non-saline whereas moderately saline and very strongly saline soils represent 17% and 14%, respectively.

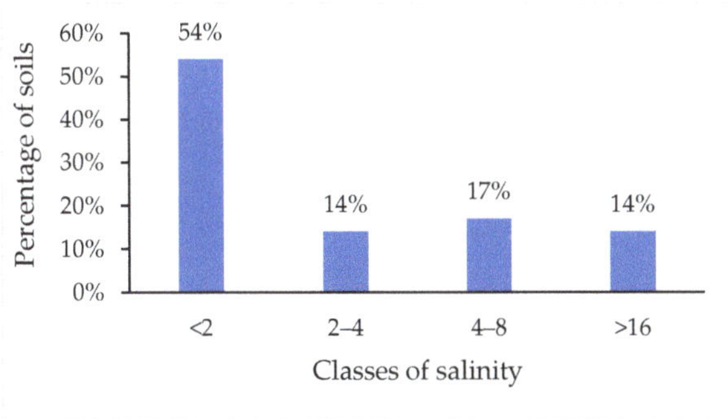

Figure 2. Distribution of soil samples according to salinity class.

Two zones of the Tarfaya province displayed the highest levels of salinity, Daoura with 19.38 and Akhfennir with 24.33 mS/cm. Still, certain species have shown a high tolerance and were present at moderate to high densities. The distribution of the species by salinity classes is displayed in Figure 3. Around 70% of the species are growing in non-saline soils, while only 15 species are found in strongly saline and very strongly saline soils. They include among others, *Aeluropus littoralis* (Gouan) Parl., *Halocnemum strobilaceum* (Pall.) M. Bieb. and the genus *Suaeda* which can be considered halophytes based on their salt-tolerating capacity and as mentioned in the literature [6–8]. *Acacia tortilis* (Forssk.) Hayne subsp. *Tortilis* for example, was present in a highly saline location but is not considered a halophyte according to others' research studies [9,10]. Conversely, the species belonging to the genus *Salsola* are considered halophytes [11], but none of *Salsola longifolia* and *S. gymnomaschala* were encountered in a saline area.

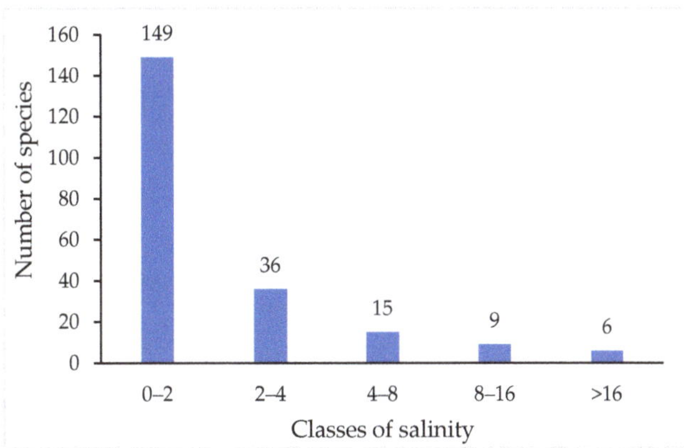

Figure 3. Distribution of the desert plant species according to salinity class.

4. Conclusions

The present study has enabled us to identify a wide array of species. A selection of them displayed a wide tolerance to high salinity levels. It would be interesting to set up and extend trials to many localities while increasing the number of repetitions to guarantee the durability of performances. Several native desert species were selected as results of several botanical expeditions conducted in 2020–2021 and will be subjected to the domestication process for valorization purposes as feed cosmetic or medicinal plants.

Author Contributions: Conceptualization, A.H.; methodology, A.H. and H.B.; software, A.E.M.; validation, A.H., H.B. and M.D.; formal analysis, A.H.; investigation, H.D.-N., M.I., Y.E.K., A.E.M., M.B. and A.H.; resources, A.H.; data curation, A.H.; writing—original draft preparation, H.D.-N.; writing—review and editing, A.H.; visualization, H.D.-N. and A.E.M.; supervision, A.H.; project administration, A.H.; funding acquisition, A.H. All authors have read and agreed to the published version of the manuscript.

Funding: This research was funded by OCP Phosboucraa Foundation within the project titled "Development of Atlas book on threatened local native plant species of the desert in Laayoune Sakia El Hamra region for various uses".

Institutional Review Board Statement: Not applicable.

Informed Consent Statement: Not applicable.

Data Availability Statement: Not applicable.

Acknowledgments: The authors gratefully acknowledge the OCP-Phosboucraa Foundation, for providing funds to conduct the project activities. We also appreciate the efforts of different collaborators including local guides.

Conflicts of Interest: The authors declare no conflict of interest.

References

1. Abrol, I.P.; Yadav, J.S.P.; Massoud, F.I. *Salt-Affected Soils and Their Management*; FAO Soils Bulletin No. 39; Food and Agriculture Organization: Rome, Italy, 1988; p. 131.
2. Arnfield, A.J. Köppen Climate Classification. 2020. Available online: https://www.britannica.com/science/Koppen-climate-classification (accessed on 29 December 2021).
3. Haut-Commissariat au Plan. Monographie de la Région Laayoune-Sakia El Hamra, 2020. Available online: https://www.hcp.ma/region-laayoune/attachment/2258177/ (accessed on 2 January 2022).
4. Haut-Commissariat au Plan. Monographie de la région Dakhla Oued Dahab, 2018. Available online: https://www.hcp.ma/region-eddakhla/attachment/1410090/ (accessed on 2 January 2022).
5. International Organization for Standardization (ISO). *ISO 11265:1996 Soil Quality—Determination of the Specific Electrical Conductivity*; ISO: Geneva, Switzerland, 1996.
6. Gulzar, S.; Khan, E. Seed germination of a halophytic grass Aeluropus logopoides. *Ann. Botany* **2001**, *87*, 319–324. [CrossRef]
7. Esmaeli, M.; Dadkhah, N.; Pirdashti, H.; Sepanloo, M. Response of two halophyte plants (*Nitraria schoberi* and *Halocnemum strobilaceum*) to potassium sulfate under saline condition. *Ukr. J. Ecol.* **2017**, *7*, 359–365. [CrossRef]
8. Fisher, D.D.; Schenk, H.J.; Thorsch, J.A.; Ferren, W.R. Leaf Anatomy and Subgeneric Affiliations of C3 and C4 Species of *Suaeda* (Chenopodiaceae) in North America. *Am. J. Bot.* **1997**, *84*, 1198–1210. [CrossRef] [PubMed]
9. Alhaddad, F.A.; Abu-Dieyeh, M.H.; ElAzazi ES, M.; Ahmed, T.A. Salt tolerance of selected halophytes at the two initial growth stages for future management options. *Sci. Rep.* **2021**, *11*, 10194. [CrossRef] [PubMed]
10. Veste, M.; Todt, H.; Breckle, S.W. Influence of halophytic hosts on their parasites-the case of *Plicosepalus acaciae*. *AoB Plants* **2014**, *7*, plu084. [CrossRef] [PubMed]
11. Taia, W.K.; Shiha, M.A.; Al-Kogali, A.A.; Abd-Almaged, A.M. Anatomical and chemo-taxonomical investigations within some *Salsola* L. species grown in the western costal region of Egypt. *Int. J. Sci. Res.* **2018**, *6*. [CrossRef]

environmental sciences proceedings

MDPI

Proceeding Paper

Mapping and Analysis of Irrigation Water Quality in the Coastal Region of Skhirat, Morocco †

Ibtissam Mariami [1,2,*], Houria Dakak [2], Abdelmjid Zouahri [2], Ahmed Douaik [2], Hamza Iaaich [2], Hinde Cherkaoui Dekkaki [3], Hasna Yachou [2] and Ahmed Ghanimi [1]

1 Department of Chemistry, Faculty of Sciences, Mohammed V University, P.O. Box 1014, Rabat 10100, Morocco; ghanimiahmed@gmail.com

2 Research Unit on Environment and Conservation of Natural Resources, Regional Center of Rabat, National Institute of Agricultural Research (INRA), P.O. Box 6356, Instituts, Rabat 10101, Morocco; houria.dakak@inra.ma (H.D.); abdelmjid.zouahri@inra.ma (A.Z.); ahmed.douaik@inra.ma (A.D.); hamza.iaaich@inra.ma (H.I.); hasna.yachou@inra.ma (H.Y.)

3 Laboratory of Research and Development in Engineering Sciences, Faculty of Sciences and Techniques, Abdelmalek Essaadi University, P.O. Box 34, Ajdir, Al Hoceima 32003, Morocco; hcherkaouidekkaki@uae.ac.ma

* Correspondence: ibtissam11smp@gmail.com

† Presented at the 1st International Online Conference on Infrastructures, 7–9 June 2022; Available online: https://ioci2022.sciforum.net/.

Abstract: This study analyzes map irrigation water quality in the region of Skhirat, Morocco. This study involves the evaluation of the physico-chemical quality of the irrigation water using Piper and ULSS diagrams and spatial mapping using GIS. The results showed significant salinization power, and demonstrated that the salinity and alkalinity classes of irrigation water dominating in the region are C3-S1 (i.e., average to poor quality; 'use with caution'), C4-S1 (i.e., poor quality; 'exclude sensitive plants and heavy soils'), and C4-S2 (i.e., poor quality; 'to be used with great care, only in light soils'). The evaluation of the Piper diagram determines two hydrochemical facies. The bathymetric map indicates a shallow level downstream and South-West. The salinity map shows high salinity downstream and upstream. In conclusion, a very alarming degradation of water in terms of salinity is noted in the region.

Keywords: Skhirat area; mapping; water quality; salinity; hydrochemistry

Citation: Mariami, I.; Dakak, H.; Zouahri, A.; Douaik, A.; Iaaich, H.; Cherkaoui Dekkaki, H.; Yachou, H.; Ghanimi, A. Mapping and Analysis of Irrigation Water Quality in the Coastal Region of Skhirat, Morocco. *Environ. Sci. Proc.* **2022**, *16*, 20. https://doi.org/10.3390/environsciproc2022016020

Academic Editor: Abdelaziz Hirich

Published: 16 June 2022

1. Introduction

Morocco is essentially a semi-arid country with limited rainfall (rainfed system), which reduces opportunities for agriculture. To feed the ever-growing population, irrigation is necessary. Groundwater is the major source of irrigation. However, the long-term use of groundwater would have negative impacts on soil quality and the sustainability of agricultural production [1].

Groundwater is the main source of drinking, agricultural, and industrial water. The increase in the exploitation of groundwater and increasingly unfavorable climatic conditions (especially the rainfall variability) have led to the reduction of the aquifer reserves and deteriorate groundwater quality [2]. These pose a danger to the sustainability of the land-use system. The most notable degradation process is the salinization of groundwater [3]. Therefore, the aim of this study was to identify the problem of water salinity, identify areas at risk, and find solutions for the management of these resources. This study consists of mapping and analyzing the quality of groundwater used for irrigation in the coastal region of Skhirat, Morocco, which is one of the most important market gardening areas in the Rabat-Salé region. This study assesses the degree of degradation and provides support for developing decision-making tools.

2. Material and Methods

2.1. Study Area

This study was carried out in the perimeter of Skhirat (270 km^2), located in the Rabat-Salé-Kenitra region. The study region lies between Wadi Ykem and Wadi Cherrat rivers and is part of the coastal Meseta. It is one of the agricultural regions with strong market gardening activities and farming based on irrigation water using groundwater. The climate of the region is dry in summer and semi-humid in winter. In the region, the average annual rainfall is 580 mm and averaged temperature around 17.5 °C. The geology of the region shows impermeable to slightly permeable primary formations of the schistose, sandstone, quartzitic, and limestone and permeable Mio-Plio-Quaternary formations. The Miocene presented by marls and limestones and the Plio-Quaternary consists of sands, gravel, and marine detrital limestone.

2.2. Water Sampling and Laboratory Analysis

In order to assess the state of the hydrochemical quality of irrigation water, a water monitoring campaign was carried out in the area. Water samples were collected from a network of 77 water points in hermetically sealed plastic bottles bearing the site code (Px), with P referring to the well and x the site number, and Lambert coordinates (X, Y) were taken using a GPS (Figure 1). The groundwater level is measured in situ using a piezometric probe. The water samples were then taken to the water analysis laboratory of the Research Unit on the Environment and the Conservation of Natural Resources—INRA-Rabat. Various physicochemical analyzes of the water (electrical conductivity, pH, and ion balance) were carried out.

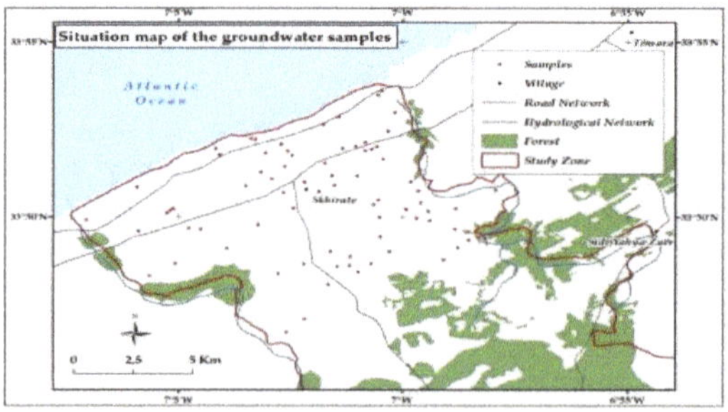

Figure 1. Location of water samples.

2.3. Mapping

The thematic maps were elaborated using spatial interpolation within Geographic Information System (GIS). We used the Geostatistical Analysis extension of the ArcGis 10 © software. The interpolation algorithm was the Inverse Distance Weighting (IDW).

3. Results and Discussion

3.1. Groundwater Depth

The bathymetric map shows that the depth of the water table oscillates between 2 and 45 m with an average of 16.5 m. The aquifer is closer to the surface in the downstream areas of the region, and to the South-West; this is due to the infiltration of water from the rivers and the sea. At the central and South-East parts, the aquifer is deeper, reaching 45 m (Figure 2).

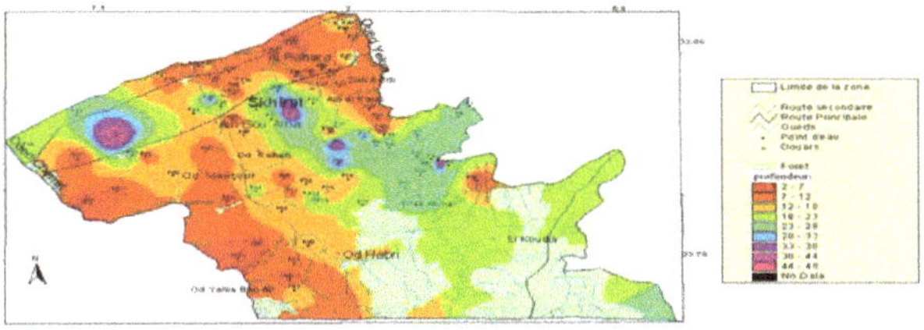

Figure 2. Spatial distribution of the water table depth.

3.2. *Chemical Facies of Waters*

The representation of the chemical results of the waters on the Piper diagram (Figure 3) made it possible to distinguish two types of facies that characterize the waters of Skhirat (sodium chloride facies in particular in the downstream and central zones and calcium chloride facies in the upstream zone).

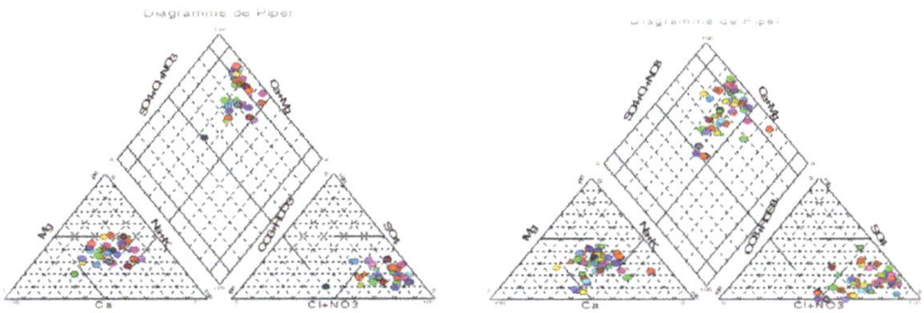

Figure 3. Piper diagram for waters downstream (**left**) and upstream (**right**).

3.3. *Water Salinity and Alkalinity*

The classification of waters, according to the American classification diagram proposed by USSL Riverside (Richards, 1954) (Figure 4), shows that the dominant classes are C3-S1 and C4-S2 (Table 1). These waters are very strongly to extremely saline and therefore have a very strong salinizing power on the soil (which can cause a reduction in the yields of sensitive crops). A total of 80% of the irrigation waters have a SAR of less than 8, and therefore these waters present a relatively low risk of alkalinization.

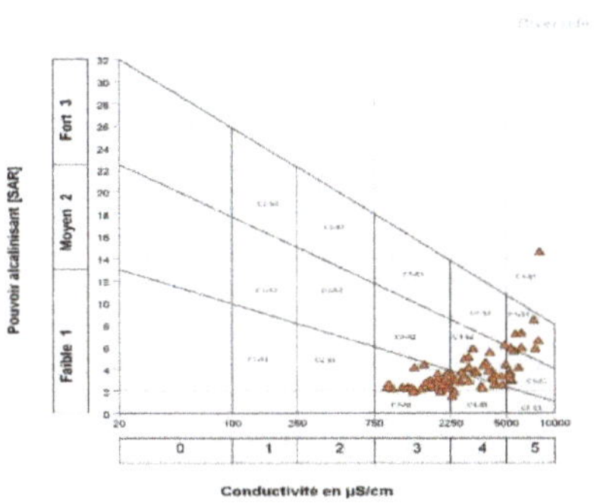

Figure 4. Classification of irrigation water (Richards, 1954).

Table 1. Classification of irrigation water according to salinity (Richards, 1954).

Salinity and Alkalinity Class	Interpretation	% of Wells
C3-S1	Average to poor quality. Use with caution.	39
C4-S1	Poor quality. Exclude sensitive plants and heavy soils.	16
C4-S2	Poor quality. To be used, with great care, only in light soils.	29
C5-S2	Very poor quality. To be used only in exceptional circumstances.	1.5
C5-S3	Very poor quality. To be used only in exceptional circumstances.	13
C5-S4	Very poor quality. To be used only in exceptional circumstances.	1.5

3.4. *Water Salinity Mapping*

The results found that the water salinity varies in a wide range from 0.9 to 8.1 dS/m with an average of 3.2 dS/m (Table 2). The map of the spatial distribution of water salinity shows that the majority of wells located in the downstream (coastal) and upstream parts of the region have high salinity. Those belonging to the middle part are relatively less saline. This can be explained by the effect of marine intrusion and the low-level depth of the water table in the littoral zone. In the upstream zone, water salinity may be due to the interaction of water with the geological formation of the shale aquifer, characterized by poor quality water. As for the middle part, the waters have a low salinity which can be attributed to the high permeability of the limestone formations of the aquifer (Figure 5).

Table 2. Classification of irrigation water based on electrical conductivity.

Salinity Class	Symbol	EC (dS/m)	Number of Wells	% of Wells
No saline	C1	<0.25	0	0
Moderately saline	C2	0.25–0.75	0	0
Strongly saline	C3	0.75–2.25	31	40
Very strongly saline	C4	2.25–5	32	42
Extremely saline	C5	>5	14	18

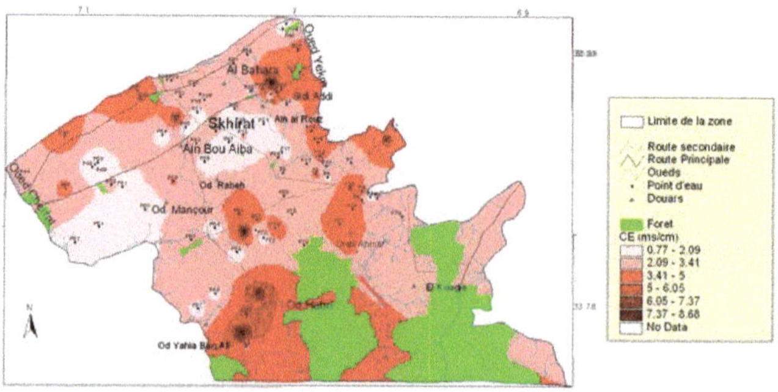

Figure 5. Groundwater salinity map from the Skhirat region.

4. Conclusions

The study analyzed and mapped the irrigation water quality (and in particular the salinity) in the Skhirat area of Morocco to assess its potential for degradation. This study showed a significant salinization power of these waters, especially in the upstream and downstream parts of the region, with relatively low risk of alkalinization. The salinity and alkalinity classes of irrigation water dominant in the region are C3-S1, C4-S2, and C4-S1. The dominant chemical facies are chloruro-sodium type in the downstream and central zones and chloruro-calcic in the upstream zone. The spatial mapping of the salinity and the depth of the water table made it possible to locate the sensitive zones of the region. Thus, the coastal and upstream areas of the region may have soil salinity problems in the future due to the high salinity of the waters in these areas.

Author Contributions: Conceptualization, H.D., A.D. and A.G.; Sampling and laboratory analyses, I.M., A.Z. and H.Y.; Software, I.M.; Orignal draft preparation, I.M.; Writing review and editing, H.D., A.D., H.I., H.C.D. and A.G. All authors have read and agreed to the published version of the manuscript.

Funding: This research was funded by INRA in the frame of Agroecology mega-project. subsidized by the general budget of INRA. The APC was funded by the LAFOBA2 organizers. Subsidized by INRA's general budget.

Institutional Review Board Statement: Not applicable.

Informed Consent Statement: This study did not include humans.

Data Availability Statement: Data can be supplied if required.

Conflicts of Interest: The authors declare no conflict of interest.

References

1. Devkota, K.P.; Devkota, M.; Rezaei, M.; Oosterbaan, R. Managing salinity for sustainable agricultural production in salt-affected soils of irrigated drylands. *Agric. Syst.* **2022**, *198*, 103390. [CrossRef]
2. Dakak, H.; Huang, J.; Zouahri, A.; Douaik, A.; Triantafilis, J. Mapping soil salinity in 3-dimensions using an EM38 and EM4 Soil inversion modelling at the reconnaissance scale in central Morocco. *Soil Use Manag.* **2017**, *33*, 553–567. [CrossRef]
3. Sehlaoui, H.; Hassikou, R.; Dakak, H.; Zouahri, A.; El Hasini, S.; Iaaich, H.; Ghanimi, A.; Douaik, A. Nitrate contamination and associated health risks of the Benslimane groundwater, Morocco. *Environ. Geochem. Health* **2022**, 1–16. [CrossRef] [PubMed]

environmental sciences proceedings

MDPI

Proceeding Paper

Mapping Soil Salinity Risk by Using an Index Approach †

Abdelwahed Chaaou [1,*], Mohamed Chikhaoui [1], Mustapha Naimi [1], Aissa Kerkour El Miad [2] and Marieme Seif-Ennasr [1]

1 Department of Natural Resources and Environment, Hassan II Institute of Agronomy and Veterinary Medicine, Rabat 10101, Morocco; mchikhaoui@gmail.com (M.C.); mnaimi2005@gmail.com (M.N.); seif.ennasr.marieme@gmail.com (M.S.-E.)

2 Department of Mathematical and Computer Sciences, Mohammed Premier University, Oujda 60000, Morocco; mid.kerkour@gmail.com

* Correspondence: abdelwahedchaaou@gmail.com; Tel.: +212-648424337

† Presented at the 2nd International Laayoune Forum on Biosaline Agriculture, 14–16 June 2022; Available online: https://lafoba2.sciforum.net/.

Abstract: Soil salinity is a worldwide problem that negatively affects soil quality and persistently spreading, especially in arid and semi-arid climate. The present study focuses on the mapping of soil salinity risk in Tadla Plain. To achieve this goal, the approach of Soil Salinity Risk Index (SSRI) has been adopted. The use of SSRI based approach reveals the occurrence of three risk classes: low, moderate and severe. The moderate risk class dominates with a coverage representing 80% of the total area. The results achieved showed the prospect of this approach to delineate areas of soils prone to salinization risk.

Keywords: salinization; soil salinity; risk index; Tadla plain; Morocco

check for updates

Citation: Chaaou, A.; Chikhaoui, M.; Naimi, M.; Kerkour El Miad, A.; Seif-Ennasr, M. Mapping Soil Salinity Risk by Using an Index Approach. *Environ. Sci. Proc.* **2022**, *16*, 24. https://doi.org/10.3390/environsciproc2022016024

Academic Editor: Abdelaziz Hirich

Published: 16 June 2022

1. Introduction

Salinization is a serious challenge for the development of modern agriculture and the preservation of environment. Globally, over one fifth of the total irrigated land is salt-effected [1]. This area could increase to above 50% in the next 30 years [2]. Human induced or secondary salinization, such as the use of water with high salt content, super-irrigation and/or lack of drainage systems, in addition to excess fertilizer [3,4], instigates the accumulation of soluble salts.

In Morocco, about 16% of irrigated lands are affected by secondary salinization [5]. In order to analyze the state of soil degradation due to salinity and evaluate the risk of its extension and aggravation, several spatio-temporal monitoring approaches have been developed and applied worldwide. The use of models allows the control of salinity by knowing its spatial distribution and its evolution over time [6,7]. However, monitoring the process and mapping the soil salinity are not carried out with sufficient reliability without the integration of several factors [8]. Given the complexity of the mechanism controlling soil salinity, the use of a multicriteria approach makes it possible to identify areas at high risk of salinity and ensure the spatiotemporal monitoring of chemical soil degradation [9]. In this context the Soil Salinity Risk Index (SSRI), which uses several factors (soil type and bedrock, climate and water table level) has been adopted for mapping soil salinity risk. This approach allows the development of a soil salinity risk map that can be used as a tool for decision-making to manage soil quality.

2. Material and Methods

2.1. Study Area

This study has been carried out in the irrigated perimeter of Tadla in central Morocco as presented in Figure 1. Geologically, the Tadla perimeter is characterized by a vast synclinal depression filled with Neogene (Miocene and Pliocene) and Quaternary deposits [10]. Soils studies carried out in thise area show the predominance of chromic Kastanozems soils [11]. The climate is arid to semi-arid with average annual temperature of about 17 °C. Mean annual rainfall varies from 200 to 600 mm during 2001–2014 [12].

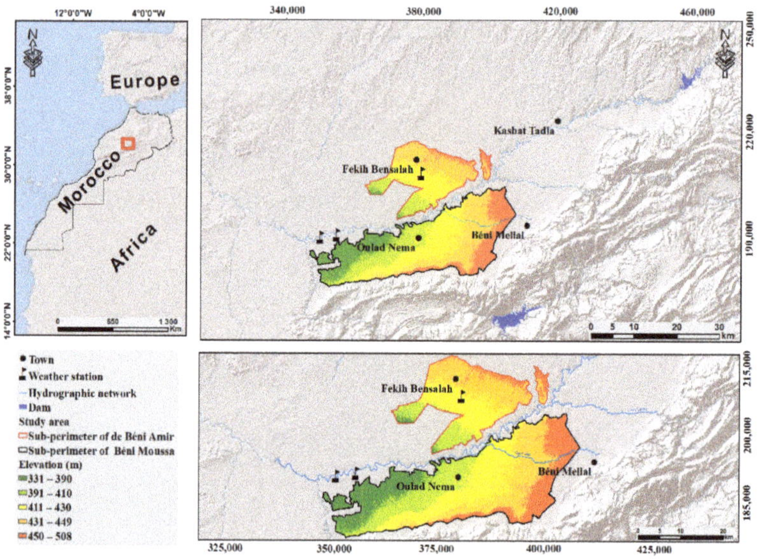

Figure 1. Location of the study area.

2.2. Soil Salinization Risk Index (SSRI)

The SSRI is an additive method [12,13], which includes 9 factors; surface water and groundwater quality, depth of groundwater, climate, aridity index, slope, texture, soil electrical conductivity and geological efficiency. The SSRI calculation is based on a 5 × 9 matrix with 2 weights (1 and 2). The computation of SSRI is based on the following equation:

$$\begin{aligned}\text{SSRI} = {} & (\text{Status of soil salinity} \times 2) + 1 \times (\text{Quality of surface water} + \\ & \text{Depth of water table} + \text{Ground water quality} + \text{Soil texture} + \text{Climate} + \\ & \text{Dry index} + \text{Slope} + \text{Efficacy of surface geology})\end{aligned}$$

The risk class index for each factor is multiplied by its respective weight. The weighted values for each factor are added together to estimate SSRI, which ranges from 10 "very low" to 50 "very high" depending on the risk class as shown in Table 1.

Table 1. Severity classes of salinity risk.

Class	None	Slight	Moderate	Severe	Very Severe
Risk score	10–15	16–25	26–35	36–45	46–50

2.3. Data Processing

The data used in this study come from soil and water quality monitoring net-works, established by the Tadla Agricultural Development Office. The IDW (Inverse Distance Weighting) module was used to interpolate the data from 2015–2016 growing season, including water table depth, electrical conductivity of groundwater and electrical conductivity of soil. The digital Elevation Model (DEM) has been generated from digitized contour lines of the Tadla topographic map (1/50,000).

3. Results and Discussion

The soil salinization risk map depicted in Figure 2a illustrates the spatial dis-tribution of salinity risk in the Tadla plain using SSRI. This approach allowed us to identify different various levels of current and potential risks by integrating the key factors influ-encing the salts accumulation process in the study area. The results showed on one hand the predominance of moderate potential risk class with an extent of 80%. This area includes soils that are very sensitive to salinity. On the other hand, area with severe risk class represents only 0.1% of the total surface of the irrigated perimeter of Tadla as shown in Figure 2b.

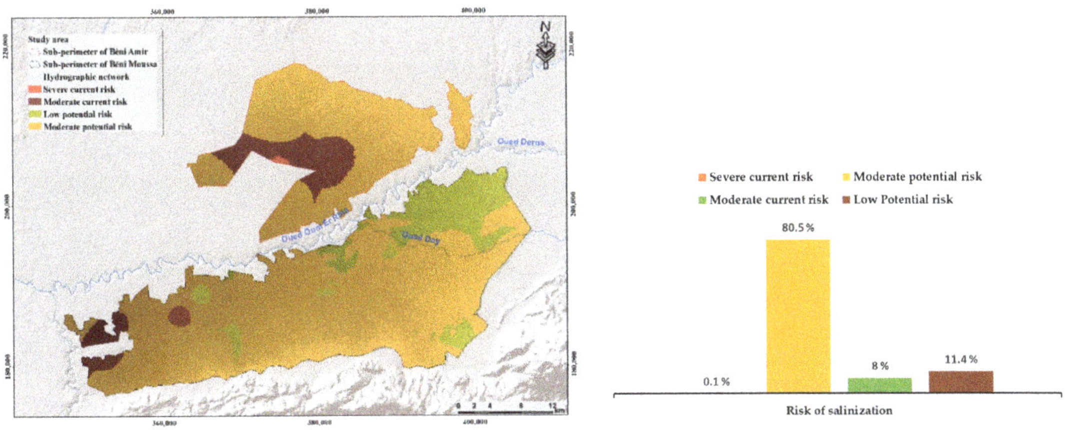

(a) (b)

Figure 2. Soil salinization risk map (**a**) and Areas of risk classes (**b**).

The Tadla plain is divided into two sub-perimeters, Béni Amir and Béni Moussa, with different hydraulic characteristics. The Béni Amir sub-perimeter is irrigated with Oued Oum Er Rbia River, classified as saline waters, while the Béni Moussa is irrigated by the non-saline waters of Bin El Ouidane dam. Furthermore, the groundwater of Béni Amir is more saline than that of Béni Moussa. Moreover, Beni Amir is characterized by an agricultural intensification accompanied by an overuse of groundwater. In addition, the use of a gravity-fed irrigation system and the absence of drainage favor the process of soil salinity. In the Béni Moussa perimeter, the installation of drip system and the irrigation with non-saline water decrease salt accumulation.

The analysis of the results shows that the soils in Béni Amir sub-perimeter are the most affected by salinity. Indeed, the downstream part of this sub-perimeter shows more concentrations of salts in soils. The spatial variability of salinity risk is linked with several factors; topography, the quality of irrigation water (surface and groundwater), the agricultural intensification, and the irrigation system.

Soil salinity is a complex process because of the spatial variability of the factors that control it. In addition, it is interesting to consider land cover as a dynamic factor as it influences soil water status.

4. Conclusions

In view of these results, we have been able to demonstrate the interest of using the SSRI approach and the integration of multisource data for the characterization of soil degradation risk by salinity in the irrigated perimeter of Tadla. The identification of current and potential risk areas allows managers to direct agricultural practices in Tadla and to alleviate salinization. The use of salinity tolerant crops, the adoption of intercropping system and the installation of a drainage system on areas at severe risk constitute some of the techniques to lessen soil degradation by salinity.

Monitoring the spatio-temporal dynamics of soil salinity is the key to understand the impact of rainfall, irrigation and agricultural practices on the evolution of soil salinity. As a consequence, agricultural land users and managers using the SSRI, can benefit from key intervention tools to mitigate the long-term effects of soil salinization on agricultural production.

Author Contributions: Conceptualization, data curation and writing-original draft preparation A.C.; methodology, supervision and project administration, M.C.; validation, supervision and writing—review and editing, M.N.; writing—review and editing, M.S.-E. and A.K.E.M. All authors have read and agreed to the published version of the manuscript.

Funding: This research received no external funding.

Institutional Review Board Statement: Not applicable.

Informed Consent Statement: Not applicable.

Data Availability Statement: The data presented in this study are available on request from the corresponding author. The data are not publicly available due to data privacy.

Acknowledgments: The authors would like to thank the National Centre for Scientific and Technical Research, the Hassan II Institute of Agronomy and Veterinary Medicine and the Hassan II Academy of Science and Technology (GISEC Project) for their manifold support. The authors would also like to extend their thanks to the Tadla Regional Office for Agricultural Development for their collaboration.

Conflicts of Interest: The authors declare no conflict of interest.

References

1. Singh, A. Soil salinity: A global threat to sustainable development. *Soil Use Manag.* **2021**, *38*, 39–67. [CrossRef]
2. Fu, Z.; Wang, P.; Sun, J.; Lu, Z.; Yang, H.; Liu, J.; Xia, J.; Li, T. Composition, seasonal variation, and salinization characteristics of soil salinity in the Chenier Island of the Yellow River Delta. *Glob. Ecol. Conserv.* **2020**, *24*, e01318. [CrossRef]
3. Elgallal, M.; Fletcher, L.; Evans, B. Assessment of potential risks associated with chemicals in wastewater used for irrigation in arid and semiarid zones: A review. *Agric. Water Manag.* **2016**, *177*, 419–431. [CrossRef]
4. Peña, A.; Delgado-Moreno, L.; Rodríguez-Liébana, J.A. A review of the impact of wastewater on the fate of pesticides in soils: Effect of some soil and solution properties. *Sci. Total Environ.* **2020**, *718*, 134468. [CrossRef] [PubMed]
5. Chaaou, A.; Chikhaoui, M.; Naimi, M.; Kerkour El Miad, A.; Achemrk, A.; Seif-Ennasr, M.; El Harche, S. Mapping soil salinity risk using the approach of soil salinity index and land cover: A case study from Tadla plain, Morocco. *Arab. J. Geosci.* **2022**, *15*, 722. [CrossRef]
6. Bouaziz, M.; Chtourou, M.Y.; Triki, I.; Mezner, S.; Bouaziz, S. Prediction of Soil Salinity Using Multivariate Statistical Techniques and Remote Sensing Tools. *Adv. Remote Sens.* **2018**, *7*, 313–326. [CrossRef]
7. Besma, Z.; Christian, W.; Didier, M.; Pierre, M.J.; Mohamed, H. Soil salinization monitoring method evolution at various spatial and temporal scales in arid context: A review. *Arab. J. Geosci.* **2021**, *14*, 283. [CrossRef]
8. Bannari, A.; Al-Ali, Z.M.; Kadhem, G.M. Effects of Topgraphic Attributes and Water-Table Depths on the Soil Salinity Accumulation in Arid Land. In Proceedings of the 2021 IEEE International Geoscience and Remote Sensing Symposium IGARSS, Brussels, Belgium, 11–16 July 2021; pp. 6548–6551. [CrossRef]
9. Castro, F.C.; Araújo, J.F.; dos Santos, A.M. Susceptibility to soil salinization in the quilombola community of Cupira—Santa Maria da Boa Vista—Pernambuco—Brazil. *CATENA* **2019**, *179*, 175–183. [CrossRef]
10. Etienne, H.; Guessab, D.; Archambault, C. The Tadla plain. *Notes et Mém. Serv. Géol. Maroc.* **1975**, *231*, 299–364.
11. FAO. *World Reference Base for Soil Resources 2014: International Soil Classification System for Naming Soils and Creating Legends for Soil Maps*; FAO: Rome, Italy, 2014.
12. Chaaou, A.; Chikhaoui, M.; Naimi, M.; El Miad, A.K.; Achemrk, A. Cartographie Du Risque De Salinité Des Sols À L'aide De L'approche Des Indices Et Des Données Multi-Sources: Cas De La Plaine De Tadla Au Maroc. *Eur. Sci. J.* **2020**, *16*, 33. [CrossRef]
13. Masoudi, M.; Patwardhan, A.; Gore, S. A new methodology for producing of risk maps of soil salinity, case study: Payab Basin, Iran. *J. Appl. Sci. Environ. Manag.* **2006**, *10*, 3. [CrossRef]

environmental sciences proceedings

MDPI

Proceeding Paper

Mapping Agronomic Suitability of Soils in Forest Savannah Transition Zone in Cameroon: A Case Study from Bokito District of the Central Region †

Olivier Leumbe [1,*], Marie Roumy Ouafo [2] and Paul Ndjigui [3]

1 Research Branch, Natural Hazards Research Laboratory, National Institute of Cartography, Yaoundé P.O. Box 157, Cameroon
2 Department of Earth Sciences, Faculty of Science, University of Douala, Douala P.O. Box 24157, Cameroon; marieroumi@yahoo.fr
3 Department of Earth Sciences, Faculty of Science, University of Yaoundé, Yaoundé P.O. Box 812, Cameroon; dndjigui@uy1.unicet.cm
* Correspondence: leumbefranck@gmail.com; Tel.: +237-677-43-1997
† Presented at the 2nd International Laayoune Forum on Biosaline Agriculture, 14–16 June 2022; Available online: https://lafoba2.sciforum.net/.

Abstract: Food security is a crucial issue in Cameroon. The Ministry of Agriculture and Rural Development, with the help of the University of Yaoundé 1 and the National Institute of Cartography, aims to produce an agronomic aptitude map of the country. The pilot site of Bokito was selected. The formula used is as follows: AA = pH × RU × K × CEC. The objective is to propose a simple, quick and inexpensive method of land evaluation that can boost the transition to second-generation agriculture. The results show that the yellowish ferrallitic soils of Bokito have good agronomic suitability.

Keywords: soils; agriculture; agronomic suitability

Citation: Leumbe, O.; Ouafo, M.R.; Ndjigui, P. Mapping Agronomic Suitability of Soils in Forest Savannah Transition Zone in Cameroon: A Case Study from Bokito District of the Central Region. *Environ. Sci. Proc.* **2022**, *16*, 38. https://doi.org/10.3390/environsciproc2022016038

Academic Editor: Abdelaziz Hirich

Published: 16 June 2022

1. Introduction

Cameroon is an agricultural country [1] and food security is a fundamental component of the sustainable development process. Numerous studies have demonstrated the importance of a soil-suitability map for the development of agriculture [2]. The "Office of Scientific and Technical Research Overseas (ORSTOM)" has previously mapped the suitability of soils in certain areas of Cameroon, but these maps are old and do not cover large areas. Cameroon aims to transition to second-generation agriculture, which requires an accurate estimate of the spatial variability in soil components. The Ministry of Agriculture and Rural Development, with the help of the University of Yaoundé 1 and the National Institute of Cartography, aims to produce a map of agronomic suitability. Its implementation requires a reliable and low-cost approach. The objective of this study is, therefore, to present and discuss a means of calculating the agronomic suitability of tropical soils, which are simple, relevant and light, and can easily be implemented in a larger territory. It aims to quantify the relationships between soil variables to map agronomic suitability using a geographic information system [3].

2. Materials and Methods

The study area was located in the Bokito district, Center region of Cameroon, with latitudes between 4°20′ and 4°40′ N and longitudes between 11°00′ and 11°20′ E (Figure 1).

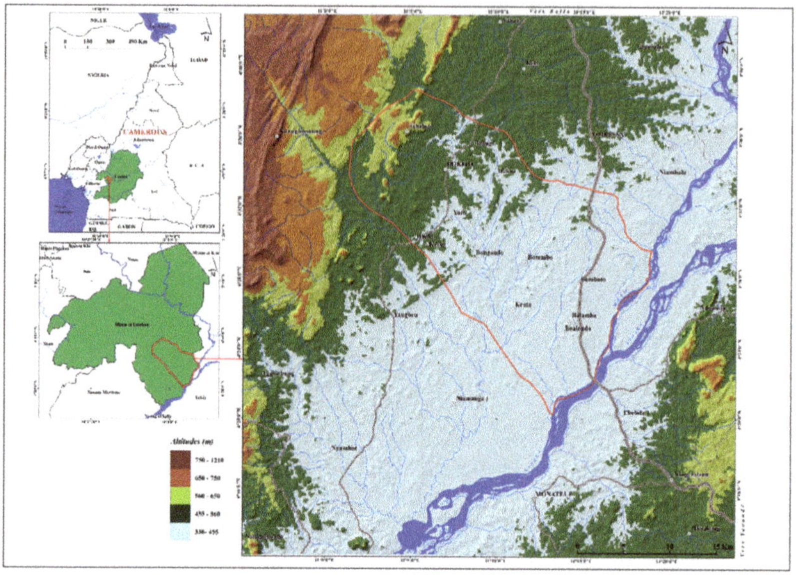

Figure 1. Location of the study area.

To map agronomic suitability, soil samples were collected from a pit ranging from 1.50 to 2.0 m deep, and laboratory analysis was carried out. The particle size and pH were determined in the soil laboratory of the Department of Soil Sciences of the University of Yaoundé 1. Organic carbon (OC) and total nitrogen content were carried out in the soil laboratory of the International Institute of Tropical Agriculture (IITA) in Yaoundé. The agronomic suitability map of the soils was produced using the following formula: AA = pH × RU × K × CEC where, AA: agronomic suitability; pH: hydrogen potential; RU: useful water reserve; K: erodibility factor; CEC: cation exchange capacity. Soil pH was spatialized using kriging. Useful soil water reserve (RU) was computed using the equation in Remy [4]: RU = H × TE × (1 − (EG/100)), where RU: useful reserve water express in millimeters; H: thickness expressed in centimeters; TE: textural index determined from the texture class; EG: coarse elements expressed as a percentage. The soil thickness (H) was measured along the profile using a tape. The texture classes were determined with the USDA texture diagram. The texture index (TE) was classified using the methodology proposed by Jamagne [5]. The quantity of coarse elements was determined using the methodology proposed by Bouma and Van Lanen [6].

Soil erodibility factor (K) was calculated according to the formula provided by Wischmeier and Smith [7]: $1000\ K = 2.8 \times 10^{-4}$ (12-%MO) $\times M^{1.4}$ + 3.25 (S-2) + 2.5 (P-3) where, MO: organic matter in percentage; M = (% sands + % silt) × (100 − % clay); S: code on the soil structure; P: infiltration capacity. Soil structure (S) was determined using the texture diagram [6]. The obtained texture classes were then transferred to the correspondence table of [7] to obtain the numerical value of the structure code S (Table 1).

Table 1. Meaning of codes on soil structure.

Code	Soil Structure (S)
1	Very fine
2	Fine
3	Medium
4	Very coarse

Soil permeability (P) was determined using the USDA diagram (Figure 2) and the correspondence was established in the table below [8] (Table 2).

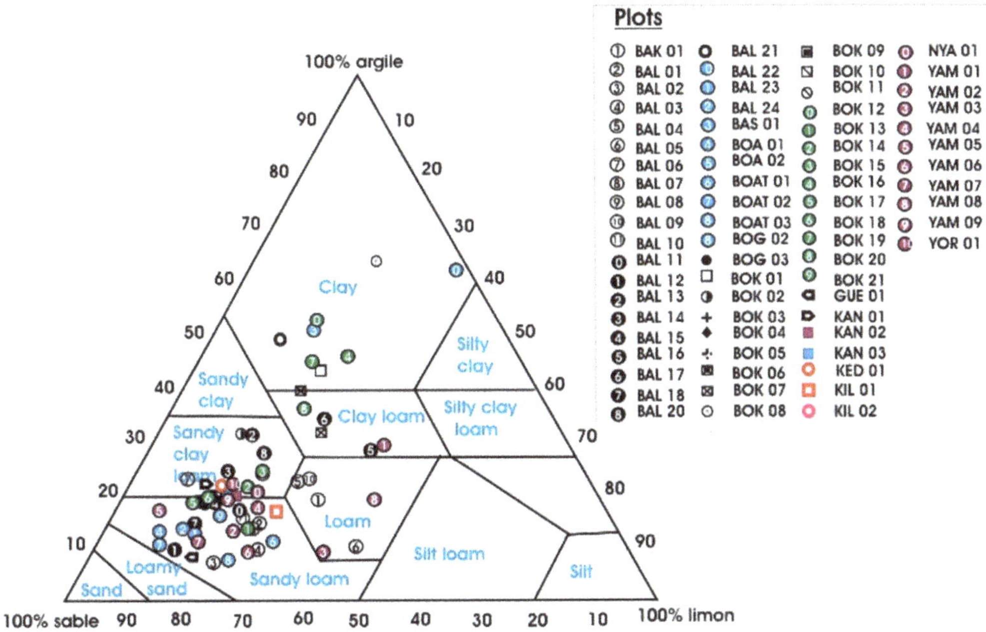

Figure 2. USDA textural diagram.

Table 2. Meaning of permeability (P) codes.

Code	Textural Class	Permeability (p)
1	Clay < 18% and Sand > 65	Fast
2	18% < Clay < 35% and Sand > 15% or 15% < Sand < 65% and Clay < 18%	Medium to fast
3	Clay < 35% and Sand < 15%	Average
4	35% < Clay < 60%	Slow to average
5	Clay > 60%	Slow

The CEC factor was calculated in the laboratory.

3. Results

Results showed that the Bokito soils range from sandy loam to sandy clay loam (Figure 2), which is dominated by coarse material (Figure 3). The yellow ferrallitic soils are moderately acidic (Figure 4A) with a good useful water reserve (Figure 4B). The CEC is maximum (Figure 4C) and the erodibility rate is high (Figure 4D). These soils have a very good agronomic suitability (Figure 4E).

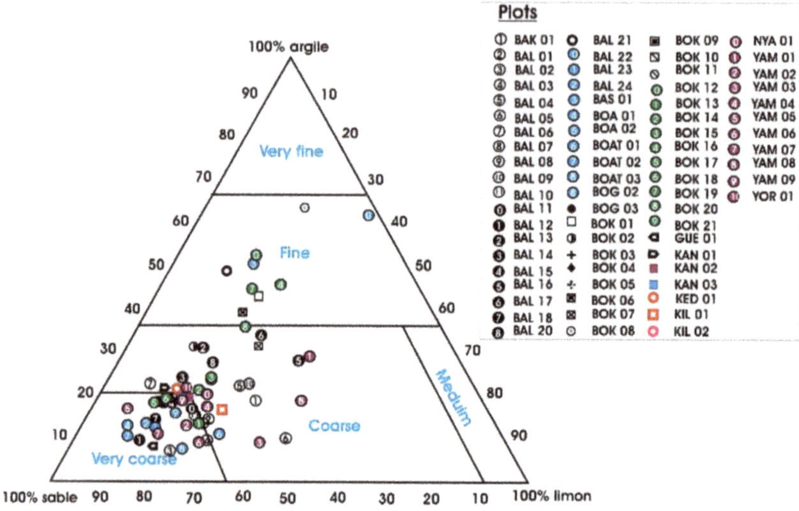

Figure 3. Textural diagram of Jamagne.

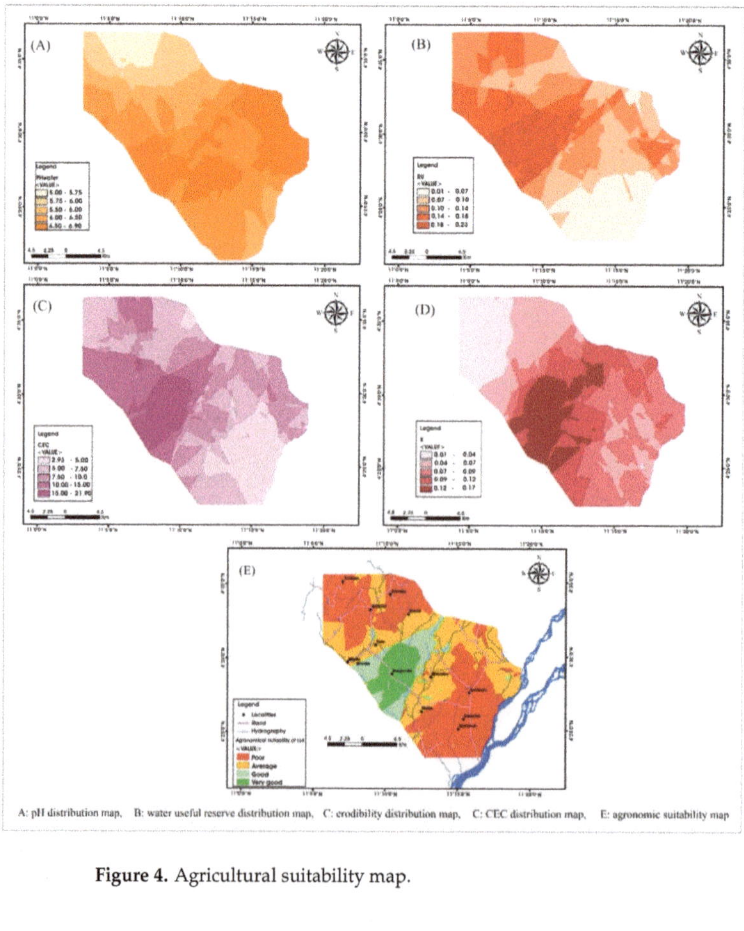

Figure 4. Agricultural suitability map.

4. Discussion

The yellow ferrallitic soils of Bokito have a very good agronomic aptitude. Their pH varies between 6 and 6.22. These thresholds are favorable for fertilizing elements and soil microorganisms' activity [9,10]. At a depth of 30 cm, the useful water reserve is at its maximum (0.229 mm) due to their high silt content [11,12]. The CEC reaches 6.65 meq/100 g soil. The very high erodibility in these soils is characteristic of humid tropical environments [13]. A field trip allowed for us to verify the obtained results. Indeed, agricultural yields are high in the yellow ferrallitic soils of Boganda village. This could support the choice of input parameters for the model.

5. Conclusions

Cameroon has agronomically suitable land that needs to be located and characterized. Their development could contribute to food security and environmental preservation. The proposed methodological approach allows for large-area mapping over a short period of time and could accelerate Cameroon's transition to second-generation agriculture.

Author Contributions: Conceptualization, O.L., M.R.O. and P.N.; Methodology, O.L.; Software, O.L.; Validation, O.L, M.R.O. and P.N.; Formal analysis, O.L., M.R.O. and P.N.; Survey, O.L. and P.N.; Resources, Ministry of Agriculture and Rural Development. Data curation, O.L. and P.N.; writing—preparation of original draft, O.L., M.R.O. and P.N.; writing—revision and editing, O.L., M.R.O. and P.N.; visualization, O.L.; supervision, O.L.; project administration, O.L. and P.N.; procurement of funding, Ministry of Agriculture and Rural Development. All authors have read and agreed to the published version of the manuscript.

Funding: This research was funded by the Ministry of Agriculture and Rural Development of Cameroon/Direction of Agricultural Development.

Institutional Review Board Statement: Not applicable.

Informed Consent Statement: Not applicable.

Data Availability Statement: Soil Science Laboratory of the University of Yaoundé.

Conflicts of Interest: The authors declare no conflict of interest.

References

1. Nguemezi, C.; Tematio, P.; Yemefack, M.; Tsozue, D.; Silatsa, T.B. Soil quality and soil fertility status in major soil groups at the Tombel area, South-West Cameroon. *Heliyon* **2020**, *6*, e3432. [CrossRef] [PubMed]
2. Tematio, P.; Tsafack, E.; Kengni, L. Effects of tillage, fallow and burning on selected properties and fertility status of Andosols in the Mounts Bambouto, West Cameroon. *Agric. Sci.* **2011**, *3*, 334–340. [CrossRef]
3. Lagacherie, P. *Digital Soil Mapping: A State of the Art*; Springer: Dordrecht, The Netherlands, 2008; Volume 2, pp. 3–14.
4. Rémy, J.-C. Méthodologie pour l'évaluation des terres: Contraintes pédologiques et facteurs limitant l'utilisation des sols. In *Sols et Environnement (Ouvrage)*; Girard, M.-C., Walter, C., Berthelin, J., Morel, J.-L., Eds.; Dunod: Paris, France, 2005.
5. PJamagne, M.; Betremieux, R.; Bégon, J.C.; Mori, A. Quelques données sur la variabilité dans le milieu naturel de la réserve en eau des sols. *Bull. Tech. Inf.* **1977**, *324*, 627–641.
6. Bouma, J.; van Lanen, H.A.J. Transfer functions and threshold values: From soil characteristics to land qualities. In Proceedings of the International Workshop on Quantified Land Evaluation Procedures, Washington, DC, USA, 27 April–2 May 1986; pp. 106–110.
7. Wischmeier, W.H.; Smith, D. Predicting Rainfall Erosion Losses, a Guide to Conservation Planning. In *Agriculture Handbook*; US Department of Agriculture: Washington, DC, USA, 1978.
8. King, D.; Le Bissonais, Y. Rôle des sols et des pratiques culturales dans l'infiltration et l'écoulement des eaux. Exemple du ruissellement et de l'érosion sur les plateaux limoneux du nord de l'Europe. *C. R. Acad. Agric. France* **1992**, *78*, 91–105.
9. Bigorre, F.; Pedro, G. Contribution des argiles et des matières organiques à la rétention de l'eau dans les sols. Signification et rôle fondamental de la capacité d'échange en cations. *C. R. Acad. Sci. Paris* **2000**, *330*, 245–250.
10. Ciesielski, H. Détermination of exchange capacity and exchangeable cations in soils by means of cobalt hexamine trichloride. Effects of expérimental conditions. *Agronomy* **1997**, *17*, 1–7. [CrossRef]
11. Ridremont, F.; Lejeune, P.; Claesens, H. Méthode pragmatique d'évaluation de la réserve en eau des stations forestières et cartographie à l'échelle régionale (Wallonie, Belgique). *Biotechnol. Agron. Société Environ.* **2011**, *15*, 727–741.
12. Duchaufour, P. *Abrégé de Pédologie*; Édition Masso: Paris, France, 1997.
13. Roose, E.; Sarrailh, J.-M. *Erodibilité de Quelques Sols Tropicaux. Vingt Années de Mesure en Parcelles D'Érosion Sous Pluies Naturelles*; Centre ORSTOM: Montpellier, France, 1990.

environmental sciences proceedings

MDPI

Proceeding Paper

Designing the Path for Soil Salinity Management: Lessons Learned and Future Perspectives in Morocco †

Marieme Seif-Ennasr [1,*], Mohamed Chikhaoui [1], Mustapha Naimi [1], Abdelwahed Chaaou [1] and Redouane Choukr-Allah [2]

1 Department of Natural Resources and Environment, Hassan II Institute of Agronomy and Veterinary Medicine, Rabat 10101, Morocco; mchikhaoui@gmail.com (M.C.); mnaimi2005@gmail.com (M.N.); abdelwahedchaaou@gmail.com (A.C.)
2 Agricultural Innovation and Technology Transfer Center (AITTC), Mohammed VI Polytechnic University (UM6P), Ben Guerir 43150, Morocco; redouane53@yahoo.fr
* Correspondence: seif.ennasr.marieme@gmail.com; Tel.: +212-665-616-271
† Presented at the 2nd International Laayoune Forum on Biosaline Agriculture, 14–16 June 2022; Available online: https://lafoba2.sciforum.net/.

Abstract: Soil salinity is a significant threat to crop sustainability and food security. This review aims to provide the basis for effective soil salinity management by examining the different solutions to develop scientifically sound guidelines for farmers, maintain profitable agricultural production in saline farmland irrigated possibly with saline groundwater, and alleviate agricultural land degradation. A lesson learned from the Moroccan experience in tackling salinity is needed for integrating soil and water management and appropriate salt-tolerant crops with innovative agricultural practices as a package of saline agriculture. Moreover, regional and global cooperation to exchange emerging challenges, successful rehabilitation studies, and innovative solutions should be considered. This review concludes that no single parameter could be suggested as the only possible way for soil salinity rehabilitation.

Keywords: salt-affected soil; climate change; strategies; water resources

Citation: Seif-Ennasr, M.; Chikhaoui, M.; Naimi, M.; Chaaou, A.; Choukr-Allah, R. Designing the Path for Soil Salinity Management: Lessons Learned and Future Perspectives in Morocco. *Environ. Sci. Proc.* **2022**, *16*, 40. https://doi.org/10.3390/environsciproc2022016040

Academic Editor: Abdelaziz Hirich

Published: 16 June 2022

1. Introduction

Under the pressure of global climate change and anthropogenic activity, salinization has been a worldwide issue that reduce severely soil quality and crop productivity and gradually reducing the cultivated area. Additionally, the salt-affected area is increasing at a rate of 10% yearly because of high evaporation, low rainfall, inadequate irrigation, and other irrational anthropogenic activities [1]. By 2050, the salinized area will exceed 50% of global arable land [2]. Researchers have identified several plant responses due to soil salinity and recommended different approaches to address the problem. However, there is a lack of coordination between the plants' physiological and biochemical responses and possible management strategies. This paper aims to provide the basis for effective soil salinity management by examining the different solutions to develop scientifically sound guidelines for farmers, maintain profitable agricultural production in saline farmland irrigated possibly with saline groundwater, and combat agricultural land degradation.

2. Material and methods

2.1. Soil Salinity Situation in Morocco

Morocco is a semi-arid Mediterranean country with scarce and irregular precipitation. Most soil and groundwater salinity problems in irrigated areas result directly from poor drainage, the rise of the saline water table, high evapotranspiration, and irrigation with water with increased risk of salinization and sodification [3]. In coastal areas, groundwater alteration is due to seawater intrusion because of over-pumping and pollution by the irrational use of fertilizers and pesticides in agriculture [4]. The areas most affected by salinity are located in the irrigated areas of Ouarzazate, Tafilalet, Haouz, Basse Moulouya, Tadla, Souss-Massa, and Gharb (Figure 1).

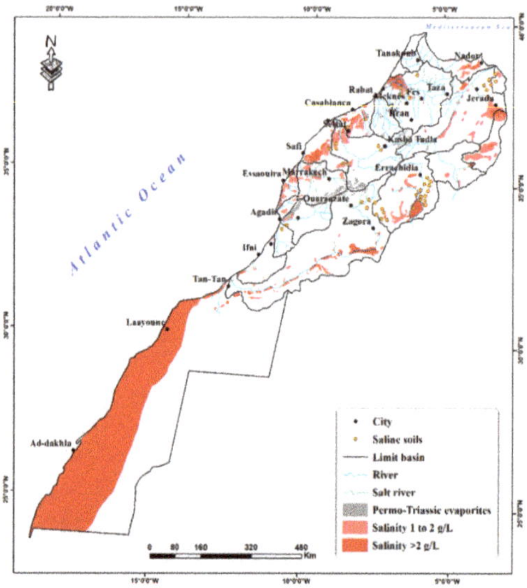

Figure 1. Spatial distribution of saline soil, aquifers and rivers in Morocco, modified from [4].

2.2. Study Approach

Many published works (n = 300) covering salt-affected soils and related salinity issues were collected to make an initial assessment. After a systematic literature review and content analysis (Figure 2), we identified the key management strategies and practices.

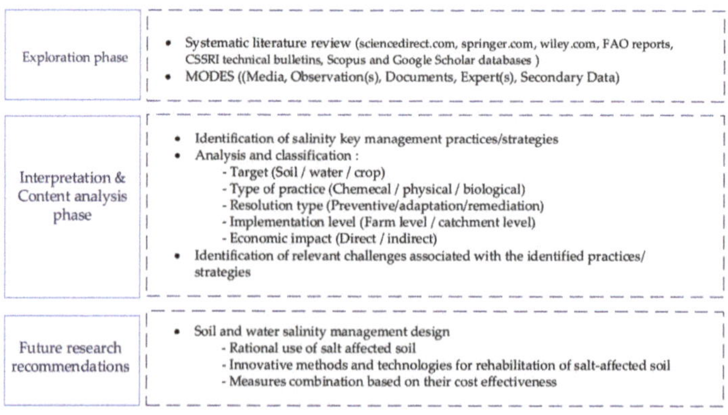

Figure 2. Methodology scheme.

3. Results and Discussion

3.1. Findings on Soil Salinity Rehabilitation in Morocco

The analysis of Moroccan regulations has shown that several actions aimed at protecting soils and combating land degradation are based on three components: (1) a better knowledge of Moroccan soils; (2) a better exploitation of soils; (3) the fight against soil degradation. In addition, the Green Generation strategy, launched in February 2020, aims to consolidate the achievements of the Green Morocco Plan, in particular through the continuation of programs to control and enhance water use and availability, which go hand-in-hand with soil preservation. This orientation would make it possible to develop and adopt a Code of Sustainable Land Management (according to the vocation), which sets the regulatory aspects, the standards, and the good practices of management, exploitation of soils, and adaptation to climate change.

Several options for managing salt-affected catchments are available. Table 1 provides a non-exhaustive list of selected examples from genetic improvement for salt-tolerance, agronomic practices and irrigation management.

Table 1. Practices and strategies adopted at the farm and territorial levels to tackle soil salinization in Morocco.

Strategies or Management Practice	Target [1]	Type [2]	Resolution [3]	Level [4]	Type of Coast [5]
Inorganic amendment (gypsum, phosphogypsum)	S	C	R	F	D
Organic amendment (Compost, humic acids, etc.)	S	B	R	F	I
Development and use of salt-tolerant cultivar	C	B	A	C	D
Introduction of alternative crops (e.g., Quinoa, Blue panicum, etc.)	CSW	B	A	F	I
Use of plant growth-promoting bacteria (PGPB) (e.g., *Robinia pseudoacacia*, *Phaseolus vulgaris* L., etc.)	S	B	A	F	D
Remote sensing and GIS in salinity mapping	S	P	R	C	D
Electromagnetic and tomographic technique in soil salinity	S	P	A	C	I
Genetic engineering (e.g., Exogenous application of compatible solutes such as proline)	C	B	A	C	I
Leaching, implementation of subsurface drainage Systems	W	P	PAR	F	D
Use of brackish water	W	P	A	C	D
Mixing saline and freshwater	W	C	A	C	D
Alternative water sources, e.g., Seawater desalination, Desalination of Irrigation water	W	C	A	C	D
Plant halophytes in high salinity areas	C	B	R	C	D
Inoculation with mycorrhizal associations	C	B	A	F	I
Incorporate agroforestry into management	S	B	A	C	D
Apply biological agents to increase crop resistance to salinity or plant growth under saline conditions (Example: *Ulva lactuca* extract)	C	B	A	F	I
Use halophilic green algaeto counter salinisation effects on the crop (e.g., *Dunaliella salina*)	C	B	A	F	I
Intervention in the nutrition of plants (e.g., *Phosphorus fertilization*)	S	C	A	F	I

[1] W: Water-based practice; C: crop-based practice; S: soil-based practice. [2] C: Chemical approach; P: physical practice; B: biological practice. [3] P: Preventive resolution; A: adaptation resolution; R: remediation resolution. [4] F: farm level; C: catchment level. [5] D: direct coast; I: indirect coast.

Under water scarcity conditions, a trade-off exists between allocating water for salinity management and production [5]. Thus, leaching practices are not relevant option in arid and semi-arid areas. El Hasini et al. [6] recommended the compost application as an amendment to alleviate soil salinity. However, compost quality needs to be very carefully checked to prevent pathogenic organisms or pollutants. Gypsum application, the most recommended amendment for alleviating sodicity stress, is available in a finite amount. In recent decades, constraints such as decreased availability of mined gypsum, deterioration in product quality, and high market prices have increasingly made gypsum use a costly and less efficient proposition in sodic land reclamation.

3.2. *How to Go Forward*

Tackling salinity requires a more scientific approach to integrate soil and water management and appropriate crop tolerance to salinity with new agricultural practices as a package of 'saline agriculture.' In addition, regional and global cooperation is of utmost importance to exchange emerging new challenges and innovative solutions.

Three considerations are necessary to improve existing farming practices and provide a solid foundation for the successful implementation of sustainable crop production intensification: (1) rational use of salt-affected soils, (2) innovative methods and technologies for the amelioration of salt-affected soils, (3) cost-effectiveness of combined measures. Combined measures for managing saline soils may be based on the following principles: (1) selection of salt-tolerant crops, (2) dilution of salts in the root zone, (3) improving soil structure with organic and biological amendments, (4) improving leaching of salts by irrigation and drainage, (5) reducing surface evaporation with mulch and/or cover crops, (6) maintaining the groundwater table at a safe depth below the root zone and (7) maintaining the crop while reclamation is underway.

4. Conclusions and Perspectives

Morocco faces enormous challenges related to the extent of salt-induced land and water degradation. Despite concerted efforts made to combat soil and water salinity, the problem persist and remains unresolved. Concepts related to the management of soils affected by salinity must be at the heart of water-saving strategies. Therefore, we suggest that all researchers and managers adopt multidisciplinary and integrated approaches to deal with the problem of soil and water salinity. These research approaches must be cost-effective and acceptable to farmers. We recommend through this attempt the creation of national regulations that would govern sustainable agriculture and encourage farmers to adopt sustainable systems based on soil health as well as capacity building to allow decision makers to explore new approaches. The Moroccan experience with salinity highlights the critical importance of high-quality scientific information to guide policy design. It also reinforces the importance of bringing together the perspectives of different disciplines to adequately address such a complex and multifaceted problem. This will require gouvernment support, particularly programs for introducing modern technologies to protect the agriculture sector, restore saline soils, and increase their fertility.

Author Contributions: Conceptualization, M.S.-E. and R.C.-A.; methodology, M.S.-E.; M.C.; M.N.; R.C.-A.; software, M.S.-E.; M.C.; M.N.; A.C.; R.C.-A.; validation, M.S.-E.; M.C.; M.N.; A.C. and R.C.-A.; formal analysis, M.S.-E.; investigation, M.S.-E. and M.C.; resources, M.S.-E.; M.C.; M.N.; R.C.-A.; data curation, M.S.-E.; M.C.; M.N.; A.C.; R.C.-A.; writing—original draft preparation, M.S.-E.; M.C.; M.N.; R.C.-A.; writing—review and editing, M.C.; M.N.; R.C.-A.; visualization, M.S.-E.; M.C.; M.N.; A.C.; R.C.-A.; supervision, R.C.-A.; project administration, R.C.-A. All authors have read and agreed to the published version of the manuscript.

Funding: This research received no external funding.

Institutional Review Board Statement: Not applicable.

Informed Consent Statement: Not applicable.

Data Availability Statement: Not applicable.

Conflicts of Interest: The authors declare no conflict of interest.

References

1. Abiala, M.A.; Abdelrahman, M.; Burritt, D.J.; Tran, L.-S.P. Salt Stress Tolerance Mechanisms and Potential Applications of Legumes for Sustainable Reclamation of Salt-Degraded Soils. *Land Degrad. Dev.* **2018**, *29*, 3812–3822. [CrossRef]
2. Sun, J.; He, F.; Zhang, Z.; Shao, H.; Pan, Y.; Yang, R.; Li, W.; Li, P.; Zheng, M. Analysis of Saline Groundwater Infiltration into Two Loam Soils. *Land Degrad. Dev.* **2018**, *29*, 3795–3802. [CrossRef]
3. Daoud, S.; Elbrik, K.; Tachbibi, N.; Bouqbis, L.; Brakez, M.; Harrouni, M.C. 9-The Potential Use of Halophytes for the Development of Marginal Dry Areas in Morocco. In *Halophytes for Food Security in Dry Lands*; Khan, M.A., Ozturk, M., Gul, B., Ahmed, M.Z., Eds.; Academic Press: San Diego, CA, USA, 2016; pp. 141–156. [CrossRef]
4. Hssaisoune, M.; Bouchaou, L.; Sifeddine, A.; Bouimetarhan, I.; Chehbouni, A. Moroccan Groundwater Resources and Evolution with Global Climate Changes. *Geosci. J.* **2020**, *10*, 81. [CrossRef]
5. Matthews, N.; Grové, B.; Barnard, J.H.; van Rensburg, L.D. Modelling the Economic Tradeoffs between Allocating Water for Crop Production or Leaching for Salinity Management. *Water SA* **2010**, *36*. [CrossRef]
6. El Hasini, S.; De Nobili, M.; El Azzouzi, M.; Azim, K.; Douaik, A.; Laghrour, M.; El Idrissi, Y.; El Alaoui El Belghiti, M.; Zouahri, A. The Influence of Compost Humic Acid Quality and Its Ability to Alleviate Soil Salinity Stress. *Int. J. Recycl. Org. Waste Agric.* **2020**, *9*, 21–31. [CrossRef]

environmental sciences proceedings

MDPI

Proceeding Paper

Groundwater Resources in Moroccan Coastal Aquifers: Insights of Salinization Impact on Agriculture †

Yassine Ez-zaouy [1,*], Lhoussaine Bouchaou [1,2], Aicha Saad [3], Mohammed Hssaisoune [2,4], Youssef Brouziyne [5], Driss Dhiba [1] and Abdelghani Chehbouni [1,6]

1 International Water Research Institute, Mohammed VI Polytechnic University (UM6P), Ben Guerir 43150, Morocco; lhoussaine.bouchaou@um6p.ma (L.B.); driss.dhiba@um6p.ma (D.D.); abdelghani.chehbouni@um6p.ma (A.C.)
2 Laboratory of Applied Geology and Geo-Environment, Ibn Zohr University, Agadir 80035, Morocco; m.hssaisoune@uiz.ac.ma
3 Polydisciplinary Faculty of Taroudant, Exploration and Management of Natural and Environmental Resources Team (EGERNE) Taroudant, Ibn Zohr University, Agadir 83000, Morocco; a.saad@uiz.ac.ma
4 Faculty of Applied Sciences, Ibn Zohr University, B.O. 6146 Azrou District, Ait Melloul 86153, Morocco
5 International Water Management Institute (IWMI), MENA Office, Giza 12661, Egypt; youssef.brouziyne@gmail.com
6 CESBIO, Université de Toulouse, CNRS, CNES, IRD, BPI 280, CEDEX 9, 31065 Toulouse, France
* Correspondence: yassine.ez-zaouy@um6p.ma; Tel.: +212-634924922
† Presented at the 2nd International Laayoune Forum on Biosaline Agriculture, 14–16 June 2022; Available online: https://lafoba2.sciforum.net/.

Abstract: Across several coastal areas in Morocco, groundwater is the strategic source of irrigation. In this work, a database of thirteen Moroccan coastal aquifers was used to assess groundwater for agriculture purposes, as well as to highlight the process responsible of the degradation of groundwater resource quality in Moroccan coastal areas. According to electrical conductivity parameter, the results show that 92% of the collected samples were not suitable for irrigation uses. This situation is due to seawater intrusion and water–rock interaction processes, in addition to intensive agriculture activities and the introduction of domestic and industrial wastewater without any treatment. In order to control the impact of groundwater salinity on agriculture, management plans are proposed.

Keywords: salinity; aquifer; water–rock interaction; seawater intrusion; agriculture; pollution

Citation: Ez-zaouy, Y.; Bouchaou, L.; Saad, A.; Hssaisoune, M.; Brouziyne, Y.; Dhiba, D.; Chehbouni, A. Groundwater Resources in Moroccan Coastal Aquifers: Insights of Salinization Impact on Agriculture. *Environ. Sci. Proc.* **2022**, *16*, 48. https://doi.org/10.3390/environsciproc2022016048

Academic Editor: Abdelaziz Hirich

Published: 16 June 2022

1. Introduction

Being a precious water resource in coastal areas, coastal aquifers are undergoing significant degradation due to overexploitation, climate change, and the implementation of several socio-economic projects in different sectors, such as tourism, industry, and agriculture [1]. Agriculture represents the main factor that requires water resources in terms of quantity and quality. Salinity risk is the main problem that threatens most of the coastal aquifers in Morocco [2]. The chemical composition of irrigation water has a direct impact on plants and agricultural soil, potentially resulting in lower productivity [3]. To ensure good agriculture performance, monitoring and assessment of water quality is needed. The main focus of this study is to assess the groundwater suitability of Moroccan coastal aquifers for irrigation purposes. Moreover, in this study, some alternative solutions are suggested in order to preserve groundwater resources against pollution and to ensure the sustainability of soil and water.

2. Materials and Methods

Chemical analyses of groundwater samples were performed in order to determine the suitability for agricultural use. A series of irrigation water indices were used: electrical conductivity (EC), sodium absorption ratio (SAR), sodium percentage (Na%), magnesium ratio

(MR), total hardness (TH), permeability index (PI), residual sodium bicarbonate (RSBC), and Kelly's ratio (KR). A database (Table 1) of 542 samples collected from thirteen Moroccan coastal aquifers [2] (Figure 1) was used. The following equations (Equations (1)–(7)) (Table 2) were used to calculate the indices. All of the ion concentrations used were converted to meq/L. TH was expressed as ppm.

Table 1. Summary of groundwater physico-chemical parameters of Moroccan coastal aquifers (*SD*: standard deviation).

Parameters	EC	Ca^{2+}	Mg^{2+}	Na^{+}	K^{+}	HCO_3^{-}	Cl^{-}	SO_4^{2-}	NO_3^{-}
	(µS/cm)	(mg/L)							
Max	21,000	37.1	65.6	200.0	6.4	21.0	220.0	46.3	521.1
Min	190	0.4	0.2	0.5	0.0	0.1	0.4	0.0	0.0
Mean	3243	7.7	8.4	17.8	0.3	5.0	21.3	7.4	64.8
SD	2600	5.8	7.6	20.9	0.6	2.6	26.1	7.6	66.2

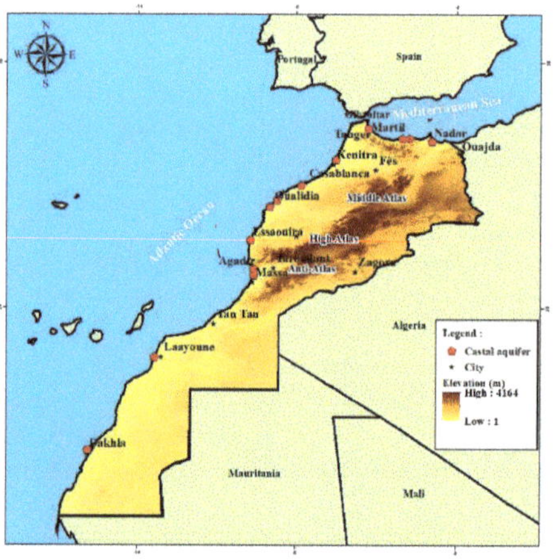

Figure 1. Location of Moroccan coastal aquifers.

Table 2. List of irrigation water indices and their equations used in this study.

Index	Index Formula	Equation Number	Source
Sodium Absorption Ratio	$SAR = Na^{+}/\sqrt{((Ca^{2+} + Mg^{2+})/2)}$	(1)	[4]
Sodium Percentage	$Na^{+}\% = ((Na^{+} + K^{+}) * 100/(Ca^{2+} + Mg^{2+} + Na^{+} + K^{+}))$	(2)	[5]
Magnesium Ratio	$MR = (Mg^{2+} * 100/(Ca^{2+} + Mg^{2+}))$	(3)	[6]
Total Hardness	$TH = (2.497\ Ca^{2+} + 4.11\ Mg^{2+})$	(4)	[7]
Permeability Index	$PI = (Na^{+} + \sqrt{HCO_3^{-}}) * 100/Ca^{2+} + Mg^{2+} + Na^{+}$	(5)	[8]
Residual Sodium Bicarbonate	$RSBC = (HCO_3^{-} - Ca^{2+})$	(6)	[9]
Kelly's Ratio	$KR = (Na^{+}/(Ca^{2+} + Mg^{2+}))$	(7)	[10]

3. Results and Discussion

Results of all indices are summarized in Table 3.

Table 3. Statistical summary of irrigation water indices used in this study.

Index	Min	Max	Mean	*SD*
SAR	0.2	45.2	6.1	6.3
Na%	5.5	94.9	45.8	17.9
MR	4.2	93.3	49.3	13.7
TH	74.1	4316.4	808.2	637.6
RSBC	−33.2	15.9	−2.7	6.1
KR	0.1	18.1	1.2	1.6
PI	10.8	101.8	57.0	15.8

The results show high values of EC reaching 21000 µS/cm (Table 1). Comparing all values with the classification adopted by [4], the results show that 8% of samples fall in the excellent class to good class. However, 92% samples fall into other classes not suitable for irrigation (permissible, doubtful, and unsuitable). This situation is explained by high concentrations of major elements in the groundwater. The SAR index shows values ranging from 0.2 to 45.2, with an average 6.1 (Table 3). According to the classification of [4], 95% of samples fall in the excellent and good classes. However, 3% and 2% of samples fall into the doubtful and unsuitable class, respectively. Concerning Na%, the results illustrate that 223 samples are within the excellent to good classes. However, 195, 101, and 23 samples fall within the permissible, doubtful, and unsuitable classes, respectively (Supplemental Table S1). Thus, 41% of the groundwater samples are suitable and 59% are unsuitable according to classification scheme of [5]. This situation reflects the abundance of Na^+ in coastal aquifers due to seawater intrusion and water–rock interaction processes (dissolution of halite rock) [2]. The MR values are between 4.2 to 93.3, with an average value of 49.3 (Table 3). Comparing these results with the classification adopted by [6], 48% of the samples have MR values <50, falling into the suitable class. However, 52% of the samples have MR values >50, falling into the unsuitable class. The TH index of groundwater samples ranged from 74.1 to 4316.4 mg/L, with an average of 808.2 mg/L (Table 3). Most of the water samples are classified as not suitable (hard and very hard). Both classes account for 97% of the samples of the study area. The results for PI show that values range from 10.8 to 101.8, with an average of 57.0 (Table 3). Nearly 14% of the samples fall into the class I category, which is considered to be the best for irrigation water. However, 85% of the samples fall into the class II category (acceptable for irrigation water), while the remaining 1% fall into the class III category, which is considered to be unacceptable for irrigation water. The RSBC index ranges from −33.2 to 15.9, with an average of −2.7 (Table 3). Moreover, 95% of the samples fall into the safe category, 4.6% of the samples fall into the marginal category, and 0.4% are in the unsatisfactory category. Results in Table 3 show that the values of KR range from 0.1 to 18.1, with an average of 1.2 (Table 3). The samples with values >1 indicate high values of sodium, and others with values < 1 are characterized by low concentrations of sodium [10]. In this study, 66% of the groundwater samples are suitable for irrigation. Nonetheless, 34% of the samples are unsuitable for irrigation purposes. These results show that the groundwater is considerably polluted and is under continued degradation. This situation is explained by over pumping in coastal areas and the occurrence of some hydrogeochemical processes that are dominant in coastal aquifers, such as evaporation, seawater intrusion, human activities (wastewater, agriculture), and water–rock interactions [2]. These processes increase the concentration of chemical elements within the water table, may result in the salinization of water, and can impact farmer activities. In addition, the return of irrigation water towards the water table and nitrate pollution represent another threat to groundwater in Morocco coastal aquifers [11,12]. In this study, the nitrate values are very high, exceeding 521.1 mg/L (Table 1). This situation is mainly due to agricultural activities (e.g., fertilizers) and anthropogenic activities (e.g., manure and septic effluents) in the coastal area. However, solutions are needed to preserve coastal aquifers against salinity risk. Pumping control, monitoring of seawater intrusion, desalination projects, bio-saline agriculture adoption, treated wastewater reuse, artificial recharge, and sustainable drip irrigation and subsurface drip irrigation can implemented as sustainable management plans to preserve groundwater resources, to ensure good water for irrigation purposes, and to increase agriculture production. The promotion of scientific research projects that bring innovative solutions for the treatment of high salinity of irrigation water is also strongly required.

4. Conclusions

The assessment of groundwater samples for irrigation uses in coastal aquifers showed that some samples were within the permissible limit and can be used with caution, for example 8% and 41% for EC and Na%, respectively. However, several water samples were observed to be inadequate for irrigation purposes. This situation is due to salinity risk (e.g., seawater), which can degrade groundwater resources in coastal environments. Some management plans have been presented in order to reduce the salinity impact on groundwater and soil resources. These proposals will serve as a reference for future water- and agriculture-related initiatives.

Supplementary Materials: The following supporting information can be downloaded at: https://www.mdpi.com/article/10.3390/environsciproc2022016048/s1, Table S1: Classification of groundwater samples of Moroccan coastal aquifers for suitability in irrigation.

Author Contributions: Conceptualization, Y.E.-z. and L.B.; methodology, Y.E.-z.; software, Y.E.-z.; validation, L.B., A.S. and M.H.; formal analysis, Y.E.-z.; investigation, Y.E.-z.; resources, Y.E.-z. and L.B.; data curation, Y.E.-z.; writing—original draft preparation, Y.E.-z.; writing—review and editing, L.B., A.S., M.H., Y.B., D.D. and A.C.; visualization, Y.B.; supervision, L.B.; project administration, Y.E.-z. and L.B.; funding acquisition, Y.E.-z. All authors have read and agreed to the published version of the manuscript.

Funding: This research was funded by Mohammed VI Polytechnic University.

Institutional Review Board Statement: Not applicable.

Informed Consent Statement: Not applicable.

Data Availability Statement: All data are from literature review.

Acknowledgments: This Research was partially supported by the grant of Mohammed VI Polytechnic University. This work was carried out within the CHARISMA Project with the assistance of the Hassan 2 Academy of Sciences and Techniques, the support of Ibn Zohr University.

Conflicts of Interest: The authors declare no conflict of interest.

References

1. Hssaisoune, M.; Bouchaou, L.; Sifeddine, A.; Bouimetarhan, I.; Chehbouni, A. Moroccan groundwater resources and evolution with global climate changes. *Geosciences* **2020**, *10*, 81. [CrossRef]
2. Ez-zaouy, Y.; Bouchaou, L.; Saad, A.; Hssaisoune, M.; Brouziyne, Y.; Dhiba, D.; Chehbouni, A. Morocco's coastal aquifers: Recent observations, evolution and perspectives. *Environ. Pollut.* **2022**, *293*, 118498. [CrossRef] [PubMed]
3. Aziane, N.; Khaddari, A.; Ebn Touhami, M.; Zouahri, A.; Nassali, H.; Elyoubi, M. Evaluation of groundwater suitability for irrigation in the coastal aquifer of Mnasra (Gharb, Morocco). *Mediterr. J. Chem.* **2020**, *10*, 197–212. [CrossRef]
4. Richards, L.A. *Diagnosis and Improvement of Saline and Alkaline Soils*; US Department of Agriculture: Washington, DC, USA, 1954; pp. 1–159.
5. Todd, D.K. *Groundwater Hydrology*, 2nd ed.; Wiley: New York, NY, USA, 1980.
6. Szaboles, I.; Darab, C. *The Influence of Irrigation Water of High Sodium Carbonate Content on Soils. 8th International Congress Soil Science Sodics Soils*; Research Institute for Soil Science and Agricultural Chemistry of the Hungarian Academy of Sciences: Budapest, Hungaria, 1964; pp. 237–246.
7. Sawyer, C.N.; Mccarty, P.L. *Chemistry for Sanitary Engineers*, 2nd ed.; McGraw Hill: New York, NY, USA, 1967; p. 518.
8. Doneen, L.D. *Notes on Water Quality in Agriculture. Published as a Water Science and Engineering*; Paper 4001; Department of Water Sciences and Engineering, University of California: Los Angeles, CA, USA, 1964.
9. Gupta, S.K.; Gupta, I.C. *Management of Saline Soils and Water*; Oxford and IBH Publication Co.: New Delhi, India, 1987; p. 399.
10. Kelley, W.P. Permissible composition and concentration of irrigation water. *Trans. Am. Soc. Civ. Eng.* **1940**, *66*, 607–613. [CrossRef]

11. Fakir, Y.; Zerouali, A.; Aboufirassi, M.; Bouabdelli, M. Potential exploitation and salinity of aquifers, Chaouia coast, Atlantic shoreline. *Morocco. J. African Earth Sci.* **2001**, *32*, 791–801. [CrossRef]
12. Re, V.; Sacchi, E. Tackling the salinity-pollution nexus in coastal aquifers from arid regions using nitrate and boron isotopes. *Environ. Sci. Pollut. Res.* **2017**, *24*, 13247–13261. [CrossRef] [PubMed]

environmental sciences proceedings

MDPI

Proceeding Paper

Salinization of Soils and Aquifers in Morocco and the Alternatives of Response †

N'gouari Gana Abdou Oumara [1] and Lahcen El Youssfi [1,2,*]

1 Natural Resources and Sustainable Development Laboratory RN2D, Faculty of Sciences, Ibn Tofail University, P.O. Box 133, Kenitra 14000, Morocco; oumaraabdou06@gmail.com
2 National School of Chemistry of Kenitra, Ibn Tofail University, P.O. Box 133, Kenitra 14000, Morocco
* Correspondence: lahcen.elyoussfi@uit.ac.ma

† Presented at the 2nd International Laayoune Forum on Biosaline Agriculture, 14–16 June 2022; Available online: https://lafoba2.sciforum.net/.

Abstract: The agriculture sector in Morocco contributes significantly to the economic development of the country; however, this sector faces several challenges. One of these important challenges is the increasing level of salinization in soils and groundwater. This has a strong impact on food security by reducing agricultural yield. The origin of salinization is usually due to marine intrusion in coastal areas, dissolution of saline aquifer rocks and infiltration of poor-quality irrigation water in the case of groundwater. In the case of soils, it is caused by irrigation with poor-quality water in poorly drained soils, and by evaporation of the water of shallow groundwater, which leads to a saline concentration in the surface layers of soils, as well as 'other' origins. Thus, many regions of Morocco are affected by this phenomenon, especially arid and semi regions with a low rainfall rate. Among the existing alternatives to contain this challenge in Morocco and in the arid and semi-arid regions in particular is the use desalination of sea water and biosaline agriculture. The adoption of the first option aims at the preservation of local production and adaptation in the context of scarcity of water resources and low quality of water for the second. The goal of this review is to present an update of the state of the salinization of Moroccan soils and aquifers and the potential alternatives to respond to these challenges.

Keywords: salinization; aquifers; soil; biosaline agriculture; water desalination

Citation: Oumara, N.G.A.; El Youssfi, L. Salinization of Soils and Aquifers in Morocco and the Alternatives of Response. *Environ. Sci. Proc.* **2022**, *16*, 65. https://doi.org/10.3390/environsciproc2022016065

Academic Editor: Abdelaziz Hirich

Published: 24 June 2022

1. Introduction

The degradation of groundwater and soil quality as a result of salinization is an increasingly worrying concern. Today, on a planetary scale, the world is losing about 10 ha/min of agricultural land and salinization accounts for 30% [1], which is 1 to 2% of irrigated land per year [2]. This is especially observed in semi and arid regions [3], where the intensive use of water resources leads to strong evaporation causing salinization of soils and groundwater [4].

This problem is also observed in Morocco, a country heavily dependent on agriculture, as it represents 19% of GDP [5], and accounts for 1/3 of national employment [6], and where the groundwater quality is very deteriorated (more than 44% is of poor quality) [7]. It is estimated that about 0.5 million ha of agricultural area is affected by salinization [8], which is 160,000 ha of irrigated areas [9]. The causes of salinization are varied: marine intrusion in coastal areas as a response to overexploitation of groundwater [4], dissolution of aquifer rocks through which water flows, or anthropogenic contamination of agricultural origin [10]. For soils, the use of saline water for irrigation is one of the main causes of salinization. In regions where the aquifers are shallow and where the waters are salinized, the rising water from the aquifers also causes soil salinization [1]. Soil salinization has been a problem of study for several years, because it is known to affect the yield of the production of different types of crops [1]. In response to the problems of salinization of water, of soil and water stress, in semi and arid regions, such as Morocco, it resorts to biosaline agriculture and water desalination (sea water or brackish) [11]. These fall within the scope of Moroccan Water Supply Management Strategies and Water Demand Management Strategies, to adapt to environmental constraints, such as the salinization of aquifers and soils, water scarcity and the increasing water demand.

The methodology consists of reviewing the studies carried out at the spatial scale on the salinization of aquifers and soils in Morocco, determining the causes of salinization, presenting the solutions of responses to these problems as well as the possible impacts that generate these solutions.

2. Salinization of Aquifers in Morocco

The salinization of aquifers can be of natural (geological) origin, by water-rock interaction, or artificial as seen in irrigation areas. It can also be of mixed origin (natural and anthropogenic) in coastal areas where overexploitation of groundwater causes marine intrusion [3]. In Morocco, in the continental zone, most aquifers are naturally salinized (by aquifer rocks) [12]. In the coastal zone, the Moroccan coastal aquifers are among the aquifers that have experienced very high salinization by marine intrusion following the overexploitation of groundwater [10]. Two regions affected by salinization are presented: Triffa (eastern Morocco) and Foum El Oued (southern Morocco).

Case of the Triffa aquifer (eastern Morocco) [13]: the Triffa plain represents the largest irrigated perimeter of the Moulaya's basin and where the main crops grown are citrus fruits [14]. The use of the region's underground water resource is dedicated to irrigation and drinking water supply. With the expansion of the irrigated perimeter for large hydraulics, groundwater in the region has been greatly exploited. This leaded to a deficit in the water balance (-32.3 Mm^3/year) in 2010 [15]. The plain has also experienced, in the past, periods of drought which have caused a strong exploitation of groundwater, and which have caused the deterioration of water quality. A study of the expansion of salinization in the aquifers of this plain showed that a good part of the aquifer is salinized following the dissolution of the aquifer rock, and where salinity values of 6g/L and 10,000 µs/cm of conductivity were observed.

Case of the Foum EL Oued aquifer (southern Morocco) [16]: this is one of a case of salinization of coastal aquifers in Morocco. It is located in limestone formations from the Cenonian and detrital from the Plio-Quaternary. The water from this aquifer is the main hydrogeological source which ensures the various supplies (agricultural, industrial and drinking water supply) of the Laayoune region. Population growths, strong agricultural activity, the nature of the Saharan-type climate increase the demand for water and lead to the overexploitation of the aquifer. Following overexploitation of the aquifer, the bevel (freshwater and saltwater interface) leans towards fresh water which therefore implies the intrusion of sea water into the aquifer. The dissolution of the aquifer rock was also a cause of salinization. The conductivity measurement shows an increasingly significant gradient towards the coastal study area.

Many studies have been made on the salinization of continental (Tafilalet [17], plain of Bahira [10], etc.)and coastal (coastal chaouia [18], Doukkala [19], Chtouka-Massa [20], Mnasra [21], Essaouira plain [22], etc.) aquifers of Morocco. All of these studies conclude that the salinization is due, in the coastal areas, tomarine intrusion, natural salinization in the continental areas and human participation in irrigated areas.

3. Soils Salinization in Morocco

Soil is an interface in the environment and also a fundamental natural resource for development [23]. It forms very slowly from the geological materials of the earth and is the loose part of the lithosphere. On the other hand, it degrades very easily in the event of indiscreet and unreserved use. It provides several fundamental functions, such as the food function, the filter function and the biological function.

Today, with population growth, the soil is subject to inappropriate use and intensive exploitation for increased agricultural production. This aggravates, modifies and deteriorates the natural state of the soil, and therefore the fertility. Salinization is one of the forms of soil degradation. It consists of a high accumulation of salt in the soils which impact negatively the agricultural productivity of the soils [24]. A soil is said to be saline when its electrical conductivity is under normal conditions exceeds 4Ds/m or 2g/L [25]. In Morocco, it is observed in almost all the irrigated perimeters of hydraulic basins (Table 1); and it affects more than half of the perimeters in certain basins (Tafilalet, Ouarzazate). It is favored by the arid climate, agricultural intensification and excessive use of chemicals (fertilizers) [7].

Table 1. Salinization in irrigated perimeters in Morocco [8].

Irrigated Area	Area Affected by Salinity, (×1000 ha)	Percentage of Irrigated Area
Gharb Low Moulouya	15.0 30.2	12.5 27.7
Haouz of Marrakech	24.6	29.9
Tafilalet	20.9	70.4
Ouarzazate	14.5	65.9
Tadla	19.3	24.5
Doukkala	0.6	1.0
Souss Massa	9.8	28.8
Loukkos	2.8	14.5
Bahira	21.0	22.8
Total	158.7	

4. Possible Alternatives of Response to the Aquifers and Soils Salinization

Several alternatives are recommended depending on the hydrodynamical conditions of the regions to mitigate and to overcome the problems of soil and aquifer salinization. For aquifers, M. S. Hussain, and al. gathered several of them [26] such as artificial recharge from surface water or treated wastewater, the reduction of water pumping, the establishment of physical barriers in the coasts (underground as an underground embankment, surface as the extension of the continent towards the coast), abstraction of brackish water, monitoring of marine intrusion, combined techniques, etc. For soils, we have drainage, leaching, the development of saline or salinity-tolerant crops, etc. [27]. However, the problem for this is, in arid and semi-arid regions (such as southern Morocco) where water resources are increasingly scarce and the soils threatened by desertification, some of these alternatives are difficult to employ. For example, artificial recharge cannot take place due to water scarcity (average rainfall of 200 mm/year [12]) and high demand; reduction of pumping without an additional water source; physical barriers are very expensive; leaching and drainage cannot take place without enough quantity of water; etc. In Morocco, two of these alternatives are now expanding: biosaline agriculture (BA) and seawater desalination (DS). (BA): Due to the impact of salinization leaving soils unvegetated or poorly covered, BA has therefore emerged as an alternative option [4]. It was introduced in 1997 by the IAEA (International Atomic Energy Agency) via a project in collaboration with INRA (National Institute of Agronomic Research) and ORMVA (Regional Office for Agricultural Development) of Tafilalet [28]. This first test study was carried out in Aïn El Atti, where 11 species of trees and shrubs, notably Ecalyptus camaldulensis and Atriplex lentiformis, were irrigated with highly salinized water (10 g/L). The results showed good salinity tolerance and good crop growth [29]. Other studies were conducted in other regions impacted by salinization. For example, the case of Foum El Oued where soil and water are salinized, alternative species are tested, such as Quinoa, Pearl millet, barley, Panicum Blue, etc., and have shown very high tolerance to salinity [28]. DS: Water salinization is observed in most hydraulic basins of Morocco. However, it is more observed in the southern regions (Souss Massa, Draa-Tafilalet, Oued Ed-Dahab-Lagouira, Laâyoune-Sakia El Hamra), which are arid with important agriculture activities. The region of Souss Massa alone provide 90% of the national export of fruits and vegetables [30]. Thus, the groundwater in this region is excessively exploited to respond to the high agriculture demand and lead to salinization by marine intrusion in the coastal zone [31]. DS for irrigation (of approximately 15,000 ha) of crops with high added value, although expensive, was the alternative chosen by the state and in agreement with the farmers, to deal with the scarcity of water, to preserve it and to ensure the agricultural productivity of the region. The use of seawater here explains the precariousness of freshwater resources (even brackish) in this region. Otherwise, desalination of groundwater and brackish surface water is more advantageous than desalination of seawater. The disadvantage of desalination is the costs (energy and maintenance) it generates. This can be offset by good irrigation practice (localized irrigation), by PV supply which is very competitive and also desalinated water can be mixed with marginal water before irrigation.

With these alternatives, salinized soils and waters are therefore exploited and not abandoned, to maintain productivity and the local economy satisfy water demand and preserve aquifers. The expansion of these alternatives could even have a great impact on the economy of the country. Nevertheless, a combination with other alternatives will contribute more. Among the combinations we have the one proposed by Hussain et al. [32] which is the abstraction of brackish water, its desalination, its use (potable water or in irrigation) and the treatment of wastewater for the recharge of aquifers. The advantage of this is that it allows not only to ensure a hydraulic gradient towards the sea and to preserve the aquifers, but it allows increasing the piezometric level of the aquifers.

5. Conclusions

Morocco as an arid and semi-arid country, the deterioration of the quality of its soils and aquifers following salinization continue to be a very worrying problem. Freshwater resources being scarce, the demand being increasing especially for agriculture, the adequate alternative is to use saline water either by using it directly for salinity-tolerant crops, or by proceeding on desalination and maintaining the production of traditional crops. This was therefore the case of many successful projects in Morocco. Biosaline Agriculture is important because it allows the use of water resources with lower quality, to adapt to local climatic conditions, to use salinized soils for the development of halophyte crops (local and introduced species) and to preserve the groundwater resource against overexploitation. In fact, it contributes significantly to food security and in insuring income for rural in arid and semi-arid regions. However, there is a need to encourage research and extension programs concerning new/local crops that showed a higher potential of productivity under salinity and water scarcity conditions. The desalination of water for irrigation, combined with a sustainable irrigation strategy (such as drip irrigation), allows preserving the know-how and the local economy. The extension of these solutions could have a great impact on the problems of water stress, demand management, desertification, maintaining agricultural productivity (with high added value), maintaining employment and other environmental constraints. It can be also a progress for the framework of the 2030 SDG agenda. However, a combination with other alternatives will have more impact to these challenges.

Author Contributions: N.G.A.O. performed writing, original draft; L.E.Y. performed writing, review and conceptualization. All authors have read and agreed to the published version of the manuscript.

Funding: This research received no external funding.

Institutional Review Board Statement: Not applicable.

Informed Consent Statement: Not applicable.

Data Availability Statement: Not applicable.

Conflicts of Interest: The authors declare no conflict of interest.

References

1. FAO; ITPS; GSBI; CBD; EC. *State of Knowledge of Soil Biodiversity. Status, Challenges and Potentialities, Report 2020*; FAO: Rome, Italy, 2020; p. 616. [CrossRef]
2. Barbouchi, M.; Lhissou, R.; Chokmani, K.; Abdelfattah, R.; El Harti, A.; Ben Aissa, N. *Caractérisation de la Salinité des Sols à L'aide de L'imagerie Radar Satellitaire: Cas de la Tunisie et du Maroc*; Centre Eau Terre et Environnement, Institut national de la recherche scientifique: Quebec City, QC, Canada, 2013; p. 99.
3. Kloppmann, W.; Bourhane, A.; Schomburgk, S. *Salinisation des Masses D'eaux en France Métropolitaine et dans l'Outre-mer*; ONEMA: Vincennes, France, 2010; p. 44.
4. Khan, M.A.; Öztürk, M.A.; Gul, B.; Ahmed, M.Z. *Halophytes for Food Security in Dry Lands*; Elsevier: Waltham, MA, USA, 2016; pp. 141–156.
5. Badran, K.; Baker, J.; Collins, B. *Water Managementand and Conservation in Rural Morocco*; Worcester Polytechnic Institute: Worcester, MA, USA, 2012; p. 80.
6. Birol, F. *Energy Policies beyond IEA Contries Morocco*; IEA: Paris, France, 2019; p. 158.
7. Nunes, A.; Rott, O.; Hirvonen, J.; Georgiadis, G.; Mosanu, V.; Sochirca, A.; Laraki, K. *Morocco Environmental Performance Reviews*; United Nations Economic Commission for Africa: Geneva, Switzerland, 2014; p. 284.
8. 3éme Rapport sur L'etat de L'environnement Maroc. 2015; p. 187. Available online: http://www.environnement.gov.ma/PDFs/Rapport-reem.pdf (accessed on 7 February 2022).
9. Badraoui, M. Connaissance-et-Utilisation-des-Ressources-en-sol-au-Maroc. 2006, pp. 91–117. Available online: http://www.abhatoo.net.ma/maalama-textuelle/developpement-economique-et-social/developpement-economique/agriculture/agriculture-generalites/connaissance-et-utilisation-des-ressources-en-sol-au-maroc (accessed on 8 February 2022).
10. El Mokhtar, M.; Fakir, Y.; El Mandour, A.; Benavente, J.; Meyer, H.; Stigter, T. Salinisation des Eaux Souterraines aux Alentours des Sebkhas de Sad Al Majnoun et Zima (plaine de la Bahira, Maroc). *Secheresse* **2012**, *23*, 48–56. [CrossRef]
11. Saadi, A.; Ouazzani, N. Perspectives of desalination of brackish water for valorisation in arid regions of Morocco. *Desalination* **2004**, *165*, 81. [CrossRef]
12. Hssaisoune, M.; Bouchaou, L.; Sifeddine, A.; Bouimetarhan, I.; Chehbouni, A. Moroccan Groundwater Resources and Evolution with Global Climate Changes. *Geosciences* **2020**, *10*, 81. [CrossRef]
13. Hssaisoune, M.; Bouchaou, L.; Sifeddine, A.; Bouimetarhan, I.; Chehbouni, A. Extension spatiale de la salinisation des ressources en eau et modèle conceptuel des sources salées dans la plaine des Triffa (Maroc nord-oriental). *Comptes Rendus Geosci.* **2020**, *338*, 768–774. [CrossRef]
14. Hadria, R.; Elmansouri, L.; Benabdelouahab, T.; Ouatiki, H.; Lebrini, Y.; Htitiou, A.; Khellouk, R. Recours aux satellites pour appuyer le management de l'eau d'irrigation. *Afr. J. Land Policy Geospat. Sci.* **2019**, *2*, 135–148.
15. ONEDD. *Etat de L'environnement, Region de L'oriental*; ONEDD: Rabat, Morocco, 2013; p. 145.
16. EL Mokhtar, M.; Chibout, M.; Kili, M.; EL Mansouri, B.; Chao, J.; EL Kanti, S.M.; Ntarmouchant, A.; Benslimane, A. *Évaluation de L'intrusion Saline dans la Nappe de Foum El Oued, Province de Laâyoune, Maroc*; Bulletin de l'Institut Scientifique: Rabat, Morocco, 2018; p. 17.
17. Aoubouazza, M.; Dakak, H.; Zouahri, A. Qualité des Aquifères de la Région de Aïn el Atti, Amont Immédiat de la Plaine du Tafilalet (Maroc). *Rev. Maroc. Des Sci. Agron. Et Vétérinaires* **2021**, *9*, 5.
18. Najib, S.; Mehdi, K.; Riss, J.; Fadili, A.; Pulido-Bosch, A.; Guessir, H. Salinisation de l'aquifère libre de la Chaouia côtière (Azemmour-Tnine Chtouka), Maroc. *Hydrol. Sci. J.* **2017**, *62*, 749–759. [CrossRef]
19. El Achheb, A.; Mania, J.; Mudry, J. Processus de Salinisation des Eaux Souterraines dans le Bassin Sahel-Doukkala (Maroc Occidental). In Proceedings of First International Conference on Saltwater Intrusion and Coastal Aquifers—Monitoring, Modeling, and Management, Essaouira, Morocco, 23–25 April 2001; p. 14.
20. Malki, M.; Bouchaou, L.; Hirich, A.; Brahim, Y.A.; Choukr-Allah, R. Impact of agricultural practices on groundwater quality in intensive irrigated area of Chtouka-Massa, Morocco. *Sci. Total Environ.* **2017**, *574*, 760–770. [CrossRef] [PubMed]
21. Batchi, M.; Karkouri, A.; El Maaqili, M.; Fenijiro, I. Cartographie de la vulnérabilité à l'intrusion marine de l'aquifère Côtier de Mnasra (Littoral Du Gharb, Maroc-Nord.-Occidental). *Mar. J.* **2014**, *11*, 16.
22. Ammar, H. Vulnérabilité des ressources en eaux des aquifères côtiers en zones semi-arides—Etude comparative entre les bassins d'Essaouira (Maroc) et de la Jeffara (Tunisie). *Comun. Geológicas* **2008**, *95*, 107–121.
23. Robert, M. *Le Sol: Interface dans L'environnement, Ressource Pour le Développement*; Masson: Paris, France, 1996; p. 241.
24. Rengasamy, P. World salinization with emphasis on Australia. *J. Exp. Bot.* **2006**, *57*, 1017–1023. [CrossRef] [PubMed]
25. US Salinity Laboratory Staff. *Diagnosis and Improvement of Saline and Alkali Soils*; USDA Handbook No. 60; US Government Printing Office: Washington, DC, USA, 1954; p. 166.
26. Hussain, M.S.; Abd-Elhamid, H.F.; Javadi, A.A.; Sherif, M.M. Management of Seawater Intrusion in Coastal Aquifers: A Review. *Water* **2019**, *11*, 2467. [CrossRef]

27. Montoroi, J.-P. La Salinisation des Ecosystèmes: De la Dégradation Insidieuse à la Remédiation Continuelle par les Hommes. 2017, p. 7. Available online: https://horizon.documentation.ird.fr/exl-doc/pleins_textes/divers17-08/010070731.pdf (accessed on 24 February 2022).
28. Hirich, A.; Choukr-Allah, R.; Ezzaiar, R.; Shabbir, S.A.; Lyamani, A. Introduction of alternative crops as a solution to groundwater and soil salinization in the Laayoune area, South. Morocco. *Euro-Mediterr. J. Environ. Integr.* **2021**, *6*, 52. [CrossRef]
29. El Allam, M.; El Khadir, M.; Zouahri, A. Evaluation de la Tolérance à la Salinité D'arbres et D'arbustes Pour la Valorisation de Terres Marginales et Eaux Salines; Actes du Symposium International sur le Développement Durable des Systèmes Oasiens. 2005, pp. 284–288. Available online: https://doczz.net/doc/139756/inra--1er-2e-partie---theme-1----institut-national-de-la- (accessed on 24 February 2022).
30. Hirich, A.; Choukr-Allah, R.; Rami, A.; El-Otmani, M. Feasibility of Using Desalination for Irrigation in the Souss Massa Region in the South of Morocco. In *Recent Progress in Desalination, Environmental and Marine Outfall Systems*; Baawain, M., Choudri, B.S., Ahmed, M., Purnama, A., Eds.; Springer International Publishing: Cham, Switzerland, 2015; pp. 189–203. [CrossRef]
31. Seminaire international sur la gestion des aquiferescotiers. *Vulnérabilité et adaptation aux impacts du changement climatique sur la rive sud du bassin méditerranéen*; GEB-Environnement: Rabat, Morocco, 2010; p. 96.
32. Hussain, M.S.; Javadi, A.A.; Sherif, M.M. Assessment of different management scenarios to control seawater intrusion in unconfined coastal aquifers. *J. Duhok Univ.* **2017**, *20*, 259–275. [CrossRef]

environmental sciences proceedings

MDPI

Proceeding Paper

Chemical Seed Priming with Zinc Sulfate Improves Quinoa Tolerance to Salinity at Germination Stage †

Mohammad Redouane Bourhim [1], Said Cheto [1], Ahmed Qaddoury [1], Abdelaziz Hirich [2] and Cherki Ghoulam [1,2,*]

[1] Center of Agrobiotechnology and Bioengineering, Research Unit Labeled CNRST, Cadi Ayyad University, Marrakech 40000, Morocco; medred1.mrb@gmail.com (M.R.B.); cheto@gmail.com (S.C.); a.qadoury@uca.ma (A.Q.)

[2] African Sustainable Agriculture Research Institute (ASARI), Mohammed VI Polytechnic University (UM6P), Laayoune 70000, Morocco; abdelaziz.hirich@um6p.ma

* Correspondence: c.ghoulam@uca.ma

† Presented at the 2nd International Laayoune Forum on Biosaline Agriculture, 14–16 June 2022; Available online: https://lafoba2.sciforum.net/.

Abstract: This study was conducted to assess the effect of seed pre-treatment "priming" with zinc sulfate ($ZnSO_4$) on the improvement in germination in three quinoa genotypes, "ICBA-Q5", "Puno" and "Titicaca", under different salinity levels and to characterize some physiological traits of seed tolerance to salinity. The germination tests were conducted to assess the priming effect on germination. Samples of 50 quinoa seeds of the 3 genotypes were soaked in 1 g/L of $ZnSO_4$ solution for 8 h and then were dried under ambient temperature. Then, each seed sample was placed in a Petri dish containing filter paper imbibed with a salt solution of 300, 400 and 500 mM NaCl. The numbers of germinated seeds were noted every 24 h and seed samples were collected for reserve mobilization analysis. The results showed that, under control conditions, ICBCA-Q5 showed the highest germination percentage, followed by Puno and then Titicaca. The salinity level of 300 and 400 mM NaCl severely inhibited the seed germination in all of the tested genotypes and the concentration of 400 mM NaCl was considered the highest threshold for germination in the quinoa genotypes tested. The priming treatment improved the germination parameters and the improvement was more evident for germination speed and the final germination percentages that were generally increased by $ZnSO_4$ priming by more than 100% for all of the genotypes.

Keywords: priming; quinoa; salinity; germination; zinc sulfate

Citation: Bourhim, M.R.; Cheto, S.; Qaddoury, A.; Hirich, A.; Ghoulam, C. Chemical Seed Priming with Zinc Sulfate Improves Quinoa Tolerance to Salinity at Germination Stage. *Environ. Sci. Proc.* **2022**, *16*, 23. https://doi.org/10.3390/environsciproc2022016023

Academic Editor: Redouane Choukr-Allah

Published: 16 June 2022

1. Introduction

Salinity is among the most harmful abiotic stresses that causes many disruptions to crop productivity [1–3] consequently affecting plant growth and water accessibility, due to the osmotic stress, the ion toxicity, the oxidative stress and a critical K+ deficiency [2].

Generally, plant tolerance to such abiotic stresses involves responses at the cellular level until the entire plant levels throughout at the phenological stage, the physiological interactions that occur between plant processes, plant organs and plants and their environment, resulting in the alleviation of the adverse stress and then improvement of crop production [4]. Germination is a crucial stage for crop establishment, and it is very sensitive to any variation in the necessary environmental conditions allowing embryo-development and seedling emergence. For many crops, ranked as tolerant to salt salinity, their seeds were revealed to be sensitive to this constraint at germination. For quinoa, in spite of its tolerance to salinity, its seed germination was severely inhibited particularly with high salinity levels [3]. To ensure high crop establishment under salinity, seed germination needs to be enhanced by some efficient and easily applied pretreatments [5].

Seed priming is a physiological technology that provides an advantage by improving the speed of seed germination, even under abiotic stresses, such as salinity, drought, very high and very low temperatures, etc. [6]. Priming is also considered as a pre- germinating stress exposure that will create a stress memory as information that improves the plant's physiological processes [2]. Priming is also known as seed preconditioning that has the capacity to modulate the effects of abiotic stresses on crop plants [4].

Our present work aims for the improvement of seeds germination under high salinity levels by some pre-germinative treatments in three genotypes of quinoa (*Chenopodium quinoa* Willd).

2. Materials and Methods

2.1. Plant Material

In this present experiment, the seeds of three quinoa genotypes "Puno", "Titicaca", and "Q5" were used in this study to test the priming effect on the improvement of seed germination using filter paper in Petri dishes.

2.2. Priming Tests and the Germination under Salinity Stress

2.2.1. Determination of Accurate Priming Duration and Zinc Sulfate Concentration

This test was carried out on the genotype with the highest germination percentage in distilled water, in order to determine the efficient concentration of the priming chemical "zinc sulfate" and the optimal time of treatment. Thus, the seeds were soaked in 3 different concentrations (0.1, 0.5 and 1 g/L of zinc sulfate) at 4 durations (4, 6, 8 and 12 h). The germination test was carried out in 9 cm Petri dishes containing filter paper at a density of 50 seeds per dish. The seeds and filter paper were imbibed with 1 mL of 200 mM NaCl solution to ensure the differences between $ZnSO_4$ concentrations can be easily observed. Additionally, a negative control with no priming treatment and a positive control consisting of germination in distilled water were included.

2.2.2. Effect of Priming with Zinc Sulfate on the Improvement of Germination under Salinity

The experiments were conducted in the laboratory and carried out on three genotypes with the selected concentration of $ZnSO_4$ (1 g/L) and the treatment time retained from the first test (8 h). The germination process was carried out in 9 cm Petri dishes, as described above. Samples of 50 seeds were placed on each dish, and the filter papers were imbibed with 1 mL of saline solutions consisting of 300, 400 and 500 mM of NaCl. Simultaneously, the negative control (without priming) was prepared for each genotype and saline concentration. Each treatment was repeated three times and the number of germinated seeds was noted each day (24 h).

Some germination parameters as the mean germination time (MGT) and final germination percentage (FGP) were calculated, as described by [7].

$$MGT = \sum(NiTi) / \sum Ni$$

where N is the number of seeds germinated at time i, and Ti is the time (day) from sowing. The value of MGT is inversely proportional to the germination speed.

$$FGP = (\text{Finalnumberofgerminatedseeds}/\text{Totalnumberofseedssown}) \times 100$$

2.3. Statistical Analysis

The statistical analysis was performed using "SPSS" program version 25. All means values and standard deviations (SD) were obtained from 3 replicates. p values of <0.05 were considered statistically significant according to the Tukey HSD test.

3. Results and Discussion

3.1. Effect of the Duration and the Concentration of Zinc Sulfate Priming on Quinoa "Q5" Germination under Salt Stress

The results showed that salinity treatment (200 mM) severely affected seed germination in the ICBA-Q5 genotype mainly after 48 h of germination. The concentrations of 1 g/L of $ZnSO_4$ presented the highest improvement in germination compared to the negative control (unprimed). We noticed that there was no significant difference between the durations of 6, 8 and 12 h of treatment. Based on these results, we retained the $ZnSO_4$ concentration of 1 g/L with 8 h as the treatment duration.

3.2. Effect of the Duration and the Concentration of Zinc Sulfate Priming on Quinoa "Q5" Germination under Salt Stress

The results from the second experiment confirmed that, under the priming condition, Puno and ICBA-Q5 presented the highest-rated results on the FGP and the variety Titicaca had the lowest rate. On the other hand, this parameter was significantly improved by the zinc sulfate priming, but only under 300 mM of NaCl for all the genotypes.

3.3. Effect of Priming with Zinc Sulfate on Germination in Three Quinoa Genotypes

The results illustrated in this part of the experiment proved that the applied salinity levels 300 and 400 mM NaCl drastically reduced the FGP in all of the tested varieties, with a variation between them (Figure 1 A). Indeed, under 300 mM NaCl, the variety Puno was the least affected, followed by ICBA-Q5 and then Titicaca. The salinity level of 500 mM NaCl was too high to completely inhibit seed germination in the tested genotypes, especially in ICBA-Q5 and Titicaca. The zinc sulfate priming treatment improved the germination in the three genotypes. This effect was evident mainly for 300 mM NaCl and was more pronounced for the Titicaca variety for which the induced increase in FGP was more than 100%. Under 400 mM NaCl, the positive effect of priming induced increases in FGP of around 100% for all of the tested varieties. The results of Figure 1B showed that salinity stress increased the MGT in all of the tested varieties but the priming treatment with zinc sulfate reduced this parameter compared to the unprimed seeds.

A: Final Germianion Percentage (FGP) **B:** Mean Germination Time (MGT)

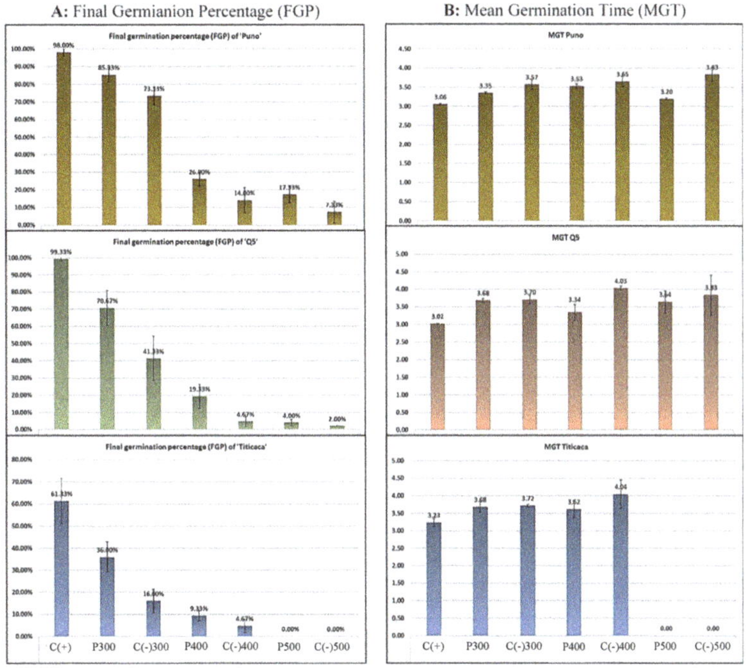

Figure 1. Effect of zinc sulfate priming on the final germination percentage (FGP) (**A**) and the mean germination time (MGT) (**B**) in three quinoa varieties, "Puno, Q5 and Titicaca", positive control "C(+)"

refers to unprimed and non-stressed seeds; negative control "C(-)300; C(-)400; C(-)500" refers to unprimed seeds germinated in 300, 400 and 500 Mm NaCl, respectively; "P300; P400; P500" refers to the primed seeds germinated in 300, 400 and 500 Mm NaCl, respectively. Results were means ± SD of the three replicates.

Under 400 mM NaCl salinity, seed germination was very low in all the tested genotypes and we can state that this NaCl concentration is the salinity threshold for germination for the studied quinoa varieties.

These results are in agreement with those reported by [6], stating that the priming of zinc sulfate had a strong positive effect on the speed and emergence percentage of wheat seeds grown under rainfed conditions. In addition, [8] reported an enhancement of biochemical parameters and seed germination of chickpeas by zinc sulfate priming.

4. Conclusions

These preliminary results allowed us to conclude that the studied quinoa genotypes presented varied seed germination capacities under high salinity levels, as for 300 and 400 mM NaCl, the Puno variety was the most tolerant as compared to ICBA-Q5 and Titicaca. The priming treatment based on soaking seeds for 8 h in 1 g/L solution of zinc sulfate is efficient for improving the germination percentage by more than 100% under salinity levels of 300 and 400 mM NaCl. The concentration of 400 mM NaCl was considered as the germination threshold for the tested quinoa genotypes, especially for ICBA-Q5 and Titicaca.

Author Contributions: Conceptualization, C.G. and M.R.B.; methodology, M.R.B. and C.G.; investigation, M.R.B. and S.C.; resources, A.H.; data curation, C.G. and A.Q.; writing—original draft preparation, M.R.B.; writing—review and editing, C.G. and A.H.; supervision, C.G. and A.Q.; project administration, C.G. and A.H.; funding acquisition, C.G. All authors have read and agreed to the published version of the manuscript.

Funding: This research was funded by SALAD project in the framework of ERANet -FOSC-056 program.

Institutional Review Board Statement: Not applicable.

Informed Consent Statement: Not applicable.

Data Availability Statement: Reported data were generated by experimental work of this study.

Conflicts of Interest: The authors declare no conflict of interest.

References

1. Derbali, W.; Manaa, A.; Goussi, R.; Derbali, I.; Abdelly, C.; Koyro, H.-W. Post-stress restorative response of two quinoa genotypes differing in their salt resistance after salinity release. *Plant Physiol. Biochem.* **2021**, *164*, 222–236. [CrossRef] [PubMed]
2. Moreno, C.; Seal, C.E.; Papenbrock, J. Seed priming improves germination in saline conditions for Chenopodium quinoa and Amaranthus caudatus. *J. Agron. Crop Sci.* **2017**, *204*, 40–48. [CrossRef]
3. Yang, A.; Akhtar, S.S.; Iqbal, S.; Qi, Z.; Alandia, G.; Saddiq, M.S.; Jacobsen, S.-E. Saponin seed priming improves salt tolerance in quinoa. *J. Agron. Crop Sci.* **2017**, *204*, 31–39. [CrossRef]
4. Azeem, M.; Iqbal, N.; Kausar, S.; Javed, M.T.; Akram, M.S.; Sajid, M.A. Efficacy of silicon priming and fertigation to modulate seedling's vigor and ion homeostasis of wheat (*Triticum aestivum* L.) under saline environment. *Environ. Sci. Pollut. Res.* **2015**, *22*, 14367–14371. [CrossRef] [PubMed]
5. Hajihashemi, S.; Skalicky, M.; Brestic, M.; Pavla, V. Cross-talk between nitric oxide, hydrogen peroxide and calcium in salt-stressed Chenopodium quinoa Willd. At seed germination stage. *Plant Physiol. Biochem.* **2020**, *154*, 657–664. [CrossRef] [PubMed]
6. Aboutalebian, M.A.; Zare Ekbatani, G.; Sepehri, A. Effects of on-farm seed priming with zinc sulfate and urea solutions on emergence properties, yield and yield components of three rainfed wheat cultivars. *Sch. Res. Libr. Ann. Biol. Res.* **2012**, *3*, 4790–4796.
7. Scott, S.J.; Jones, R.A.; Williams, W.A. Review of Data Analysis Methods for Seed Germination. *Crop Sci.* **1984**, *24*, 1192–1199. [CrossRef]
8. Tajlil, A.H.; Pazoki, A.; Asli, D.E. Effects of seed priming by mannitol and zinc sulfate on biochemical parameters and seed germination of chickpea. *Int. J. Farming Allied Sci.* **2014**, *3*, 294–298.

environmental sciences proceedings

MDPI

Proceeding Paper

The Emergence of a Governance Landscape for Saline Agriculture in Europe, the Middle East and North Africa †

Katarzyna Negacz [1,*], Pim van Tongeren [1], Lucia Ferrone [2], Federico Martellozzo [2] and Filippo Randelli [2]

[1] Department of Environmental Policy Analysis, Institute for Environmental Studies (IVM), Vrije Universiteit Amsterdam, De Boelelaan 1111, 1081 HV Amsterdam, The Netherlands; p.van.tongeren@vu.nl

[2] Department of Economics and Management , University of Florence, Via delle Pandette 27, 50127 Firenze, Italy; lucia.ferrone@unifi.it (L.F.); federico.martellozzo@unifi.it (F.M.); filippo.randelli@unifi.it (F.R.)

* Correspondence: k.e.negacz@vu.nl

† Presented at the 2nd International Laayoune Forum on Biosaline Agriculture, 14–16 June 2022; Available online: https://lafoba2.sciforum.net/.

Abstract: Salinization is one of the main challenges of contemporary agriculture affecting food security and sustainability. Climate change with more persistent droughts, floods and sea-level rise is expected to increase this challenge, making it one of the most common land degradation processes. At the same time, an increasingly complex institutional landscape has emerged across multiple issue areas of global environmental governance related to salinization. This can be seen in a myriad of public, private, and hybrid actors coming together by creating initiatives to address the issue of growing salinization through saline agriculture. Therefore, the aim of this paper is to characterize the development of a governance landscape of cooperative initiatives for saline agriculture in Europe, North Africa, and the Middle East, and to discuss how to harness their potential and orchestrate their efforts. The preliminary findings suggest that the fragmented landscape of initiatives is predominated by public actors and research institutions. This potentially hampers benefit sharing and upscaling efforts. Operational activities are most frequently the governance function, followed by information and networking efforts thereafter. Thematically, initiatives focus on the development of new crop varieties and water and soil management practices. Linkages to the Sustainable Development Goals suggest saline agriculture is connected to policy debates on sustainable food systems, climate change, water security, and land degradation.

Keywords: salinization; governance; international cooperative initiatives; policy; saline agriculture

Citation: Negacz, K.; van Tongeren, P.; Ferrone, L.; Martellozzo, F.; Randelli, F. The Emergence of a Governance Landscape for Saline Agriculture in Europe, the Middle East and North Africa. *Environ. Sci. Proc.* **2022**, *16*, 33. https://doi.org/10.3390/environsciproc2022016033

Academic Editor: Abdelaziz Hirich

Published: 16 June 2022

1. Introduction

Salinization of water and soil resources is a substantial driver of land degradation and stresses freshwater provisions, particularly in arid and semi-arid regions [1]. It depletes the soil of pivotal nutrients, decreases the water quality and is a significant constituent of desertification processes [2,3]. This potentially threatens global food security and nutrition needs [4]. Moreover, it can trigger the collapse of local fishery industries, reduce biodiversity, and change local climatic conditions [3]. Salinization thus poses a significant barrier to ensure food security under the pressures of population growth and climate change [5]. Anthropogenic salinization has occasionally contributed to the destruction of formerly successful agrarian societies, such as ancient Mesopotamia and the Tigris–Euphrates valley [6]. Effective management can mitigate factors accounted both to high environmental and social costs [3].

The extent of salinized land is difficult to determine accurately. One billion hectares of land, divided over more than 100 countries, can be classified as salt-affected [7]. Around 10% of the global arable land [8] and 20% of irrigated lands [9] are salt-affected. However, estimates vary widely among different countries. In some countries, up to 50% of irrigated land is salt-affected as mapped through remote sensing [10].

Current academic research focuses on the identification and exploration of strategies that mitigate or adapt to salinization. Saline agriculture often translates to mitigation techniques that aim to either move soluble salts to lower soil depths through leaching, natural or artificial drainage systems, or by removing salt through mechanical or biological means [11]. Moreover, halophytic plants with desalinizing properties have been utilized successfully for soil reclamation [12]. Furthermore, humic substances [13] and microorganisms [14] may also mitigate salinity stress. Next to saline mitigation techniques, there is a growing field of knowledge and practice on saline adaptation. Saline agriculture seems feasible for crops that can withstand relatively large amounts of salts that have been built up in root zones [15]. This could be achieved by using salt resistant rootstocks, either by genetic modification or classical breeding [15].

In order to mitigate salinization and freshwater shortage, a sustainable transition needs to occur in the environmental policy landscape. A sustainable transition can be set in motion by a collective effort of experimental niche initiatives [16]. Transitions require system innovations that transcend individual actors and construct relationships between private and public entities [17]. Sustainable development can occur in innovative environments through these initiatives. One of the earliest examples of an initiative investigating the potential of saline agriculture dates back to 1954, when the U.S. Salinity laboratory was launched. The number of initiatives exploring saline agriculture has exponentially increased over the past two decades, highlighted by examples such as the International Centre for Biosaline Agriculture (ICBA), founded in 2000 in the United Arab Emirates.

Salinization on a global scale is a problem positioned in a complex and fragmented institutional landscape [18]. No clear pathway or solution can solve the issues caused by salinity, threatening global food and water security. Many stakeholders are involved, with volatile perceptions regarding associated problems and solutions. These characteristics make this a vital problem [19] which must be resolved by incorporating design thinking in the process [20]. To achieve this, a systematic, structured, and interdisciplinary approach must be taken. This short paper aims to structure and explore the field of saline agriculture by mapping and analyzing suitable initiatives in a comprehensive way. At the time of writing, this approach has never been adopted with regard to saline agriculture.

2. Materials and Methods

Using a systematic approach, we created a database of cooperative initiatives for saline agriculture by internet snowballing and expert interviews. Further, we applied a semi-automated content analysis to the mission and vision statements and about sections of initiatives websites in order to validate their link with saline agriculture.

To describe the evolving institutional landscape and make policy recommendations, we code characteristics of each initiative including *inter alia*, their members, governance functions, goals, and geographic coverage. We analyzed the characteristics of these initiatives using descriptive statistics to illustrate the patterns across the sample. The data were collected from publicly available online data sources.

3. Preliminary Results

The preliminary results show a few overarching trends in the sample of ca. 100 initiatives selected for the analysis. There is an increasing number of cooperative initiatives focusing on saline agriculture over time, particularly in years 2019–2020. The majority of initiatives lasted for a finite period of time, with an average duration of 3 years. The initiatives are often led by diversified sets of actors, but the public actors are predominant. A significant number of initiatives are related to research institutions. Their main governance functions focus on operational activities followed by information sharing and networking. The main themes for the initiatives are cultivation of conventional crops under saline conditions and water management practices. Around 20% of the initiatives incorporate halophytic plants for saline mitigation within their operations. The key sustainable development goals (SDGs), i.e., global goals for sustainable development constructed by the United Nations [21], addressed are SDG2 "Zero hunger", SDG13 "Climate action", SDG6 "Clean water and sanitation", and SDG8 "Decent work and economic growth". Our preliminary results indicate that most of the initiatives do not report publicly, but those with reports exhibit high verification rates.

4. Discussion and Conclusions

Our findings suggest that the governance landscape for saline agriculture is dominated by public and research initiatives. This may indicate a high interest of national government and international organizations, as well as good access to the public funding. This could indicate a low involvement of other stakeholders such as farmers, distributors, or environmental associations can influence access and benefit sharing and potentially hamper upscaling efforts. Further research in a form of network analysis could help in understanding connections between main actors.

The main governance functions of saline agriculture initiatives indicate high interest in practical application, field trials, and experiments. This is consistent with the main themes discussed in the literature on saline agriculture, which focuses on the development of new crop varieties and soil and water management practices [22,23]. Although of great importance in the field of saline agriculture, and often tackled with a scenario-based approach [24], the analysis of salinity parameters within the governance landscape falls outside the scope of our contribution. The second most common function, information sharing and networking, points to an increased interest in building a saline agriculture community [25], which is also reflected in a rising number of initiatives in recent years.

The key SDGs partially correspond with previous research [26]. However, climate change adaptation is found to have a more prominent role in our analysis of initiatives. This could serve as a link to increase the presence of saline agriculture as a point on the policy agenda. As a multifaceted topic, saline agriculture is connected to policy debates on climate change, sustainable food systems, water security, and land degradation. Integration of saline agriculture in these policy domains can contribute to addressing multiple SDGs through synergistic actions and instruments. Furthermore, the orchestration of this fragmented landscape potentially led by international organizations, such as FAO, could provide a pathway to harness the potential of international initiatives for saline agriculture for addressing climate change, water, and food security.

Author Contributions: Conceptualization: K.N.; methodology: K.N. and P.v.T.; Software: P.v.T.; validation: K.N., P.v.T. and L.F.; formal analysis: K.N. and P.v.T.; investigation: P.v.T.; data curation, K.N. and P.v.T.; writing—original draft preparation: K.N. and P.v.T.; writing—review and editing: K.N., P.v.T., L.F., F.M. and F.R.; project administration: K.N. and P.v.T.; funding acquisition: K.N., F.M. and F.R. All authors have read and agreed to the published version of the manuscript.

Funding: This project is part of the programme of the ERA-NET Cofund FOSC that has received funding from the European Union's Horizon 2020 research and innovation programme under grant agreement No. 862555.

Institutional Review Board Statement: Not applicable.

Informed Consent Statement: Not applicable.

Data Availability Statement: For the data supporting reported results, please contact the corresponding author.

Conflicts of Interest: The authors declare no conflict of interest.

References

1. Vengosh, A. Salinization and Saline Environments. *Treatise Geochem.* **2003**, *9*, 612.
2. Thomas, D.S.G.; Middleton, N.J. Salinization: New Perspectives on a Major Desertification Issue. *J. Arid Environ.* **1993**, *24*, 95–105. [CrossRef]

3. Williams, W.D. Anthropogenic Salinisation of Inland Waters. In *Saline Lakes*; Springer: Berlin/Heidelberg, Germany, 2001; pp. 329–337.
4. Ingram, J. A Food Systems Approach to Researching Food Security and Its Interactions with Global Environmental Change. *Food Secur.* **2011**, *3*, 417–431. [CrossRef]
5. Rojas, R.V.; Achouri, M.; Maroulis, J.; Caon, L. Healthy Soils: A Prerequisite for Sustainable Food Security. *Environ. Earth Sci.* **2016**, *75*, 180. [CrossRef]
6. Gelburd, D.E. Managing Salinity Lessons from the Past. *J. Soil Water Conserv.* **1985**, *40*, 329–331.
7. Ivushkin, K.; Bartholomeus, H.; Bregt, A.K.; Pulatov, A.; Kempen, B.; De Sousa, L. Global Mapping of Soil Salinity Change. *Remote Sens. Environ.* **2019**, *231*, 111260. [CrossRef]
8. Shahid, S.A.; Zaman, M.; Heng, L. Soil Salinity: Historical Perspectives and a World Overview of the Problem. In *Guideline for Salinity Assessment, Mitigation and Adaptation Using Nuclear and Related Techniques*; Springer: Berlin/Heidelberg, Germany, 2018; pp. 43–53.
9. Ghassemi, F.; Jakeman, A.J.; Nix, H.A. *Salinisation of Land and Water Resources: Human Causes, Extent, Management and Case Studies*; CAB International: Wallingford, UK, 1995.
10. Metternicht, G.I.; Zinck, J.A. Remote Sensing of Soil Salinity: Potentials and Constraints. *Remote Sens. Environ.* **2003**, *85*, 1–20. [CrossRef]
11. Qadir, M.; Schubert, S.; Noble, A.D.; Saqib, M. Amelioration Strategies for Salinity-Induced Land Degradation. *CAB Rev.* **2006**, *1*, 12. [CrossRef]
12. Saddhe, A.A.; Manuka, R.; Nikalje, G.C.; Penna, S. Halophytes as a Potential Resource for Phytodesalination. In *Handbook of Halophytes: From Molecules to Ecosystems towards Biosaline Agriculture*; Springer: Cham, Switzerland, 2020; pp. 1–21.
13. Ouni, Y.; Ghnaya, T.; Montemurro, F.; Abdelly, C.; Lakhdar, A. The Role of Humic Substances in Mitigating the Harmful Effects of Soil Salinity and Improve Plant Productivity. *Int. J. Plant Prod.* **2014**, *8*, 353–374.
14. Banerjee, A.; Sarkar, S.; Cuadros-Orellana, S.; Bandopadhyay, R. Exopolysaccharides and Biofilms in Mitigating Salinity Stress: The Biotechnological Potential of Halophilic and Soil-Inhabiting Pgpr Microorganisms. In *Microorganisms in Saline Environments: Strategies and Functions*; Springer: Berlin/Heidelberg, Germany, 2019; pp. 133–153.
15. Cuevas, J.; Daliakopoulos, I.N.; del Moral, F.; Hueso, J.J.; Tsanis, I.K. A Review of Soil- Improving Cropping Systems for Soil Salinization. *Agronomy* **2019**, *9*, 295. [CrossRef]
16. Rotmans, J.; Loorbach, D.; Kemp, R. Complexity and Transition Management. In *Complexity and Planning*; Routledge: London, UK, 2016; pp. 195–216.
17. Rotmans, J. Societal Innovation: Between Dream and Reality Lies Complexity. 2005. Available online: https://papers.ssrn.com/sol3/papers.cfm?abstract_id=878564 (accessed on 22 February 2022).
18. Pattberg, P.; Widerberg, O.; Isailovic, M.; Dias Guerra, F. Mapping and Measuring Fragmentation in Global Governance Architectures: A Framework for Analysis. 2014. Available online: https://ssrn.com/abstract=2484513 (accessed on 22 February 2022).
19. Rittel, H.W.J.; Webber, M.M.; Huppatz, D.J. Dilemmas in a General Theory of Planning. *Policy Sci.* **1973**, *4*, 155–169. [CrossRef]
20. von Thienen, J.; Meinel, C.; Nicolai, C. How Design Thinking Tools Help to Solve Wicked Problems. In *Design Thinking Research*; Springer: Berlin/Heidelberg, Germany, 2014; pp. 97–102.
21. Desa, U.N. Transforming Our World: The 2030 Agenda for Sustainable Development. 2016. Available online: https://stg-wedocs.unep.org/bitstream/handle/20.500.11822/11125/unepswiosm1inf7sdg.pdf?sequence=1 (accessed on 22 February 2022).
22. Vargas, R.; Pankova, E.I.; Balyuk, S.A.; Krasilnikov, P.V.; Khasankhanova, G.M. *Handbook for Saline Soil Management*; FAO/LMSU: Washington, DC, USA, 2018.
23. Hamdy, A.; Abdel-Dayem, S.; Abu-Zeid, M. Saline Water Management for Optimum Crop Production. *Agric. Water Manag.* **1993**, *24*, 189–203. [CrossRef]
24. Lee, D.J.; Howitt, R.E. Modeling Regional Agricultural Production and Salinity Control Alternatives for Water Quality Policy Analysis. *Am. J. Agric. Econ.* **1996**, *78*, 41–53. [CrossRef]
25. Negacz, K.; Vellinga, P.; Barrett-Lennard, E.; Choukr-Allah, R.; Elzenga, T. Saline Agriculture: A Call to Action. In *Future of Sustainable Agriculture in Saline Environments*; CRC Press: Boca Raton, FL, USA, 2021; pp. 3–12.
26. Negacz, K.; Bruning, B.; Vellinga, P. Achieving Multiple Sustainable Development Goals through Saline Agriculture. *Future Sustain. Agric. Saline Environ.* **2021**, *13*, 187.

environmental sciences proceedings

MDPI

Proceeding Paper

Four Species with Crop Potential in Saline Environments: The SALAD Project Case Study †

Giulia Atzori

Dipartimento di Scienze e Tecnologie Agrarie, Alimentari, Ambientali e Forestali (DAGRI), Università degli Studi di Firenze, Viale delle Idee 30, 50019 Sesto Fiorentino, Italy; giulia.atzori@unifi.it

† Presented at the 2nd International Laayoune Forum on Biosaline Agriculture, 14–16 June 2022; Available online: https://lafoba2.sciforum.net/.

Abstract: With sea levels rising due to climate change, salinity intrusion will increase and new crops, specifically appropriate to such particular ecological conditions, are needed. In the project "SALAD—Saline Agriculture as a Strategy to Adapt to Climate Change", the possibility of growing tomato (*Solanum lycopersicum*), potato (*Solanum tuberosum*), quinoa (*Chenopodium quinoa*), and New Zealand spinach (*Tetragonia tetragonioides*) in saline conditions is explored, together with their market upscaling opportunity. The crops are described in terms of their origin and distribution, botanical description and edible use. Moreover, the state of the art of the four crops' response under saline conditions is reviewed.

Keywords: saline agriculture; salt-tolerant crops; tomato; potato; quinoa; New Zealand spinach; SALAD project

Citation: Atzori, G. Four Species with Crop Potential in Saline Environments: The SALAD Project Case Study. *Environ. Sci. Proc.* **2022**, *16*, 54. https://doi.org/10.3390/environsciproc2022016054

Academic Editor: Abdelaziz Hirich

Published: 16 June 2022

1. Introduction

The current food system feeds the great majority of the world population and supports the livelihoods of over one billion people. Yet, it is both a major driver of climate change and increasingly vulnerable to it because of its intrinsic dependency on natural resources. With sea levels rising due to climate change, salinity intrusion will increase and not only affect agricultural production, but also the living conditions of farmers, the quality of natural resources and whole ecosystems. In salt-affected areas, the world's major crops are not adequate to supply the calories, proteins, fats, and nutrients people need. New crops are needed that are specifically appropriate to such particular ecological conditions. In the recent funded project "SALAD—Saline Agriculture as a Strategy to Adapt to Climate Change", the possibility of growing four crops in saline conditions is explored, together with their market upscaling opportunity. The main objective of the SALAD project is to improve the climate resilience of agricultural production in the current and potential salt-affected areas and promote the use of innovative salt-tolerant crops and cultivation techniques to improve food security and sustainability in the context of climatic changes, by boosting transnational cooperation between knowledge institutes, farmers, and entrepreneurs, consumers, and the public sector, around and beyond partner regions. The case study species to be studied in the project are tomato (*Solanum lycopersicum*), potato (*Solanum tuberosum*), quinoa (*Chenopodium quinoa*), and New Zealand spinach (*Tetragonia tetragonioides*). The four crops are currently not at the same stage of development and awareness among the general public, with quinoa and potato crops already extensively screened for their performance in saline conditions, tomato showing very promising quality improvements of the fruits under saline conditions, and New Zealand spinach is still at an early stage of research. Regarding the salt tolerance of the four species, potato and tomato crops are glycophytes, characterized by a high difference among varieties, whereas quinoa and New Zealand spinach are halophytes, or "salt-loving" species. The four crops are here introduced in terms of their origin and distribution, their botanical description, and edible use. Moreover, the state of the art of the four crops' response under saline conditions is reviewed.

2. Tomato Crop

2.1. Origin, Distribution, and Botanical Description

Tomato (*Solanum lycopersicum* L.) is an important food and industrial crop, cultivated all over the world. The species originated in western South America and Central America and numerous varieties of the tomato plant are widely grown in temperate climates across the world, with greenhouses allowing for their production during all seasons of the year. Tomatoes are fruits, botanically classified as berries. They are commonly used culinarily as a vegetable ingredient.

2.2. Tomato Response to Salinity

This species is classified as a moderate-salinity-tolerant species [1], and its salt tolerance has been studied since the 1970s [2]. Salt-sensitive and salt-tolerant genotypes (i.e., ecotypes from the Galapagos Islands) were investigated by [3] assessing a far stronger salt resistance in the Galapagos ecotypes, which were surviving in a full-strength seawater nutrient solution, EC roughly corresponding to 50 dS m^{-1}. The salt-resistant genotypes were firstly used in breeding programs with the aim of transferring genetic information to cultivated tomato, but the process did not give the expected results and, to date, creating new salt-tolerant tomato cultivars with breeding programs faces many obstacles, mainly because traits related to salt tolerance are not combined in a single donor genotype [4]. Several trials were also performed to determine the seawater tolerance of cultivated tomato, assessing, in the meantime, the characteristics of fruits; even if seawater irrigation generally reduced the crop yield, according to the tested variety, in particular, 10 to 20% seawater (roughly EC of 8 dSm^{-1} to 14 dS m^{-1}) increased the nutritional value of the product. In particular, fruit dry matter and total soluble solids were reported by several authors to increase with the use of seawater concentrations of 10–12% compared to control. In addition to that, the concentration of reducing sugars (RS) and titratable acidity (TA) also increased in the berries exposed to seawater irrigation, resulting in tastier fruits than control ones [5]. Similarly, glucose, fructose, and citric and ascorbic acid increased proportionally to salinity, with glucose concentrations up to 139% and fructose up to 101% higher compared to the control treatment [6].

3. Potato Crop

3.1. Origin and Distribution, and Botanical Description

The potato is a starchy tuber used as a root vegetable, native to the Americas. The plant was introduced to Europe in the 16th century by the Spanish and today, potatoes are a staple food in many parts of the world and an integral part of much of the world's food supply, being a rich source of carbohydrates, proteins, dietary fiber, ascorbic acid, riboflavin, and minerals [7]. Following millennia of selective breeding, there are now over 5000 different types of potatoes.

3.2. Potato Response to Salinity

The potato crop is classified as a moderately salt-sensitive crop since the threshold value of soil salinity of saturated soil extract (EC_e) is 1.7 dS m^{-1} and irrigation water salinity (EC_W) is 1.1 dS m^{-1} [7]. Yet, the results of field experiments suggest that there is more potential for conventional crop production under "moderate saline" conditions than is generally assumed, with numerous cultivars (i.e., 'Miss Mignonne', 'Achilles', 'Foc', 'Met', and '927') not showing yield losses at EC up to 4.1 dS m^{-1} [8], with some varieties showing greater levels of salt tolerance compared to others. Moreover, 'Désirée' was irrigated with saline water (EC 6.6 dS m^{-1}) in the Negev Desert without loss of yield [9], even if the combination of salinity and heat instead led to a reduced tuber yield. Under saline conditions, leaf water and osmotic potentials were observed to decline significantly, even if a positive turgor was maintained by plants throughout the entire growth period, thanks to the fact that plants mainly adjusted osmotically due to chloride and proline [10]. Potato leaves are very sensitive (especially to salt applied at the beginning of tuber formation) and are severely damaged by overhead irrigation with saline water that may induce toxicity, exhibited as leaf burn along the margins [9]. Especially on tubers, high salinity levels (EC greater than 10 dS m^{-1} in the root zone) may cause coarse russeting and furrowing of the tubers, accompanied by severe browning of the surface [9].

4. Quinoa Crop

4.1. Origin and Distribution, and Botanical Description

The species originated in the Andean region of northwestern South America. Today, its cultivation has spread to more than 70 countries, including Kenya, India, the United States, and several European countries. It is an herbaceous annual and it is grown as a crop primarily for its edible seeds, which are rich in protein, dietary fiber, B vitamins, and dietary minerals.

4.2. Quinoa Response to Salinity

Quinoa is a facultative halophytic plant species with the most tolerant varieties being able to cope with salinity levels as high as those present in seawater [11]. Of the almost 2500 existing quinoa accessions, more than 200 have been tested under saline conditions, and shown to differ in their response to salinity [11]. The natural variability in a number of traits (i.e., inflorescence type, seed color, seed size, life-cycle duration, salinity tolerance, saponin content, and nutritional value) allows quinoa to adapt to diverse environments [12]. Generally, optimal plant growth has been observed at NaCl concentrations of 100–200 mM (~10–20 dS m^{-1}) and biomass production, seed yield, and harvest index were proved to be higher under moderately saline conditions (10–20 dS m^{-1}) than under non-saline conditions [13,14]. Several studies to identify mechanisms of salt tolerance in quinoa have been conducted over the last few years, identifying many factors, among which are the efficient control of Na^+ sequestration in leaf vacuoles, a higher ROS tolerance, a better K^+ retention, and the accumulation of compatible solutes, such as proline and total phenolics [12].

5. New Zealand Spinach Crop

5.1. Origin and Distribution, and Botanical Description

Tetragonia, or New Zealand spinach, is a widespread species native to eastern Asia, Australia, and New Zealand that has been introduced in many parts of Africa, Europe, North America, and South America. Sandy shorelines represent its natural habitat. It is a halophyte that grows well in saline ground and it is often cultivated as a leafy vegetable.

5.2. New Zealand Spinach Response to Salinity

Several trials have shown that Tetragonia may withstand an EC of the growing medium of 10 dS m^{-1} [15,16]. Other studies identified a salt-induced growth response at salinity levels of 50–100 mM NaCl (EC 5–10 dS m^{-1}) [17,18] and even higher salinity thresholds (i.e., 18 dS m^{-1}) have been assessed in hydroponic conditions [18], regardless of the type of salinity tested (i.e., NaCl solution vs seawater at comparable EC) [19]. Salinity was also correlated to an increased nutritional value of the edible leaves due to the enhanced accumulation of mineral elements and a decrease in nitrates in saline conditions compared to control ones [18].

Funding: This project is part a program of the ERA-NET Co-fund FOSC that has received funding from the European Union's Horizon 2020 research and innovation program, under grant agreement No. 862555.

Institutional Review Board Statement: Not applicable.

Informed Consent Statement: Not applicable.

Acknowledgments: All partners of the ERA-NET Co-fund FOSC Project "SALAD—Saline Agriculture as a Strategy to Adapt to Climate Change".

Conflicts of Interest: The author declares no conflict of interest.

References

1. Maas, E.V.; Hoffman, G.J. Crop Salt Tolerance–Current Assessment. *J. Irrig. Drain. Div. Am. Soc. Civ. Eng.* **1977**, *103*, 115–134. [CrossRef]
2. Atzori, G.; Mancuso, S.; Masi, E. Seawater Potential Use in Soilless Culture: A Review. *Sci. Hortic.* **2019**, *249*, 199–207. [CrossRef]
3. Rush, D.W.; Epstein, E. Genotypic Responses to Salinity. *Plant Physiol.* **1976**, *57*, 162–166. [CrossRef] [PubMed]
4. Cuartero, J.; Fernandez-Munos, R. Tomato and Salinity. *Sci. Hortic.* **1999**, *78*, 83–125. [CrossRef]
5. Sgherri, C.; Kadlecová, Z.; Pardossi, A.; Navari-Izzo, F.; Izzo, R. Irrigation with Diluted Seawater Improves the Nutritional Value of Cherry Tomatoes. *J. Agric. Food Chem.* **2008**, *56*, 3391–3397. [CrossRef]
6. Kawai, Y.; Hiratsuka, S.; Tashiro, T.; Kunoh, H. Effects of Deep-Sea Water Application on Fruit Qualities of Satsuma Mandarin [Citrus Unshiu] and Tomato [Lycopersicon Esculentum]. *Hortic. Res.* **2002**, *1*, 179–182. [CrossRef]
7. Chourasia, K.N.; Lal, M.K.; Tiwari, R.K.; Dev, D.; Kardile, H.B.; Patil, V.U.; Kumar, A.; Vanishree, G.; Kumar, D.; Bhardwaj, V.; et al. Salinity Stress in Potato: Understanding Physiological, Biochemical and Molecular Responses. *Life* **2021**, *11*, 545. [CrossRef] [PubMed]
8. De Vos, A.; Bruning, B.; van Straten, G.; Oosterbaan, R.; Rozema, J.; van Bodegom, P. *Crop Salt Tolerance under Controlled Field Conditions in The Netherlands, Based on Trials Conducted at Salt Farm Texel*; Salt Farm Texel: Den Hoorn, The Netherlands, 2016.
9. Levy, D.; Veilleux, R.E. Adaptation of Potato to High Temperatures and Salinity—A Review. *Am. J. Potato Res.* **2007**, *84*, 487–506. [CrossRef]
10. Heuer, B.; Nadler, A. Physiological Response of Potato Plants to Soil Salinity and Water Deficit. *Plant Sci.* **1998**, *137*, 43–51. [CrossRef]
11. Adolf, V.I.; Jacobsen, S.E.; Shabala, S. Salt Tolerance Mechanisms in Quinoa (Chenopodium Quinoa Willd.). *Environ. Exp. Bot.* **2013**, *92*, 43–54. [CrossRef]
12. Hinojosa, L.; Gonzalez, J.A.; Barrios-Masias, F.H.; Fuentes, F.F.; Murphy, K.M. Quinoa Abiotic Stress Responses: A Review. *Plants* **2018**, *7*, 106. [CrossRef] [PubMed]
13. Hariadi, Y.; Marandon, K.; Tian, Y.; Jacobsen, S.E.; Shabala, S. Ionic and Osmotic Relations in Quinoa (Chenopodium Quinoa Willd.) Plants Grown at Various Salinity Levels. *J. Exp. Bot.* **2011**, *62*, 185–193. [CrossRef] [PubMed]
14. Jacobsen, S.E.; Mujica, A.; Jensen, C.R. The Resistance of Quinoa (Chenopodium Quinoa Willd.) to Adverse Abiotic Factors. *Food Rev. Int.* **2003**, *19*, 99–109. [CrossRef]
15. Neves, M.A.; Miguel, M.G.; Marques, C.; Panagopoulos, T.; Beltrao, J. The Combined Effects of Salts and Calcium on Growth and Mineral Accumulation of *Tetragonia tetragonioides*—A Salt Removing Species. *WSEAS Trans. Environ. Dev.* **2008**, *4*, 1–5.
16. Wilson, C.; Lesch, S.M.; Grieve, C.M. Growth Stage Modulates Salinity Tolerance of New Zealand Spinach (*Tetragonia tetragonioides*, Pall.) and Red Orach (*Atriplex hortensis* L.). *Ann. Bot.* **2000**, *85*, 501–509. [CrossRef]
17. Yousif, B.S.; Liu, L.Y.; Nguyen, N.T.; Masaoka, Y.; Saneoka, H. Comparative Studies in Salinity Tolerance between New Zealand Spinach (*Tetragonia tetragonioides*) and Chard (Beta Vulgaris) to Salt Stress. *Agric. J.* **2010**, *5*, 19–24. [CrossRef]
18. Atzori, G.; Nissim, W.G.; Macchiavelli, T.; Vita, F.; Azzarello, E.; Pandolfi, C.; Masi, E.; Mancuso, S. *Tetragonia tetragonioides* (Pallas) Kuntz. as Promising Salt-Tolerant Crop in a Saline Agriculture Context. *Agric. Water Manag.* **2020**, *240*, 106261. [CrossRef]
19. Guidi Nissim, W.; Masi, E.; Pandolfi, C.; Mancuso, S.; Atzori, G. The Response of Halophyte (*Tetragonia tetragonioides* (Pallas) Kuntz.) and Glycophyte (*Lactuca sativa* L.) Crops to Diluted Seawater and Nacl Solutions: A Comparison between Two Salinity Stress Types. *Appl. Sci.* **2021**, *11*, 6336. [CrossRef]

environmental sciences proceedings

MDPI

Proceeding Paper

How Does Organic Amendment Improve Quinoa Growth and Productivity under Saline Conditions? †

Ayoub El Mouttaqi [1], Ihssane Mnaouer [1], Mohamed Belcaid [1], Mohamed Ibourki [1], Lamine Diatta [1], Krishna Prasad Devkota [1], Abdelaziz Nilahyane [1], Jamaa Zim [2], Cherki Ghoulam [3], Redouane Choukr-Allah [4], Lamfeddal Kouisni [1] and Abdelaziz Hirich [1,*]

[1] Biosaline Agriculture Research Program, African Sustainable Agriculture Research Institute (ASARI), Mohammed VI Polytechnic University (UM6P), Laayoune 70000, Morocco; ayoub.elmouttaqi@um6p.ma (A.E.M.); ihssane.mnaouer@um6p.ma (I.M.); med.belcaid80@gmail.com (M.B.); mohamed.ibourki@um6p.ma (M.I.); diattalamine29@gmail.com (L.D.); krishna.devkota@um6p.ma (K.P.D.); abdelaziz.nilahyane@um6p.ma (A.N.); lamfeddal.kouisni@um6p.ma (L.K.)

[2] Hassan II Institute of Agronomy and Veterinary Medicine, Rabat 10101, Morocco; j.zim@iav.ac.ma

[3] Faculty of Sciences and Techniques (FST), Cadi Ayyad University, Marrakech 40000, Morocco; c.ghoulam@uca.ma

[4] Agricultural Innovation and Technology Transfer Center (AITTC), Mohammed VI Polytechnic University (UM6P), Ben Guerir 43150, Morocco; redouane53@yahoo.fr

* Correspondence: abdelaziz.hirich@um6p.ma

† Presented at the 2nd International Laayoune Forum on Biosaline Agriculture, 14–16 June 2022; Available online: https://lafoba2.sciforum.net/.

Abstract: Nowadays, salinization is becoming a serious problem affecting several agricultural areas, especially irrigated ones. A field experiment was conducted in Foum El Oued, south of Morocco, testing quinoa responses to three irrigation-water salinities (4, 12, and 20 dS/m) combined with nine organic amendments. The obtained results indicate that most of the growth and productivity parameters were negatively affected by salinity, while the effect of organic amendment varied from one salinity level to another. Under high salinity, sheep manure, compost, and insects improved quinoa seed yield by 157, 110, and 83%, respectively, compared to the control. The findings of this study recommend that organic amendment could be a judicious practice to improve quinoa yield under saline conditions.

Keywords: compost; manure; biomass; seed yield; irrigation

check for updates

Citation: El Mouttaqi, A.; Mnaouer, I.; Belcaid, M.; Ibourki, M.; Diatta, L.; Devkota, K.P.; Nilahyane, A.; Zim, J.; Ghoulam, C.; Choukr-Allah, R.; et al. How Does Organic Amendment Improve Quinoa Growth and Productivity under Saline Conditions? *Environ. Sci. Proc.* **2022**, *16*, 59. https://doi.org/10.3390/environsciproc2022016059

Academic Editor: Bas Bruning

Published: 17 June 2022

Publisher's Note: MDPI stays neutral with regard to jurisdictional claims in published maps and institutional affiliations.

1. Introduction

Agriculture in Morocco is facing several challenges, such as climate change, desertification, water scarcity and salinity, that limit its productivity. Nowadays, salinization is becoming a serious problem affecting several agricultural areas, especially irrigated ones. It is estimated that more than 50% of groundwater in Morocco has an EC value greater than 3 dS/m, causing a decline in crop production; thus, several agricultural areas have already been abandoned due to the salinity problem. Biosaline agriculture can be a judicious solution to bring salt-affected land back to production through the introduction of salt-tolerant and halophytic crops. Crops such as blue panicum, halophytic grasses, quinoa, sesbania, and barley showed great performance, and have already been adopted by farmers [1]. The application of organic amendments is one of the best practices used to cope with salinity and water stress [2]. The objective of this study was to evaluate the impacts of organic amendments and salinity on quinoa growth and productivity.

2. Materials and Methods

This research was conducted on the experimental farm at the National Institute of Agricultural Research in the Foum El Oued area, Laayoune, south of Morocco (latitude:

27.176°; longitude: −13.349°; Z = 37 m). The soil at the experimental site was sandy loam (61% sand, 18% silt, and 18% clay), and was moderately saline and poor in organic matter and nutrients.

Quinoa (*Chenopodium quinoa* Willd.) of the variety ICBA-Q5 was used in this experiment. The area of each plot was 7.5 m^2 and each consisted of five rows with 50 cm between them. The crop was sown manually on 13 April 2021. Crop spacing was kept at 50 cm row-to-row. Irrigation was applied twice a week with an amount of 8.33 mm per irrigation. The irrigation system consisted of drip irrigation with a flow rate of 2 L/h, and the distance between drippers was equal to 20 cm. Three levels of irrigation-water salinity, including 4, 12 and, 20 dS/m, were applied. The low salinity level (4 dS/m) corresponded to the lowest groundwater salinity level in the region. The salinity level of 12 dS/m corresponded to the groundwater salinity level at the experimental site; and 20 dS/m was reached by adding NaCl to low-saline groundwater (4 dS/m).

Nine different organic amendments in addition to a non-treated control were tested, including:

- Cow manure, 30 T/ha;
- Sheep manure, 30 T/ha;
- Goat manure, 30 T/ha;
- Chicken manure, 30 T/ha;
- Compost, 5 T/ha;
- Biochar, 5 T/ha;
- Lombricompost, 5 T/ha;
- Insect frass of *Tenebrio molitor* (mealworm), 5 T/ha; and
- Insect frass of *Hermetia illucens* (black soldier fly), 5 T/ha.

The adopted experimental design was a split-plot design whereby we had two levels of factors; the first level, representing salinity level, was applied in the large plot, and the second level, representing soil amendments, was applied in the sub-plots.

3. Results

3.1. Biomass Accumulation

The effects of irrigation-water salinity and organic amendments on the final fresh and dry biomass are presented in Figure 1. The presented data clearly indicate that biomass was greatly affected by increased salinity. Under high-salinity conditions (20 dS/m) both sheep manure and black soldier fly had the highest biomass accumulation.

3.2. Seed Yield

Figure 2 presents the seed yield variation in quinoa subjected to both salinity and organic amendment effects. It is obvious from the presented data that seed yield decreased with increased salinity. Under high salinity levels (20 dS/m), sheep manure, compost and the insect frass of *Hermetia illucens* showed the highest performances, and improved quinoa seed yield by 157, 110, and 83%, respectively, compared to the control. Under medium salinity levels (12 dS/m), cow manure, the insect frass of *Tenebrio monitor*, and biochar resulted in the highest productivity, and improved seed yield by 44, 38, and 20%, respectively, compared to the control. Under low salinity levels (4 dS/m), the insect frass of *Tenebrio molitor*, chicken, and the insect frass of *Hermetia illucens* showed the highest performance, and increased seed yield by 47, 35, and 31%, respectively, compared to the control. The harvest index was stable under all tested salinity levels, as both biomass and seed yield responded in the same way to both salinity and organic amendment.

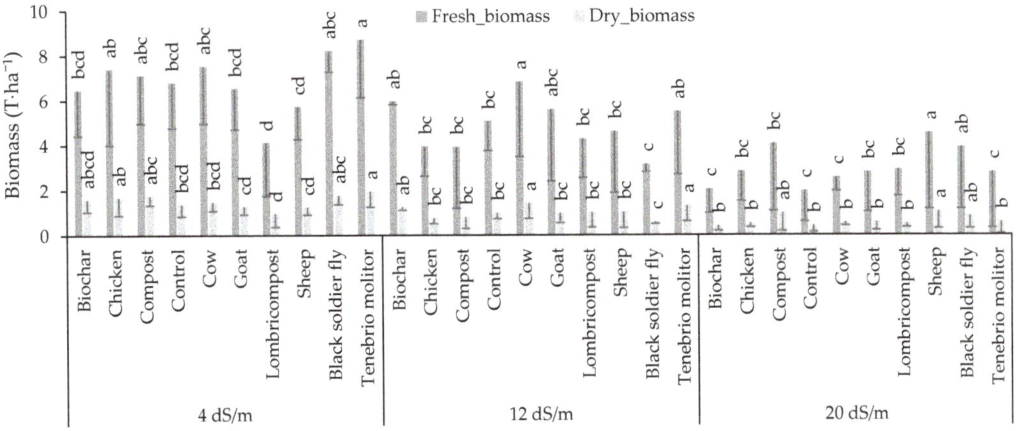

Figure 1. Variation in final fresh and dry biomass as response to irrigation-water salinity and organic amendments. Error bars indicate the standard deviation. a, b, c, and d present Tuckey's test at $p = 0.05$.

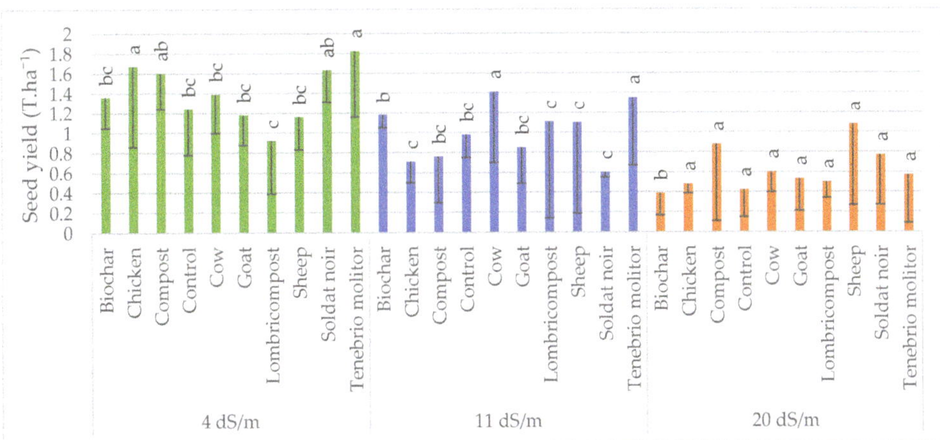

Figure 2. Variation in seed yield as response to irrigation-water salinity and organic amendments. Error bars indicate the standard deviation. a, b, and c present Tuckey's test at $p = 0.05$.

4. Discussion

It is well known that organic amendments can have an important role in alleviating the adverse effects of salinity and drought by improving organic matter content, nutrient availability, water-holding capacity, soil porosity, etc. [3], thus increasing crop productivity. In a similar study testing other types of organic amendment, Alcívar et al. [4] showed that quinoa seed yield in the AZ-51 genotype presented significant increases of 116 and 85% for gypsum and humic substances with gypsum treatments, while biochar had no significant effect on seed yield under saline conditions; this is in agreement with our findings. Similarly, Yang et al. [5] found that the seed yield of quinoa subjected to full irrigation under saline conditions was not affected by biochar amendment, while the latter had a significant effect under deficit irrigation and alternate root-zone-drying irrigation regimes.

5. Conclusions

The findings of this study recommend that organic amendment could be a judicious practice to improve quinoa yield under saline conditions. Furthermore, new organic amendments such as insect frass showed great potential compared to traditional manure.

Author Contributions: Conceptualization, A.H., A.E.M. and R.C.-A.; methodology, A.H. and A.E.M.; software, A.E.M.; validation, A.H. and K.P.D.; investigation, A.E.M., I.M., M.B., M.I., L.D. and J.Z.; resources, A.H.; data curation, A.E.M.; writing—original draft preparation, A.H. and A.E.M.; writing—review and editing, A.N. and L.K.; visualization, A.H and A.E.M.; supervision, A.H.; project administration, A.H. and C.G.; funding acquisition, A.H. and C.G. All authors have read and agreed to the published version of the manuscript.

Funding: This research was funded by the Moroccan Ministry of National Education, Professional Training, Higher Education and Scientific Research within the SALAD project of the ERA-Net Cofund on Food Systems and Climate.

Institutional Review Board Statement: Not applicable.

Informed Consent Statement: Not applicable.

Data Availability Statement: Not applicable.

Acknowledgments: We would like to thank the National Agricultural Research Institute in Foum El Oued, Laayoune for providing a field facility in which to undertake this research.

Conflicts of Interest: The authors declare no conflict of interest.

References

1. Hirich, A.; Choukr-Allah, R.; Ezzaiar, R.; Shabbir, S.A.; Lyamani, A. Introduction of alternative crops as a solution to groundwater and soil salinization in the Laayoune area, South Morocco. *Euro Med. J. Environ Integr.* **2021**, *6*, 52. [CrossRef]
2. Leogrande, R.; Vitti, C. Use of organic amendments to reclaim saline and sodic soils: A review. *Arid. Land Res. Manag.* **2019**, *33*, 1–21. [CrossRef]
3. Bello, S.K.; Alayafi, A.H.; AL-Solaimani, S.G.; Abo-Elyousr, K.A.M. Mitigating Soil Salinity Stress with Gypsum and Bio-Organic Amendments: A Review. *Agronomy* **2021**, *11*, 1735. [CrossRef]
4. Alcívar, M.; Zurita-Silva, A.; Sandoval, M.; Muñoz, C.; Schoebitz, M. Reclamation of Saline–Sodic Soils with Combined Amendments: Impact on Quinoa Performance and Biological Soil Quality. *Sustainability* **2018**, *10*, 3083. [CrossRef]
5. Yang, A.; Akhtar, S.S.; Li, L.; Fu, Q.; Li, Q.; Naeem, M.A.; He, X.; Zhang, Z.; Jacobsen, S.-E. Biochar Mitigates Combined Effects of Drought and Salinity Stress in Quinoa. *Agronomy* **2020**, *10*, 912. [CrossRef]

environmental sciences proceedings

MDPI

Proceeding Paper

Soil Salinity Prediction and Mapping Using Electromagnetic Induction and Spatial Interpolation †

Houria Dakak [1,*], Hinde Cherkaoui Dekkaki [2], Abdelmjid Zouahri [1], Rachid Moussadek [1,3], Hamza Iaaich [1], Hasna Yachou [1], Ahmed Ghanimi [4] and Ahmed Douaik [1]

1 Research Unit on Environment and Conservation of Natural Resources, Regional Center of Rabat, National Institute of Agricultural Research (INRA), P.O. Box 6356, Instituts, Rabat 10101, Morocco
2 Laboratory of Research and Development in Engineering Sciences, Faculty of Sciences and Techniques, Abdelmalek Essaadi University, P.O. Box 34, Ajdir, Al Hoceima 32003, Morocco
3 International Center for Research in Dry Areas (ICARDA), Rabat P.O. Box 1302, Morocco
4 Department of Chemistry, Faculty of science, Mohammed V University, Rabat P.O. Box 1014, Morocco
* Correspondence: houria.dakak@inra.ma

† Presented at the 2nd International Laayoune Forum on Biosaline Agriculture, 14–16 June 2022, Available online: https://lafoba2.sciforum.net/.

Abstract: For better soil management in terms of salinization to ensure sustainable agriculture, a soil salinity mapping and prediction study based on the measurement of apparent electrical conductivity using an electromagnetic instrument (EM38) combined with geostatistical interpolation (kriging) is proposed herein for the soil of the semi-arid region of Beni Amir (2060 ha), Tadla, Morocco. This solution is efficient and quick to use, with the ability to provide reliable information for assessing the spatial distribution of soil salinity. The results of measurements through a spatial analysis offered us the possibility of identifying several classes of salinity that can be used for the sustainable management of land and water.

Keywords: digital soil mapping; electromagnetic induction; geostatistics; irrigated perimeter; salinity

Citation: Dakak, H.; Dekkaki, H.C.; Zouahri, A.; Moussadek, R.; Iaaich, H.; Yachou, H.; Ghanimi, A.; Douaik, A. Soil Salinity Prediction and Mapping Using Electromagnetic Induction and Spatial Interpolation. *Environ. Sci. Proc.* **2023**, *16*, 76. https://doi.org/10.3390/environsciproc2022016076

Academic Editor: Abdelaziz Hirich

Published: 3 January 2023

1. Introduction

Intensive agricultural development under irrigation in arid areas has improved agricultural production, but has led to degradation of soil quality, which poses a serious danger to the sustainability of the land use system [1,2]. In the Beni Amir irrigated sub-perimeter of Tadla (Morocco), where our study area is located, land irrigated with salt water from the Oum Er Rbia River is affected by salinization. The overall soil salinity varies between 0.35 and 13 dS/m. It is important to characterize the saline state of the land before any exploitation as well as a regular monitoring of its evolution, in order to arrive at strategies for its use. This study was carried out in agricultural soil, located upstream of the irrigated perimeter of Beni Amir. It consists of the prediction and mapping of soil salinity by implementing rapid, efficient, and economical methods; in particular, the technique of electromagnetic induction combined with spatial interpolation methods of soil properties allowed us to draw reliable maps for diverse soil types [3].

2. Material and Methods

2.1. Study Area

The study area is located upstream of the irrigated perimeter of Beni Amir, on the right bank of the irrigated perimeter of Tadla (Figure 1). It is 12 km far from Fkih Ben Salah. The irrigated area is 2060 ha. The topography is generally regular, with an altitude of 400 m. The climate is of the arid to semi-arid Mediterranean type, with an annual rainfall of 350 mm.

Figure 1. Study area.

2.2. Soil Sampling and Laboratory Analysis

A dataset to be calibrated, made up of measurements of the apparent electrical conductivity (ECa) at 91 sites, was created using the EM38 electromagnetic induction technique in the vertical position [4] (Figure 2). The calibration dataset corresponds to 12 sites, chosen from among the 91 sites intended for the measurement of ECa. The points were taken with a manual auger at a depth of 90 cm for three horizons (0–30, 30–60, and 60–90 cm). At these 12 sites, electrical conductivity of the saturated past extract (ECe) was evaluated in the laboratory of the Environment and Natural Resources Unit (INRA, Rabat). The average ECe over the 3 depths is used in this study.

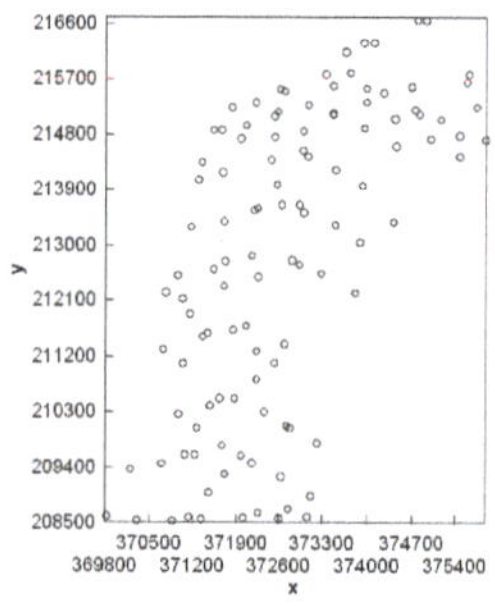

Figure 2. Network of soil measurement points by EM38.

2.3. Calibration and Mapping

Soil apparent electrical conductivity (ECa) was converted to electrical conductivity of soil paste extract (ECe) using the calibration equation based on a simple linear regression. Afterwards, estimated soil salinity data were interpolated using a geostatistical method [5] to identify the spatial variations of the salts using the variogram and to map the different levels of soil salinity using kriging.

3. Results and Discussion

ECe and ECa are strongly correlated, as confirmed by the Pearson coefficient of determination, which is 0.88 with the equation of calibration as ECe = 0.58 + 4.22 * ECa. The histogram and some statistics about ECe are given in Figure 3.

The data generally follows a normal distribution, with a few large values. The average value is 1.92 dS/m and values range between 0.8 and 4.38 dS/m, with a coefficient of variation of 39%.

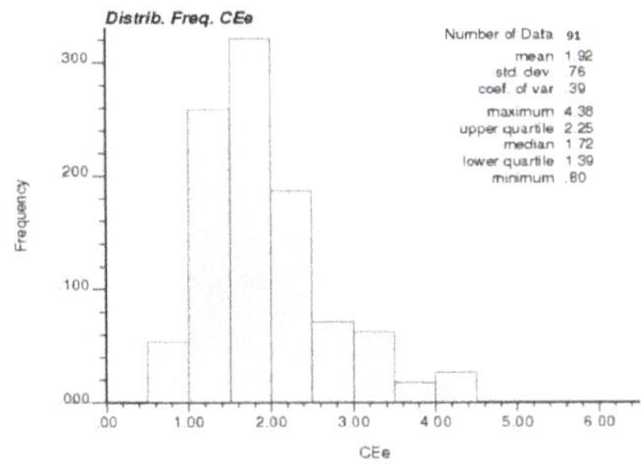

Figure 3. Histogram and some descriptive statistics for ECe.

The variogram of ECe is given in Figure 4 and its corresponding parameters are reported in Table 1. A spherical model was fitted to the experimental variogram with the following equation: $\gamma(\mathbf{h}) = \mathrm{Sph}(\mathbf{h}/a) = c_0 + c_1[(3/2)\,(\mathbf{h}/a) - (1/2)(\mathbf{h}/a)^3]$, where **h** is spatial lag (separation distance), c_0 is the variance of the nugget effect, c_1 is the partial sill variance, and a is the range.

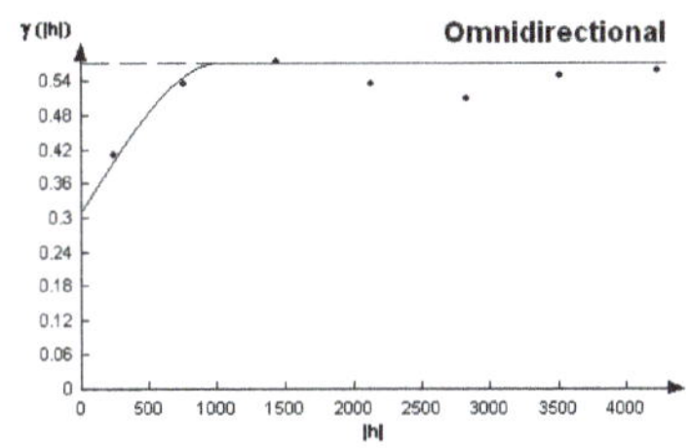

Figure 4. Experimental (dots) and fitted (curve) variograms for ECe.

Table 1. Parameters of the fitted model for the experimental variogram of ECe.

Nugget Effect (dS/m)^2	Range (m)	Partial Sill (dS/m)^2	Relative Nugget Effect (%)
0.312	1030	0.264	54

The range of the spatial dependence of ECe is 1030 m, with a relative nugget effect of 54%, meaning that more than half of the variation in ECe is not spatially structured, i.e., it is random.

Soil salinity (ECe) was predicted by kriging (Figure 5 left) using a regular 30 m × 30 m grid. The predicted values are between 1 and 3 dS/m. Compared to the observed values, the predicted minimum value is higher (0.8 dS/m) and the predicted maximum value is lower (4.38 dS/m); this property is known as the kriging smoothing effect. The map reflects the spatial distribution of ECe in the same way as the observed values (Figure 5). However, there is a complete spatial coverage of the study area, whereas the observed sampling was limited to only 91 locations.

Kriging prediction errors for soil salinity (ECe) are given in Figure 5 (right). This map has a model that is a function of sampling intensity. Normally, at sampled locations, these errors should be zero and will increase as we move further away from these sampled locations. On this map, the errors are between 0.6 and 0.8 dS/m, which means that, depending on the choice of the interpolation grid, none of the predicted locations coincided with the sampled ones.

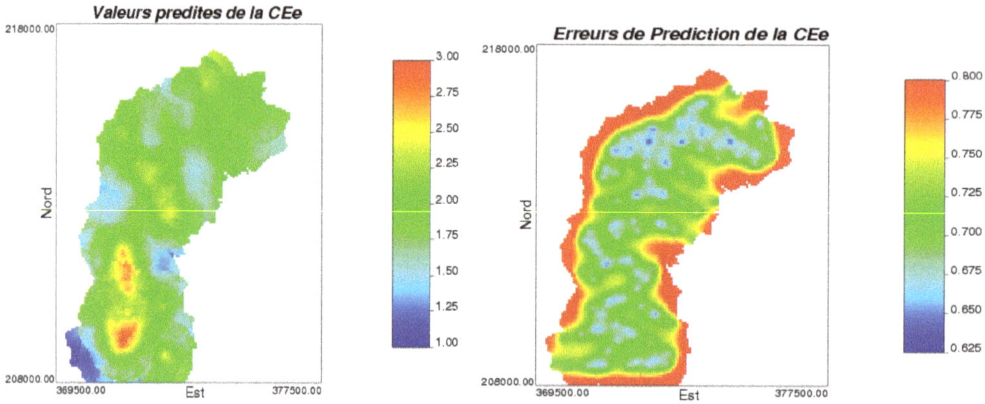

Figure 5. Spatial interpolation (**left**) and prediction errors (**right**) of soil salinity (ECe).

4. Conclusions

Apparent electrical conductivity (ECa) measured in the field using electromagnetic induction appears to be a fast and reliable method to assess soil salinity efficiently and cost-effectively. Additionally, spatial interpolation by kriging contributes to the description and modeling of soil salinity to spatially predict at unsampled locations. This will map the entire study area. This work has shown the usefulness of using indirect and inexpensive measurements of soil salinity for more reliable mapping than conventional laboratory measurements which are laborious and expensive.

Author Contributions: Conceptualization, H.D., A.D., and A.G.; sampling and laboratory analyses, H.D., A.Z., and H.Y.; software, H.D.; original draft preparation, H.D.; writing—review and editing, H.D., H.C.D., A.D., H.I., R.M., and A.G. All authors have read and agreed to the published version of the manuscript.

Funding: This research was funded by INRA in the frame of the Agroecology mega-project, subsidized by the general budget of INRA. The APC was funded by the LAFOBA2 organizers.

Institutional Review Board Statement: Not applicable.

Informed Consent Statement: This study did not include humans.

Data Availability Statement: Data can be supplied if required.

Conflicts of Interest: The authors declare no conflict of interest.

References

1. Huang, J.; Kilminster, T.; Barrett-Lennard, E.G.; Triantafilis, J. Characterization of field-scale dryland salinity with depth by quasi-3d inversion of DUALEM-1 data. *Soil Use Manag.* **1990**, *33*, 205–215. [CrossRef]
2. Dakak, H.; Huang, J.; Zouahri, A.; Douaik, A.; Triantafilis, J. Mapping soil salinity in 3-dimensions using an EM38 and EM4 Soil inversion modelling at the reconnaissance scale in central Morocco. *Soil Use Manag.* **2017**, *33*, 553–567. [CrossRef]
3. Douaik, A. Evaluation of the Space-Time Variability of Soil Salinity by Statistical, Geostatistical and Bayesian Maximum Entropy Methods. Ph.D. Thesis, Ghent University, Ghent, Belgium, 2005.
4. McNeill, J.D. *Electrical Conductivity of Soils and Rocks*; Tech. Note TN-6; Geonics Ltd.: Mississauga, ON, Canada, 1980.
5. Webster, R.; Oliver, M. *Geostatistics for Environmental Scientists*, 2nd ed.; Wiley: New York, NY, USA, 2007.

MDPI
St. Alban-Anlage 66
4052 Basel
Switzerland
Tel. +41 61 683 77 34
Fax +41 61 302 89 18
www.mdpi.com

Environmental Sciences Proceedings Editorial Office
E-mail: environsciproc@mdpi.com
www.mdpi.com/journal/environsciproc

www.ingramcontent.com/pod-product-compliance
Lightning Source LLC
Chambersburg PA
CBHW061556220926
51850CB00005B/353
* 9 7 8 3 0 3 6 5 6 1 6 9 1 *